AF735375

Smart Hydrogel Functional Materials

Preface

Inspired by the natural stimuli-responsive materials, artificial smart hydrogels are designed to be able to respond to physical and chemical stimuli in the environment, such as temperature, pH, ionic strength, magnetic field, electrical field, photo irradiation, glucose concentration, oxidoreduction, and/or chemical or biological species. Such smart hydrogels have been considered to be highly potential in versatile applications in numerous fields, such as controlled drug and gene delivery systems, chemical-/bio-separations, adaptive liquid microlenses, soft machines, chemical sensors and/or actuators, chemical valves, scaffolds for tissue engineering, and so on. Therefore, smart hydrogel functional materials have attracted great interests and ever-increasing attention from various fields in recent years. Nowadays, a lot of scientists and technologists all over the world are devoting themselves to designing and developing novel smart hydrogel functional materials for various applications. Thus, it is believable that smart hydrogel functional materials will definitely bring better technology and better quality of life for human beings in the future.

The authors' group at Sichuan University < group website: http://teacher.scu.edu.cn/ftp_teacher0/cly/> has been devoted to the development of smart hydrogel functional materials since the beginning of this century. In the last decade, they have successfully developed various environmental stimuli-responsive smart hydrogel functional materials, including thermo-responsive, pH-responsive, thermo-/pH-dual-responsive, ethanol-responsive, glucose-responsive, ion-recognizable, and molecular-recognizable ones, for myriad applications ranging from controlled release to chemical valves, chemical-/bio-separations, soft machines, chemical sensors and actuators, etc.

This book, entitled *Smart Hydrogel Functional Materials*, comprehensively and systematically describes modern understanding of smart or intelligent hydrogel functional materials with environmental stimuli-responsive functions. The contents range from hydrogels (including hydrogel-functionalized membranes) to microgels (including hydrogel-functionalized microcapsules) with various response properties, such as thermo-response, pH-response, pH-/thermo-dual-response, glucose-response, ethanol-response, ion-recognition, molecular-recognition, and so on. While chapters may be read as stand-alone, together they clearly describe design concepts, fabrication strategies and methods, microstructures, and performances of smart hydrogel functional materials. Vivid schematics and illustrations throughout the book enhance accessibility to the theories and technologies of these materials. This book is directed primarily to those readers who are interested in smart hydrogel functional materials, and aims to be a definitive reference book for a wide general readership including chemists, materials researchers, chemical engineers, pharmaceutical scientists, and biomedical researchers in the field of smart functional materials.

The book is composed of 16 chapters in seven parts. Because in many cases the environmental temperature fluctuations can occur naturally and the temperature stimuli can be easily designed and artificially controlled, *Part I* is focused on the thermo-responsive hydrogel functional materials, in which the structure-function relationship of thermo-responsive hydrogels (Chap. 1), the preparation and properties of monodisperse thermo-responsive microgels (Chap. 2), flow and aggregation characteristics of thermo-responsive microgels during phase transition (Chap. 3), polyphenol-induced phase transition of thermo-responsive hydrogels (Chap. 4), functional membranes with thermo-responsive hydrogel gates (Chap. 5), and functional microcapsules with thermo-responsive hydrogel shells (Chap. 6) are introduced systematically. In *Part II*, pH-responsive hydrogel functional materials, especially the preparation and properties of monodisperse pH-responsive microgels (Chap. 7) and pH-responsive membranes and microcapsules for controlled release (Chap. 8), are introduced. In *Part III*, the contents are focused on thermo-/pH-dual-responsive hydrogel functional materials, especially the thermo-/pH-dual-responsive hydrogels with rapid response properties (Chap. 9). In *Part IV*, ethanol-responsive hydrogel functional materials, especially smart functional membranes with alcohol-responsive characteristics (Chap. 10), are introduced. In *Part V*, glucose-responsive hydrogel functional materials, including hydrogels with rapid response to glucose concentration change at physiological temperature (Chap. 11) and glucose-responsive membranes and microcapsules for controlled release (Chap. 12), are introduced. In *Part VI*, the contents are focused on the ion-recognizable hydrogel functional materials, especially the preparation and properties of ion-recognizable smart hydrogels (Chap. 13) and functional microcapsules with ion-recognizable properties (Chap. 14). Finally, *Part VII* introduces the molecular-recognizable hydrogel functional materials including the preparation and properties of molecular-recognizable smart hydrogels (Chap. 15) and functional membranes with molecular-recognizable properties (Chap. 16).

Most of the contents in this book are the fresh achievements of the authors' group on smart hydrogel functional materials since the beginning of this century. Professor Liang-Yin Chu wrote Chaps. 2, 3, 4, and 9; Prof. Rui Xie wrote Chaps. 5, 6, 10, 15, and 16; Prof. Xiao-Jie Ju wrote Chaps. 1, 7, 8, 13, and 14; and Dr. Wei Wang wrote Chaps. 11 and 12. The authors would like to thank all the former and current group members who contributed to the investigations on smart hydrogel functional materials, especially Dr. Xin-Cai Xiao, Dr. Yan Li, Dr. Hai-Dong Wang, Dr. Jie Zhang, Dr. Mei Yang, Dr. Ming-Yu Zhou, Dr. Chang-Jing Cheng, Dr. Peng-Fei Li, Dr. Tao Meng, Dr. Gang Chen, Dr. Lie-Wen Xia, Dr. Li-Li Yue, Dr. Li Liu, Dr.

Jian-Bo Qu, Dr. Dan Xu, Lin Hu, Peng Mi, Wen-Chuan Yang, Xiao-Li Zhu, Hao Zhang, Ping-Wei Ren, Jian-Ping Yang, Shi-Bo Zhang, Yong-Chao Chen, Jing Jiang, Han-Guang Wu, Gang Wei, Ji-Yun Wang, Shuo-Wei Pi, Bao Zhang, Cheng-Jing Wu, Guan Wang, Xi Lin, Ya-Lan Yu, Li Mei, Nan-Nan Deng, Zhuang Liu, Chuan-Lin Mu, Jie Wei, Ying-Mei Liu, Hai-Rong Yu, Tao Luo, Hong-Bo Wei, Mao-Jie Zhang, Xiao-Heng He, Ming-Yue Jiang, Zhi-Jun Meng, Chen Yao, Wei-Gang Liang, Xiao-Lu Song, Yi-Meng Sun, Yan-Ni Wang, Chao Yang, and Mei Yuan, for their hard work and creative research on developing smart hydrogel functional materials. The authors gratefully acknowledge all the professors, friends, and colleagues who helped them in carrying out investigations in the field of smart hydrogel functional materials, and thank all the organizations who financially supported the study.

Finally, the authors would like to acknowledge the kind help of Mr. Jinxiang Du at Chemical Industry Press and the editorial staff at Springer during the preparation and publication of this book.

Chengdu, China
October, 2012

Liang-Yin Chu

Contents

Part I
Thermo-responsive Hydrogel Functional Materials

Chapter 1
Structure-Function Relationship of Thermo-responsive Hydrogels

Abstract In this chapter, the structure-function relationship of thermo-responsive hydrogel materials with various internal microstructures is introduced. The fabrication, microstructure control, and performance of thermo-responsive hydrogels are described. The relationships between the internal microstructures of the thermo-responsive hydrogels and their thermo-responsive and mechanical properties, such as equilibrium thermo-responsive phase transition, dynamic thermo-responsive phase transition, thermo-responsive controlled-release characteristics, and mechanical strength, are discussed systematically.

1.1 Introduction

Because in many cases environmental temperature fluctuations can occur naturally and the environmental temperature stimuli can be easily designed and controlled, thermo-responsive hydrogels have attracted much attention and are the most studied smart hydrogel functional materials.

Poly(*N*-isopropylacrylamide) (PNIPAM) is a typical kind of temperature-sensitive polymer and has been widely studied. PNIPAM exhibits a temperature-induced phase transition in aqueous solution at a lower critical solution temperature (LCST) around 32 °C due to the presence of both hydrophilic amide groups and hydrophobic isopropyl groups in its side chains [1, 2]. Since first reported by Hirokawa and Tanaka in 1984 [3], cross-linked PNIPAM hydrogels have attracted increasing interest due to their excellent thermo-responsive properties and potential applications in many fields, such as controlled drug and gene delivery, enzyme and cell immobilization, separation of proteins, and recyclable adsorption. As the environmental temperature varies across its corresponding volume-phase transition temperature (VPTT), PNIPAM hydrogels undergo a reversible volume-phase change (Fig. 1.1). Below the VPTT, the hydrogel is swollen and absorbs a significant amount of water, while above the VPTT, the hydrogel dramatically releases free water and is in shrunken state.

L.-Y. Chu et al., *Smart Hydrogel Functional Materials*,
DOI 10.1007/978-3-642-39538-3_1,

© Chemical Industry Press, Beijing and Springer Berlin Heidelberg 2013

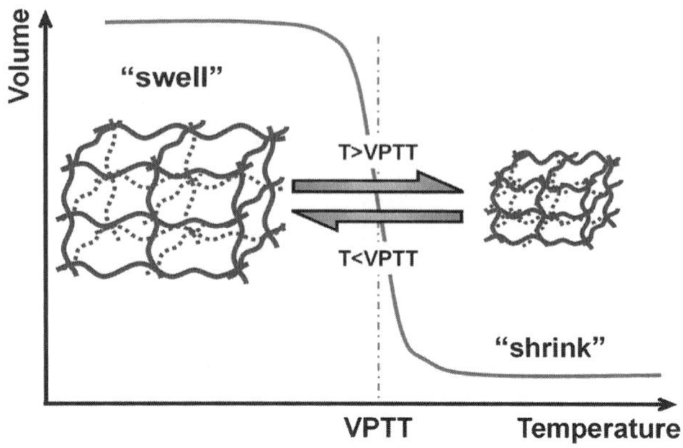

Fig. 1.1 Schematic illustration of thermo-responsive "swell/shrink" volume transition of the PNIPAM hydrogel

Recently, the authors' group has studied the effects of internal microstructures and chemical structures of PNIPAM hydrogels on their thermo-responsive phase transitions and mechanical strength. PNIPAM thermo-responsive hydrogels with different internal microstructures and chemical structures have distinctly different thermo-responsive and mechanical properties.

1.2 Effect of Internal Microstructure on the Equilibrium Thermo-responsive Phase Transition

PNIPAM hydrogels, which are fabricated at two different temperatures, 25 °C (below the LCST of PNIPAM) and 60 °C (above the LCST of PNIPAM), by free-radical cross-linking polymerization with *N,N′*-methylenebisacrylamide (MBA) as cross-linker, are found to have two distinctly different internal microstructures [4]. The preparation recipes and conditions are listed in Table 1.1. The polymerization reaction carried out at 60 °C is a thermally initiated free-radical polymerization using ammonium persulfate (APS) as an initiator. The initiator in the polymerization reaction at 25 °C (below the LCST) is a redox initiator system which is composed of APS and *N,N,N′,N′*-tetramethylethylenediamine (TEMED).

Observed by scanning electron microscopy (SEM), PNIPAM hydrogels prepared at 25 °C (below the LCST of PNIPAM) have a homogeneous porous microstructure like the honeycomb, whereas PNIPAM hydrogels prepared at 60 °C (above the LCST of PNIPAM) have a heterogeneous internal microstructure composed of numerous small gel particles with sub-micrometer dimensions, as shown in Fig. 1.2 [4]. The formation of different microstructures of PNIPAM hydrogels could be mainly affected by the preparation temperature. When the preparation temperature is below the LCST of PNIPAM, water remains in the polymer network phase throughout the whole reaction process, resulting in the formation of a homogeneous internal microstructure. However, when the temperature is above the LCST of

Table 1.1 Preparation recipes and conditions of PNIPAM hydrogels with different internal microstructures (Reproduced with permission from Ref. [4], Copyright (2006), IOP)

Hydrogel sample no.	Monomer [NIPAM] (mol/L)	Cross-linker [MBA]/[NIPAM] (mol/mol)	Preparation temperature (°C)	Preparation time (h)
1-1	1.5	1/100	25	6
1-2	1.5	1/50	25	6
1-3	1.0	1/100	25	6
2-1	1.5	1/100	60	10
2-2	1.5	1/50	60	10
2-3	1.0	1/100	60	10

Note: Initiator dosage [APS]/[NIPAM] = 1/200 (mol/mol)

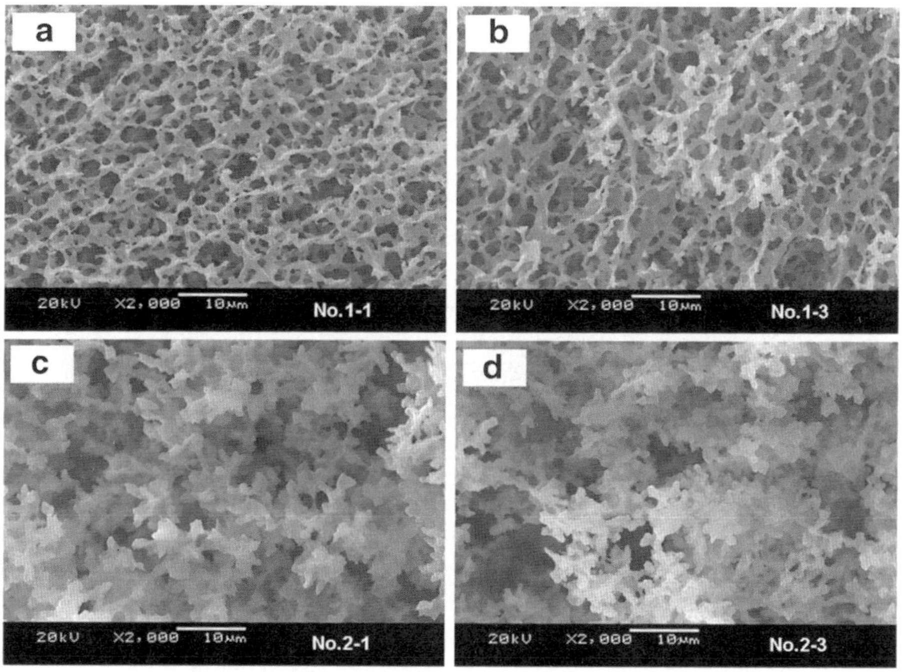

Fig. 1.2 SEM images of the internal microstructures of PNIPAM hydrogels: (**a**, **b**) with homogeneous microstructure (samples no. 1-1 and no. 1-3) and (**c**, **d**) with heterogeneous microstructure (samples no. 2-1 and no. 2-3). Scale bars are 10 μm (Reproduced with permission from Ref. [4], Copyright (2006), IOP)

PNIPAM, water separates out of the polymer network phase during the reaction. Thus, the polymerization system tends to separate into polymer-rich and polymer-poor regions. Due to the high concentration of vinyl groups in polymer-rich regions, one might expect enhanced rates of cross-linking and multiple cross-linking reactions, which would lead to the formation of highly cross-linked, microgel-like

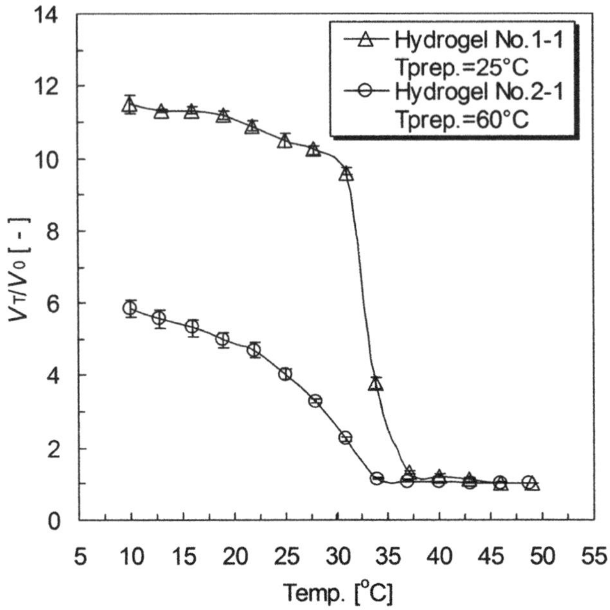

Fig. 1.3 Temperature dependence of the equilibrium volume deswelling ratios of PNIPAM hydrogels with different internal microstructures in pure water (Reproduced with permission from Ref. [4], Copyright (2006), IOP)

regions in the final gels. On the other hand, the polymer-poor regions became slightly cross-linked due to the high content of water; as a result, they constitute the interstices between the microgels in the final polymer hydrogel networks. In addition, the cross-linker is more soluble in water at higher temperature, but the PNIPAM macroradicals would obviously come out of the solution phase as a consequence of the LCST. This might be one of the reasons for the nonuniform microstructure.

The relationships between the internal microstructure of PNIPAM hydrogels and the equilibrium thermo-responsive volume-phase transition are experimentally investigated [4]. As shown in Fig. 1.3, it can be clearly seen that the degree of volume-phase transition (change of equilibrium volume deswelling ratio versus ambient temperature around the LCST) of PNIPAM hydrogels with homogeneous internal microstructures prepared below the LCST is remarkably greater than that of hydrogels with heterogeneous internal microstructures prepared above the LCST. The reason for the above phenomenon should be that there are larger interspaces in the internal structure of hydrogels when prepared below the LCST. The volume change of hydrogels is considered to be related to the amount of water in hydrogels. Larger interspaces could contain more water inside the polymer networks and will result in a greater change in volume when the ambient temperature changes across the LCST.

1.3 Effect of Internal Microstructure on the Dynamic Thermo-responsive Phase Transition

For some potential applications of thermo-responsive hydrogels, a fast response to temperature change is necessary for their practical applications. For instance, some targeted drug delivery systems need hydrogels to release drugs immediately in a specific time and location, while an acting actuator requires an instantaneous feedback after receiving signals.

The relationships between the internal microstructure of PNIPAM hydrogels and the dynamic thermo-responsive volume-phase transition are also investigated [4]. PNIPAM hydrogels are first equilibrated in deionized water at 25 °C. The dynamic experiment is carried out by changing the temperature of deionized water around PNIPAM hydrogels from 25 °C to 40 °C at a rate of 1.5 °C/min using a thermostatic unit, and the instantaneous volume of the hydrogel is obtained at a regular time interval of 1 min. Interestingly, it is found that the PNIPAM hydrogel with a heterogeneous internal microstructure exhibits a more rapid thermo-responsive volume-phase transition speed than that with a homogeneous internal microstructure does, as shown in Fig. 1.4. The reason might be that the volume deformation of the PNIPAM hydrogel with a homogeneous honeycomb-like network might be restricted due to three-dimensional stereo restraint. Therefore, the volume-phase transition of the homogeneous hydrogel could not be very fast. On the other hand, the microstructure of the PNIPAM hydrogel prepared above the LCST is heterogeneous with numerous microgel-like particles of sub-micrometer dimensions. These microgel-like particle clusters have numerous free ends; as a result, the volume-phase transition of the hydrogel could be very rapid, because these free ends inside the hydrogel could flex without any restriction and consequently the deformation could be fast in respond to the change of ambient temperature. The situation for the PNIPAM hydrogel with a heterogeneous internal microstructure, i.e., with numerous free-end microgel-like clusters, is very similar to that of the comb-type grafted PNIPAM hydrogels [5, 6]. Because the grafted chains in the comb-type grafted PNIPAM hydrogels have free ends, the comb-type grafted PNIPAM hydrogels are reported to have a rapid deswelling response to temperature changes. Like the grafted chains of the comb-type grafted hydrogels, the microgel-like particle clusters in the heterogeneous hydrogel could dehydrate immediately when the ambient temperature increases across the LCST; as a result, the volume change of the hydrogel could be rapid.

As mentioned above, an effective method for improving the response rate of thermo-responsive hydrogels is demonstrated by fabricating a comb-type grafted hydrogel with freely mobile chains grafted onto its backbone network [5, 6]. The authors' group successfully prepared a novel dual thermo- and pH sensitive comb-type grafted hydrogel by grafting PNIPAM macromonomer chains with freely

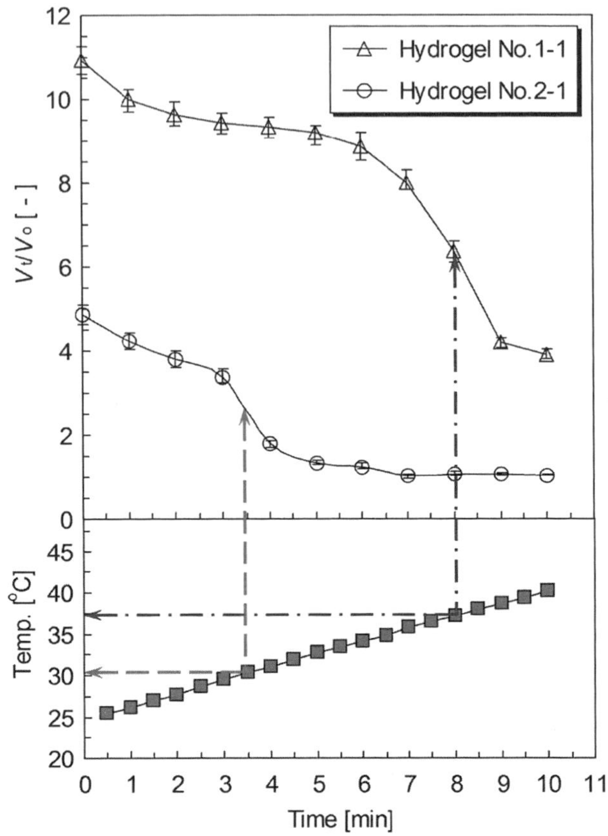

Fig. 1.4 Dynamic volume deswelling behaviors of PNIPAM hydrogels with different internal microstructures in pure water (Reproduced with permission from Ref. [4], Copyright (2006), IOP)

mobile ends onto the backbone of the cross-linked poly(*N*-isopropylacrylamide-*co*-acrylic acid) (P(NIPAM-*co*-AAc)) network [7]. Schematic structures of normal-type and comb-type grafted P(NIPAM-*co*-AAc) hydrogels are shown in Fig. 1.5. The grafted chains within the comb-type grafted hydrogel have freely mobile ends. That is, the network structure of the comb-type grafted hydrogel is distinct from that of the normal-type hydrogel, in which the ends of chains are cross-linked and relatively immobile.

The PNIPAM macromonomer is synthesized by radical telomerization of NIPAM monomer using 2-hydroxyethanethiol (HESH) as a chain transfer agent as previously described by the literature [8]. The normal-type P(NIPAM-*co*-AAc) hydrogel is prepared by free-radical copolymerization with NIPAM and AAc as comonomers, MBA as cross-linker, APS as redox initiator, and TEMED as accelerator. As illustrated in Fig. 1.6, the comb-type grafted P(NIPAM-*co*-AAc) hydrogel is synthesized by free-radical copolymerization of PNIPAM macromonomer with NIPAM and AAc [7]. The backbone networks of the comb-type grafted P(NIPAM-*co*-AAc) hydrogel are made up of NIPAM and AAc components, and the linear PNIPAM polymers served as the freely mobile chains are grafted onto the backbone by fixing

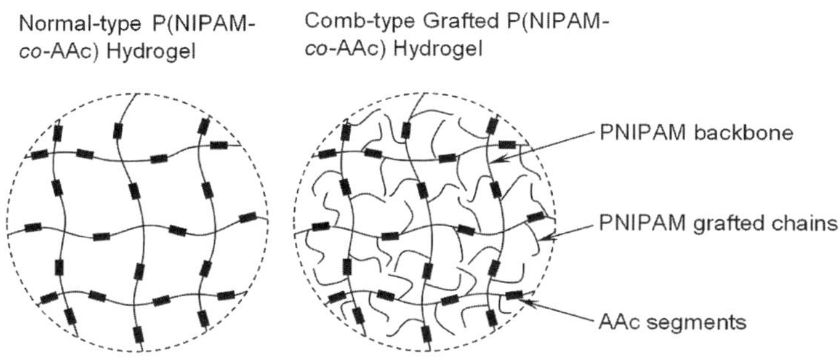

Fig. 1.5 Schematic structures of normal-type and comb-type grafted P(NIPAM-*co*-AAc) hydrogels (Reproduced with permission from Ref. [7], Copyright (2007), Elsevier)

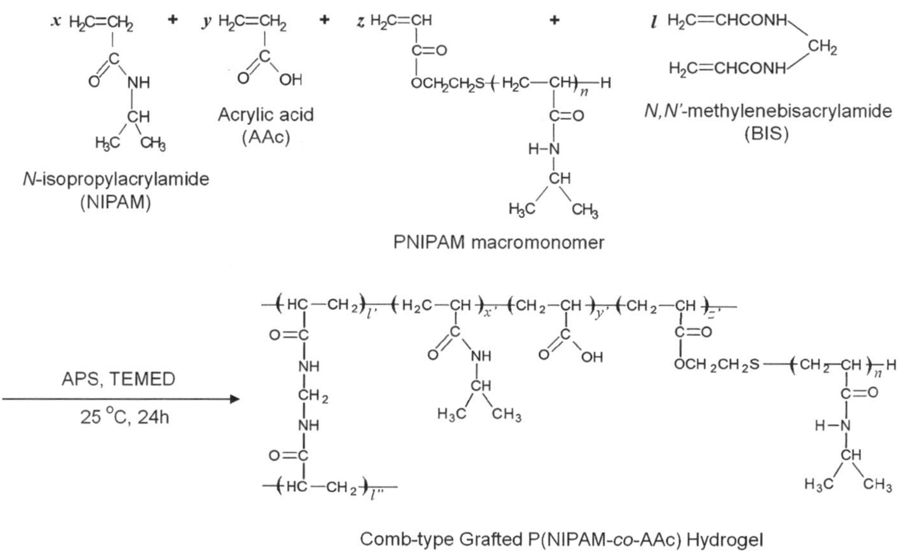

Fig. 1.6 Synthetic scheme for the preparation of comb-type grafted P(NIPAM-*co*-AAc) hydrogels by radical copolymerization (Reproduced with permission from Ref. [7], Copyright (2007), Elsevier)

one end structurally. The feed compositions for preparation of normal-type and comb-type grafted P(NIPAM-*co*-AAc) hydrogels are listed in Table 1.2. The comb-type grafted P(NIPAM-*co*-AAc) hydrogels constructed with different contents of NIPAM and macromonomer are designated as GNA30-1 and GNA30-2, and normal-type P(NIPAM-*co*-AAc) hydrogel without grafted chains is designated as NNA30.

The prepared comb-type grafted P(NIPAM-*co*-AAc) hydrogels exhibit a more rapid deswelling rate than normal-type P(NIPAM-*co*-AAc) hydrogel in deionized water in response to abrupt temperature change from 25 °C to 60 °C, as shown in Fig. 1.7a [7]. And the quick response is ranked in the order of the feed weight of

Table 1.2 Feed compositions for the preparation of normal-type and comb-type grafted P(NIPAM-*co*-AAc) hydrogels (Reproduced with permission from Ref. [7], Copyright (2007), Elsevier)

Component	NNA30	GNA30-1	GNA30-2
NIPAM (g)	1.0	0.6	0.5
Macromonomer (g)	0	0.4	0.5
AAc (μL)	30	30	30

Note: Solution (ultrapure water) = 20 mL, MBA = 0.02 g, TEMED = 50 mL. 5 wt% APS = 0.2 mL

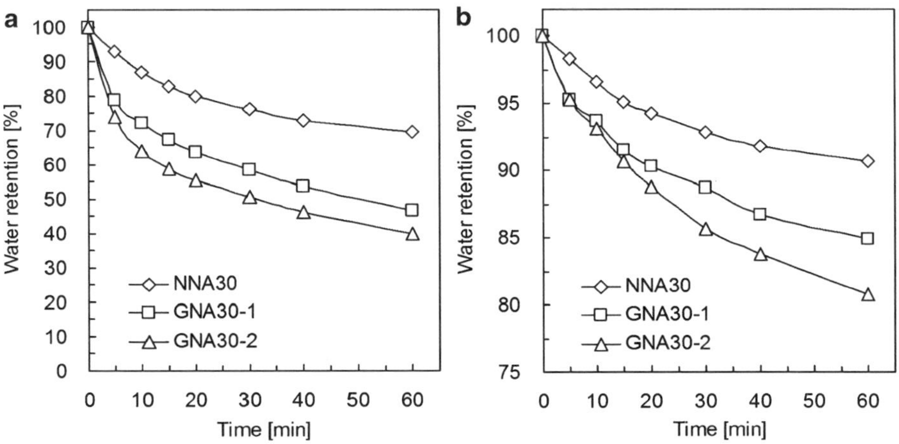

Fig. 1.7 Deswelling kinetics of different P(NIPAM-*co*-AAc) hydrogels at 60 °C as measured from an equilibrium swelling condition at 25 °C in deionized water (**a**) and that in buffer at pH 2.0 as measured from an equilibrium swelling condition in buffer at pH 7.4 at 25 °C (**b**) (Reproduced with permission from Ref. [7], Copyright (2007), Elsevier)

PNIPAM macromonomer in response to temperature changes. Such rapid shrinking of the comb-type grafted hydrogel is due to the immediate dehydration of freely mobile grafted chains, followed by subsequent hydrophobic interactions between dehydrated grafted chains preceding shrinkage of the whole network. On the other hand, dense shrunken skin layers are not formed on the surface of the hydrogel due to decreased hydrophobic aggregation forces by hydrophilic AAc component. The same is true in buffer solution with a pH value jumping from 7.4 to 2.0 at 25 °C, as shown in Fig. 1.7b [7]. This deswelling with pH change is due to the pH sensitivity of the AAc incorporated into the backbone of hydrogels. With the same AAc content, comb-type grafted hydrogels deswell faster than normal-type hydrogel, and the deswelling rate increases with an increasing amount of grafted chains. For comb-type grafted hydrogels, the AAc units in the backbone may have been closer, owing to the higher density of pH-sensitive comonomers. Therefore, the ability of the hydrogel to respond to pH changes is enhanced, thereby reducing the response time. As expectedly, when temperature and pH simultaneously change

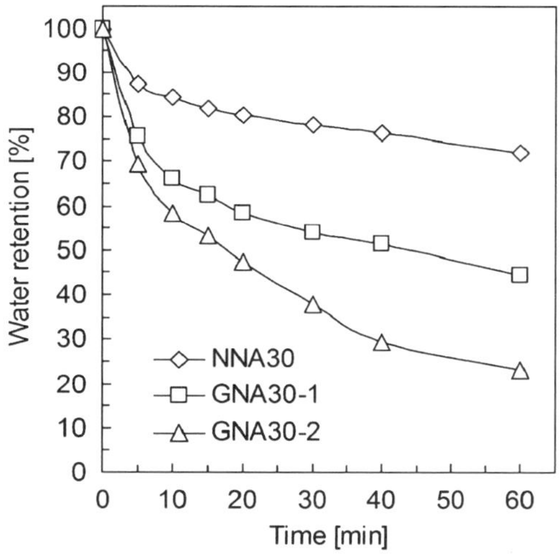

Fig. 1.8 Deswelling kinetics of different P(NIPAM-*co*-AAc) hydrogels in buffer at pH 2.0 and 60 °C as measured from an equilibrium swelling condition in buffer at pH 7.4 and 25 °C (Reproduced with permission from Ref. [7], Copyright (2007), Elsevier)

from pH 7.4/25 °C to pH 2.0/60 °C, much quicker shrinkage of comb-type grafted P(NIPAM-*co*-AAc) hydrogels is observed because of the cooperation of thermo- and pH responses, as shown in Fig. 1.8 [7].

The authors' group developed a novel type of microgel-cross-linked hydrogel (MCG) with both fast response rate and large volume-change ratio to temperature change [9]. As shown in Fig. 1.9, the proposed hydrogel is featured with a novel microstructure that is constructed by cross-linking thermo-responsive P(NIPAM-*co*-AAc) linear chains with thermosensitive P(NIPAM-*co*-AAc) microgels as the cross-linker and fabricating grafted P(NIPAM-*co*-AAc) thermosensitive linear chains on the microgel surfaces. The proposed MCG P(NIPAM-*co*-AAc) hydrogel is fabricated in a three-step process. Firstly, P(NIPAM-*co*-AAc) microgels with carboxyl acid groups on the surface are prepared by precipitation copolymerization of NIPAM and AAc at 60 °C (Fig. 1.9a). Secondly, P(NIPAM-*co*-AAc) microgels are treated with cold hydrogen peroxide (H_2O_2) containing a certain amount of concentrated H_2SO_4 at 25 °C to generate peracid groups on the surface (Fig. 1.9b) [10]. Thirdly, P(NIPAM-*co*-AAc) microgels with peracid groups on the surface are used as both cross-linker and initiator to prepare the MCG P(NIPAM-*co*-AAc) hydrogel by free-radical copolymerization of NIPAM and AAc at 60 °C (Fig. 1.9c, d).

The internal microstructures, the equilibrium, and the dynamic phase transitions of the MCG P(NIPAM-*co*-AAc) hydrogels are studied systematically [9]. The MCG hydrogels prepared with different concentrations of microgel dispersions (1.9 wt%, 3.8 wt%, and 7.6 wt%) are labeled as MCG1.9, MCG3.8, and MCG7.6. The normal P(NIPAM-*co*-AAc) hydrogel prepared at 0 °C using KPS and TEMED as the redox initiator system is labeled as NG0, and the normal P(NIPAM-*co*-AAc) hydrogel prepared at 60 °C using KPS as the initiator is labeled as NG60.

As shown in SEM images in Fig. 1.10 [9], P(NIPAM-*co*-AAc) microgels prepared by the precipitation copolymerization are quite monodisperse (Fig. 1.10a), and the chemical treatment on microgels with H_2O_2 does not affect their sphericity

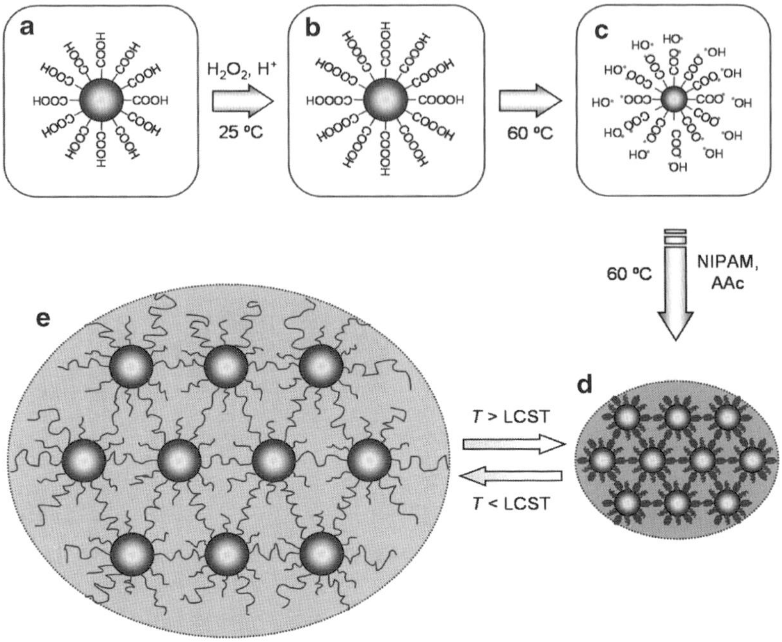

Fig. 1.9 Fabrication illustration of the MCG P(NIPAM-*co*-AAc) hydrogel. (**a**) P(NIPAM-*co*-AAc) microgel with carboxyl acid groups on the surface, (**b**) P(NIPAM-*co*-AAc) microgel with peracid groups on the surface, (**c**) P(NIPAM-*co*-AAc) microgel with free radicals on the surface, (**d**) MCG P(NIPAM-*co*-AAc) hydrogel showing shrunken state at temperature above the LCST, (**e**) MCG P(NIPAM-*co*-AAc) hydrogel showing swollen state at temperature below the LCST (Reproduced with permission from Ref. [9], Copyright (2010), Elsevier)

and monodispersity (Fig. 1.10b). The MCG P(NIPAM-*co*-AAc) hydrogel presents a heterogeneous internal microstructure (Fig. 1.10c), and the microgel-cross-linked polymeric network can be clearly seen in Fig. 1.10d. Such heterogeneous internal microstructure resulted from the formation process of the hydrogel. When P(NIPAM-*co*-AAc) microgels with peracid groups heated to 60 °C in water, the peracid groups on the microgel surfaces decompose to generate a mass of free radicals COO· on the microgel surfaces and ·OH in solvent (Fig. 1.9c). Subsequently, COO· initiates the graft polymerization of NIPAM and AAc monomers onto microgels, and ·OH initiates the copolymerization of NIPAM and AAc in solvent. On one hand, the bulk hydrogel is formed by coupling reaction between the grafted polymer free radicals on one microgel and the grafted polymer free radicals on neighboring microgels and/or termination reaction of the grafted polymer free radicals on one microgel with COO· on neighboring microgels. On the other hand, P(NIPAM-*co*-AAc) dangling chains on microgel surfaces are formed by termination reaction between the grafted polymer free radicals on microgels and free radicals existed in solvent and/or terminators unavoidably in the system as impurity and/or by the chain transfer reaction.

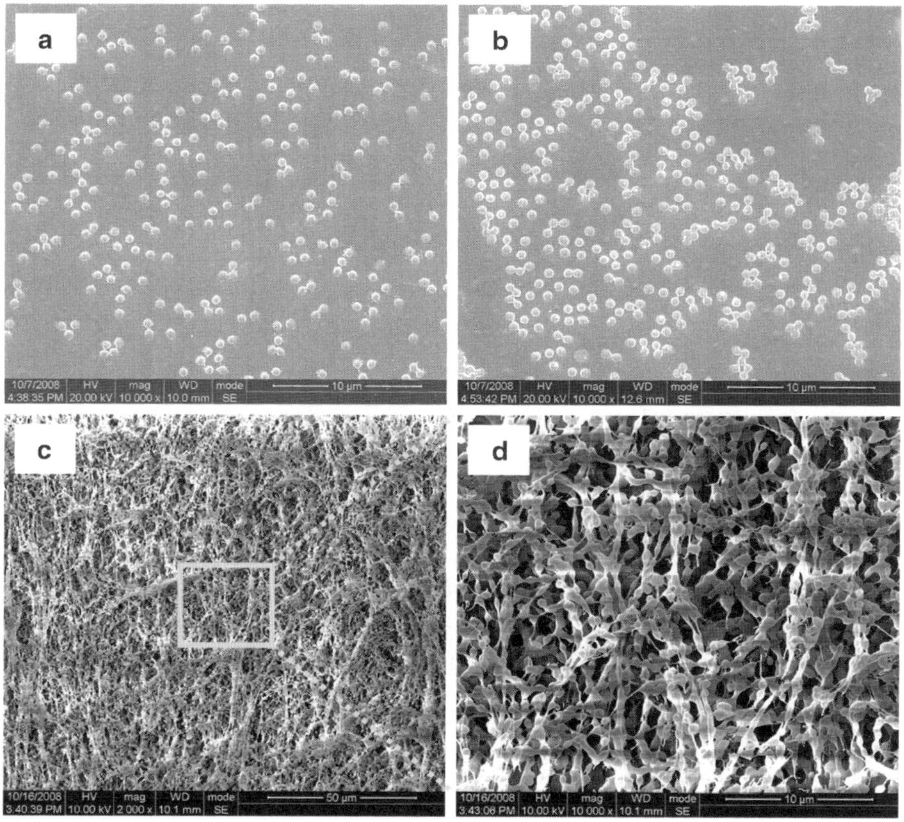

Fig. 1.10 SEM images of P(NIPAM-*co*-AAc) microgels before (**a**) and after (**b**) being treated with H_2O_2 and SEM images of internal microstructures of MCG3.8 hydrogel (**c**, **d**). The *scale bars* in (**a**), (**b**), and (**d**) are 10 μm and that in (**c**) is 50 μm (Reproduced with permission from Ref. [9], Copyright (2010), Elsevier)

MCG P(NIPAM-*co*-AAc) hydrogels exhibit larger thermo-responsive volume change than normal hydrogels (NG0 and NG60) as shown in Fig. 1.11 [9]. The functional microgel content in the MCG hydrogel has remarkable influence on the swelling ratio of MCG hydrogels. The lower the microgel content in the MCG hydrogel is, the larger swelling ratio the MCG hydrogel shows. When lower concentration of microgel dispersion is used in MCG hydrogel preparation, longer P(NIPAM-*co*-AAc) chain should form between microgels because the polymeric radicals on one microgel need to be long enough to meet radicals on another microgel. Longer P(NIPAM-*co*-AAc) chain can stretch its ends to longer distance and leads to higher volume change of MCG hydrogels. In other words, under the condition of lower microgel content, the cross-linking degree is lower and the swelling ratio is larger. Thus, the MCG1.9 hydrogel presents largest swelling ratio at 10 °C and has largest thermo-responsive volume-phase transition degree.

Fig. 1.11 Temperature dependence of the equilibrium swelling ratio of normal and MCG P(NIPAM-*co*-AAc) hydrogels (Reproduced with permission from Ref. [9], Copyright (2010), Elsevier)

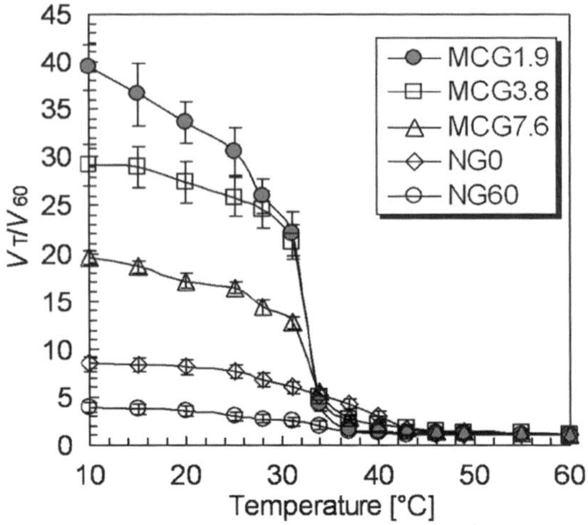

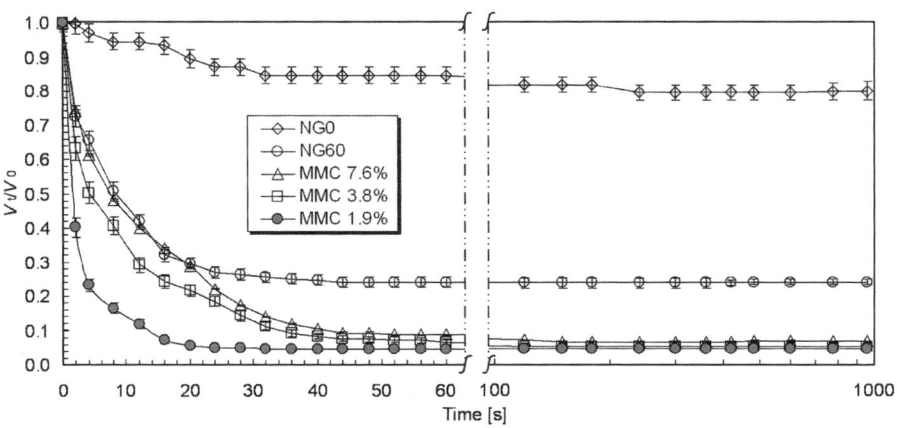

Fig. 1.12 Dynamic deswelling behaviors of normal and MCG P(NIPAM-*co*-AAc) hydrogels after the environmental temperature jumping abruptly from 15 °C to 55 °C (Reproduced with permission from Ref. [9], Copyright (2010), Elsevier)

Figure 1.12 shows the dynamic deswelling behaviors of normal and MCG P(NIPAM-*co*-AAc) hydrogels after the environmental temperature jumping abruptly from 15 °C to 55 °C [9]. It can be clearly seen that MCG hydrogels present much faster deswelling rates than NG hydrogels. The MCG hydrogel presents interconnected 3-D micro-spaces between polymeric networks, which are beneficial for water transport during the volume-phase transition process. The cooperative function of all the fast shrinking of microgels, the fast collapse of the grafted dangling chains on the microgel surfaces, and the 3-D interconnected water transportation channels between the microgels and polymeric networks inside the proposed hydrogels make the MCG hydrogels respond very fast to the environmental temperature change. The MCG hydrogel with improved thermo-responsive phase transition property proposed provides a novel strategy for developing efficient stimuli-responsive hydrogel materials for sensors, actuators, bioseparation absorbents, and so on.

1.4 Effect of Internal Microstructure on the Thermo-responsive Controlled-Release Characteristics

For the purpose of thermo-responsive controlled release, a rapid response rate to ambient temperature is expected. Therefore, PNIPAM-based hydrogels with a heterogeneous internal microstructure would be suitable to be used for thermo-responsive controlled-release systems.

In order to study the thermo-responsive controlled-release characteristics of PNIPAM hydrogels with different internal microstructures, sodium chloride (NaCl) is chosen as the model solute in the release experiments [4]. The NaCl-loaded hydrogel is transferred into deionized water at 25 °C, and the temperature of the ambient water is immediately increased from 25 °C to 40 °C at a rate of 1.5 °C/min using a thermostatic unit. During the period of increasing temperature, the instantaneous volume of the hydrogel and the NaCl concentration of the ambient solution are measured simultaneously at regular time intervals.

Figure 1.13 shows the thermo-responsive controlled-release characteristics of NaCl-loaded PNIPAM hydrogels with different internal microstructures and the corresponding dynamic volume deswelling behaviors of these PNIPAM hydrogels [4]. Obviously, the thermo-responsive release of NaCl from hydrogels is a result of the dynamic volume change of hydrogels. Because of the rapid response of volume-phase transition to the environmental temperature, PNIPAM hydrogel no. 2-1 with heterogeneous internal microstructure exhibits a much more satisfactory thermo-responsive controlled-release property than hydrogel no. 1-1 does. For PNIPAM hydrogel no. 2-1, a rapid NaCl release is found just corresponding to the abrupt volume deswelling; however, such result is not found for hydrogel no. 1-1 with a homogeneous internal microstructure.

1.5 Effect of Internal Microstructure on the Mechanical Strength of Thermo-responsive Hydrogels

Mechanical strength is also an important factor for thermo-responsive hydrogel materials used in industrial and biomedical fields. The mechanical properties are highly dependent on the polymer network structure of the hydrogel. Many efforts have been focused on increasing the mechanical strength of the hydrogels with different network structure [11].

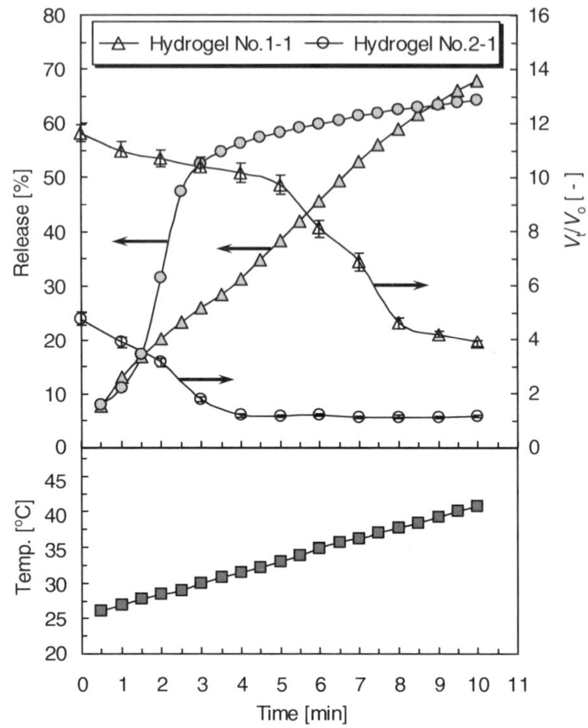

Fig. 1.13 Thermo-responsive controlled-release characteristics of solutes from PNIPAM hydrogels with different internal microstructures and the corresponding dynamic volume deswelling behaviors of NaCl-loaded PNIPAM hydrogels (Reproduced with permission from Ref. [4], Copyright (2006), IOP)

The authors' group also developed another kind of thermo-responsive MCG hydrogel, which could bear repeated elongations and compressions. This second kind of MCG hydrogels is prepared by copolymerization of modified poly(acrylamide) (mPAM) microgel and NIPAM in water (Fig. 1.14). The mPAM microgel with unsaturated vinyl bonds on its surface is prepared by treating PAM microgel with *N*-methylolacrylamide. Table 1.3 lists the preparation recipes for these mPAM microgel-cross-linked MCG hydrogels.

As shown in Fig. 1.15, the internal microstructure of mPAM microgel-cross-linked PNIPAM hydrogel is similar to the MCG hydrogels constructed by H_2O_2-treated P(NIPAM-*co*-AAc) microgels as mentioned above. Such MCG PNIPAM hydrogel also exhibits larger degree of volume change and faster response rate to the ambient temperature change compared to normal PNIPAM hydrogel due to the similar microgel-cross-linked internal microstructure.

Figure 1.16 shows the compression tests of MCG2-1.5 and NG2-1.5 hydrogels at temperature lower than their LCSTs. It can be clearly seen that MCG hydrogel exhibits better compression resistance than normal hydrogel NG2-1.5. Furthermore, such MCG hydrogel cross-linked with mPAM microgels also shows good tensile properties. Figure 1.17 shows the typical elongation process of MCG3-1.5 and NG2-1.5 hydrogels at temperature lower than their LCSTs. The MCG

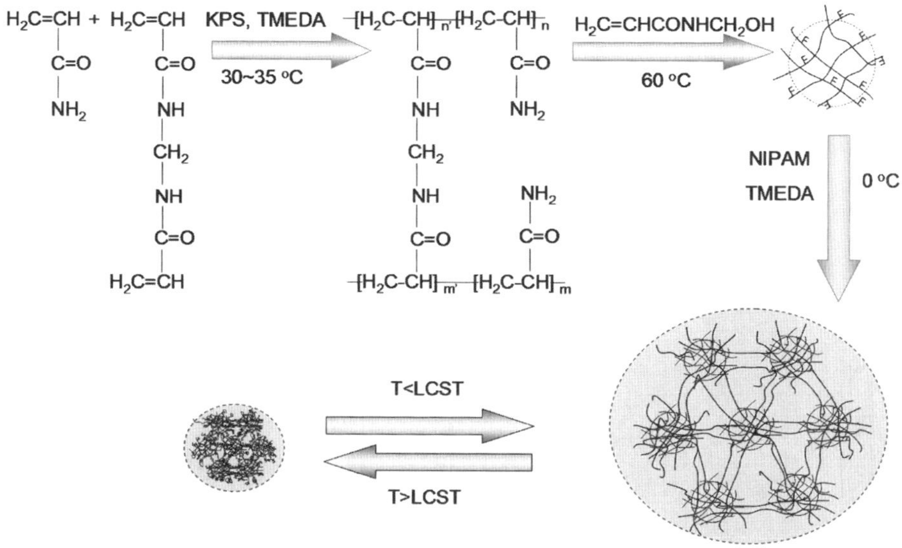

Fig. 1.14 The schematic illustration of the preparation of MCG PNIPAM hydrogel cross-linked by mPAM microgels with unsaturated vinyl bonds

Table 1.3 Preparation recipes of mPAM microgel-cross-linked PNIPAM hydrogels

Hydrogel sample no.	Water (g)	Cross-linker mPAM microgel (g)	Monomer NIPAM (g)	Cross-linker MBA (g)	Initiator KPS (g)	Accelerator TEMED (μL)
MCG2-0.5	19.6	0.4	1.13	0	0.027	30
MCG2-1.5	19.6	0.4	3.39	0	0.081	70
MCG3-1.5	19.6	0.6	3.36	0	0.081	70
NG2-1.5	20	0	3.39	0.0924	0.081	70

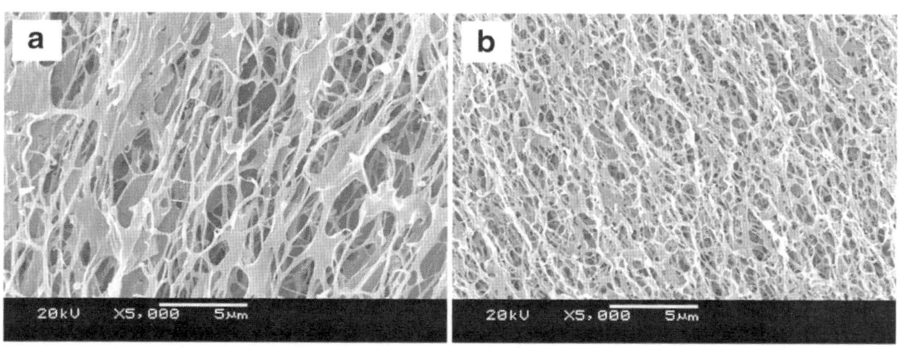

Fig. 1.15 SEM images of the internal microstructures of (**a**) MCG2-1.5 and (**b**) MCG2-0.5 PNIPAM hydrogels. The scale bars are 5 μm

Fig. 1.16 Compression tests of MCG2-1.5 and NG2-1.5 hydrogels at temperature lower than their LCSTs: (**a**, **d**) before compression, (**b**, **e**) under compression, and (**c**, **f**) after compression. (**a**, **b**, **c**) NG2-1.5 hydrogel and (**d**, **e**, **f**) MCG2-1.5 hydrogel. The scale bars are 10 mm

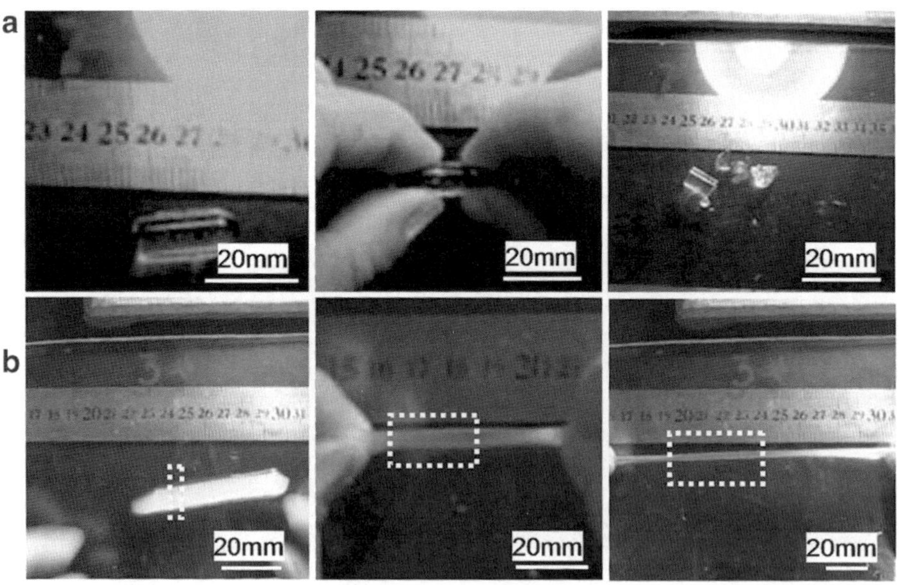

Fig. 1.17 The elongation process of MCG3-1.5 and NG2-1.5 hydrogels at temperature lower than their LCSTs. (**a**) NG2-1.5 hydrogel and (**b**) MCG3-1.5 hydrogel. The scale bars are 20 mm

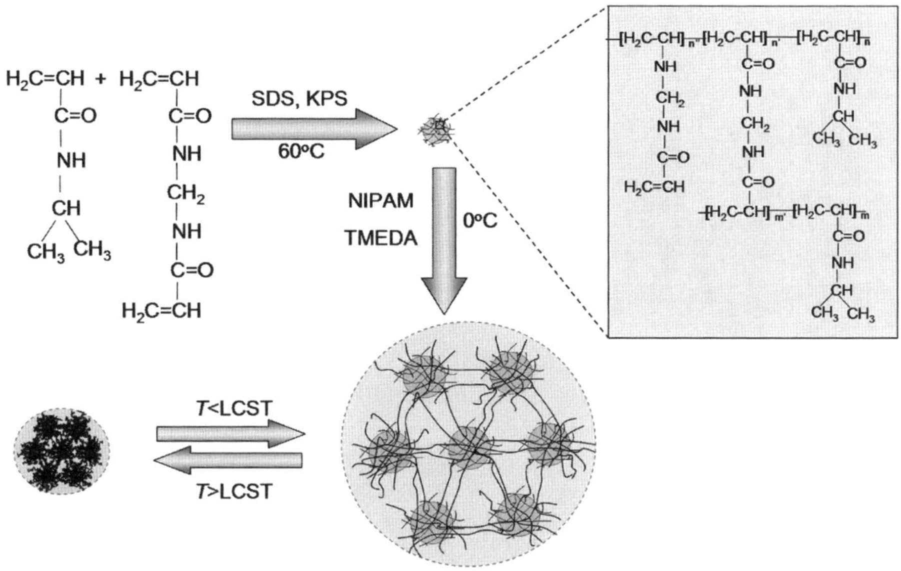

Fig. 1.18 Scheme for the preparation of microgel composite PNIPAM hydrogel using active PNIPAM microgels as the cross-linker

Table 1.4 Preparation recipes of the third kind of MCG hydrogels

Hydrogel sample no.	NIPAM (g)	Water (g)	aPN-xx dispersion (g)
MCGxx-1.5	0.8475 (1.5 mmol/gH_2O)	0.5	4.5
MCGxx-1.0	0.565 (1.0 mmol/gH_2O)	0.5	4.5
MCGxx-0.5	0.2825 (0.5 mmol/gH_2O)	0.5	4.5
MCGxx-0.25	0.1413 (0.25 mmol/gH_2O)	0.5	4.5
MCGxx-1.5-0.5C	0.8475 (1.5 mmol/gH_2O)	2.5	2.5

Note: Preparation temperature = 0 °C, TEMED = 20 μL. The MCG PNIPAM hydrogels are labeled as MCGxx-yy, in which "*aPN-xx*" refers to the polymerization time in active PNIPAM microgel preparation and *yy* refers to the NIPAM concentration

hydrogel could be elongated so much without fracture. More preferably, the mPAM microgel-cross-linked hydrogel can be elongated and compressed repeatedly and can restore to its original shape and size after experienced large strain.

To further improve the mechanical strength of thermo-responsive PNIPAM hydrogels, the authors' group developed the third kind of MCG hydrogels, which is prepared by copolymerization of NIPAM and active PNIPAM microgels. The schematic illustration of the preparation process is shown in Fig. 1.18, and the active PNIPAM microgels with unsaturated carbon bonds are prepared by precipitation polymerization of NIPAM in water at 60 °C using sodium dodecyl sulfate (SDS)as space obstructor. Table 1.4 lists the preparation recipes of the third kind of MCG hydrogels. The content of unsaturated carbon bonds in active PNIPAM microgels decreases with the increase of polymerization time.

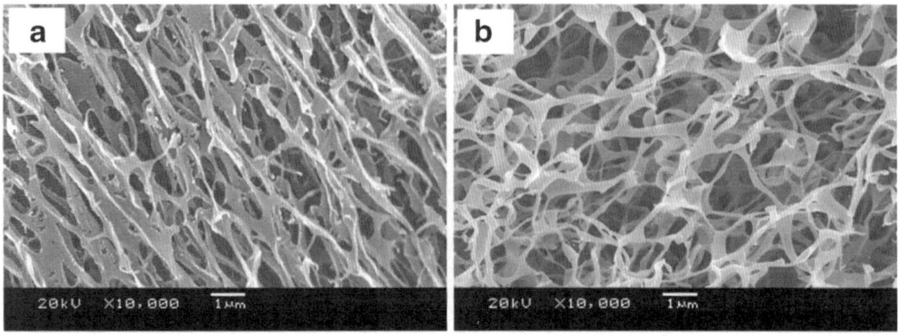

Fig. 1.19 SEM images of the internal microstructures of MCG hydrogels prepared with different aPN microgel concentrations: (**a**) MCG40-1.5 and (**b**) MCG40-1.5-0.5C. The scale bars are 1 μm

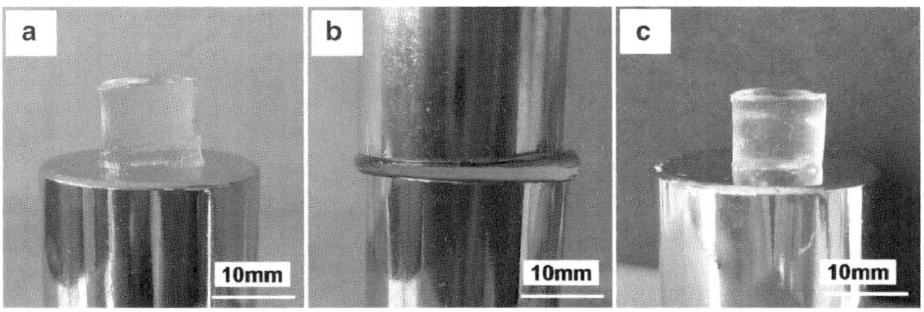

Fig. 1.20 Photos of MCG20-1.5-0.5C hydrogel compressed at temperature lower than its LCST: (**a**) before compression, (**b**) under compression, and (**c**) after compression. The scale bars are 10 mm

The third kind of MCG hydrogels using active PNIPAM microgels as cross-linker presents heterogeneous microstructure like cross-linked threads, as shown in Fig. 1.19. So, such MCG hydrogels also exhibit larger volume change and faster response rate to the ambient temperature change compared to normal PNIPAM hydrogel. Especially, this kind of MCG hydrogels is able to restore to their initial shape and size after being compressed, chopped, bent, contorted, knotted, or elongated with large strain, as shown in Figs. 1.20, 1.21, and 1.22. The high mechanical strength of MCG hydrogels is from the internal microstructure. There exist a large number of long PNIPAM bridge chains between active PNIPAM microgels, which can disperse the stress added on MCG hydrogels. The long PNIPAM bridge chains have sufficient space for configuration change, which could be available to withstand the stress even at large strain.

There exists a hysteresis circle when the third kind of MCG hydrogels restores after being elongated, as shown in Fig. 1.23. The degree of hysteresis for the MCG hydrogels is affected by cross-linking density and decreases when the concentrations of initial NIPAM and aPN microgels increase or the polymerization time in the preparation of aPN microgels decreases.

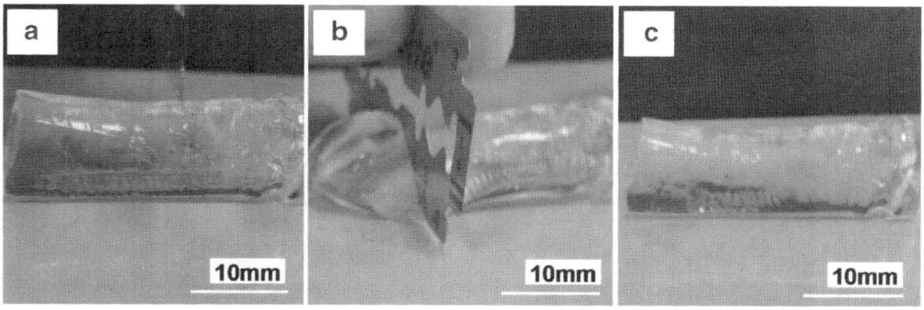

Fig. 1.21 Photos of MCG20-1.5-0.5C hydrogel cut by a blade at temperature lower than its LCST: (**a**) before cutting, (**b**) under cutting, and (**c**) after cutting. The scale bars are 10 mm

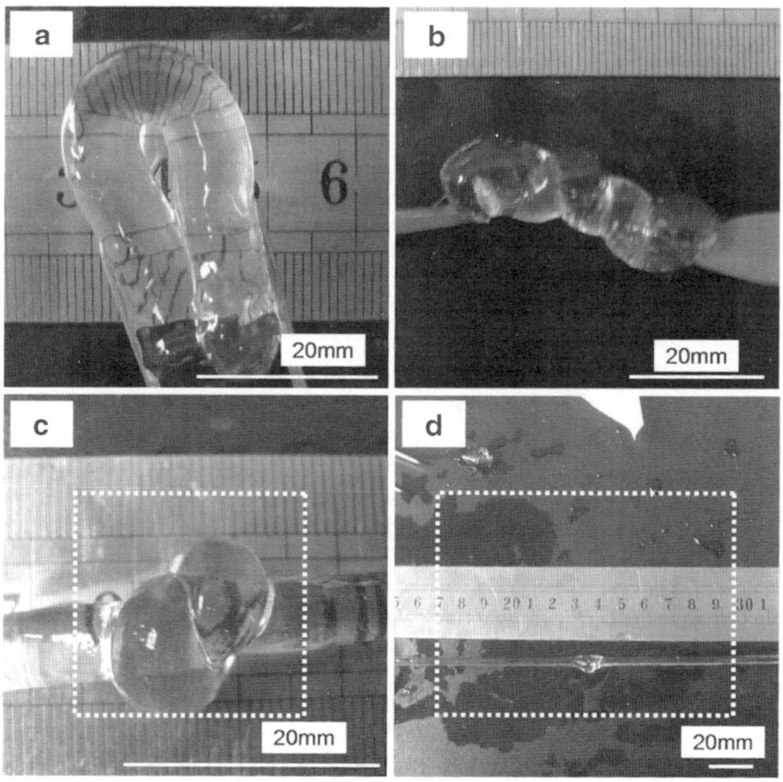

Fig. 1.22 Photos of MCG20-1.5-0.5C hydrogel demonstrating its elasticity characteristics at temperature lower than its LCST. (**a**) bending, (**b**) torsion, (**c**) knotting, and (**d**) elongation after knotting. The scale bars are 20 mm

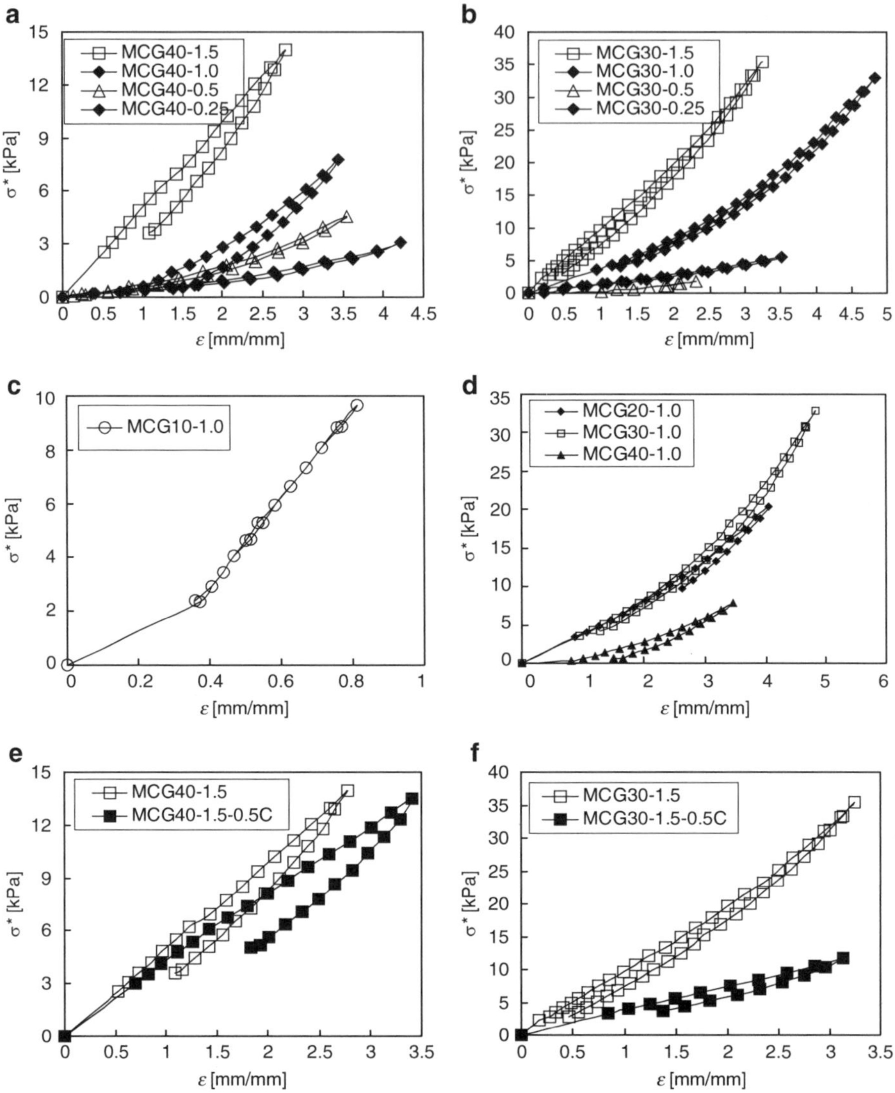

Fig. 1.23 The elongation and restoration curve of MCG PNIPAM hydrogels at temperature lower than their LCSTs. (**a**, **b**) MCG hydrogels prepared with different NIPAM concentrations, (**c**, **d**) MCG hydrogels prepared with different aPN microgel, and (**e**, **f**) MCG hydrogels prepared with different aPN microgel concentrations

1.6 Summary

According to the above results, the PNIPAM-based thermo-responsive hydrogels with different internal microstructures and chemical structures have distinctly different thermo-responsive and mechanical properties. The PNIPAM hydrogel with a homogeneous internal microstructure shows a remarkably greater volume change; on the other hand, the PNIPAM hydrogel with a heterogeneous internal microstructure exhibits a more rapid response rate to ambient temperature and correspondingly has a more rapid thermo-responsive controlled-release property. Comb-type grafted P(NIPAM-*co*-AAc) hydrogels with freely mobile grafted chains exhibit rapid responses to both temperature and pH stimuli. Thermo-responsive MCG hydrogel with fast response rate and large thermo-responsive volume change and/or improved mechanical strength is achieved using macromolecular microgels as the cross-linker. These properties of a specific hydrogel are extremely important in selecting which materials are suitable for a given application. For instance, for pulse controlled-release systems, a fast response rate to ambient temperature is required; on the other hand, for some separation and extraction systems, a large volume change is needed. So, the thermo-responsive hydrogel materials should be designed and fabricated with suitable internal microstructure and molecular structure for different purposes.

References

1. Heskins M, Guillet JE (1968) Solution properties of poly(*N*-isopropylacrylamide). J Macromol Sci Chem A2:1441–1455
2. Schild HG (1992) Poly(*N*-isopropylacrylamide): experiment, theory and application. Prog Polym Sci 17:163–249
3. Hirokawa Y, Tanaka T (1984) Volume phase transition in a nonionic gel. J Chem Phys 81:6379–6380
4. Ju XJ, Chu LY, Zhu XL et al (2006) Effects of internal microstructures of poly(*N*-isopropylacrylamide) hydrogels on thermo-responsive volume phase-transition and controlled-release characteristics. Smart Mater Struct 15:1767–1774
5. Yoshida R, Uchida K, Kaneko Y et al (1995) Comb-type grafted hydrogels with rapid de-swelling response to temperature changes. Nature 374:240–242
6. Kaneko Y, Sakai K, Kikuchi A et al (2006) Influence of freely mobile grafted chain length on dynamic properties of comb-type grafted poly(*N*-isopropylacrylamide) hydrogel. Macromolecules 28:7717–7723
7. Zhang J, Chu LY, Li YK et al (2007) Dual thermo- and pH-sensitive poly(*N*-isopropylacrylamide-co-acrylic acid) hydrogels with rapid response behaviors. Polymer 48:1718–1728
8. Annaka M, Tanaka C, Nakahira T et al (2002) Fluorescence study on the swelling behavior of comb-type grafted poly(*N*-isopropylacrylamide) hydrogels. Macromolecules 35:8173–8179
9. Xia LW, Ju XJ, Liu JJ et al (2010) Responsive hydrogels with poly(*N*-isopropylacrylamide-*co*-acrylic acid) colloidal spheres as building blocks. J Colloid Interface Sci 349:106–113
10. Greenspan FP (1946) The convenient preparation of per-acids. J Am Chem Soc 68:907
11. Tanaka Y, Gong JP, Osada Y (2005) Novel hydrogels with excellent mechanical performance. Prog Polym Sci 30:1–9

Chapter 2
Preparation and Properties of Monodisperse Thermo-responsive Microgels

Abstract Monodispersity is very important for the stimuli-responsive microspheres to improve their performance in various applications. In this chapter, several strategies including surfactant-free emulsion polymerization, membrane emulsification, and microfluidics are introduced to prepare monodisperse thermo-responsive microgels. The type of microgels includes both negatively and positively submicron-sized monodisperse thermo-responsive core-shell hydrogel microspheres and micro-sized monodisperse thermo-responsive microspheres and microcapsules.

2.1 Introduction

Environmental stimuli-responsive polymeric hydrogels have attracted increasing attention due to their potential applications in numerous fields [1–5], including controlled drug delivery [6–11], chemical separations [12, 13], sensors [14–16], catalysis [17], enzyme immobilization [18], and color-tunable crystals [19]. For some potential applications of these materials, such as "smart" actuators and on-off switches, a fast response is needed. To increase the response dynamics, several strategies have been explored. Besides improving the internal architecture or structure of the hydrogels [6, 20, 21] and introducing linear-grafted hydrogel chain configurations with freely mobile ends [22–26], another main strategy is developing microgels or hydrogel microspheres [1–5, 7, 27–34], because the characteristic time of gel swelling has been reported to be proportional to the square of a linear dimension of the hydrogels [35]. Furthermore, a small dimension is necessary for certain applications of the stimuli-responsive hydrogels. For example, there is a size limit for the drug delivery systems (DDS) to traverse certain organs [36], and a small particle size minimizes any potential irritant reaction at the injection site [37]. Therefore, the fabrication of small-sized environmental stimuli-responsive microspheres is of both scientific and technological interest. As there are many cases in which environment temperature fluctuations occur naturally and in which the

L.-Y. Chu et al., *Smart Hydrogel Functional Materials*,
DOI 10.1007/978-3-642-39538-3_2,

© Chemical Industry Press, Beijing and Springer Berlin Heidelberg 2013

environment temperature stimuli can be easily designed and artificially controlled, in recent years, much attention has been focused on thermo-responsive hydrogel microspheres [1–5, 7, 27–34].

Monodispersity is very important for the stimuli-responsive microspheres to improve their performance in various applications. For example, a uniform microsphere particle size is important for DDS. Both the distribution of the microspheres within the body and the interaction with biological cells are greatly affected by the particle size [38]. In addition, if monodispersed microspheres are available, the drug release kinetics can be manipulated, thereby making it easier to formulate more sophisticated intelligent DDS.

In this chapter, several strategies including surfactant-free emulsion polymerization, membrane emulsification, and microfluidics are introduced to prepare monodisperse thermo-responsive microgels. The type of microgels includes both negatively and positively submicron-sized thermo-responsive core-shell hydrogel microspheres [27, 39, 40] and micro-sized thermo-responsive microspheres and microcapsules [41–46].

2.2 Submicron-Sized Monodisperse Thermo-responsive Core-Shell Hydrogel Microspheres Fabricated via Surfactant-Free Emulsion Polymerization

In this section, preparation of submicron-sized thermo-responsive core-shell hydrogel microspheres with poly(*N*-isopropylacrylamide-*co*-styrene) [P(NIPAM-*co*-St)] cores and poly(*N*-isopropylacrylamide) (PNIPAM) shells will be introduced, in which the core seeds are prepared by a surfactant-free emulsion polymerization method and shell layers are fabricated by a seed polymerization method [39].

2.2.1 *Preparation of P(NIPAM-*co*-St) Seeds*

The P(NIPAM-*co*-St) seeds are prepared by an emulsifier-free emulsion polymerization method. A mixture of styrene and *N*-isopropylacrylamide is dissolved in 185 ml of deionized water in a 250 ml four-necked round-bottom flask equipped with a condenser, a nitrogen inlet, a thermometer, and a stirrer. Nitrogen is bubbled into the solution, and the mixture is stirred for 30 min to remove oxygen from the monomer solution. Polymerization is initiated by adding 15 ml of aqueous solution containing a certain amount of $K_2S_2O_8$ at 70 °C. The reaction is allowed to proceed for 24 h at 70 °C under stirring. The resulting P(NIPAM-*co*-St) microsphere samples are dialyzed and purified by repetitive centrifugation, decantation, and redispersion and then freeze-dried.

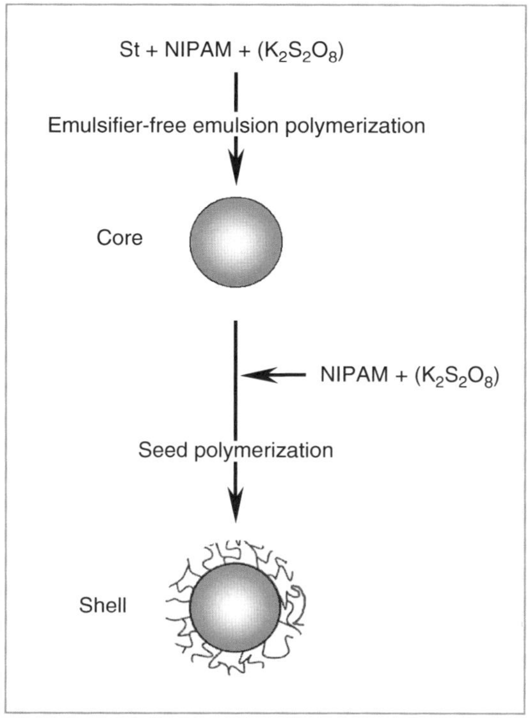

Fig. 2.1 Schematic illustration of preparation procedure of the core-shell microspheres with PNIPAM shell layers (Reproduced with permission from Ref. [39], Copyright (2004), American Chemical Society)

2.2.2 Preparation of Core-Shell Microspheres with PNIPAM Shell Layers

The PNIPAM shell layers are fabricated on the as-prepared core seeds by a seed polymerization method. The preparation procedure of the core-shell microspheres is schematically illustrated in Fig. 2.1. When the above-mentioned reaction for the P(NIPAM-*co*-St) seeds has not been stopped, another 15 ml of aqueous solution containing certain amount of NIPAM is added, and the polymerization is allowed to continue for 22 h under stirring at 200 rpm. By this method, the reaction manner is graft polymerization because of some living radicals on the surfaces of core particles [28, 29], and the resulting shell layers of the core-shell microspheres are hairy because no cross-linker is used. The prepared core-shell microsphere samples are also dialyzed and purified by repetitive centrifugation, decantation, and redispersion and then freeze-dried. The morphology of both core seeds and core-shell microspheres is observed by a scanning electron microscope (SEM S-450, Hitachi). All specimens for SEM observations are sputtered with gold at fixed conditions (time 150 s, current 20 mA, voltage 2 kV).

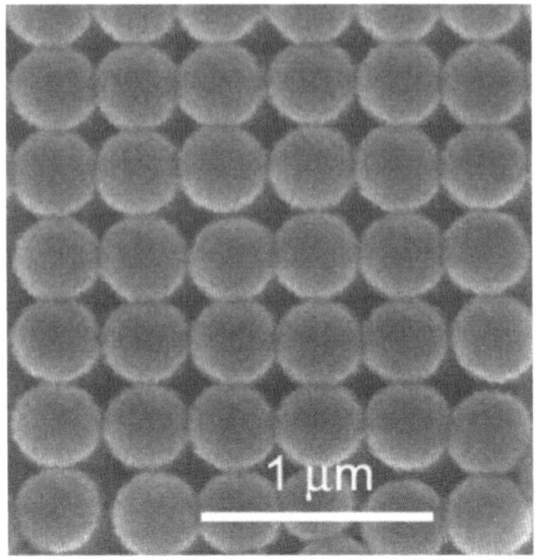

Fig. 2.2 SEM photographs of the core-shell microspheres with P(NIPAM-*co*-St) cores and PNIPAM shell layers (Reproduced with permission from Ref. [39], Copyright (2004), American Chemical Society)

2.2.3 Monodispersity of Core-Shell Microspheres with P(NIPAM-co-St) Cores and PNIPAM Shell Layers

Figure 2.2 shows the SEM photographs of the monodisperse core-shell microspheres with P(NIPAM-*co*-St) cores and PNIPAM shell layers. The PNIPAM shell layers are fabricated on the core seeds by a seed polymerization method. By this method, the reaction manner is graft polymerization because of some living radicals on the surfaces of core particles. Because the number of core seeds per unit volume is constant and monomer NIPAM contributes only to the PNIPAM shell formation on the seeds, the mean diameters of the core-shell microspheres became larger simply with increasing the NIPAM dosage. On the other hand, the free-radical density of smaller particles is relatively larger, which is helpful for the smaller particles to absorb more monomers or polymers with low molecular weight onto their surfaces to form larger particles. Therefore, the core-shell microspheres with P(NIPAM-*co*-St) cores and PNIPAM shell layers are highly monodisperse.

2.2.4 Thermo-responsive Characteristics of the Core-Shell Microspheres with PNIPAM Shell Layers

The hydrodynamic diameters of the prepared core-shell hydrogel microspheres at different temperatures are determined by temperature-programmed photon correlation spectroscopy (TP-PCS; Brookhaven BI-9000AT). This technique has been applied extensively to the characterization of such material, as it allows for in situ size characterization of soft material that cannot be reliably sized by electron microscopes due to deformation and/or dehydration under vacuum [2]. The dispersed

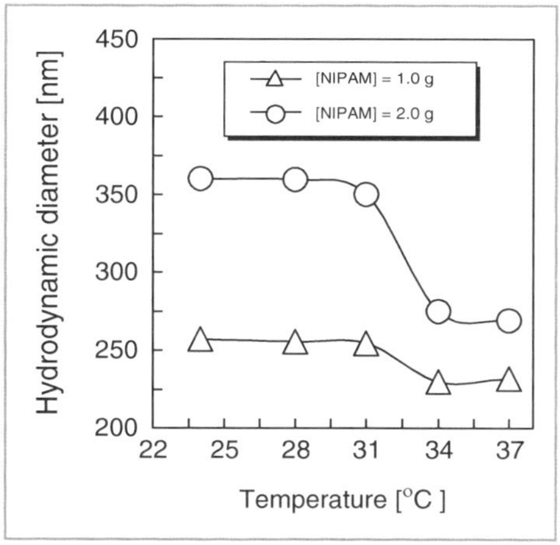

Fig. 2.3 Effect of the NIPAM dosage in the preparation of shells on thermo-responsive swelling characteristics of the core-shell microspheres (Reproduced with permission from Ref. [39], Copyright (2004), American Chemical Society)

particles in water are allowed to equilibrate thermally for 10–15 min before measurements are taken at each temperature. The hydrodynamic diameters of particles are calculated from diffusion coefficients by the Stokes-Einstein equation, and all correlogram analyses are performed using the manufacturer-supplied software. In the data presented here, each data point at a given temperature represents the average valve of 15–20 measurements, with a 20 s integration time for each measurement.

Figure 2.3 shows the effect of NIPAM dosage in the preparation of shell layers on the thermo-responsive swelling characteristics of the core-shell microspheres. With increasing the NIPAM dosage in the fabrication of the shell layers, the thermo-responsive swelling ratio of the hydrodynamic diameters of the core-shell microspheres at temperatures below the LCST of PNIPAM to those above the LCST increases. The hydrophilic groups of hairy PNIPAM chains on the core-shell microsphere surfaces form hydrating layers by hydrogen bond with water. The longer the PNIPAM chains result from increasing NIPAM dosage, the thicker the hydrating layer and then the larger the hydrodynamic diameter. The thickness of the hydrating layer decreases because of the breakage of hydrogen bonds with increasing temperature. When temperature approaches to the LCST, hydrogen bonds are broken seriously, which leads the thickness of hydrating layer to decrease rapidly, and then the linear PNIPAM polymer chains collapse quickly, resulting in a rapid decrease in the hydrodynamic diameters of the core-shell microspheres.

2.3 Positively Thermo-responsive Submicron-Sized Monodisperse Core-Shell Hydrogel Microspheres

Up to now, most of the thermosensitive hydrogel microspheres have been featured with negatively thermo-responsive volume-phase transition characteristics, i.e., the particles deswell with increasing environment temperature, because most of them have been prepared with PNIPAM and some with poly(*N*-vinylcaprolactam) (PVCL). In certain applications, however, an inverse mode of the thermo-responsive phase transition behavior is preferred. In this chapter, a strategy will be introduced to prepare monodisperse thermosensitive core-shell hydrogel microspheres featured with attractive monodispersity and positively thermo-responsive volume-phase transition characteristics, i.e., the particle swelling is induced by an increase rather than a decrease in temperature [27, 40].

2.3.1 Preparation of Positively Thermo-responsive Submicron-Sized Monodisperse Core-Shell Hydrogel Microspheres

The schematic illustration of the preparation of the positively thermosensitive core-shell hydrogel microsphere is illustrated in Fig. 2.4. The microsphere is composed of a poly(acrylamide-*co*-styrene) (poly(AAM-*co*-St)) core and an interpenetrating polymer network (IPN) shell of poly(acrylamide) (PAAM) and poly(acrylic acid) (PAAC). It is known that PAAM and PAAC form polycomplexes in solution through hydrogen bonding. By the cooperative "zipping" interactions between the molecules that result from hydrogen bonding, a positively thermosensitive volume-phase transition characteristic of PAAM-/PAAC-based IPN macroscopic hydrogels in water has been found [47]. When the environment temperature is lower than the upper critical solution temperature (UCST) of the PAAM-/PAAC-based IPN gel, PAAC forms intermolecular hydrogen bonds with PAAM, and the IPN hydrogels keep shrinking state by the interaction between two polymer chains or the so-called chain-chain zipper effect; on the other hand, when the environment temperature is higher than the UCST of the IPN gel, PAAC dissociates intermolecular hydrogen bonds with PAAM, and the IPN hydrogels keep swelling state by the relaxation of the two polymer chains. Therefore, the microspheres shrink at temperatures below the UCST due to the complex formation by hydrogen bonding and swell at temperatures above the UCST due to PAAM/PAAC complex dissociation by the breakage of hydrogen bonds.

The core-shell microspheres with PAAM-/PAAC-based IPN shells are fabricated in a three-step process [27]. In the first step, monodisperse poly(AAM-*co*-St) seeds are prepared by emulsifier-free emulsion polymerization. In the second step, PAAM or poly(acrylamide-*co*-butyl methacrylate) (poly(AAM-*co*-BMA)) shells are fabricated on the microsphere seeds by free-radical polymerization. In the third step, the core-shell microspheres with PAAM-/PAAC-based IPN shells are finished by a method of sequential IPN synthesis. After repetitive centrifugation, decantation, and redispersion with well-deionized water, the core-shell seeds with PAAM or poly(AAM-*co*-BMA) shells are immersed in aqueous acrylic acid (AAC) solution containing potassium persulfate (KPS) and methylenebisacrylamide (MBA) as

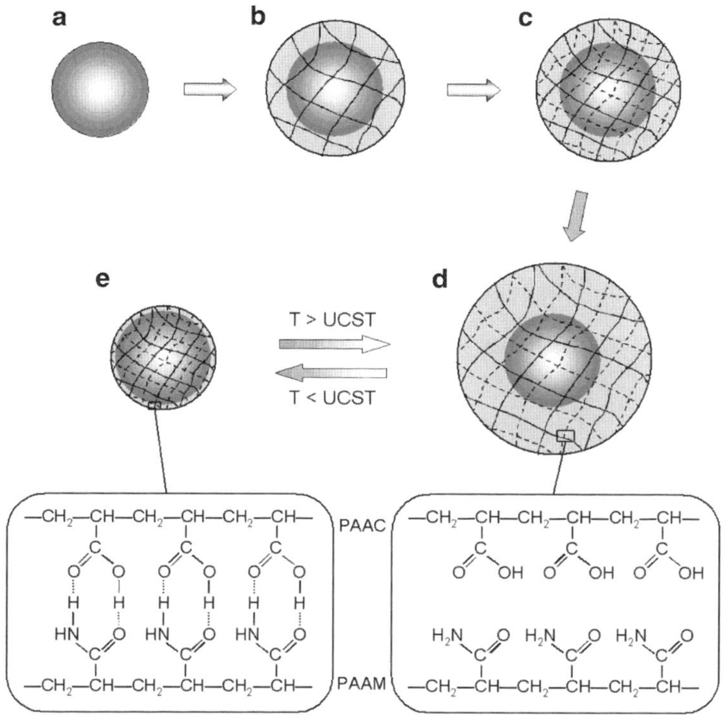

Fig. 2.4 Schematic illustration of the preparation of the positively thermosensitive core-shell hydrogel microsphere. (**a**) Monodisperse poly(acrylamide-*co*-styrene) (poly(AAM-*co*-St)) seeds prepared by emulsifier-free emulsion polymerization, (**b**) microsphere seeds with PAAM shells prepared by free-radical polymerization, (**c**) microspheres with PAAM-/PAAC-based IPN shells prepared by a method of sequential IPN synthesis, (**d**) thermosensitive PAAM-/PAAC-based IPN microspheres in the swollen state at temperatures above UCST due to PAAM/PAAC complex dissociation by the breakage of hydrogen bonds, and (**e**) microspheres in the shrunken state at temperatures below the UCST due to the complex formation by hydrogen bonding (Reproduced with permission from Ref. [27], Copyright (2003), Wiley-VCH Verlag GmbH & Co. KGaA)

initiator and cross-linker, respectively, to swell for 24 h, and the monomer AAC is subsequently polymerized and cross-linked within the initial PAAM matrix gels of the seed shells to form PAAM-/PAAC-based IPN.

2.3.2 Morphological Analyses of the Microspheres

Figure 2.5 shows SEM images of the poly(AAM-*co*-St) seeds with PAAM shells and corresponding microspheres with PAAM-/PAAC-based IPN shells, from which it can be seen that both the core-shell seeds and the resulting microspheres are highly

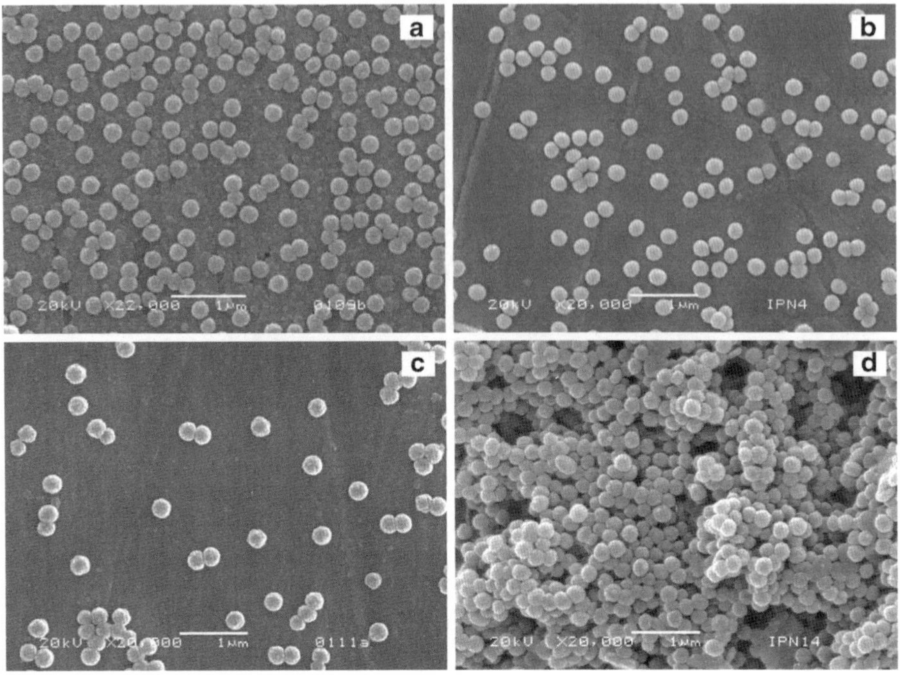

Fig. 2.5 SEM images of seeds with PAAM shells and the resulting microspheres with PAAM-/PAAC-based IPN shells. (**a**, **c**) seeds and (**b**, **d**) corresponding microspheres. (**a**, **b**) No butyl methacrylate (BMA) is added in the preparation of PAAM shells on seed cores (the second step), and [MBA]/[AAC] = 1 wt% in the synthesis of IPN; (**c**, **d**) [BMA]/([BMA] + [AAM]) = 34.4 wt% in the preparation of PAAM shells on seed cores, and [MBA]/[AAC] = 1.5 wt% in the synthesis of IPN. Scale bar 1 μm (Reproduced with permission from Ref. [27], Copyright (2003), Wiley-VCH Verlag GmbH & Co. KGaA)

monodisperse. Because the seeds with PAAM shells are the initial matrix for the IPN synthesis, the monodispersity of the final IPN microspheres is dependent on that of the core-shell seeds consequently. Another interesting phenomenon is that the SEM images show that the mean particle size of the IPN microspheres appears to be a little smaller than that of the core-shell seeds. This is due to the freeze-drying method for the preparation of SEM samples. Because the environment temperature is lower than the UCST of the PAAM-/PAAC-based IPN gel, the IPN microspheres shrink when they are treated by freeze-drying, but the seeds do not. Therefore, the IPN microspheres appear to be a little smaller than the seeds in the SEM micrographs.

2.3.3 Positively Thermosensitive Swelling Characteristics

Temperature-programmed photon correlation spectroscopy (TP-PCS) is used to determine the temperature dependence of the hydrodynamic diameter of

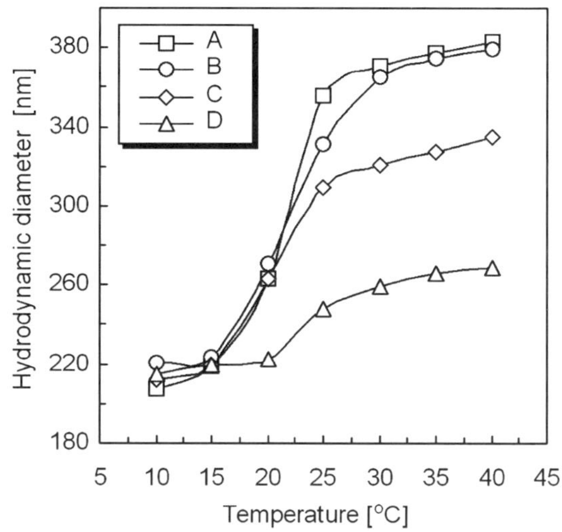

Fig. 2.6 Temperature dependence of hydrodynamic diameter of microspheres with PAAM-/PAAC-based IPN shells. The sample code is defined in Table 2.1 (Reproduced with permission from Ref. [27], Copyright (2003), Wiley-VCH Verlag GmbH & Co. KGaA)

Table 2.1 Experimental recipe for the preparation of microspheres with PAAM-/PAAC-based IPN shell (Reproduced with permission from Ref. [27], Copyright (2003), Wiley-VCH Verlag GmbH & Co. KGaA)

	In the second step: fabrication of PAAM or poly(AAM-*co*-BMA) shell on the seed core		In the third step: PAAM/PAAC IPN synthesis	
Sample code	[BMA]/([BMA] + [AAM]) [wt%]	[MBA]/([St] + [AAM]) [wt%]	[AAC]/([AAC] + [AAM]) [mol/mol]	[MBA]/[AAC] [wt%]
A	0	0	1.0	1
B	0	1.2	0.5	1
C	34.4	0	0.5	1
D	34.4	0	0.5	1.5

microspheres with PAAM-/PAAC-based IPN shells. Figure 2.6 shows the temperature dependence of hydrodynamic diameter of the prepared microspheres with PAAM-/PAAC-based IPN shells, in which the sample code is defined in Table 2.1. All of the microspheres exhibit positively thermosensitive volume-phase transition characteristics. The hydrodynamic diameters of the PAAM-/PAAC-based IPN microspheres are about 210 nm in the temperature range 10–15 °C and increase to 260–380 nm in the temperature range 30–40 °C. A sharp transition of the hydrodynamic diameters occurs from 15 °C to 25 °C, which corresponds to the UCST of PAAM-/PAAC-based IPN hydrogels [47]. Below the UCST, PAAM/PAAC intermolecular complexes form by hydrogen bonding, and the chain-chain zipper effect makes the IPN microspheres shrunken; as a result, the mean hydrodynamic diameter is small. In contrast, the IPN microspheres are in the swollen state at temperatures above the UCST due to PAAM/PAAC complex dissociation by the breakage of hydrogen bonds, and therefore, a larger hydrodynamic diameter is the result (as illustrated in Fig. 2.4).

The swelling ratios of the hydrodynamic diameters of IPN microspheres decreased when adding hydrophobic monomer butyl methacrylate (BMA) into the experimental recipe for the preparation of PAAM shells on the seeds in the second step. The reason exists in two aspects as follows: (1) The hydrophobic monomer BMA contributes to the mechanical properties of the IPNs, and (2) the hydrophobic interactions stabilize the hydrogen-bonding complexation of the IPNs. In the second step for fabricating poly(AAM-*co*-BMA) shell on the seed core, hydrophobic monomer BMA and hydrophilic monomer AAM had "equal chance" to participate in the free-radical copolymerization reaction. Consequently, certain contents of hydrophobic polymer are randomly distributed in the prepared poly(AAM-*co*-BMA) shell layer and therefore in the resulting IPN shell layer. The mechanical strength of PAAM-/PAAC-based IPN macroscopic hydrogels increases with increasing BMA content in the IPNs. The increase in the mechanical strength of the IPN shell restricts the swelling and shrinkage of the microsphere. In addition, hydrophobic interactions are proposed to increase with increasing BMA content in the IPNs. The hydrophobic interactions prevent the polymer chain complexes from dissociating, i.e., stabilize the hydrogen-bonding complexation of the IPNs. As a result, the swelling of the microsphere is depressed.

With increasing cross-linker MBA dosage in the IPN synthesis or adding MBA in the second step for PAAM shell preparation, the results show that the swelling ratios of the hydrodynamic diameters of IPN microspheres also decrease to some extent. Although cross-linker is helpful to prepare the IPNs, an overfull cross-linker dosage results in a decrease of the thermosensitive swelling ratio. By adding cross-linker MBA in the PAAM shell preparation or increasing MBA dosage in the IPN synthesis, the network of PAAM chains or that of PAAC chains becomes more rigid, and the polymer chains restrict to each other more strongly; therefore, the deformation of the IPNs becomes more difficult. As a result, the swelling ratio of the hydrodynamic diameter of the microspheres is depressed. That is to say, the positively thermosensitive swelling ratio of the prepared core-shell microspheres could be adjusted by regulating the hydrophobic monomer BMA dosage and the cross-linker MBA dosage.

2.4 Monodisperse Thermo-responsive Hydrogel Microspheres and Microcapsules Prepared via Membrane Emulsification

In this chapter, a simple two-step method, which includes Shirasu porous glass (SPG) membrane emulsification and UV-induced free-radical polymerization, is introduced to prepare monodisperse PNIPAM microspheres and hollow microcapsules [41].

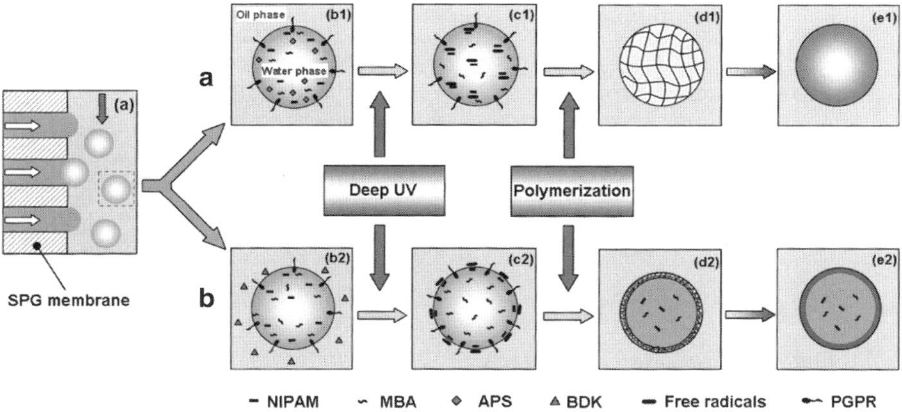

Fig. 2.7 Schematic illustration of preparation of monodisperse PNIPAM microspheres in situ (**a**: *b1* → *e1*) and microcapsules at water/oil interface (**b**: *b2* → *e2*) by SPG membrane emulsification and UV irradiation polymerization (Reproduced with permission from Ref. [41], Copyright (2008), Elsevier)

2.4.1 Strategies for Preparation of Monodisperse PNIPAM Microspheres and Microcapsules via Membrane Emulsification

Figure 2.7 shows the schematic illustration of preparation processes of monodisperse PNIPAM microspheres in situ and microcapsules at water/oil interface by SPG membrane emulsification and UV irradiation polymerization. First, monodisperse monomer-contained water-in-oil (W/O) emulsions are prepared by SPG membrane emulsification method [41]. SPG membrane emulsification is based on injecting a disperse phase through a porous membrane, with the resulting droplets forming at the end of pores on the membrane surface after coming into contact with the continuous phase. Thus, using this technique, it is easier to control the droplet size and size distribution [23, 41]. Then, monomers dissolved in the aqueous droplets are initiated by initiator ammonium persulfate (APS) dissolved in the disperse phase or 2,2-dimethoxy-2-phenylacetophenone (BDK) dissolved in the continuous phase to polymerize and form microspheres or hollow microcapsules under the UV radiation at 20 °C.

2.4.2 Morphology of Prepared Monodisperse PNIPAM Microspheres

To evaluate the monodispersity of the particles quantitatively, particle size dispersal coefficient, δ, is defined as [23, 41]

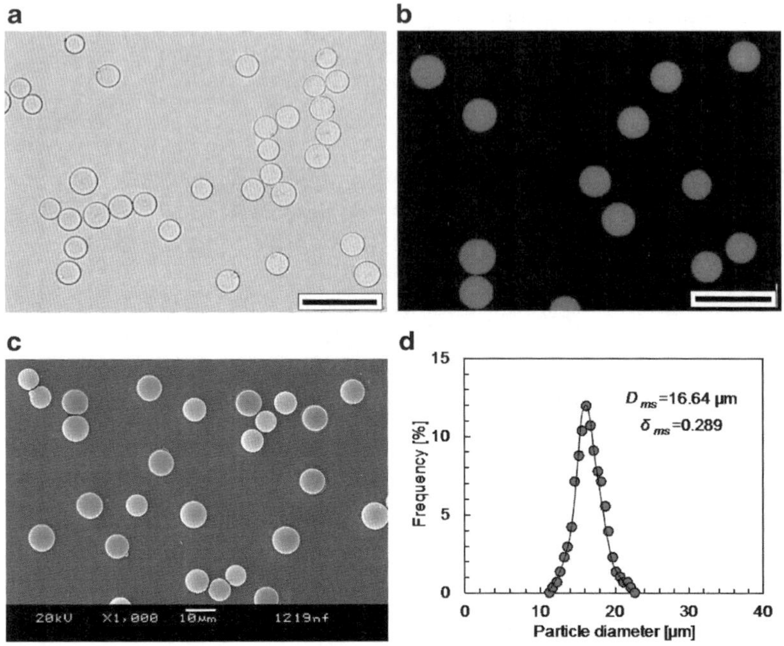

Fig. 2.8 (**a**) Optical microgram of PNIPAM microspheres in water, (**b**) CLSM image of Rd B-dyed PNIPAM microspheres, (**c**) SEM image of air-dried PNIPAM microspheres, and (**d**) size distributions of the resultant PNIPAM microspheres dispersed in water. $T = 20$ °C. Scale bar: (**a**) 60 μm, (**b**) = 40 μm, and (**c**) = 10 μm (Reproduced with permission from Ref. [41], Copyright (2008), Elsevier)

$$\delta = \frac{D_{90} - D_{10}}{D_{50}} \tag{2.1}$$

where D_n ($n = 10, 50, 90$) denotes the cumulative number percentage of particles with diameter up to D_n and n stands for the percentage. The smaller the value of δ is, the narrower the size distribution. Usually, if the coefficient $\delta \leq 0.4$, the particles can be considered to be monodisperse.

Figure 2.8 shows the optical microgram, CLSM and SEM images, as well as the size distribution of the PNIPAM microspheres prepared by SPG membrane emulsification and UV-induced free-radical polymerization. The prepared microspheres have a solid structure (Fig. 2.8b), good sphericity (Fig. 2.8a–c), narrow size distribution (Fig. 2.8d), and almost the same δ value as that of the corresponding monomer-contained W/O emulsions. Moreover, the mean diameter of microspheres in water at 20 °C is much larger than that of air-dried. This is ascribed to that the polymer chains of microspheres stretch fully in water when environment temperature is lower than the LCST of PNIPAM (about 32 °C); on the other hand, microspheres become dense and impact owing to the shrinkage of polymer chains,

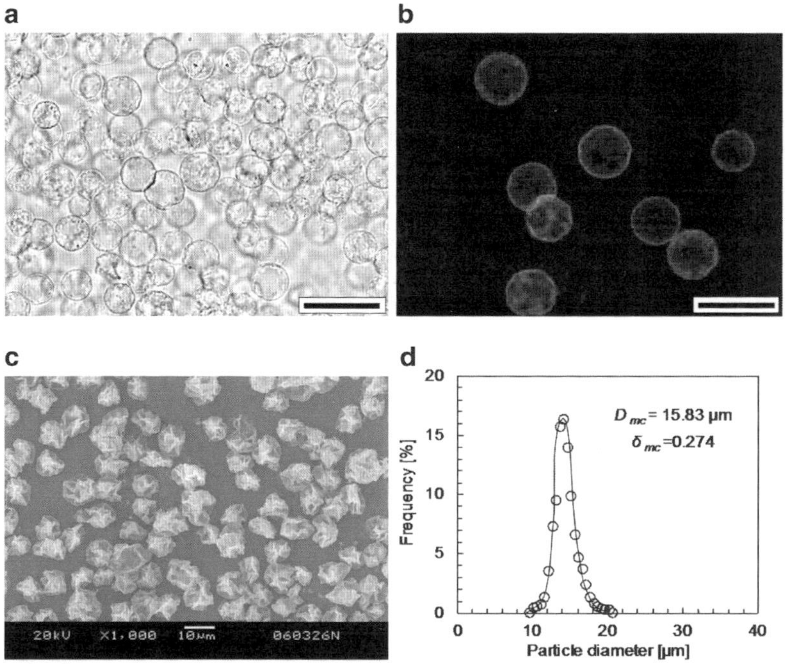

Fig. 2.9 (**a**) Optical microgram of PNIPAM microcapsules in water, (**b**) CLSM image of Rd B-dyed PNIPAM microcapsules, (**c**) SEM image of air-dried PNIPAM microcapsules, and (**d**) size distributions of the resultant PNIPAM microcapsules dispersed in water. $T = 20$ °C. Scale bar: (**a**) = 40 μm, (**b**) = 30 μm, and (**c**) = 10 μm (Reproduced with permission from Ref. [41], Copyright (2008), Elsevier)

and water is extruded from the hydrated microspheres as a result of the disruption of hydrogen bonds and hydrophobic interactions of isopropyl groups of neighboring polymer chains.

2.4.3 Morphology of Prepared Monodisperse PNIPAM Microcapsules

Figure 2.9 shows the optical microgram, CLSM and SEM images, as well as the size distribution of the PNIPAM microcapsules prepared by SPG membrane emulsification and UV-induced free-radical polymerization. The resulting PNIPAM microcapsules are also characterized with good sphericity and narrow size distribution. The hollow structure of PNIPAM microcapsules with thin membranes can be obviously observed in Fig. 2.9b. In the SEM image of dried PNIPAM microcapsules without washing with isopropyl alcohol and water (Fig. 2.9c), collapsed and crinkled structures resulted from the drying process with oil surrounding. By making a comparison, the structure of PNIPAM microcapsules is definitely different from that of PNIPAM microspheres. This result verifies once again that the resulting microcapsules are featured with hollow structures.

Different polymerization mechanisms in the reaction are responsible for the distinct structures of microspheres and microcapsules. When aqueous-soluble initiator APS is introduced into the disperse phase before membrane emulsification (Fig. 2.7b1), APS molecules dissociate to generate a great deal of reactive free radicals within the aqueous droplets under the UV irradiation (Fig. 2.7c1), and then the monomer NIPAM molecules dissolved in water phase are initiated in situ to polymerize and form many long-chain macromolecules (Fig. 2.7d1). Meanwhile, those macromolecules are cross-linked under the action of MBA and then form the solid microspheres (Fig. 2.7e1). On the other hand, when oil-soluble initiator BDK is added into the continuous phase after membrane emulsification (Fig. 2.7b2), under the UV irradiation, BDK also decomposes to produce many active free radicals in the oil phase. The free radicals then diffuse across the oil/water interface to the water phase of W/O emulsions (Fig. 2.7c2). Meantime, monomer NIPAM molecules also diffuse toward the water/oil interface and are initiated to polymerize by the active radicals. Subsequently, cross-linked PNIPAM layer forms at the interface owing to the function of MBA within the aqueous droplets (Fig. 2.7d2). When the cross-linked PNIPAM layer at the interface becomes thicker and thicker, fewer and fewer active free radicals can diffuse across it and enter into the aqueous phase; as a result, the rate of polymerization becomes slower and slower. Finally, when the PNIPAM shell becomes too thick for the free radicals to diffuse across, the polymerization reaction stops and hollow PNIPAM microcapsules result (Fig. 2.7e2).

2.4.4 Effect of Freeze-Drying and Rehydrating Treatment on the Thermo-responsive Characteristics of PNIPAM Microspheres

Figure 2.10 shows the monodisperse PNIPAM microspheres prepared by SPG membrane emulsification and UV-induced polymerization. Figure 2.10a, b, respectively, show the typical optical micrograms of PNIPAM microspheres dispersed in deionized water before and after freeze-drying, and Fig. 2.10c, d show the typical SEM images of air-dried and freeze-dried PNIPAM microspheres, respectively. Figure 2.10e shows the typical SEM image of air-dried PNIPAM microspheres that have experienced a freeze-drying and rehydrating treatment before air-drying at room temperature. No matter whether the PNIPAM microspheres have experienced the freeze-drying and rehydrating treatment or not, the air-dried PNIPAM microspheres always exhibit a satisfactory sphericity and a smooth and compact surface (Fig. 2.10c, e). However, a marked change takes place in the morphology

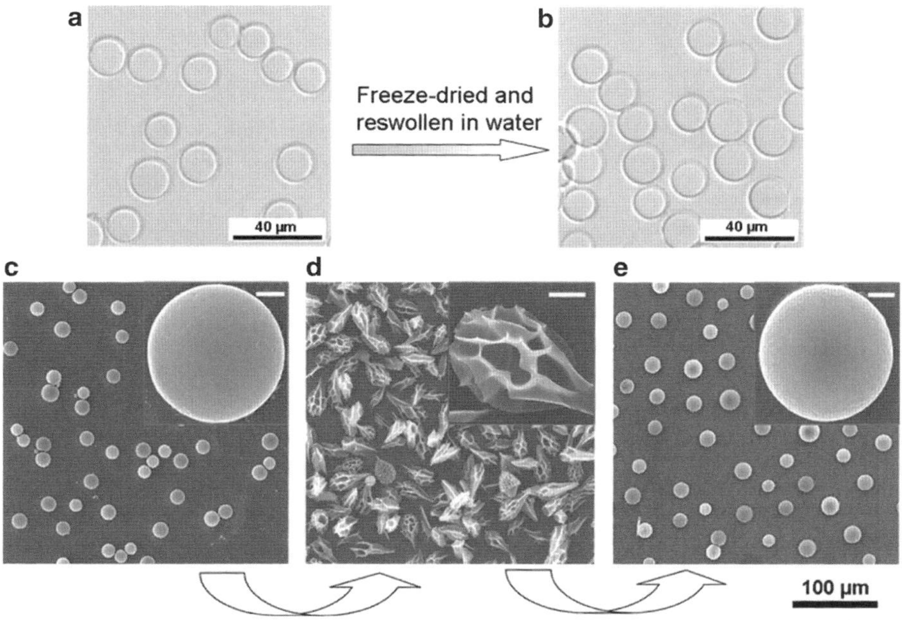

Fig. 2.10 (**a**) Optical micrograms of PNIPAM microspheres dispersed in deionized water at 20 °C. (**b**) Optical micrograms of PNIPAM microspheres that are firstly freeze-dried and then reswollen in deionized water at 20 °C. (**c**, **d**) SEM micrograms of the air-dried and freeze-dried PNIPAM microspheres; (**e**) SEM micrograms of air-dried PNIPAM microspheres that are firstly freeze-dried and then reswollen in deionized water. In the inserted pictures, scale bar (**c**, **e**) 2 μm; (**d**) 5 μm. $[C_{NIPAM}] = 1.0$ mol/L, $[C_{MBA}] = 0.05$ mol/L, and $[C_{APS}] = 8.8$ mmol/L (Reproduced with permission from Ref. [42], Copyright (2008), Springer)

and structure of microspheres after freeze-drying treatment (as shown in Fig. 2.10d). After freeze-drying, the microspheres unexpectedly become a flower-like porous structure and produce a tip-shaped tail at one end of microsphere, and a large number of micropores with a size of several microns are generated throughout the microsphere.

The micropores within the microspheres after freeze-drying resulted from the ice crystals presented in the swollen PNIPAM microspheres acting as a template for pore generation [48]. Water inside swollen microspheres is dispersed in interconnected polymeric network and allows the polymer network to enlarge and swell in water. So, the ice crystals form rapidly within the network of microspheres upon immerging them into liquid nitrogen. The frozen microspheres are dried by sublimation of ice crystals under vacuum at a temperature below the ice freezing point. The direction of micropores resulted from the orientation of ice crystals [48]. When ice crystals are subjected to an external stress such as rapid cooling, for instance, immerged into liquid nitrogen in this case, they will be oriented due to a temperature gradient and subsequently lead to the formation of ordered direction of micropores. As for the formation of tip-shaped tails, a possible reason is that

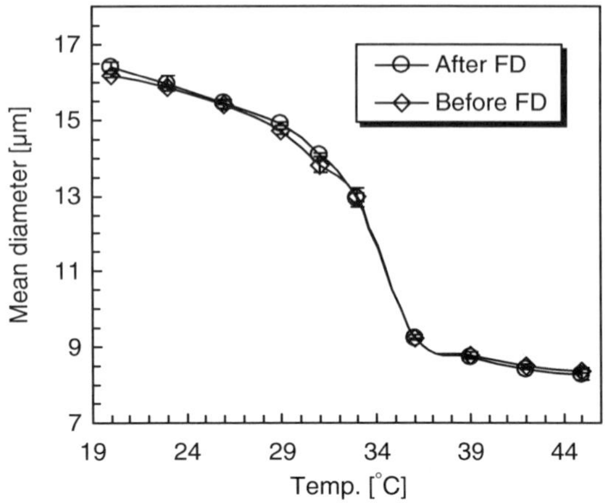

Fig. 2.11 Temperature-dependent equilibrium volume change of PNIPAM microspheres before and after freeze-drying treatment. [C_{NIPAM}] = 1.0 mol/L, [C_{MBA}] = 0.05 mol/L, and [C_{APS}] = 8.8 mmol/L (Reproduced with permission from Ref. [42], Copyright (2008), Springer)

water within the top of microsphere network is pressed to flow downwards during the frozen process of microsphere by a driving force. The top of microsphere is first frozen, and the water streamed from the top of microsphere is also gradually frozen. Finally, a tip-shaped tail resulted. In view of this phenomenon, a conclusion can be drawn that such a freeze-drying treatment is ineffective for hydrogel microspheres to maintain their good spherical shapes.

It is particularly interesting that the freeze-dried porous microspheres can recover the spherical morphology after reswelling in deionized water (Fig. 2.10b, e). This is mainly ascribed to the elastic and flexible characteristics of PNIPAM polymeric networks. Stretched polymer networks can occur to contract, the pore regions are stepwise occupied by water when the porous microspheres are reswollen, and then the rehydrated microspheres can return to the surface morphology without lyophilizing treatment. Furthermore, the size of freeze-dried and reswollen microspheres is almost the same as that without freeze-drying treatment.

Figure 2.11 shows the temperature-dependent equilibrium volume change of monodisperse PNIPAM microspheres before and after freeze-drying (FD) treatment. The volume of both microspheres changes slowly when the environment temperature is below 31 °C, then suffers from a sharp decrease while the temperature is raised across 32 °C, and finally keeps almost constant upon heating above 36 °C, which is in good agreement with the bulky PNIPAM hydrogel. The results show that the freeze-drying treatment nearly has no effect on the LCST and the equilibrium volume changes of the PNIPAM microspheres. The volume of microspheres nearly does not change after the lyophilizing treatment, except that the size of PNIPAM microspheres after freeze-drying and reswelling with deionized water at 20 °C is slightly larger than that without freeze-drying.

Figure 2.12 shows the time-dependent diameter change and volume change of monodisperse PNIPAM microspheres with different cross-linkages before and after freeze-drying treatment when environment temperature is suddenly increased from 20 °C to 40 °C and 50 °C, respectively. When other preparation conditions, such as

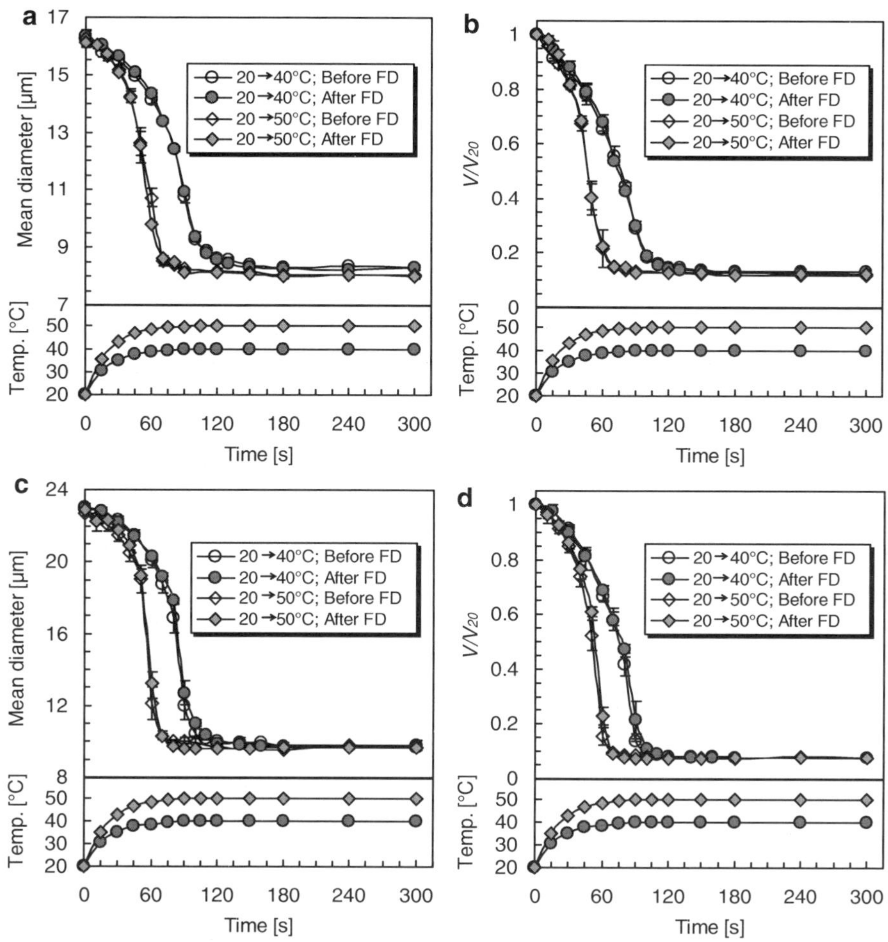

Fig. 2.12 Time-dependent diameter changes (**a**, **c**) and volume-change rate (**b**, **d**) of monodisperse PNIPAM microspheres with different cross-linkages before and after freeze-drying treatment. $[C_{NIPAM}] = 1.0$ mol/L and $[C_{APS}] = 8.8$ mmol/L. (**a**, **b**) $[C_{MBA}] = 0.05$ mol/L; (**c**, **d**) $[C_{MBA}] = 0.01$ mol/L (Reproduced with permission from Ref. [42], Copyright (2008), Springer)

the monomer NIPAM and initiator APS concentrations and the transmembrane pressure, are kept unvaried, the initial mean diameter of swollen PNIPAM microspheres in water (at 20 °C) increases with decreasing the cross-linker MBA concentration in preparation. The lower the MBA concentration is, the lower the cross-linkages of the polymer networks inside the microgels, i.e., the looser the polymer networks inside the PNIPAM microspheres, the larger the microspheres swelled in water at a temperature below the LCST of PNIPAM.

The deswelling rate of PNIPAM microspheres increases with increasing the upper temperature limit. The higher the environment temperature is, the quicker the heat transferred into the PNIPAM microspheres; therefore, the hydrogen bonds between the polymer chains and water molecules are ruptured more rapidly upon heating; as a result, the deswelling rate of microgels is faster. On the other hand, when the solution temperature is increased from 20 °C to 40 °C or to 50 °C (above the LCST of PNIPAM), the treatment of freeze-drying and rehydration nearly does not affect the deswelling rate of PNIPAM microspheres no matter how the cross-linkage changed. Because the size of microspheres (micron-sized scale) is much smaller in comparison with that of bulky PNIPAM hydrogel (at least millimeter scale), the microsphere itself has a quick deswelling rate before freeze-drying treatment, since the characteristic time of gel deswelling is proportional to the square of a linear dimension of hydrogel. The deswelling rate of a hydrogel depends mainly on the diffusion velocity of water passing through the polymer network and the skin layer. The thickness of skin layer of micron-sized PNIPAM microspheres should be much thinner than that of bulky hydrogel, and the freeze-drying treatment for the microsphere does not distinctly reduce the thickness of skin layer. Therefore, the accelerating extent that resulted from freeze-drying treatment for the deswelling rate of microgels is negligible compared with that resulted from the effect of the small dimension of the microspheres. That is, when the microgel is treated by frozen in only liquid nitrogen as rapid cooling, freeze-drying treatment has very little effect on the deswelling rate of micron-sized PNIPAM microspheres when the environment temperature is increased from 20 °C to 40 °C or to 50 °C.

2.5 Monodisperse Thermo-responsive Hydrogel Microspheres and Microcapsules Fabricated with Microfluidics

Recently, microfluidic methods have provided a very promising approach to prepare monodisperse polymeric microspheres, because the microfluidic technique can control the fluid flow precisely and thus ensure high monodispersity of the prepared particles. Furthermore, both the emulsification and the polymerization can be carried out in the same microfluidic device. In this section, microfluidic strategies for preparation of monodisperse thermo-responsive hydrogel microspheres and microcapsules will be introduced [43–46].

2.5.1 Microfluidic Fabrication of Monodisperse Thermo-responsive Microgels with Tunable Volume-Phase Transition Kinetics

The kinetics of swelling and deswelling of stimuli-responsive polymeric hydrogels are typically governed by diffusion-limited transport of water in and out of the polymeric networks [35]. Hence, if a facile way to control this transport of water is devised, the volume-phase transition kinetics of the microgels can be precisely tuned. Here, this is accomplished by introducing different void structures inside the microgels. An internal structure with voids offers less resistance to the transport of water compared to a voidless core. Thus, by varying the size and number of voids inside a microgel, its swelling and shrinking dynamics can be effectively controlled. The microgels with spherical voids are prepared by using a two-step method. First, monodisperse microgels with different numbers/sizes of solid polystyrene microspheres are synthesized in a microfluidic device. Second, the embedded microspheres are chemically dissolved to form spherical voids inside the microgels.

The microfluidic device used to synthesize the microgels consists of cylindrical glass capillary tubes nested within square glass capillary tubes, as shown in Fig. 2.13. It features a coaxial co-flow geometry ensured by matching the outside diameters of the cylindrical tubes to the inner dimensions of the square tubes. An aqueous suspension containing the monomer NIPAM, a cross-linker MBA, an initiator ammonium persulfate (APS), and polystyrene microspheres is pumped into the left end of the left square tube. The continuous phase, kerosene containing a surfactant, polyglycerol polyricinoleate (PGPR 90), is pumped through the outer coaxial region between the left square tube and a tapered round microcapillary tube. The aqueous phase breaks into droplets at the entrance orifice of the tapered tube, forming monodisperse emulsion drops in the tube. The accelerator, *N,N,N',N'*-tetramethylethylenediamine (TEMED), dissolved in the continuous phase, is pumped through the outer coaxial region between the right square tube and the right end of the tapered round tube. When the accelerator meets the initiator, it starts a redox reaction that polymerizes the monomers, thus forming monodisperse microgels with embedded polystyrene microspheres. The addition of the reaction accelerator downstream of the other chemicals delays the polymerization process long enough to allow the formation of monodisperse emulsion drops of these chemicals in the continuous phase and eliminates the possibility of the entrance orifice of the round tube becoming clogged by untimely polymerization of the monomers. Once the polymerization process is completed, the PNIPAM microgels are washed with isopropanol and subsequently immersed in xylene for a fixed period of time to dissolve the polystyrene beads leaving behind holes in the microgels. The microgels with spherical voids thus formed are again rinsed with isopropanol and are finally dispersed in water. Specific comparisons are made between the thermosensitive behavior of voidless microgels, without any voids, and those with hollow void structures, having either one large void of 75 μm diameter or different numbers of smaller voids.

Microgels containing different numbers of 25 μm diameter polystyrene beads and the microgels with spherical voids formed by dissolving these beads from similar microgels are shown in Fig. 2.14. The optical micrographs suggest that the size and number of spherical voids in a microgel depend on the size and number of the encapsulated polystyrene beads. The precise circular periphery of the voids and unchanged size of the microgels also suggest that the chemical dissolution of the beads does not affect the structural integrity of the surrounding microgel

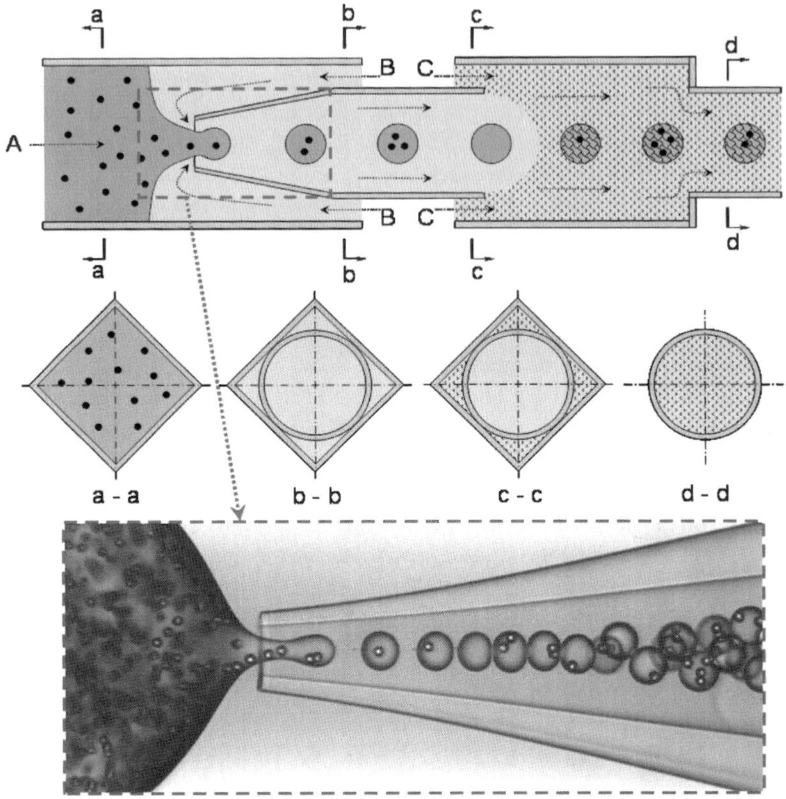

Fig. 2.13 Microfluidic device for preparing monodisperse microgels with embedded solid particles. Fluid **A** is an aqueous suspension containing the monomer, initiator, cross-linker, and solid polystyrene particles; fluid **B** is oil containing a surfactant; and fluid **C** is a mixture of an oil and a reaction accelerator. Illustrations *a-a*, *b-b*, *c-c*, and *d-d* are cross-sectional images of the capillary microfluidic device in relevant positions, which clearly show how the square capillary tubes and cylindrical tubes are assembled in the device (Reproduced with permission from Ref. [43], Copyright (2007), Wiley-VCH Verlag GmbH & Co. KGaA)

structure. The equilibrium diameter of the microgel with four voids is compared to that of the microgel with no voids at various temperatures between 20 °C and 47 °C in Fig. 2.15. The samples are equilibrated for 30 min at each temperature before the measurements are made. It appears that the induced void structure has no effect on the equilibrium size change of the microgels. For same size microgels, the equilibrium volume change is a function of the homogeneity of the internal microstructure. Since NIPAM is polymerized under identical conditions in both cases, the two microgels should have the same homogenous microstructures. The overlapping of the two curves confirms that the presence of polystyrene microspheres during the NIPAM polymerization process does not affect the homogeneous net-like microstructure of the microgel.

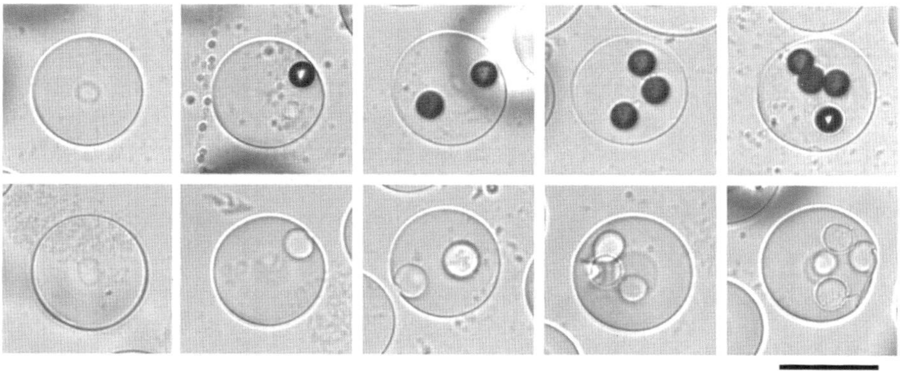

Fig. 2.14 (*top*) Microgels with different numbers of embedded polystyrene beads; (*bottom*) microgels with spherical voids formed by dissolving the embedded beads from such microgels. Scale bar = 100 μm (Reproduced with permission from Ref. [43], Copyright (2007), Wiley-VCH Verlag GmbH & Co. KGaA)

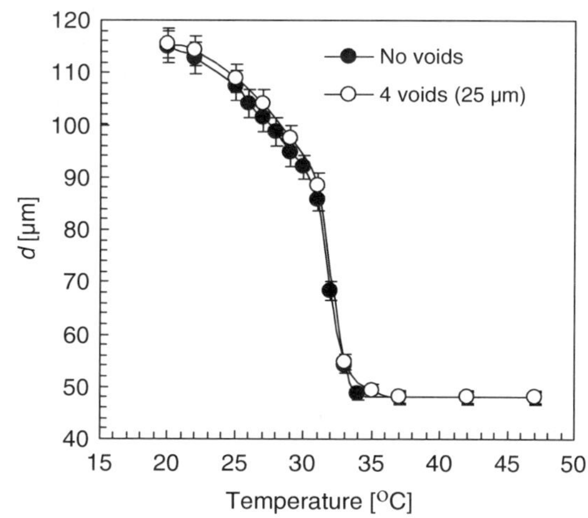

Fig. 2.15 Equilibrium diameters of the voidless microgel and the microgel with 4 voids (25 μm dia. each) plotted as a function of temperature. The samples are held for 30 min at each temperature before measuring their diameters (Reproduced with permission from Ref. [43], Copyright (2007), Wiley-VCH Verlag GmbH & Co. KGaA)

The dynamics of shrinking and swelling of the microgels are studied by heating them together with water from 23 °C to 47 °C and subsequently cooling them back to 23 °C in a transparent sample holder placed on a microscope-mounted thermal stage (Fig. 2.16). The same sample holder and volume of water are used for microgels with different internal structures. Although the equilibrium volume change of the microgels is unaffected by the internal structure, their shrinking and swelling kinetics are affected by the induced void structure. When the temperature is increased from 23 °C to 47 °C, the microgel with multiple voids and the one with a hollow shell structure (one large void) shrink faster than the voidless microgel as shown in Fig. 2.16a. For example, at $t = 49$ s, the diameter of the voidless microgel is the largest, while that of the one with multiple voids is the smallest. Similar behavior is observed when the microgels are cooled back to 23 °C: The microgel

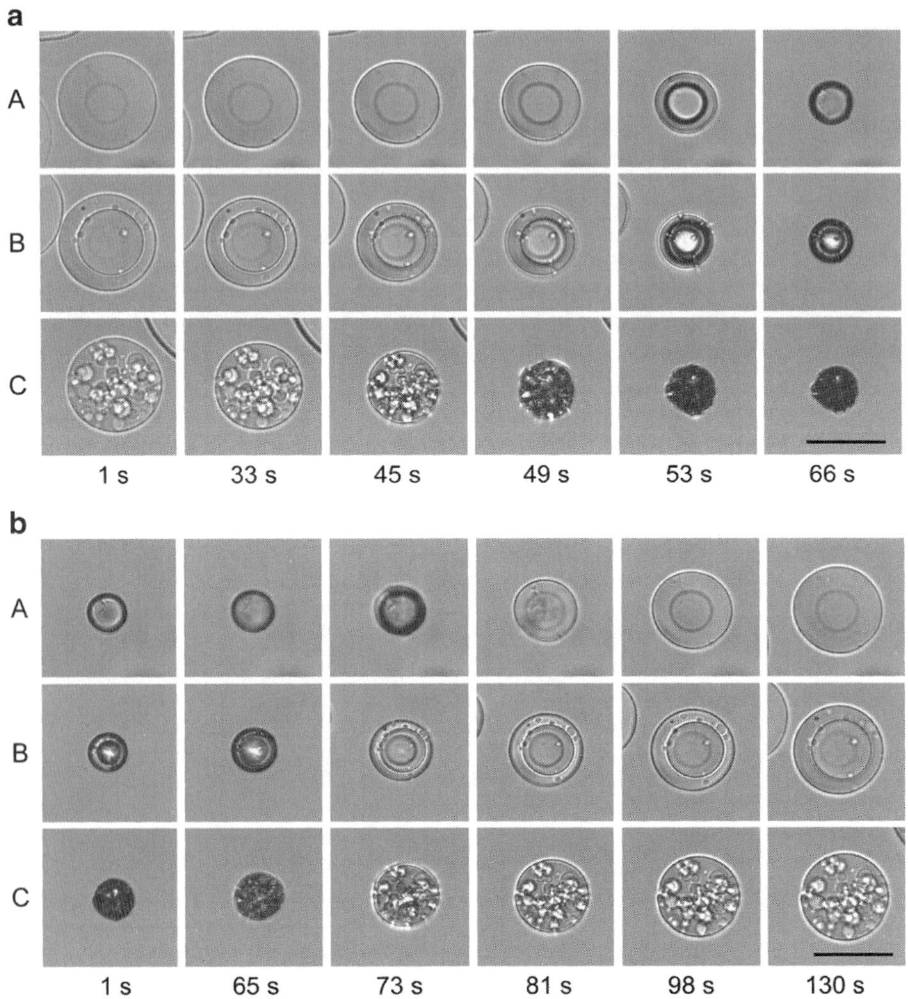

Fig. 2.16 Effect of internal structure on the dynamic volume shrinking and swelling behavior of microgels upon (**a**) heating from 23 °C to 47 °C and (**b**) cooling from 47 °C to 23 °C. *A*: Voidless microgel; *B*: microgel with hollow shell structure; *C*: microgel with multiple voids. Scale bar = 100 μm (Reproduced with permission from Ref. [43], Copyright (2007), Wiley-VCH Verlag GmbH & Co. KGaA)

with multiple voids and the one with a hollow shell structure swell faster than the voidless microgel (Fig. 2.16b). A quantitative comparison of the temporal evolution of sizes of the three microgels during heating and cooling is presented in Fig. 2.17 where we have plotted the instantaneous diameters of the microgels as a function of time. The temperature of the microgels is shown in the lower panel, and $t = 0$ refers to when the temperature change is initiated. The different behavior for each sample is evidenced by the separation between the curves. The inset in Fig. 2.17b

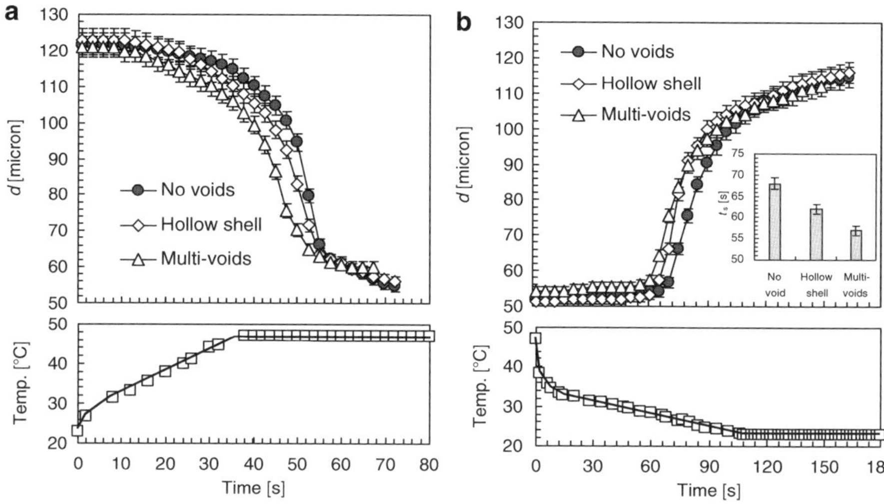

Fig. 2.17 Time-dependent diameter of microgels with different internal structures upon (**a**) heating from 23 °C to 47 °C and (**b**) cooling from 47 °C to 23 °C; t_s is the time elapsed before the microgels begin to swell (Reproduced with permission from Ref. [43], Copyright (2007), Wiley-VCH Verlag GmbH & Co. KGaA)

reveals that the inception of swelling in the microgel with multi-voids, indicated by a sudden change in its diameter, precedes that in the voidless microgel by as much as 10 s. The swelling of the microgels is perceptible only after a significant amount of water is transferred into the microgel. Thus, the result suggests that the transport of water in the microgels with voids occurs much before the voidless microgel.

The response rate of PNIPAM microgels to changes in temperature is governed by diffusion-limited transport of water in and out of the polymeric networks. Voids within a microgel offer much less resistance to the transport of water compared to the three-dimensional PNIPAM network; hence, the microgels with voids respond faster to changes in temperature than the voidless one. Furthermore, in the microgel with multi-voids, the voids can form continuous channels from the core to the surface of the microgel, further expediting transport of water in and out of the microgel. This is presumably why the microgel with multi-voids responds faster to temperature changes than even the microgel with the hollow shell structure.

To evaluate the effects of a progressive increase in the number of voids inside the microgels on their response kinetics, we prepared monodisperse PNIPAM microgels with one, two, three, and four voids of 25 μm diameter each. The dynamics of shrinking and swelling of these microgels are compared to that of a same-sized voidless microgel in Fig. 2.18. When temperature is increased from 23 °C to 47 °C, the microgels with four voids shrink faster than all other samples, whereas the voidless microgel is the slowest to respond. The differences are more distinct during swelling. The inset in Fig. 2.18b clearly reveals a systematic decrease in the time required for the inception of swelling with an increase in the number of voids.

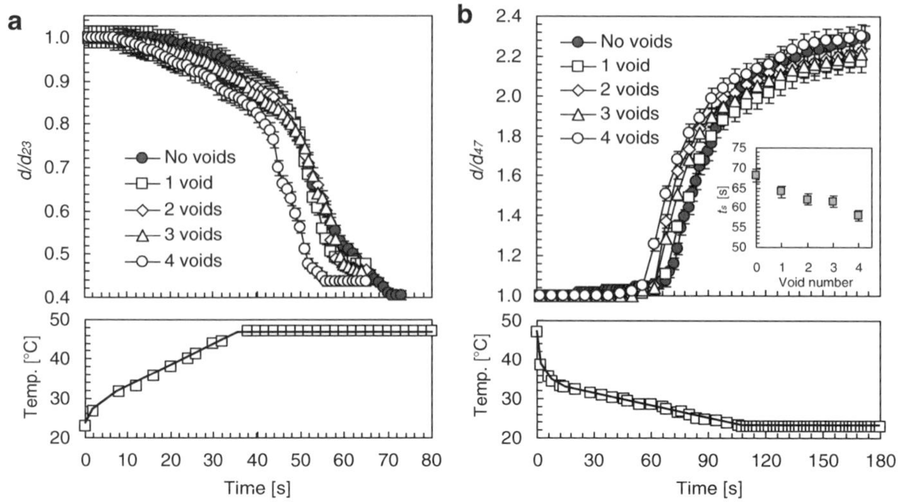

Fig. 2.18 Effect of the number of spherical voids on the dynamic volume shrinking and swelling behavior of microgels. The diameter of spherical voids is 25 μm. The samples are (**a**) heated from 23 °C to 47 °C and (**b**) cooled from 47 °C to 23 °C; d_{23} and d_{47} are the microgel diameters at 23 and 47 °C, respectively, and t_s is the time elapsed before the microgels begin to swell (Reproduced with permission from Ref. [43], Copyright (2007), Wiley-VCH Verlag GmbH & Co. KGaA)

This suggests that the volume-phase transition kinetics of microgels can be tuned simply by varying the number of voids in them. To gain such a precise control, it is necessary that all microgels in a given sample should not only be monodisperse, as shown above, but also contain the exact same number of voids.

2.5.2 Fabrication of Monodisperse Thermo-responsive Microgels in a Microfluidic Chip

In this section, an on-chip fabricating technique is introduced for the preparation of highly monodisperse and homogenous thermosensitive PNIPAM microgels. Instead of using UV irradiation, a redox reaction approach is designed in the microfluidic chip to initiate the polymerization of NIPAM monomer. Because the polymerization is carried out in the chip below the LCST, the resultant PNIPAM microgels are highly homogeneous. Another advantage of this approach is that all fabrication processes are completed in one chip and without any other supplementary instrument, e.g., UV lamp; thus, this microfluidic reactor is more compact, which makes it more easily scalable.

The microfluidic chip is made of poly(dimethylsiloxane) (PDMS) by using a standard soft-lithography method [49], which allows rapid replication of an integrated microchannel prototype. The flow-focusing geometry is introduced in

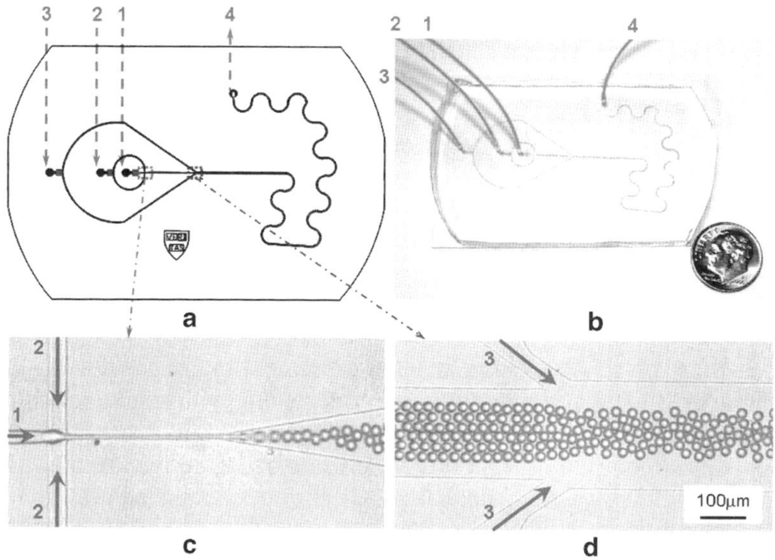

Fig. 2.19 (**a**) Schematic illustration of the channel design in the microfluidic chip. (**b**) A photograph of the PDMS microfluidic system compared with a one-dime coin of the USA. The polyethylene tubing and microchannels are filled with a dye-labeled aqueous solution to increase the contrast of image. The channels in the PDMS device have a height of 15 μm. The throat channel has a width of 10 μm and a length of 270 μm. Fluid 1 is an aqueous solution containing monomer, initiator, and cross-linker and is pumped through an inlet channel with width of 20 μm. The viscosity of fluid 1 is ~1 mPa·s. Fluid 2 is a kerosene solution containing surfactant and is pumped through two flanking channels with width of 30 μm for the continuous phase. The viscosity of fluid 2 is ~4 mPa·s. Fluid 3 is a kerosene solution containing surfactant and accelerator. (**c**) Optical microscope image of the drop formation in the emulsification step. (**d**) Optical microscope image of adding accelerator solution in the downstream of emulsification (Reproduced with permission from Ref. [44], Copyright (2008), Royal Society of Chemistry)

the device to generate monodisperse emulsion droplets, as shown in Fig. 2.19. In the emulsification-polymerization approach to prepare PNIPAM hydrogels, the polymerization of NIPAM monomer is usually initiated by adding accelerator into the monomer solution to start the redox reaction, by heating the monomer solution to a temperature above the LCST, or by irradiating with UV light. Generally, the redox reaction generates homogeneous internal microstructure inside the hydrogel, while the latter two methods result in heterogeneous internal microstructure. The PNIPAM microgel with homogeneous microstructure is reported to have a much larger thermo-responsive volume-change ratio than that with heterogeneous microstructure. To prepare homogeneous microgels in the microfluidic chip, if the accelerator is added into the monomer solution, the polymerization will occur in a very short time; thus, the microchannel will be clogged by the polymerized PNIPAM hydrogels and microgels cannot be gotten. To prevent this problem from happening in our device, here a novel approach is designed to add the accelerator. The accelerator is put into the oil phase, which is added into the downstream of emulsion solution (Fig. 2.19). Because the accelerator TEMED is more soluble in water, when it is added into the oil phase (Fig. 2.19d), it will diffuse into the water phase. When the accelerator meets the initiator APS inside the monomer emulsion droplets, redox reaction is initiated to polymerize the NIPAM monomer. Thus, not only the channel clogging problem is completely avoided, but also homogeneous internal microstructure is generated inside the microgels.

In the system shown in Fig. 2.19, fluid 1 is an aqueous solution containing monomer NIPAM (11.3 % w/v), an initiator APS (1.13 % w/v), and a cross-linker MBA (0.77 % w/v); fluid 2 is a kerosene solution containing surfactant PGPR 90 (8 % w/v); and fluid 3 is a kerosene solution containing both PGPR 90 (8 % w/v) and an accelerator TEMED (10 % v/v). The solutions are supplied to the microfluidic device through polyethylene tubing attached to syringes operated by syringe pumps. A Phantom high-speed camera is used to record the drop formation processes.

Another feature of this microfluidic approach is that it is able to precisely control the drop sizes inside the channel and meanwhile maintain the size monodispersity, because a thin and long throat channel is designed for the drop formation (Fig. 2.19c). In the experiments, monodisperse monomer droplets over the size range from ~10 to ~3 μm by radius can be generated.

After the polymerization, the microgels are washed 5 times with isopropanol by centrifuge to remove the oil on the microgel surface and then immersed into pure water. To test the thermosensitive volume-phase transition behavior, the PNIPAM microgels together with pure water are put into a transparent sealed holder on the slide glass, which is put on a heating and cooling stage for microscope. The actual temperature inside the sample holder is measured by an infrared thermometer. A digital camera is used to record the thermo-responsive behavior of microgels. The prepared PNIPAM microgels illustrate homogeneous structures and excellent thermosensitivity (Fig. 2.20). At 22 °C, the diameter of the PNIPAM microgels is about 2.7 times as that at 40 °C. The temperature-dependent diameter change of the microgel is satisfactorily reversible and shows a sharp transition near the LCST, as shown in Fig. 2.20c.

2.5.3 Fabrication of Monodisperse Microspheres with PNIPAM Core and Poly(2-Hydroxyethyl Methacrylate) (PHEMA) Shell

Generally, the thermo-response rate of PNIPAM microspheres is fast, and the reversible volume-change ratio is large, which enable PNIPAM microspheres highly promising in numerous applications. However, the biocompatibility of PNIPAM has been disputed for a long period of time, which hinders the application of PNIPAM microspheres in the biomedical field to a large degree. Recently, some investigations have been carried out to try to improve the biocompatibility of PNIPAM-based microspheres by coating biocompatible materials onto the outer

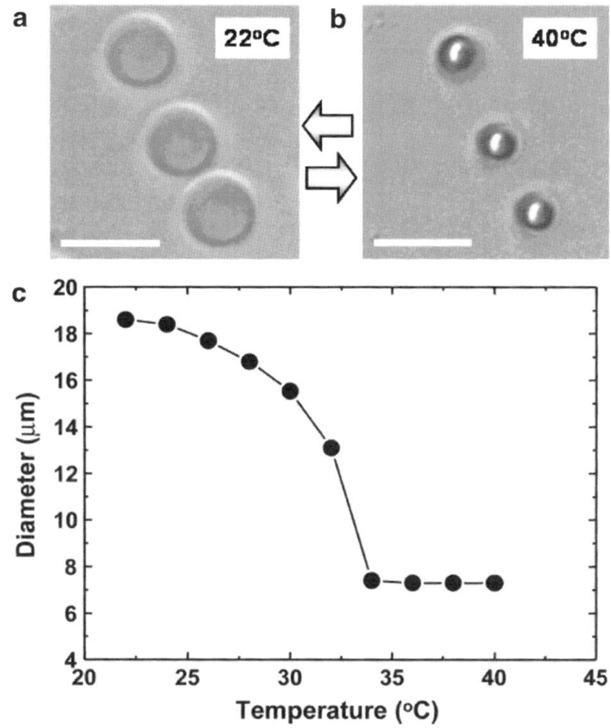

Fig. 2.20 (**a**) Optical microscope image of the microgels in pure water at 22 °C. (**b**) Optical microscope image of the same microgels in pure water at 40 °C. The scale bar is 25 μm. (**c**) Temperature-dependent diameters of PNIPAM microgels with temperature changes (Reproduced with permission from Ref. [45], Copyright (2011), Wiley-VCH Verlag GmbH & Co. KGaA)

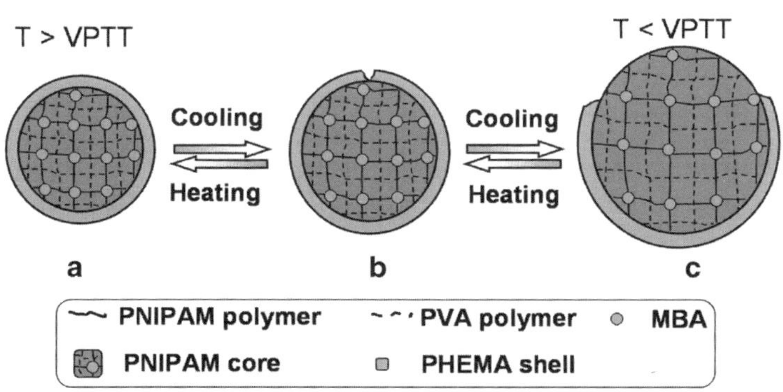

Fig. 2.21 Schematic illustration of the thermo-responsive behavior of the microsphere with PNIPAM core and PHEMA shell (Reproduced with permission from Ref. [46], Copyright (2010), Elsevier)

surface of PNIPAM microspheres. In this section, a microfluidic strategy for preparation of monodisperse microspheres with PNIPAM core and PHEMA shell is introduced [46].

The schematic illustrations of the structure and the thermo-responsive behavior of the microspheres with PNIPAM core and PHEMA shell are shown in Fig. 2.21.

The monodisperse PNIPAM cores are fabricated via microfluidic emulsification and free-radical polymerization methods, and the PHEMA shells are grafted onto the shrunken PNIPAM cores with atom-transfer radical polymerization (ATRP) method. At temperatures above the VPTT, the PNIPAM core shrinks and the PHEMA shell entirely covers the PNIPAM core (Fig. 2.21a). When the environment temperature decreases below the LCST, the PNIPAM core swells dramatically and tries to fill to the point of bursting inside the PHEMA shell. Because the PHEMA shell is not deformable but the inside PNIPAM core swells to a large extent, a visible crack comes into being on the PHEMA shell very soon (Fig. 2.21b), then the crack gets larger and larger, and finally the PHEMA shell ruptures a large area along the crack (Fig. 2.21c). Such thermo-responsive behavior of the microspheres with PNIPAM core and PHEMA shell is reversible and prompt when the temperature changes across the VPTT, which is adjustable by adding hydrophilic or hydrophobic comonomer for copolymerization of PNIPAM-based copolymeric cores.

Preparation of the core-shell microspheres includes two steps, namely, fabricating PNIPAM core and PHEMA shell subsequently. In the first step, monodisperse water-in-oil (W/O) emulsion droplets are prepared by a microfluidic approach (Fig. 2.22a), and the inner water phase contained NIPAM monomer and polyvinyl alcohol (PVA). The prepared W/O emulsion droplets act as templates for synthesizing PNIPAM/PVA semi-interpenetrating polymer network (semi-IPN) microspheres in the subsequent free-radical polymerization (Fig. 2.22b). PVA on the surface of the semi-IPN microspheres provides active groups (i.e., hydroxyl groups) to graft PHEMA shell in the next step. In the second step, hydroxyl groups react with 2-bromoisobutyryl bromide (BIBB) in order to introduce atom-transfer radical polymerization (ATRP) initiator (–Br) onto the microspheres firstly, and then PHEMA shell is fabricated by ATRP method (Fig. 2.22c).

Figure 2.23 shows the optical micrographs of PNIPAM cores in soybean oil and microspheres with PNIPAM core and PHEMA shell at room temperature and the corresponding size distributions. The diameters of both the PNIPAM core and microspheres with PNIPAM core and PHEMA shell are uniform, and their size distributions are narrow. The mean diameters of the original PNIPAM core microspheres and the core-shell microspheres are 140 and 202 μm, respectively. The calculated coefficient of variation (*CV*) values of them are 4.84 % and 3.18 %, respectively [46], which demonstrate that both the PNIPAM cores and the core-shell microspheres are of good monodispersity.

As mentioned above, the microfluidic emulsification is an intriguing emulsification method for preparing highly monodisperse emulsions. The monodispersity of W/O emulsions generated by microfluidic emulsification method is good. With the monodisperse W/O emulsion droplets as templates, monodisperse microspheres are subsequently synthesized. The reaction rate of free-radical polymerization is fast, which is beneficial to remain the monodispersity of the templates. The activator TEMED which can be dissolved in water better than in oil is fully dissolved in the outer oil phase in the experiments. After the single W/O emulsions form at the orifice, TEMED molecules uniformly and quickly diffuse into the inner monomer water phase and initiate the free-radical polymerization as soon as they encounter

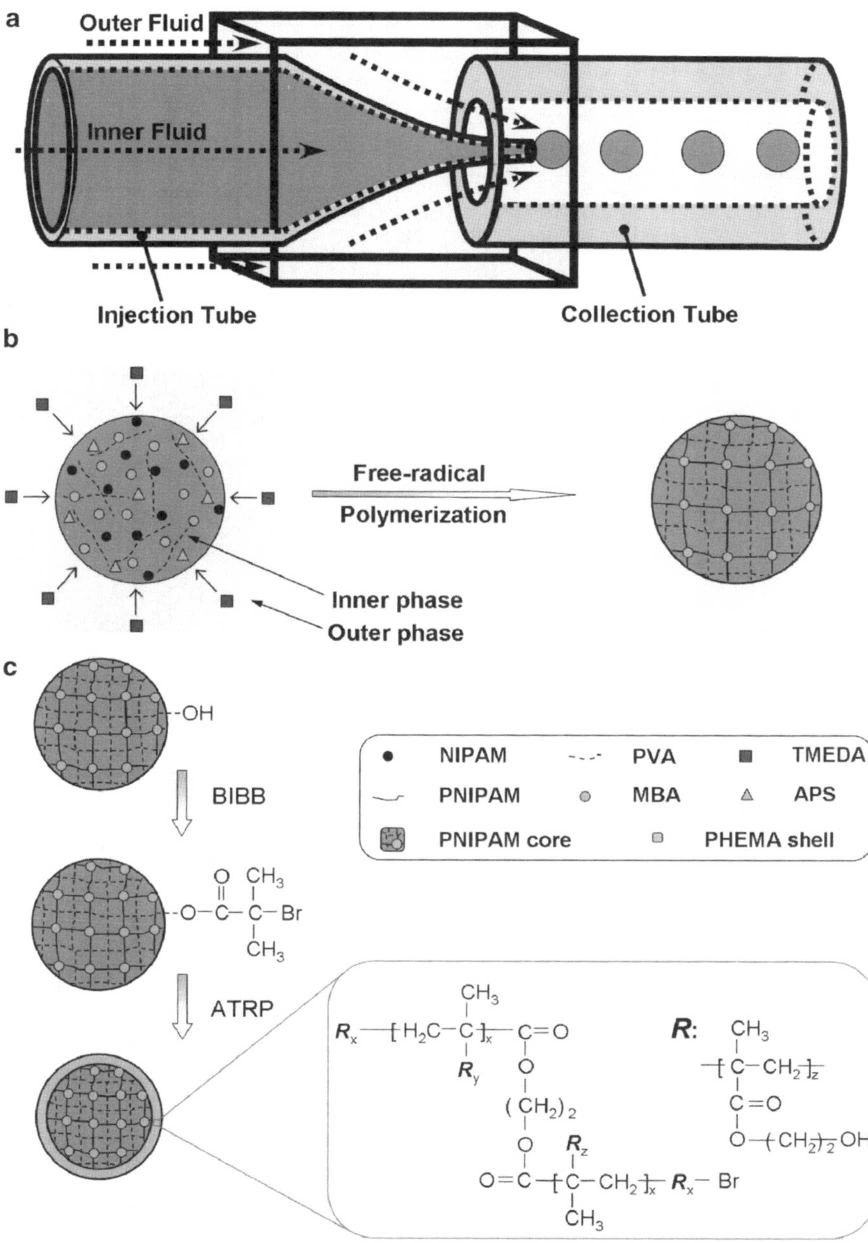

Fig. 2.22 Schematic illustration of the preparation procedure of the microsphere with PNIPAM core and PHEMA shell. (**a**) Microfluidic preparation of monodisperse emulsion droplets containing NIPAM monomer and PVA polymer, (**b**) polymerization of PNIPAM core with the emulsion droplet as synthesis template, and (**c**) fabrication of PHEMA shell on the PNIPAM core via ATRP method (Reproduced with permission from Ref. [46], Copyright (2010), Elsevier)

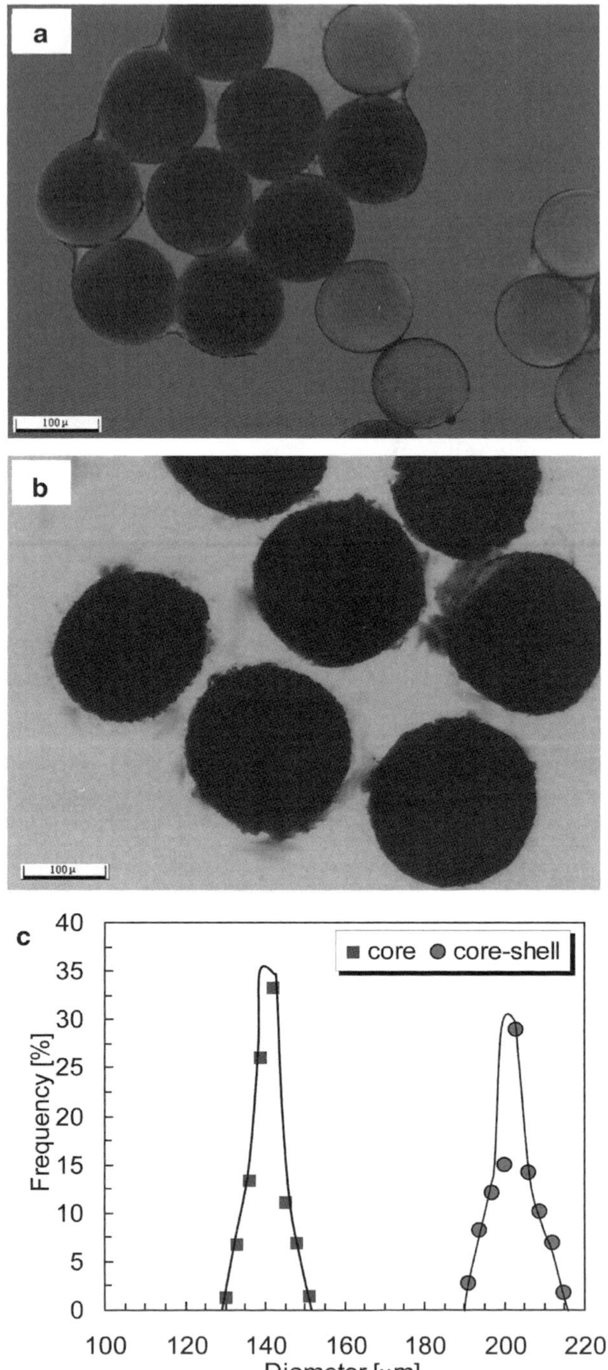

Fig. 2.23 (**a**, **b**) Optical micrographs of PNIPAM cores in soybean oil (**a**) and microspheres with PNIPAM core and PHEMA shell (**b**) at room temperature and (**c**) the corresponding size distributions (Reproduced with permission from Ref. [46], Copyright (2010), Elsevier)

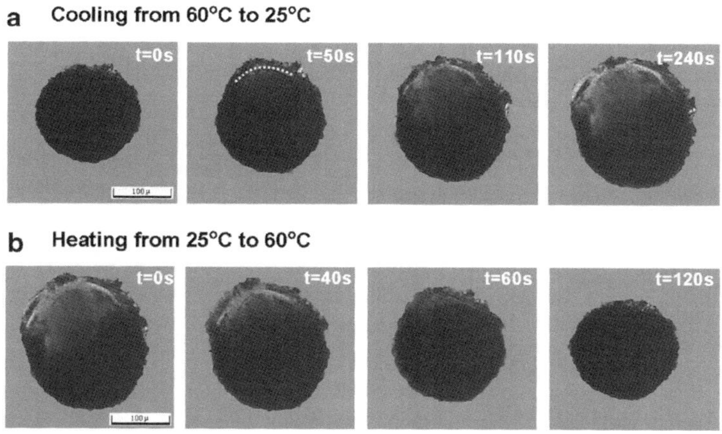

Fig. 2.24 Dynamic thermo-responsive characteristics of prepared microspheres with PNIPAM core and PHEMA shell during cooling-down process (**a**) and heating-up process (**b**). The scale bars are 100 μm (Reproduced with permission from Ref. [46], Copyright (2010), Elsevier)

with initiator APS. With high concentration of initiator APS (4 % molar ratio to monomer), the free-radical polymerization process is prompt and accomplished within several minutes. During the ATRP, PNIPAM core microspheres are uniformly distributed in the well-mixed HEMA monomer solution by gently magnetic stirring. Therefore, PHEMA polymers are uniformly grafted onto the PNIPAM cores to form the monodisperse core-shell microspheres with homogeneous thickness.

The dynamic thermo-responsive characteristics of core-shell microspheres prepared with PVA concentration of 2 % and grafting time of 2 h are shown in Fig. 2.24. During the cooling-down process (60 °C → 25 °C), the core-shell microsphere swells gradually, and at 50 s, there appears an obvious crack on the PHEMA shell of the microsphere (see the white dotted line in Fig. 2.24a). With the PNIPAM core microsphere swells further, the crack becomes larger and larger, and eventually the PHEMA shell of the core-shell microsphere ruptures a large area about 30 % of the whole surface area at 240 s. The microsphere with PNIPAM core and PHEMA shell rapidly reaches to the swelling equilibrium state at 240 s when the temperature is cooling down from 60 °C to 25 °C. The thermo-responsive PNIPAM core begins to swell at the temperature below the LCST, which acts as an expansive force on the outer PHEMA shell. The PHEMA shell is stretched to some degree until it cannot afford the larger and larger expansive force from the expanding PNIPAM core. Then, the thinnest area of the PHEMA shell breaks at first and loses the restraint for the further swelling of PNIPAM core. Therefore, the ruptured area becomes larger and larger along the crack on the surface of PHEMA shell till the PNIPAM core reaches to the swelling equilibrium.

Inversely, during the heating-up process (25 °C → 60 °C), the core-shell microsphere rapidly shrinks, and the ruptured PHEMA shell recovers to the state before the cooling-down process within 120 s (Fig. 2.24b). With the temperature heating up, the inner PNIPAM core shrinks, and the PHEMA shell promptly recovers to the original covering state on the PNIPAM core surface in absence of the expansive force. The as-prepared microspheres with PNIPAM core and PHEMA shell exhibit the thermo-responsive swelling/shrinking and more importantly the "open/close" switching characteristics during the cooling-down/heating-up processes, and the thermo-responsive function of the proposed microspheres is highly reversible by changing the temperature across the LCST (Fig. 2.24).

The thermo-responsive swelling/shrinking behavior of the PNIPAM core and the corresponding opening/closing behavior of the PHEMA shell crack of the microspheres enable such microspheres to be competent to deliver water-soluble drugs in a controllable way. Before delivering, the encapsulated drugs in PNIPAM core are protected by the biocompatible PHEMA shell at temperatures above the LCST, in which case the crack of the PHEMA shell is closed. On request to deliver the encapsulated drugs at specific site or programmed time, the local temperature is decreased below the LCST, so that the PNIPAM core swells and the crack of the PHEMA shell opens to provide unblocked channels for the encapsulated drugs to release.

2.6 Summary

Highly monodispersed thermo-responsive hydrogel microspheres with submicron size can be effectively fabricated by using a surfactant-free emulsion polymerization method, and monodispersed thermo-responsive hydrogel microspheres or microcapsules with diameters ranging from several microns to several hundreds of microns can be effectively prepared by using membrane emulsification or microfluidic emulsification method. Core-shell structures of microspheres or microcapsules with either thermo-responsive shells or thermo-responsive cores can be achieved by using either seed polymerization method, free-radical polymerization method, or ATRP method. The monodisperse thermo-responsive hydrogel microspheres with diverse microstructures are highly attractive for a wide range of applications in pharmaceutics and cosmetics and as sensors and actuators.

References

1. Pelton R (2000) Temperature-sensitive aqueous microgels. Adv Colloid Interface Sci 85:1–33
2. Gan D, Lyon LA (2001) Tunable swelling kinetics in core-shell hydrogel nanoparticles. J Am Chem Soc 123:7511–7517
3. Gan D, Lyon LA (2001) Interfacial nonradiative energy transfer in responsive core-shell hydrogel nanoparticles. J Am Chem Soc 123:8203–8209
4. Jones CD, Lyon LA (2000) Synthesis and characterization of multiresponsive core-shell microgels. Macromolecules 33:8301–8306

5. Jones CD, Lyon LA (2003) Shell-restricted swelling and core compression in poly(*N*-isopropylacrylamide) core-shell microgels. Macromolecules 36:1988–1993
6. Kwon IC, Bae YH, Kim SW (1991) Electrically credible polymer gel for controlled release of drugs. Nature 354:291–293
7. Ichikawa H, Fukumori Y (2000) A novel positively thermosensitive controlled-release microcapsule with membrane of nano-sized poly(*N*-isopropylacrylamide) gel dispersed in ethylcellulose matrix. J Control Release 63:107–119
8. Jeong B, Bae YH, Lee DS et al (1997) Biodegradable block copolymers as injectable drug-delivery systems. Nature 388:860–862
9. Leobandung W, Ichikawa H, Fukumori Y et al (2003) Monodisperse nanoparticles of poly(ethylene glycol) macromers and *N*-isopropylacrylamide for biomedical applications. J Appl Polym Sci 87:1678–1684
10. Murthy N, Thng YX, Schuck S et al (2002) A novel strategy for encapsulation and release of proteins: hydrogels and microgels with acid-labile acetal cross-linkers. J Am Chem Soc 124:12398–12399
11. Vihola H, Laukkanen A, Hirvonen J et al (2002) Binding and release of drugs into and from thermosensitive poly(*N*-vinyl caprolactam) nanoparticles. Eur J Pharm Sci 16:69–74
12. Kawaguchi H, Fujimoto K (1998) Smart latexes for bioseparation. Bioseparation 7:253–258
13. Kondo A, Kaneko T, Higashitani K (1994) Development and application of thermo-sensitive immunomicrospheres for antibody purification. Biotechnol Bioeng 44:1–6
14. Hu ZB, Chen YY, Wang CJ et al (1998) Polymer gels with engineered environmentally responsive surface patterns. Nature 393:149–152
15. Panchapakesan B, DeVoe DL, Widmaier MR (2001) Nanoparticle engineering and control of tin oxide microstructures for chemical microsensor applications. Nanotechnology 12:336–349
16. van der Linden H, Herber S, Olthuis W (2002) Development of stimulus-sensitive hydrogels suitable for actuators and sensors in microanalytical devices. Sens Mater 14:129–139
17. Bergbreiter DE, Case BL, Liu YS et al (1998) Poly(*N*-isopropylacrylamide) soluble polymer supports in catalysis and synthesis. Macromolecules 31:6053–6062
18. Guiseppi-Elie A, Sheppard NF, Brahim S et al (2001) Enzyme microgels in packed-bed bioreactors with downstream amperometric detection using microfabricated interdigitated microsensor electrode arrays. Biotechnol Bioeng 75:475–484
19. Debord JD, Eustis S, Debord SB et al (2002) Color-tunable colloidal crystals from soft hydrogel nanoparticles. Adv Mater 14:658–662
20. Yoshida R, Uchida K, Kaneko Y et al (1995) Comb-type grafted hydrogels with rapid de-swelling response to temperature changes. Nature 374:240–242
21. Wu XS, Hoffman AS, Yager PJ (1992) Synthesis and characterization of thermally reversible macroporous poly(*N*-isopropylacrylamide) hydrogels. J Polym Sci Polym Chem 30:2121–2129
22. Chu LY, Yamaguchi T, Nakao S (2002) A molecular recognition microcapsule for environmental stimuli-responsive controlled-release. Adv Mater 14:386–389
23. Chu LY, Park SH, Yamaguchi T et al (2002) Preparation of micron-sized monodispersed thermo-responsive core-shell microcapsules. Langmuir 18:1856–1864
24. Chu LY, Niitsuma T, Yamaguchi T et al (2003) Thermo-responsive transport through porous membranes with grafted PNIPAM gates. AIChE J 49:896–909
25. Chu LY, Park SH, Yamaguchi T et al (2001) Preparation of thermo-responsive core-shell microcapsules with a porous membrane and poly(*N*-isopropylacrylamide) gates. J Membr Sci 192:27–39
26. Yamaguchi T, Ito T, Sato T et al (1999) Development of a fast response molecular recognition ion gating membrane. J Am Chem Soc 121:4078–4079
27. Xiao XC, Chu LY, Chen WM et al (2003) Positively thermo-sensitive monodisperse core-shell microspheres. Adv Funct Mater 13:847–852
28. Matsuoka H, Fujimoto K, Kawaguchi H (1998) Monodisperse microspheres exhibiting discontinuous response to temperature change. Polym Gels Netw 6:319–332

29. Matsuoka H, Fujimoto K, Kawaguchi H (1999) Stimuli-response of microsphere having poly(*N*-isopropylacrylamide) shell. Polym J 31:1139–1144
30. Zhu PW, Napper DH (2000) Effect of heating rate on nanoparticle formation of poly(*N*-isopropylacrylamide)-poly(ethylene glycol) block copolymer microgels. Langmuir 16:8543–8545
31. Varga I, Gilanyi T, Meszaros R et al (2001) Effect of cross-link density on the internal structure of poly(*N*-isopropylacrylamide) microgels. J Phys Chem B 105:9071–9076
32. Gao J, Hu Z (2002) Optical properties of *N*-isopropylacrylamide microgel spheres in water. Langmuir 18:1360–1367
33. Zha L, Zhang Y, Yang W et al (2002) Monodisperse temperature-sensitive microcontainers. Adv Mater 14:1090–1092
34. Bouillot P, Vincent B (2000) A comparison of the swelling behaviour of copolymer and interpenetrating network microgel particles. Colloid Polym Sci 278:74–79
35. Tanaka T, Fillmore DJ (1979) Kinetics of swelling of gels. J Chem Phys 70:1214–1218
36. Thews G, Mutschler E, Vaupel P (1980) Anatomie, physiologie, pathophysiologie des menschen. Wissenschaftl Verlagsges, Stuttgart
37. Little K, Parkhouse J (1962) Tissue reactions to polymers. Lancet II(7261):857–861
38. Shiga K, Muramatsu N, Kondo T (1996) Preparation of poly(D, L-lactide) and copoly-(lactide-glycolide) microspheres of uniform size. J Pharm Pharmacol 48:891–895
39. Xiao XC, Chu LY, Chen WM et al (2004) Preparation of submicron-sized monodispersed thermo-responsive core-shell hydrogel microspheres. Langmuir 20:5247–5253
40. Xiao XC, Chu LY, Chen WM et al (2005) Monodispersed thermo-responsive hydrogel microspheres with a volume phase transition driven by hydrogen bonding. Polymer 46:3199–3209
41. Cheng CJ, Chu LY, Zhang J et al (2008) Preparation of monodisperse poly(*N*-isopropylacrylamide) microspheres and microcapsules via Shirasu-porous-glass membrane emulsification. Desalination 234:184–194
42. Cheng CJ, Chu LY, Zhang J et al (2008) Effect of freeze-drying and rehydrating treatment on the thermo-responsive characteristics of poly(*N*-isopropylacrylamide) microspheres. Colloid Polym Sci 286:571–577
43. Chu LY, Kim JW, Shah RK et al (2007) Monodisperse thermo-responsive microgels with tunable volume-phase transition kinetics. Adv Funct Mater 17:3499–3504
44. Shah RK, Kim JW, Agresti JJ et al (2008) Fabrication of monodisperse thermosensitive microgels and gel capsules in microfluidic devices. Soft Matter 4:2303–2309
45. Kim JW, Chu LY (2011) New functional microgels from microfluidics. In: Fernandez-Nieves A, Wyss HM, Mattsson J, Weitz DA (eds) Microgel suspensions: fundamentals and applications. Wiley-VCH, Weinheim, pp 53–69
46. Yu YL, Xie R, Zhang MJ et al (2010) Monodisperse microspheres with poly(*N*-isopropylacrylamide) core and poly(2-hydroxyethyl methacrylate) shell. J Colloid Interface Sci 346:361–369
47. Ilmain F, Tanaka T, Kokufuta E (1991) Volume transition in a gel driven by hydrogen bonding. Nature 349:400–401
48. Kang HW, Tabata Y, Ikada Y (1999) Fabrication of porous gelatin scaffolds for tissue engineering. Biomaterials 20:1339–1344
49. Xia YN, Whitesides GM (1998) Soft lithography. Angew Chem Int Ed 37:551–575

Chapter 3
Flow and Aggregation Characteristics of Thermo-responsive Microgels During Phase Transition

Abstract The flow characteristics and kinetic characteristics during the phase transition may directly affect the performance of thermo-responsive microspheres in many applications. In this chapter, to probe into the flow and aggregation behaviors of thermo-responsive microspheres in microchannel during the phase transition, the flow characteristics of monodisperse poly(*N*-isopropylacrylamide) (PNIPAM) microspheres in microchannel with local heating are introduced systematically. Furthermore, the effects of microchannel surface wettability and roughness on the flow behaviors of thermo-responsive microspheres during the phase transition are also introduced. The phenomena show that the microspheres can aggregate together during the phase transition and stop automatically at a desired position in the microchannel by local heating if designed properly, which is what the targeting drug delivery systems expected.

3.1 Introduction

In the last decades, stimuli-responsive microspheres have attracted widespread interest from both scientific and technological aspects because they may find many applications in the fields of controlled drug delivery, chemical separation, enzyme immobilization, catalysis, sensors, and so on. Such microspheres are able to change their physical-chemical properties and colloidal properties in response to fluctuations in environmental conditions, alone or in combination, including temperature, pH, magnetic field, and other stimuli signals. There are many cases where environment temperature fluctuations occur naturally and in which the temperature stimuli can be easily designed and artificially controlled. Therefore, much attention has been focused on thermo-responsive microspheres recently.

Most of previous researches on the thermo-responsive microspheres are focused on improving their monodispersity, thermo-responsiveness, and controlled-release property. Nevertheless, investigations on the flow characteristics of such

L.-Y. Chu et al., *Smart Hydrogel Functional Materials*,
DOI 10.1007/978-3-642-39538-3_3,
© Chemical Industry Press, Beijing and Springer Berlin Heidelberg 2013

microspheres during the phase transition are very few. The reason may exist in that such investigations fall into an interdisciplinary field, which might be easily neglected. However, if the thermo-responsive microspheres are applied to targeting drug delivery systems and other applications, their flow characteristics and kinetic characteristics during the phase transition may directly affect the site-specific targeting performance. Therefore, it is essential and very important to understand the flow characteristics of thermo-responsive microspheres in microchannels during the phase transition process. For the patients with diabetes or hyperlipidemia [1, 2], hydrophobic fattiness in the blood can easily deposit onto the inner surface of blood vessels, which turns the inner surface of blood vessels from hydrophilic and smooth to hydrophobic and rough. Therefore, for thermo-responsive microspheres to achieve applications as thermo-responsive drug carriers specifically targeting such pathologically changed sites in blood vessels, it is necessary and essential to study the flow and aggregation behaviors of such thermo-responsive microspheres in microchannels with different surface wettabilities and roughness.

In this chapter, the flow and aggregation characteristics of the poly(*N*-isopropylacrylamide) (PNIPAM) spheres during the phase transition in a transparent glass pipe with hydrophilic inner wall [3] will be introduced first. Then, the flow and aggregation behaviors of monodisperse PNIPAM microspheres in glass capillary microchannel during their phase transition triggered by local heating [4] will be introduced. Finally, the effects of microchannel surface wettability and roughness on the flow behaviors of thermo-responsive microspheres during the phase transition [5] will be discussed systematically.

3.2 Flow and Aggregation Characteristics of Thermo-responsive Spheres During the Phase Transition

Compared with microgels, spheres with diameters of several millimeters are easier to be observed with the naked eye; therefore, flow and aggregation characteristics of thermo-responsive spheres during the phase transition are introduced in this section.

3.2.1 Preparation of Monodisperse PNIPAM Hydrogel Spheres

PNIPAM spheres are prepared by the cross-linked polymerization [3]. First, NIPAM (1.13 g) and MBA (0.0154 g) are dissolved in 10 ml of deionized water (solution A), and 0.2 ml TEMED is dissolved in 4 ml of deionized water (solution B). Then, 0.3 ml of 10 % APS solution and 0.3 ml of solution B are added in succession to solution A. The solution is quickly stirred and then added dropwise into paraffin oil by glass syringe. The solution forms spherical droplets on the bottom surface of the dish. The droplets become gels within 5 min. Then, the hydrogel spheres are put into deionized water with room temperature (about 20 °C), in which they swell with the water. In the preparation, the temperature of the paraffin oil is kept at a constant 50 °C using a thermostatic unit. The diameter of the sphere is controlled by the size of needle and the extruding speed. The PNIPAM spheres with the same diameter are picked out for the measurement of flow and aggregation characteristics.

3.2.2 Thermo-responsive Volume-Phase Transition Characteristics of PNIPAM Hydrogel Spheres

The prepared PNIPAM spheres are taken out of the paraffin oil and put into deionized water for a week, exchanging with fresh deionized water every 12–24 h to remove surplus oil and impurities until they reached the swelling/deswelling equilibrium state with deionized water. Then, PNIPAM spheres are put into thermostated water bath, in which the water temperature is changed from 10 °C to 49 °C. The diameter of the sphere is measured every 3 °C, and the volume of the sphere is calculated as a function of temperature. The volume-change ratio is defined as the ratio of the volume of the sphere at every fixed temperature to that at 49 °C. Thermo-responsive volume-phase transition behavior of PNIPAM hydrogel spheres is shown in Fig. 3.1a. The maximal volume-change ratio is about 7, and the LCST of PNIPAM sphere is 30 °C or so. Because of the boundary tension between water and paraffin oil, the PNIPAM hydrogels had satisfying spherical shape.

3.2.3 Flow Characteristics of PNIPAM Hydrogel Spheres During the Phase Transition in a Transparent Glass Pipe

The inner wall of the transparent Pyrex glass pipe is pretreated to be hydrophilic [3]. In the measuring section between point A and point B (Fig. 3.1b), the temperature range is designed to include the LCST of PNIPAM. The interior diameter of the pipe is 6.4 mm, and the length of section AB is 400 mm. Digital pickup camera is used to record flow behaviors of PNIPAM spheres inside the pipe during the phase transition. Because the operating temperature of the rotameter flowmeter should be the room temperature (about 20 °C), much lower than outlet temperature (about 43 °C), a cooling bottle is used. Before the experiment of measuring the flow characteristics of the spheres, the following considerations should be confirmed. First, it is necessary to confirm that the range of Reynolds number falls in laminar flow, since the character of blood flow is viscous laminar flow in micro-blood circulation system. The pressure drops are measured at different flow rates; the results show that the Reynolds number for laminar flow should be lower than 547. Therefore, the flow rates in the subsequent experiments are selected from 20 to

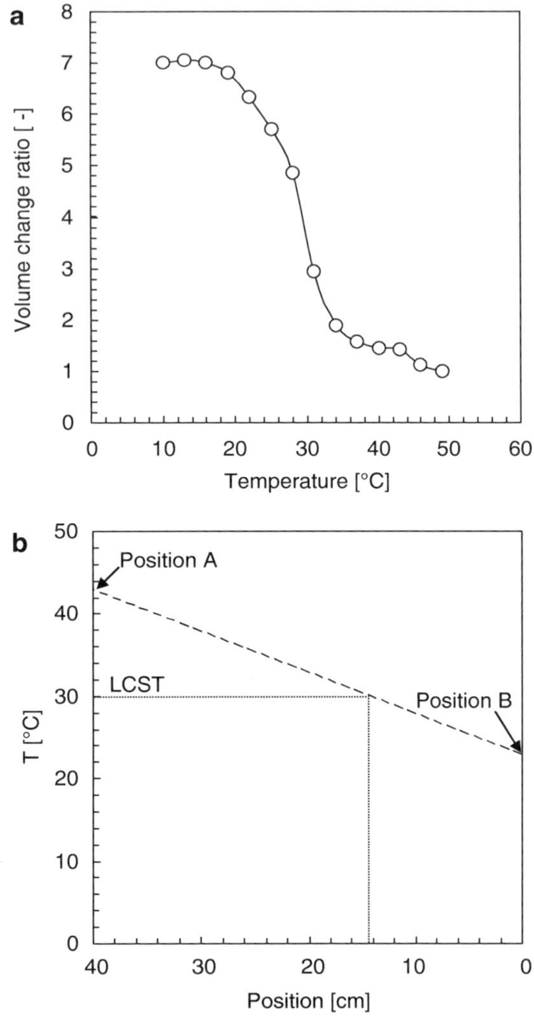

Fig. 3.1 Thermo-responsive volume-phase transition characteristics of PNIPAM hydrogel spheres (**a**) and temperature distribution along the glass tube (**b**) (Reproduced with permission from Ref. [3], Copyright (2006), Elsevier)

60 ml/min. At each flow rate, PNIPAM hydrogel spheres with the same diameter are put into the pipe and the motions are recorded. Second, the temperature distribution along the testing tube should be confirmed. Considering the classical principle of convection heat transfer in the counterflow heat exchanger, the boundary conditions of constant thermal flow density is used to indicate that average temperature of liquid varies linearly along the testing tube from inlet. Figure 3.1b shows the temperature distribution along the testing tube, in which “position” is the position fixed on the tube section AB, “0 cm” stands for the inlet point B, and “40 cm” meant the outlet point A. The average velocities of PNIPAM spheres in the pipe are calculated through the distance and time interval. A series of corresponding figures are snatched by media software from the videos of flow behaviors of the spheres when they are replayed.

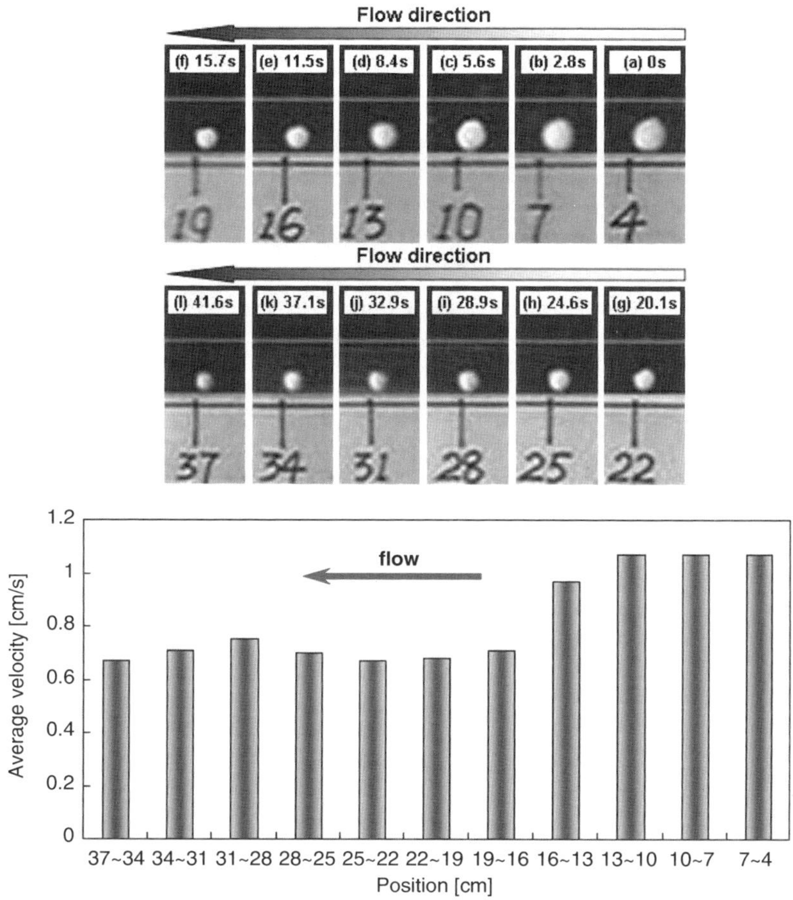

Fig. 3.2 Process of phase transition of a PNIPAM hydrogel sphere (**a**) and velocity variation of the PNIPAM hydrogel sphere during the process of phase transition in glass pipe (**b**). The fluid flow rate is 20 ml/min, and the unit for the position scale in (**a**) is cm (Reproduced with permission from Ref. [3], Copyright (2006), Elsevier)

Many interesting phenomena about the flow behaviors of PNIPAM hydrogel spheres are found during the phase transition. For one PNIPAM sphere (see Fig. 3.2a), the process of phase transition is obvious along the pipe when the water temperature is increased across the LCST of PNIPAM. From the position scale of 10 cm ($T = 28$ °C) to 19 cm ($T = 32.5$ °C), the diameter of the sphere decreased rapidly while the velocity of it slowed down sharply (see Fig. 3.2b). On the other hand, before and after this section, the diameter of the sphere had little change as well as the velocity. Because the smaller the sphere became, the closer it came to the bottom of laminar flow, the slower the average velocity of fluid became, and then the less the impetus force on the sphere became (see Fig. 3.3a, b). The forces acted on the sphere are illustrated in Fig. 3.3a, b. The drag force (F) and lift force

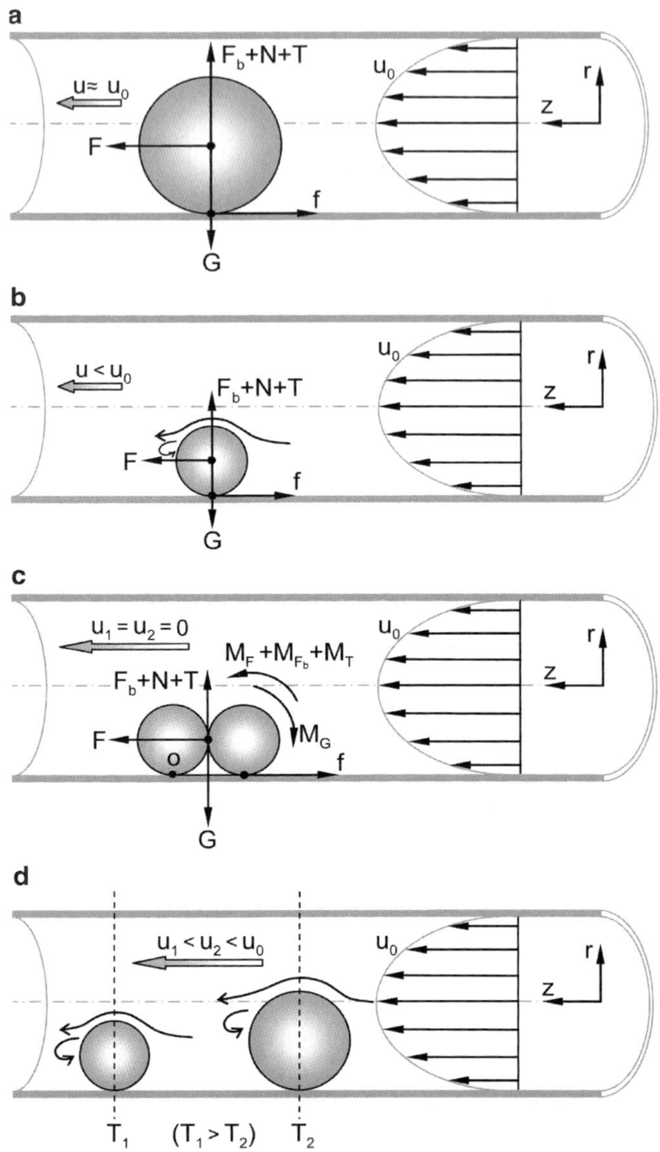

Fig. 3.3 (**a**, **b**) Forces acted on a PNIPAM hydrogel sphere in horizontal pipe before the phase transition (T < LCST) (**a**) and after the phase transition (T > LCST) (**b**). u_0 is the average velocity of fluid. (**c**) Forces and moments acted on two PNIPAM hydrogel spheres (with no initial distance) just after the phase transition in horizontal pipe and (**d**) flow and volume-change behavior of two PNIPAM hydrogel spheres (with a initial distance) during the process of phase transition in horizontal pipe (Reproduced with permission from Ref. [3], Copyright (2006), Elsevier)

(T) are the result of pressure and shear stress that can be obtained by the integration of pressure and shear stress across the surface of the sphere [6]. F_b is the buoyancy force, N is the supporting force from the pipe wall, and f is the friction force that can be calculated as

$$f = \mu N = \mu (G - F_b - T) \tag{3.1}$$

where μ is the coefficient of friction.

From Fig. 3.3a, b and Eq. 3.1, it can be concluded that when the gravity of agglomerate of PNIPAM spheres (after aggregation) is so large that the friction force became bigger than the drag force ($f > F$), they would stop moving.

For two PNIPAM spheres, three conditions are introduced as follows: There is no initial distance between them at the entrance of pipe section AB, and the initial distances are 5.5 and 8.5 mm, respectively. Under the first condition, two spheres with no initial distance move forward steadily side by side; however, when they come to the position scale of about 15 cm, at which the temperature is 30.5 °C, their diameters decreased dramatically and the velocity slows down. When they come to the position scale of 18 cm, something interesting happens, i.e., the two spheres start to overturn and subsequently roll forward with the two spheres aggregating together. At high temperature ($T >$ LCST), the hydrophobic effect of the PNIPAM hydrogel sphere surface makes them aggregate together. Because the velocity of fluid upside the sphere is bigger than that downside, as well as the existence of friction of pipe wall, when the friction force is larger than drag force ($f > F$) and the sum of moments M_F, M_{F_b}, and M_T is also larger than moment M_G ($M_F + M_{F_b} + M_T > M_G$), an anticlockwise resultant moment occurred and made the two spheres overturn (see Fig. 3.3c).

When the initial distance between the spheres is 5.5 mm at the entrance of pipe section AB (as shown in Fig. 3.4a), the two spheres move forward with the distance keeping 5.5 mm at first. However, as phase transition goes on, the distance between them becomes closer and closer, especially from the position scale of 10 cm ($T = 28$ °C) to 14 cm ($T = 30$ °C). At the same time, the velocity of the two spheres decreases sharply in this section, while the difference between the two velocities is bigger than anytime else (see Fig. 3.4b). From the position scale of 7 to 14 cm, the average velocities of sphere 2 are always larger than that of sphere 1; therefore, the distance between them becomes closer and closer. Finally they aggregate together after the phase transition due to the hydrophobic effect of the PNIPAM hydrogel spheres. Like the condition mentioned above, the two spheres roll forward subsequently. Before the phase transition, the velocity of fluid is almost the same as the velocity of the sphere; so at that time, there is little wake flow. With the temperature increasing, since the velocity of sphere decreases, the velocity of fluid is larger than the velocity of the sphere, the fluid would flow around the sphere, and wake flow of sphere 2 enhances and makes the sphere 1 slow down.

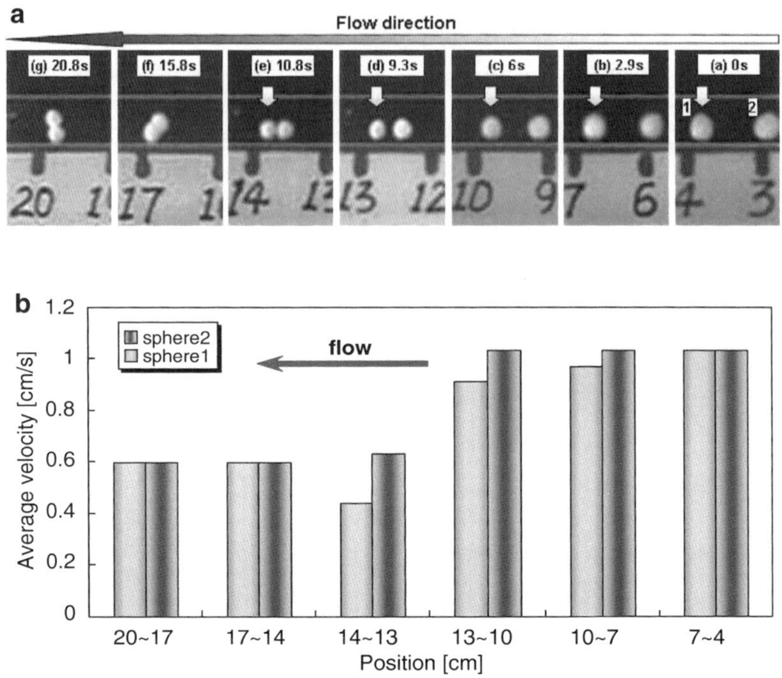

Fig. 3.4 Flow and aggregation characteristics (**a**) and velocity variation (**b**) of two PNIPAM hydrogel spheres (with initial distance of 5.5 mm) during the process of phase transition in horizontal pipe. The fluid flow rate is 20 ml/min, and the unit for the position scale in (**a**) is cm (Reproduced with permission from Ref. [3], Copyright (2006), Elsevier)

On the other hand, sphere 1 is in front of sphere 2, so the phase transition of sphere 1 as well as the decrease of velocity would be earlier than sphere 2. Because of the two effects, the two spheres come together, while the latter plays a more important role (see Fig. 3.3d).

When the initial distance between the spheres at the entrance of pipe section AB (as shown in Fig. 3.5a) comes to 8.5 mm, with the temperature increasing, the distance becomes close at first and then becomes far after the phase transition, and the two spheres do not aggregate together during the phase transition. The analysis of the average velocity of two spheres is shown in Fig. 3.5b. It indicates that the average velocity of sphere 1 slows gently, whereas the average velocity of sphere 2 slows dramatically from the position scale of 13 cm ($T = 29.5$ °C) to 19 cm ($T = 32.5$ °C). It is also found that the two PNIPAM spheres would never aggregate together during the phase transition when the initial distance between them at the entrance is larger than 8.5 mm.

If PNIPAM spheres aggregate together during the phase transition, different aggregation configurations for different numbers of spheres would be formed and then roll forward. For three PNIPAM spheres, if they aggregate together during the phase transition, they would form a regular triangle and then roll forward. A regular

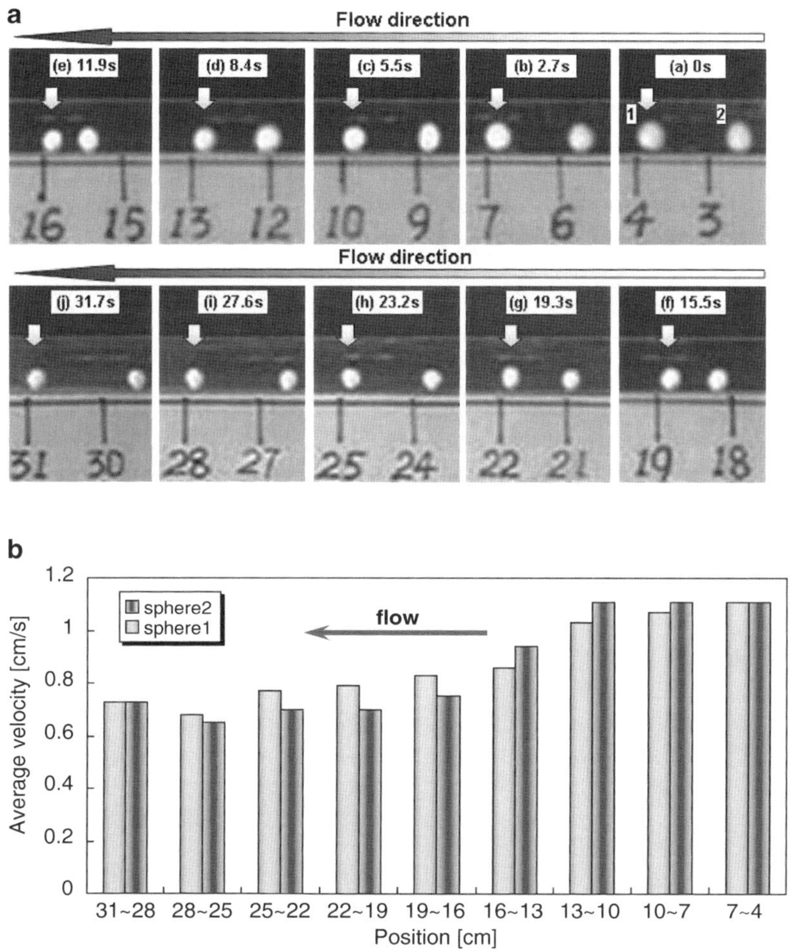

Fig. 3.5 Flow characteristics (**a**) and velocity variation (**b**) of two PNIPAM hydrogel spheres (with initial distance of 8.5 mm) during the process of phase transition in horizontal pipe. The fluid flow rate is 20 ml/min, and the unit for the position scale in (**a**) is cm (Reproduced with permission from Ref. [3], Copyright (2006), Elsevier)

tetrahedron would be formed for four PNIPAM spheres, a regular hexahedron would be formed with five spheres, and a regular octahedron would be formed with six spheres. At flow rate of 20 ml/min, a series of regular aggregation configurations of PNIPAM spheres during the phase transition are found. Because the fluid impetus on the spheres decreases as the phase transition goes on, when the impetus decreases to equal the friction of inner wall of the pipe, the spheres themselves would aggregate together owing to the impetus and then roll forward as rolling friction is smaller than breakaway friction. Furthermore, because the spherical particles with the same diameter gather together according to the best thick fill, i.e., the least interstice ratio,

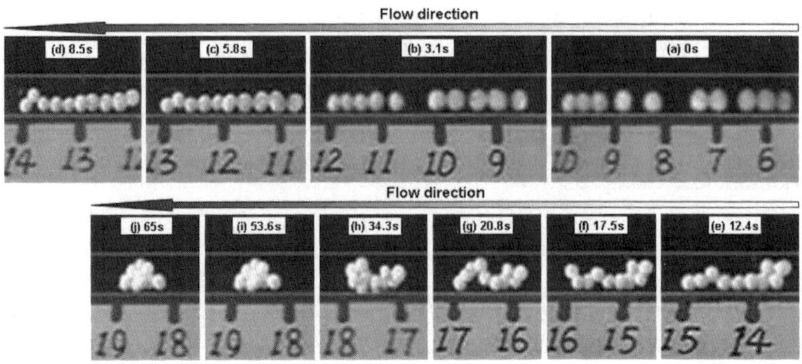

Fig. 3.6 Flow and aggregation characteristics of ten PNIPAM hydrogel spheres (with initial distances less than 5.0 mm) during the process of phase transition in horizontal pipe. The fluid flow rate is 20 ml/min, and the unit for the position scale is cm (Reproduced with permission from Ref. [3], Copyright (2006), Elsevier)

these arrays are the most stable among all cases. When the number comes to ten, different phenomena occurred (see Fig. 3.6). Since the agglomerate of PNIPAM spheres becomes too big, according to Eq. 3.1, friction between pipe wall and the agglomerate is much bigger than fluid impetus on the agglomerate ($f > F$), and because the size of the agglomerate of the spheres is too big to roll inside the pipe or the sum of moments M_F, M_{F_b}, and M_T is smaller than moment M_G ($M_F + M_{F_b} + M_T < M_G$), the whole agglomerate stops at the position scale of 19 cm where the temperature is about 32.5 °C after the phase transition. The above-mentioned phenomenon indicates that local heating on pathological part could be used to make thermo-responsive drug carriers slow down and even stop and release drugs there to achieve site-specific targeting therapeutic effects.

3.3 Flow Characteristics of Thermo-responsive Microspheres in Microchannel During the Phase Transition

3.3.1 Synthesis of Microspheres in a Simple Microfluidic Device

Monodisperse PNIPAM microspheres are prepared using a simple microfluidic device [4]. The dispersed phase is de-ionized water containing NIPAM (11.3 % (w/v)), MBA (0.154 % (w/v)) and ammonium persulfate (APS, 0.2 % (w/v)). The continuous phase is kerosene containing PGPR 90 (14.125 % (w/v)) and TEMED (3.5 % (v/v)). The dispersed and continuous phase solutions are pumped into the simple microfluidic device with adjustable flow rates by syringe pumps. The continuous phase flows through poly(vinyl chloride) (PVC) tubing (1 mm inner diameter and 3 mm outer diameter), and the dispersed phase is pumped into the PVC tubing via a 30 gauge needle (150 μm inner diameter) inserted through the wall of the PVC tubing with the needle tip located in the middle of the channel. The aqueous phase breaks into droplets at the tip of the needle, and the resultant droplets are carried away by the continuous phase flow, forming monodisperse W/O emulsion drops in the tubing. TEMED is both oil and water soluble; therefore, it can diffuse from kerosene to the emulsified aqueous phase and initiate the radical polymerization at room temperature. The PVC tubing with a length of 12 m ensures the complete polymerization of microspheres in the tube. The PNIPAM microspheres with different diameters are prepared at different flow rates of dispersed phase (4 ml/h and 2 ml/h, respectively) while at the same flow rate of continuous phase (120 ml/h).

3.3.2 Flow Characteristics of PNIPAM Microspheres in Horizontal Microchannel at Low Reynolds Number of Fluid

The microchannel device (Fig. 3.7a) with the function of local heat exchange is prepared using poly(methyl methacrylate) (PMMA) and glass capillary [4]. In the observation section, the temperature range is designed to across the lower critical solution temperature (LCST) of PNIPAM microspheres in SDS solution. Before the experiments on flow behaviors, temperature distribution of the observation section along the testing microchannel is measured by utilizing an infrared thermodetector (MiniTemp MT6, Raytek, USA) (Fig. 3.7b). The temperature measurement errors are in control of $\pm$ 1 °C. The interior diameter of the capillary is 520 μm, and the length of observation section is about 46 mm. In order to not only avoid the slack movement of PNIPAM microspheres but also prevent the LCST of PNIPAM microspheres from changing a lot, 0.02 wt% SDS solution is introduced into the microchannel as the fluid. The average velocity of the fluid is obtained by dividing the flow rate of fluid by cross-sectional area of capillary. The flow and aggregation behaviors of PNIPAM microspheres during the phase transition are observed by an optical microscope from top, and a digital pickup camera equipped on the microscope is used to record the video. The average velocities of PNIPAM microspheres in the microchannel are calculated through the moving distance and relevant time interval. A series of corresponding optical micrographs are snatched by Windows Movie Maker from the videos to analyze the flow behaviors of PNIPAM microspheres. For measuring the velocity of microspheres, 3–5 parallel measurements are carried out to verify the repeatability, and the average data are used to plot the velocity figures. The Reynolds numbers of fluid are tested in order to ensure the range of Reynolds number of fluid flow in the microchannel fall in laminar flow, since the character of blood flow is viscous laminar flow in microcirculation system [7]. Many interesting phenomena about the flow behaviors of PNIPAM microspheres with $d_s/D_{mc} = 0.913$ (where d_s stands for the diameter of PNIPAM microspheres and D_{mc} for the inner diameter of the microchannel) are found during the phase transition.

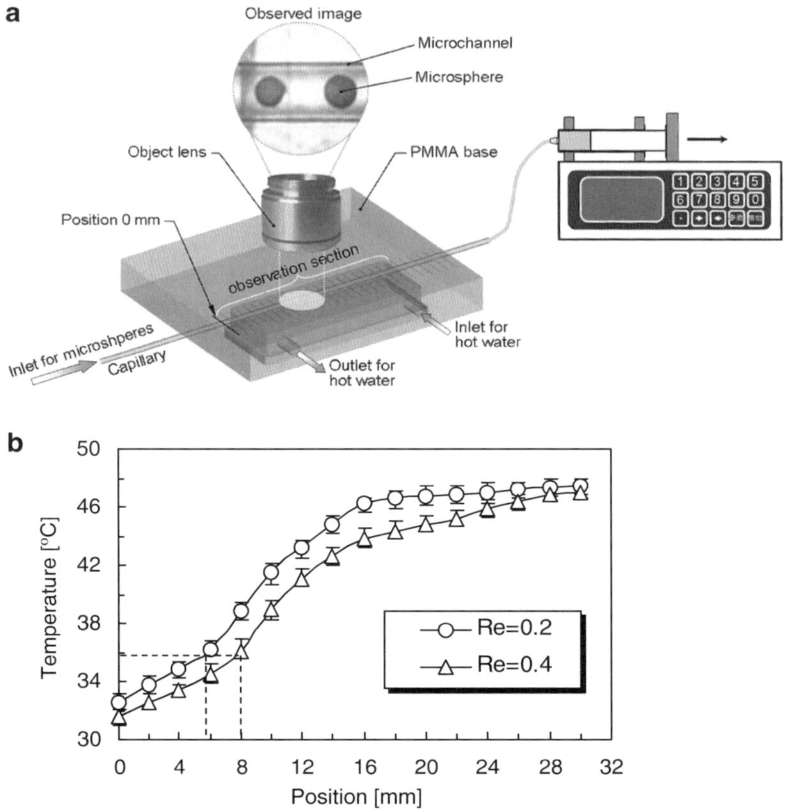

Fig. 3.7 (**a**) Schematic illustration of experiment apparatus for investigating flow characteristics of PNIPAM microspheres and (**b**) the temperature distribution along the observation section of the microchannel at different Reynolds numbers (Reproduced with permission from Ref. [4], Copyright (2009), John Wiley and Sons)

For one PNIPAM microsphere (Fig. 3.8a), the process of phase transition is obviously observed along the microchannel when the fluid temperature increases across the LCST of PNIPAM microsphere in 0.02 wt% SDS solution. As is known, PNIPAM has both isopropyl (hydrophobic) and amidogen (hydrophilic) groups, and PNIPAM hydrogels can shrink as the environment temperature increases from below the LCST to above the LCST and swell as the temperature decreases from above the LCST to below the LCST [3]. When the microsphere comes to the position scale of 8 mm, the environment temperature equals to the LCST of PNIPAM ($T \approx 35.7\,°C$), so the whole polymer network starts to shrink, and the diameter of the microsphere starts to decrease from this position and then decreases rapidly as the temperature increasing. The response time of the PNIPAM microsphere for the phase transition process is less than 10 s. Before the phase transition, the microsphere moves forward steadily with a constant velocity, and afterwards, the velocity of microsphere keeps

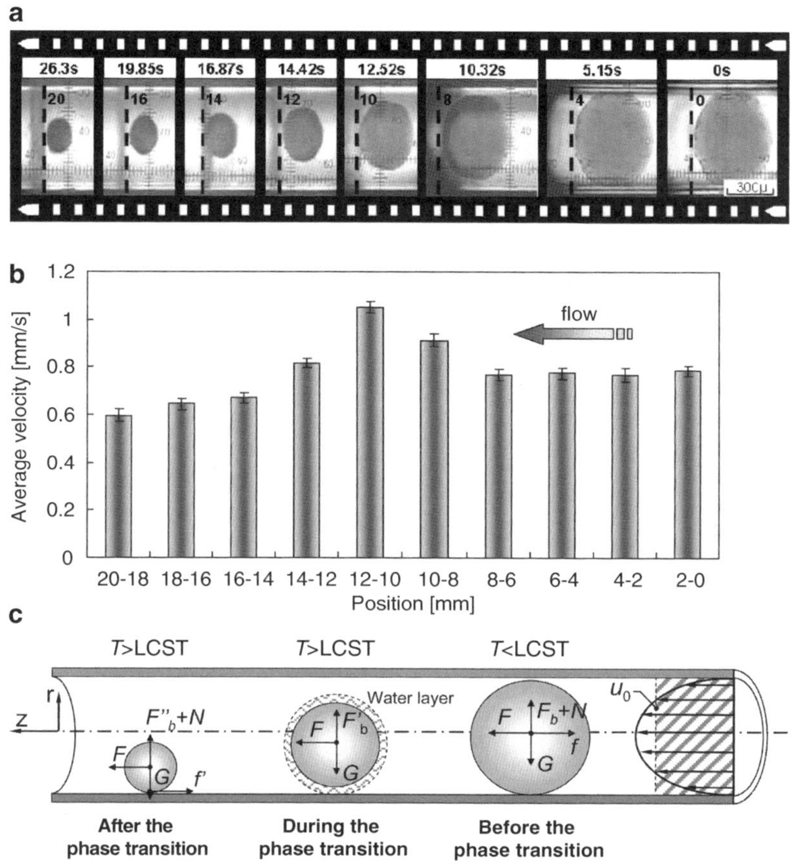

Fig. 3.8 Flow characteristics of a PNIPAM microsphere ($d_s/D_{mc} = 0.913$) during the process of phase transition in horizontal microchannel. The Reynolds number of fluid is 0.4. (**a**) Optical micrograph time series showing the movement of a PNIPAM microsphere during the phase transition process. The *dashed line* in the pictures is the position scale, and the number near the *dashed line* is the position (unit: mm). (**b**) The average velocity distribution of the microsphere varied with position. (**c**) Forces acted on a PNIPAM microsphere in horizontal microchannel during the phase transition ($T >$ LCST) and after the phase transition ($T >$ LCST). u_0, F, N, and G are the average velocity of fluid, the drag force, the supporting force from the microchannel inner wall, and gravity force, respectively. f and f' are the friction forces before the phase transition and after the phase transition. F_b, F'_b, and F''_b are the buoyancy forces before the phase transition, during the phase transition, and after the phase transition, respectively (Reproduced with permission from Ref. [4], Copyright (2009), John Wiley and Sons)

increasing during the phase transition and then decreases after the phase transition (Fig. 3.8b). Before the phase transition, as the microsphere is in the force balance (Fig. 3.8c), it moves forward steadily with a constant velocity because of no acceleration. During the phase transition, the microsphere shrinks dramatically and rapidly, the liquid inside PNIPAM polymer network is squeezed out rapidly due to the large volume-change ratio (about 22). As a result, a water layer forms around the microsphere, which removes the friction force (f) between the microsphere and the inner wall of capillary. Therefore, the force balance is broken, and an acceleration generates due to the drag force (F). Thus, the microsphere speeds up during this period. However, after the phase transition (i.e., the process of prompt deswelling finishes), the water layer surrounding the microsphere disappears, and the gravity (G) of the microsphere is larger than the buoyancy force (F''_b) after the phase transition. Hence, the microsphere locates in the bottom of the microchannel, where the local velocity is lower than the average velocity of the fluid. In addition, the existence of friction force (f') also prevents the microsphere from moving forward. Therefore, the microsphere after the phase transition slows down.

For two PNIPAM microspheres, two situations are investigated as follows: One is that there is no initial distance between two microspheres at the entrance of observation section (position scale of 0 mm shown in Fig. 3.7a) and the other is that the initial distance is 140 μm (Fig. 3.9a). In the first situation, two microspheres move forward steadily one after the other at the beginning. When they come to the position scale of about 6 mm at which the temperature is 36.2 °C, the temperature is above the LCST of PNIPAM, so the whole polymer networks of PNIPAM microspheres shrink dramatically, and the sphere diameters decrease dramatically and the velocities slow down. When they come to the position scale of 10 mm ($T = 41.5$ °C), the deswelling process of PNIPAM microspheres has almost finished, since the interval time for the microspheres moving from position scale 6 to 10 mm (about 11.3 s) is longer than the phase transition time (less than 10 s). Thus, the surface of PNIPAM microsphere turns hydrophobic. Because of the hydrophobic interaction of the surface of PNIPAM microspheres at high temperature ($T >$ LCST), the two microspheres aggregate together, and they start to overturn. After the phase transition, two aggregated microspheres may roll forward or swing forward randomly.

When the initial distance between two PNIPAM microspheres is 140 μm (Fig. 3.9a) at the entrance of observation section, the two microspheres move forward with the distance maintained as 140 μm at first. However, during the phase transition, with the temperature increasing, the surface distance between them becomes farther at first and then becomes closer; meanwhile, the center distance decreases all along during the phase transition, and finally the microspheres aggregate together. From position scale 6 mm ($T = 34.5$ °C) to 10 mm ($T = 39$ °C), the temperature increases across the LCST of PNIPAM microspheres in 0.02 wt% SDS solution. Because temperature difference between two microspheres with initial distance of 140 μm could be neglected, the phase transitions of the two microspheres occur almost isochronously. As a result, there is no difference of velocities between the two microspheres during this period (Fig. 3.9b). Therefore, the central distance (d) between the centers of two microspheres does not change. However, when the microspheres flow toward downstream, the sizes of two microspheres decrease, and consequently the surface distance (d_2) between the two microspheres increases (it seems that the distance of the two microspheres becomes longer)

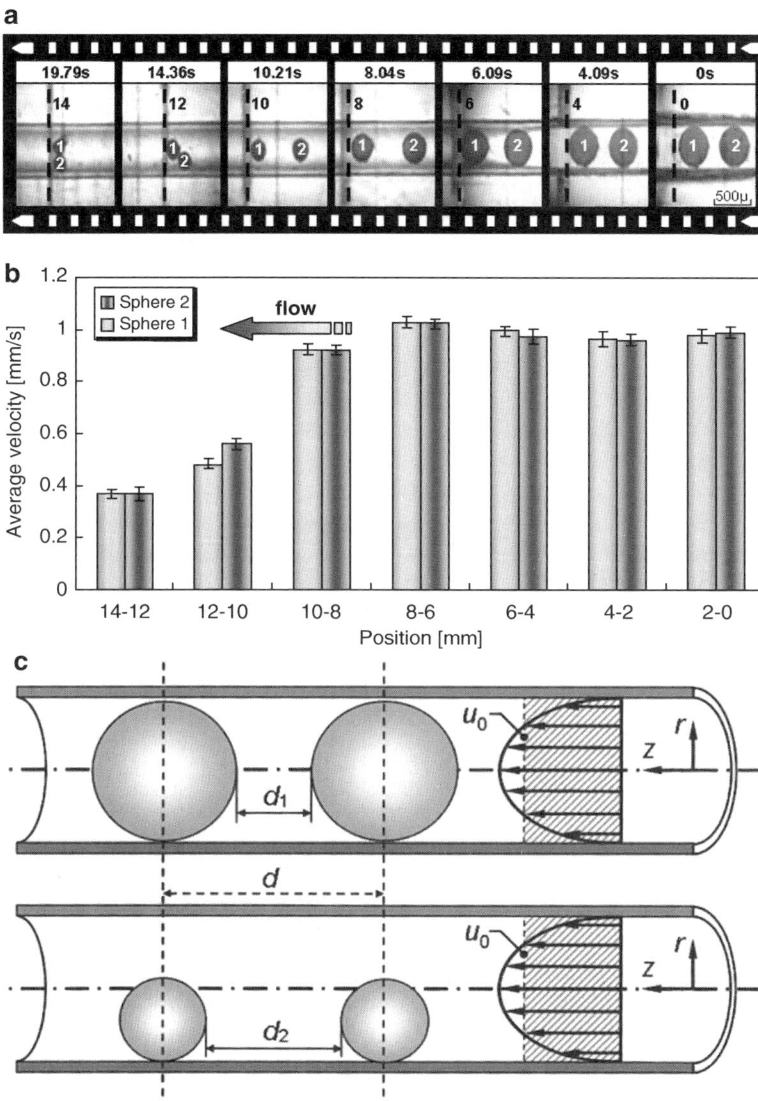

Fig. 3.9 Flow characteristics of two PNIPAM microspheres ($d_s/D_{mc} = 0.913$) with initial distance of 140 μm during the process of phase transition in horizontal microchannel. The Reynolds number of fluid is 0.4. (**a**) Optical micrograph time series showing the movement of two PNIPAM microspheres in the microchannel. (**b**) The average velocity distribution of two microspheres varied with position. (**c**) Distance variation between two microspheres during the phase transition. d and d_1 are the distances between the centers and between the surfaces of two microspheres before phase transition, respectively, and d_2 is the distance between the surfaces of two microspheres after phase transition (Reproduced with permission from Ref. [4], Copyright (2009), John Wiley and Sons)

(Fig. 3.9a, c). When the microspheres flow further toward downstream, the center distance decreases as well as the surface distance until two PNIPAM microspheres aggregate together. Even when the initial distance of two microspheres at the entrance of the observation section is as far as about 1,000 μm, the microspheres can still get aggregated during the phase transition (relevant pictures are not shown here because the eyeshot of microscope is too narrow to catch the two microspheres with such a far distance in the same view picture). The anterior microsphere always shrinks and slows down earlier than the posterior one, and the distance between two microspheres decreases continuously till zero. Since the aggregation of the microspheres makes the contact area between the microspheres and the inner wall of microchannel increase, the frictional resistance on the movement of microspheres increases. Therefore, the velocities of aggregated microspheres decrease more dramatically than that of a single microsphere during the phase transition. That is, the aggregation of PNIPAM microspheres makes them more easily to slow down or even stop at specific site when they suffer from local heating, which is what site-specific targeting drug delivery systems needed.

There are still some other factors that affect their site-specific targeting effectiveness in the microchannel. When the number of microspheres reaches 14 (Fig. 3.10a), all of the microspheres aggregate together during the phase transition. When the phase transitions of anterior PNIPAM microspheres occur, their velocities decrease. The phase transitions of anterior microspheres occur prior to those of the posterior microspheres; as a result, the posterior ones collide with the anterior ones and press them. Since the d_s/D_{mc} (i.e., 0.913) of the microspheres is too large, it is hard for the microspheres to overturn across one another in the microchannel. As a result, the microspheres can only move one by one and press each other, the chain of aggregated PNIPAM microspheres transforms due to the definite flexibility, and consequently the aggregated PNIPAM microspheres move forward with an “S” aggregation shape after the phase transition. For the 14 PNIPAM microspheres flowing in a line in the microchannel, when the phase transitions of anterior PNIPAM microspheres occur, their velocities decrease, which makes all the PNIPAM microspheres slow down during the phase transition. On the other hand, after the phase transition, the “S” aggregation shape can lead to the increase of effective area of fluid impetus. Therefore, unlike one or two microspheres, the aggregated 14 microspheres move faster rather than slow down after the phase transition (Fig. 3.10b).

3.3.3 Effect of the Diameter Ratio of PNIPAM Microsphere to Microchannel on the Flow Characteristics

To investigate the effect of the diameter ratio of PNIPAM microsphere to microchannel on the flow characteristics, PNIPAM microspheres with $d_s/D_{mc} = 0.596$ are also prepared besides the above-mentioned microspheres with $d_s/D_{mc} = 0.913$.

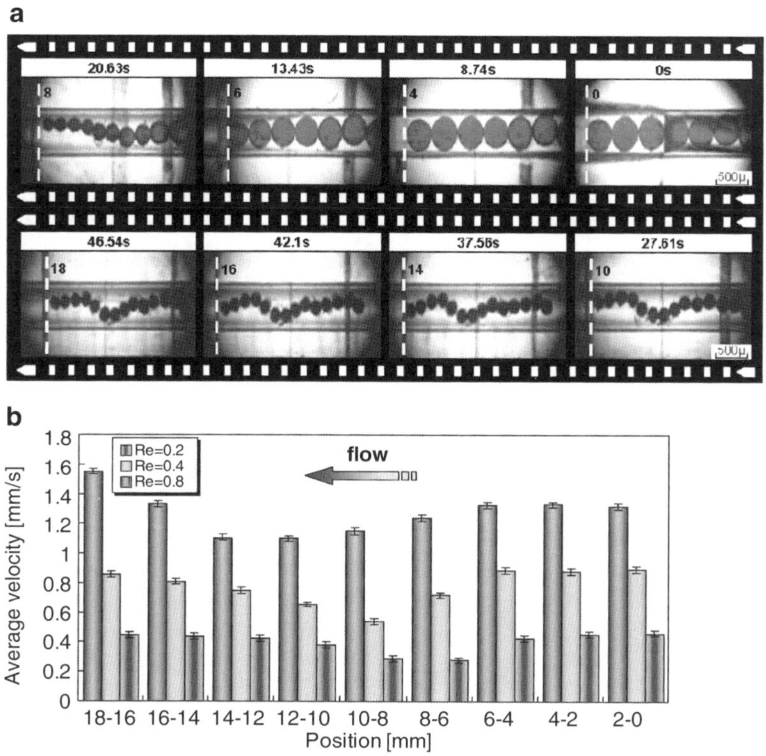

Fig. 3.10 Flow and aggregation characteristics of 14 PNIPAM microspheres ($d_s/D_{mc} = 0.913$) during the process of phase transition in horizontal microchannel. (**a**) Optical micrograph time series showing the movement of 14 PNIPAM microspheres in the microchannel. The Reynolds number of fluid is 0.2. (**b**) The average velocity distribution of microspheres varied with position at different Reynolds numbers of fluid (Reproduced with permission from Ref. [4], Copyright (2009), John Wiley and Sons)

For one PNIPAM microsphere with $d_s/D_{mc} = 0.596$ (Fig. 3.11a), the process of phase transition in the microchannel is almost the same as that of one PNIPAM microsphere with $d_s/D_{mc} = 0.913$, but the velocity change during the phase transition is very different. Unlike the case of microspheres with $d_s/D_{mc} = 0.913$, the velocity of the PNIPAM microsphere with $d_s/D_{mc} = 0.596$ decreases during the phase transition instead of increasing, which is just similar to the velocity variation of millimeter-scale PNIPAM hydrogel sphere ($d_s/D_{mc} = 0.608$) during the phase transition [3]. Since the maximal volume-change ratio (about 10 for PNIPAM microsphere with $d_s/D_{mc} = 0.596$ and 7 for millimeter-scale PNIPAM hydrogel sphere with $d_s/D_{mc} = 0.608$) is not so large, the effect of water layer surrounding the microsphere during the thermo-responsive phase transition is not so obvious during the phase transition. Compared with the PNIPAM microsphere with $d_s/D_{mc} = 0.913$, the velocity of microsphere in this study decreases to a much larger extent, and

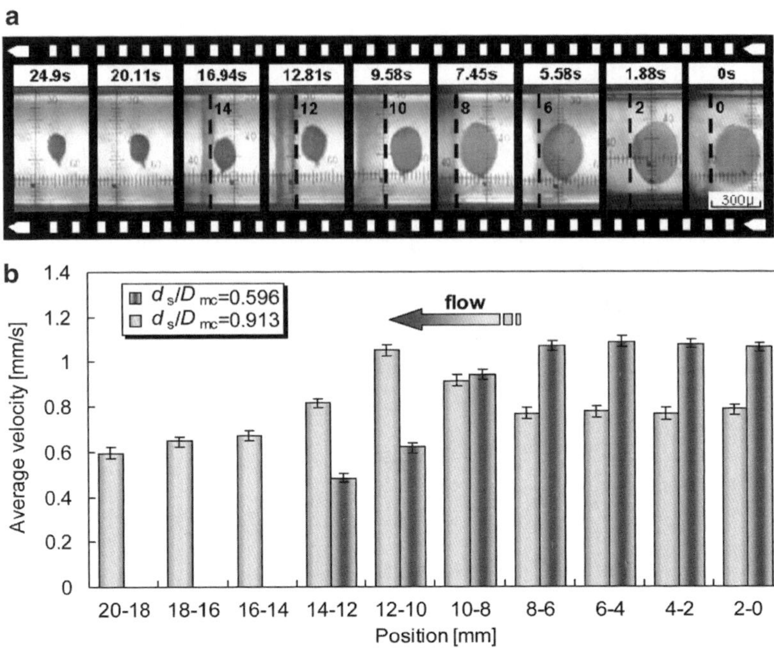

Fig. 3.11 Flow characteristics of a PNIPAM microsphere ($d_s/D_{mc} = 0.596$) during the process of phase transition in horizontal microchannel. (**a**) Optical micrograph time series showing the movement of a PNIPAM microsphere in the microchannel. The Reynolds number of fluid is 0.4. (**b**) The average velocity variation of microspheres with different ratios of d_s/D_{mc} along the position (Reproduced with permission from Ref. [4], Copyright (2009), John Wiley and Sons)

at last, the microsphere even stops just after the phase transition (Fig. 3.11). The microsphere with $d_s/D_{mc} = 0.596$ can shrink smaller than the microsphere with $d_s/D_{mc} = 0.913$ after the phase transition; on the other hand, the local velocity of fluid decreases from the central axis to the bottom of the microchannel. When the microsphere shrinks and comes to the bottom of the microchannel, the smaller the shrunken microsphere is, the lower the local velocity of fluid is. Therefore, the microsphere with smaller size ($d_s/D_{mc} = 0.596$) can slow down or even stop after the phase transition more easily than the bigger one ($d_s/D_{mc} = 0.913$).

When ten PNIPAM microspheres with $d_s/D_{mc} = 0.596$ move along the microchannel (Fig. 3.12), they also aggregate together during the phase transition. Unlike the aggregation of the larger microspheres with $d_s/D_{mc} = 0.913$ mentioned above (Fig. 3.10a), the PNIPAM microspheres with $d_s/D_{mc} = 0.596$ can overturn across one another easily during the phase transition, and finally ten PNIPAM microspheres aggregate to a large agglomerate. The aggregated microspheres stop at a certain position after the phase transition, which is almost the same as the phenomenon of ten millimeter-scale PNIPAM hydrogel spheres ($d_s/D_{mc} = 0.608$) [3]. The results indicate that the value of d_s/D_{mc} can determine the aggregation

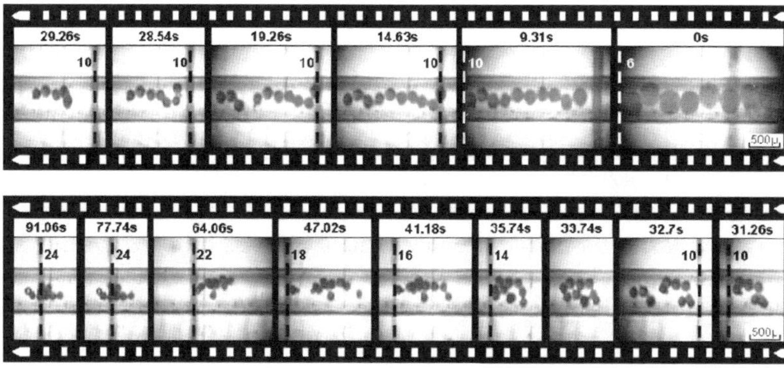

Fig. 3.12 Flow and aggregation characteristics of ten PNIPAM microspheres ($d_s/D_{mc} = 0.596$) during the process of phase transition in horizontal microchannel. The Reynolds number of fluid is 0.2 (Reproduced with permission from Ref. [4], Copyright (2009), John Wiley and Sons)

configuration of PNIPAM microspheres during the phase transition and then the velocity change of the aggregated microspheres. That is, the value of d_s/D_{mc} is a key parameter affecting the flow characteristics of PNIPAM microspheres during the phase transition. Therefore, to design thermo-responsive targeting drug delivery systems, d_s/D_{mc} should be taken into account properly to achieve site-specific stop of microsphere carriers by local heating.

3.4 Effects of Microchannel Surface Property on Flow Behaviors of Thermo-responsive Microspheres During the Phase Transition

3.4.1 Modification of Inner Surface of Glass Microchannel

The inner surfaces of glass microchannels are modified using the following three methods: by hydroxylation treatment to achieve hydrophilic surface, by self-assembly of chlorotrimethylsilane to realize hydrophobic surface, and by coating with silica nanoparticles to generate rough surface via sol-gel method [5]. Hydroxylation treatment is carried out by immersing unmodified glass microchannels into piranha solution (70 vol.% of sulfuric acid and 30 vol.% of hydrogen dioxide) at 90 °C for 30 min. Then, the glass microchannels are picked out, washed with deionized water to remove residual piranha solution, and dried by blowing with nitrogen gas. After the hydroxylation treatment, the inner surfaces of glass microchannels are modified to be hydrophilic and smooth. In order to obtain hydrophobic and smooth inner surface, the self-assembly of chlorotrimethylsilane is employed to modify the inner surface of glass microchannel. Briefly, one of the microchannels pretreated by hydroxylation treatment is immersed into 0.2 M of chlorotrimethylsilane (CTMS) hexane solution at room temperature for 2 h and then cleaned and dried as in the hydroxylation treatment. To generate rough inner surface of glass microchannel, sol-gel method is used to coat the surface with silica nanoparticles. The sol is prepared in two steps: firstly, adding 3 ml of NH_4OH (30 % in water) into 50 ml ethanol and stirring the mixture vigorously at 60 °C for 30 min and, secondly, adding 3 ml tetraethylorthosilicate (TEOS) dropwise into the NH_4OH/ethanol mixture and stirring for another 90 min. The prepared sol consists of monodisperse spherical silica nanoparticles with diameter of about 100 nm.[8] The prepared sol is injected into some of the microchannels pretreated by hydroxylation treatment, and then the channels are heated at 210 °C for 1 h. The heating helps remove the solvent in injected sol, and as a result, a layer of SiO_2 nanoparticles is formed on the inner surfaces of microchannels. The microchannels are subsequently immersed into H_2O_2 solution at room temperature for 30 min in order to generate the hydroxyl groups on their inner surfaces. Thus, the inner surfaces of the glass microchannels have been modified to be hydrophilic and rough. To prepare hydrophobic and rough inner surface, the self-assembly of chlorotrimethylsilane is employed. Briefly, one of the glass microchannels with hydrophilic and rough inner surfaces is further immersed into 0.2 M of CTMS hexane solution at room temperature for 2 h.

3.4.2 Characterization of Wettability and Roughness of Modified Glass Microchannels

The wettability of modified glass microchannels can be determined by characterizing aqueous static contact angle (θ) of their inner surfaces. The aqueous static contact angle is typically calculated by using following formula [9]:

$$\cos\theta = \frac{rh\rho g}{2\sigma} \tag{3.2}$$

where r is inner radius of the modified microchannel (m), h is liquid-meniscus height (m), ρ is liquid density (kg/m^3), and σ is liquid surface tension coefficient (N/m). At constant temperature, water has constant ρ (1,000 kg/m^3) and σ (0.073 N/m). By assuming r to be the same as the inner radius of unmodified microchannel (0.00026 m), the aqueous static contact angles of microchannel inner surfaces can be determined by measuring water-meniscus heights within the microchannels. Inner surface roughness of the modified microchannel is characterized using optical microscope (BX 61, Olympus) and scanning electron microscope (SEM, JSM 5900LV).

After being coated with a monolayer of SiO_2 nanoparticles, the smooth inner surface of the unmodified microchannel turns into a rough one. If the inner surface is coated with multilayers of SiO_2 nanoparticles, the roughness of inner

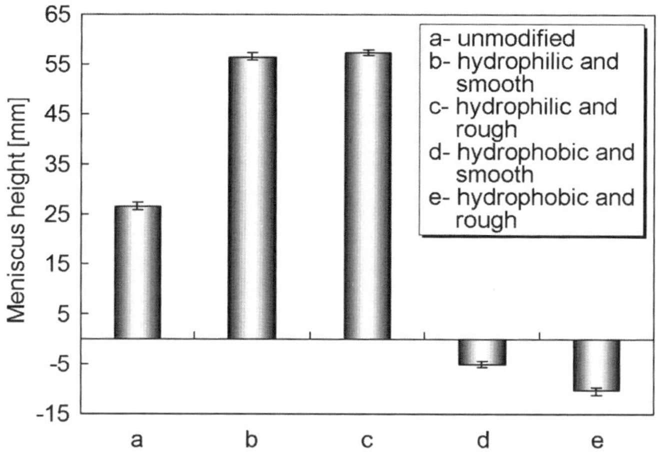

Fig. 3.13 Meniscus heights of water in unmodified and modified microchannels. The minus sign means that the meniscus height inside the microchannel is lower than the height of water level outside the microchannel. Three independent measurements of the meniscus height in microchannels are carried out (Reproduced with permission from Ref. [5], Copyright (2009), Elsevier)

surface will increase. However, highly rough inner surface can cause PNIPAM microspheres to undergo undesired slack movement in the microchannel. For that reason, only a monolayer of SiO_2 nanoparticles is deposited onto the inner surface of microchannel. The inner diameter of the modified microchannel is almost the same as that of unmodified microchannel. Therefore, the inner radius of modified microchannel and that of unmodified microchannel can be reasonably assumed to be the same when calculating the static contact angles by using Eq. 3.2.

Figure 3.13 shows the measured meniscus heights of water in both modified and unmodified microchannels. In unmodified glass microchannel, the meniscus height of water is measured to be 26.5 mm. Using Eq. 3.2, the static contact angle θ is calculated to be about 62.4°. In modified microchannels, however, the water-meniscus heights are much different. The meniscus heights of water in modified microchannel with hydrophilic and smooth inner surface and that in microchannel with hydrophilic and rough inner surface are 56.4 mm ($\theta \approx 17.5°$) and 57.2 mm ($\theta \approx 2.7°$), respectively. In addition, the meniscus heights of water in microchannel with hydrophobic and smooth inner surface and that in microchannel with hydrophobic and rough inner surface are −5 mm ($\theta \approx 95°$) and −10.3 mm ($\theta \approx 100.4°$), respectively. The minus value means that the meniscus height of water inside the microchannel is lower than the height of water level outside the microchannel. Compared with the unmodified microchannel, the modified microchannels with hydrophilic inner surfaces have smaller static contact angles, while those with hydrophobic inner surfaces have larger static contact angles. The results indicate that hydrophilic and hydrophobic modifications of inner surfaces of microchannels are successful.

3.4.3 Effects of Surface Wettability and Roughness of Microchannel on the Average Velocity of Fluid in Microchannel

In microchannels, PNIPAM microspheres move together with the fluid therein, and as a result, flow characteristics of microspheres are closely related with the average velocity of fluid. Previous reports show that the average velocity of fluid can be influenced by the wettability and roughness of channel surface [10, 11]. Therefore, before studying the flow characteristics of PNIPAM microspheres in microchannels, it is necessary to first investigate the effect of inner surface wettability and roughness of microchannel on the average velocity of fluid flow in microchannels. Such an investigation is performed using a self-made capillary microchannel device with the function of local heating [4, 5]. The inner diameter of the capillary used for microchannel device is 520 μm. Since blood flow in blood vessels is viscous laminar flow in microcirculation system [7], to ensure the fluid flow in our microchannel to also fall into laminar flow, the Reynolds numbers of fluid flow need to be tested and the fluid flow rate needs to be adjusted to ensure the Reynolds number to stay in the laminar flow range. The fluid flow rates in all measurements are fixed at 20 μl/min. Moreover, in order to study the effect of surface wettability and roughness of microchannel on the flow characteristics of PNIPAM microspheres during the phase transition (e.g., stop probabilities of PNIPAM microspheres after the phase transition), other flow rates that are either less than or larger than 20 μl/min, such as 10 μl/min and 30 μl/min, have also been tested. However, the difference of the stop probabilities at other flow rates is not as distinct as that at the flow rate of 20 μl/min. For instance, as the flow rate increases, the impetus force of fluid increases as well; as a result, it is more difficult for a PNIPAM microsphere to stop after the phase transition. When the flow rate is too large, the stop probabilities of PNIPAM microspheres in microchannels with different surface wettabilities and roughness after the phase transition are all 0 %. Therefore, 20 μl/min is an optimum flow rate to study the stop probabilities of PNIPAM microspheres in microchannels during the phase transition in the study.

The fluid phase in the experiments is 0.02 wt% SDS solution. The flow rate at which fluid is injected into a microchannel is well controlled by the control panel of syringe pump. When the fluid contains no PNIPAM microspheres and its flow rate is fixed, the fluid demonstrates different average velocities in microchannels modified to have different surface wettabilities and roughness. Figure 3.14 shows that, at same flow rate, the average velocity of fluid in microchannel with hydrophobic surface is higher than that in microchannel with hydrophilic surface. This observation can be explained using the boundary slip phenomenon. In microchannel with hydrophobic surface, the surface hydrophobicity weakens the energy transfer between the fluid and the surface, which decreases flow resistance and fluid slips on the hydrophobic surface [10, 11]. On the contrary, little boundary slip occurs in microchannel with hydrophilic surface. Therefore the fluid flows slower in hydrophilic microchannel

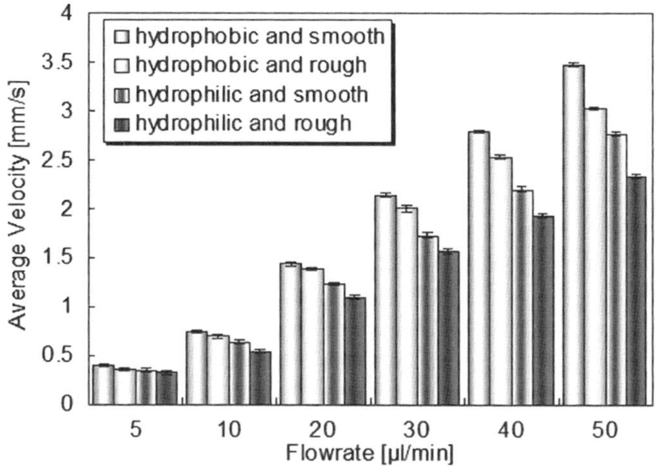

Fig. 3.14 Average velocities of fluid (0.02 wt% SDS solution) in the glass microchannels with different inner surfaces at different flow rates. Three independent measurements are carried out (Reproduced with permission from Ref. [5], Copyright (2009), Elsevier)

than in hydrophobic microchannel. As the flow rate increases, the difference between the average velocities in hydrophobic and hydrophilic microchannels increases.

Besides surface wettability, surface roughness also affects the average fluid velocity in the microchannel. Figure 3.14 also shows that when the flow rate and the surface wettability are the same, fluid in microchannel with rough surface demonstrates a lower average velocity than that in microchannel with smooth surface. The observed decrease of average fluid velocity in rough surfaced microchannels may come from different sources depending on the surface wettability. In microchannel with hydrophobic surface, rough surface restrains boundary slip [12, 13], which results in the decrease in average velocity. However, in microchannel with hydrophilic surface, rough surface increases the friction resistance, and therefore, the fluid flows more slowly.

3.4.4 Effect of Surface Wettability and Roughness of Microchannel on Aggregation Behaviors of PNIPAM Microspheres During the Phase Transition

To investigate the effect of the surface wettability and roughness of microchannel on aggregation behaviors of PNIPAM microspheres during the phase transition, the same amount of microspheres is injected into each modified microchannel. The PNIPAM microspheres injected into modified microchannels can aggregate during their phase transition, which is similar to what they do in unmodified microchannel. When three PNIPAM microspheres are injected into a microchannel, all of them form one aggregation during the phase transition which eventually evolves to a triangle and rolls forward after the phase transition; this trend persists in all the tested microchannels independent of the surface wettability and roughness. The same trend is observed when four and five PNIPAM microspheres are injected; the only difference is that, after the phase transition, aggregations formed by four and five PNIPAM microspheres evolve to tetrahedron and hexahedron, respectively. The results indicate that the surface wettability and roughness of microchannels do not influence the aggregation behaviors of PNIPAM microspheres during the phase transition. As described above, such aggregation may contribute to the site-specific targeting performance of thermo-responsive drug carriers.

3.4.5 Effect of Surface Wettability of Microchannel on Flow Characteristics of PNIPAM Microspheres

To study the effect of channel surface wettability on flow characteristics of PNIPAM microspheres, the average velocity of an individual PNIPAM microsphere flowing in smooth microchannels with either hydrophilic or hydrophobic inner surfaces is measured. In smooth microchannel with hydrophilic inner surface (Fig. 3.15a), the velocity of the microsphere displays a three-stage profile accompanying its phase transition. Before the phase transition starts, the microsphere moves forward with a constant velocity; during the phase transition, the velocity of microsphere keeps increasing; and after the phase transition completes, the velocity of microsphere starts to decrease. Before the phase transition, forces exerting on the microsphere (namely, drag force from flowing fluid and friction force between microsphere and channel surface) are well balanced. So, the microsphere moves forward at a constant velocity. Once the phase transition starts, the microsphere undergoes dramatic and rapid deswelling, which makes the microsphere squeeze water out from its PNIPAM polymer network. Water rapidly squeezed out from the polymer network forms a water layer around the microsphere, which prevents the microsphere from touching the channel surface and thus eliminates the friction force between them. Therefore, the previously achieved force balance is broken, and the microsphere speeds up by the unbalanced drag force from the flowing fluid. Once the phase transition completes, no more water can be squeezed out from the PNIPAM network. Because the microsphere has a density larger than that of the surrounding fluid, the microsphere regains touch with the bottom of microchannel, and the friction force between them comes back to work. It is known that, in the microchannel, the local velocity of fluid is the biggest along channel axial line and decreases when approaching the channel inner surface. Lower local velocity of surrounding fluid generally exerts smaller drag force. Therefore, because of the lower local velocity of surrounding fluid at bottom of the microchannel and the friction force, the microsphere slows down. Similar velocity profile is observed for microsphere in the smooth microchannel with hydrophobic inner surface (Fig. 3.15a).

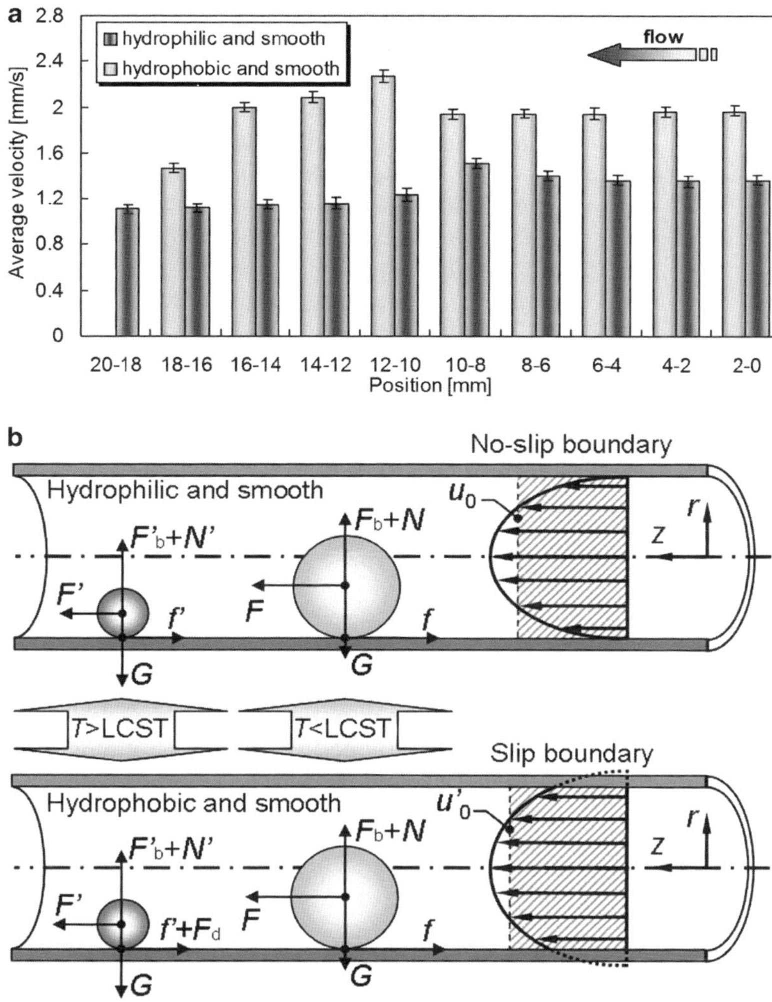

Fig. 3.15 The average velocity distribution of a PNIPAM microsphere varied with position (**a**) and boundary conditions of fluid (**b**) in smooth microchannels with different wettabilities. The fluid flow rate for (**a**) is 20 μl/min. u_0 and u'_0 are average velocities of the fluid in hydrophilic and hydrophobic microchannels, respectively. F, F_b, N, f, and G are, respectively, the drag force, buoyancy force, supporting force, friction force from the microchannel surface, and gravity force before the phase transition. F', F'_b, N', f', and G are, respectively, the drag force, buoyancy force, supporting force, friction force from the microchannel surface, and gravity force after the phase transition. F_d is the resistance resulting from the interaction of hydrophobic association. Twenty independent measurements of the microsphere velocities in microchannels are carried out (Reproduced with permission from Ref. [5], Copyright (2009), Elsevier)

Although velocity profiles for microspheres in smooth microchannels with different wettabilities exhibit qualitatively similar trend, their average velocities and post-phase-transition stop probability show quantitatively significant difference. In hydrophobic microchannel where boundary slip occurs, the local fluid velocity near the bottom is relatively high compared with that in hydrophilic microchannel (Fig. 3.15b), which leads to a relatively higher average fluid velocity in hydrophobic smooth microchannel than that in hydrophilic smooth microchannel (Fig. 3.14). As a result, average velocity of PNIPAM microsphere in hydrophobic smooth microchannel has a bigger magnitude than that in hydrophilic smooth microchannel. Besides difference in magnitude of average velocity, microspheres in microchannels with different wettabilities also exhibit different post-phase-transition stop probabilities. The microsphere keeps moving forward during the whole process of phase transition in the microchannel with hydrophilic and smooth inner surface (Fig. 3.15a); however, it may stop at a certain position after the phase transition in hydrophobic and smooth microchannel (Fig. 3.15a). The phase transition process of PNIPAM microsphere is not only a process of its volume shrinking but also a process in which the hydrophilic polymer network turns hydrophobic. After the phase transition, the whole PNIPAM polymer network shrinks and becomes hydrophobic. The hydrophobic association occurs between hydrophobic microsphere and hydrophobic inner surface of microchannel, and as a result, PNIPAM microsphere seems to be grasped by the inner surface of microchannel after the phase transition. Thus, a resistance (F_d in Fig. 3.15b) opposite to the flow direction makes the microsphere possibly stop at a certain position after the phase transition. However, not all samples can stop after the phase transition due to the interaction of hydrophobic association. In 20 parallel measurements, the stop probability of PNIPAM microsphere after the phase transition is about 57 % in hydrophobic and smooth microchannel.

The relatively high post-phase-transition stop probability of PNIPAM microspheres in hydrophobic microchannel indicates that hydrophobic inner surface of microchannel can contribute to the site-specific targeting of PNIPAM microspheres during the phase transition. The information demonstrates that thermo-responsive drug carriers could be used for site-targeting local therapy at certain pathologically changed sites.

3.4.6 Effect of Surface Roughness of Microchannel on Flow Characteristics of PNIPAM Microspheres

Figure 3.16 shows the variation of average velocity of an individual PNIPAM microsphere flowing in hydrophilic microchannels with different surface roughness. In the hydrophilic microchannel with smooth inner surface, the microsphere keeps moving forward during the whole process of phase transition (Fig. 3.16). The microsphere velocity is constant before the phase transition, keeps increasing during the phase transition, and decreases after the phase transition. Similar variation of average velocity is observed for PNIPAM microsphere in hydrophilic microchannel with rough inner surface. Since rough inner surface provides bigger friction force than smooth inner surface does, PNIPAM microsphere flows slower in rough hydrophilic microchannel than in smooth hydrophilic microchannel. As a result,

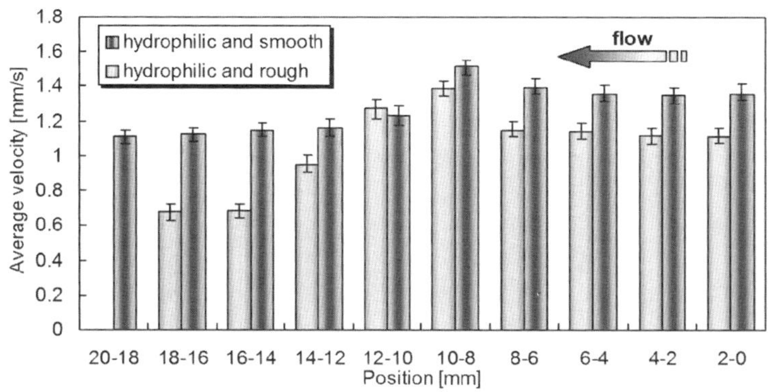

Fig. 3.16 The average velocity distribution of a PNIPAM microsphere varied with position in hydrophilic microchannels with different surface roughness. The fluid flow rate is 20 μl/min. Twenty independent measurements of the microsphere velocities in microchannels are carried out (Reproduced with permission from Ref. [5], Copyright (2009), Elsevier)

in the hydrophilic microchannel with rough inner surface, the microsphere stops at a certain position after the phase transition (Fig. 3.16). However, not all samples can stop after the phase transition in the microchannel with hydrophilic and rough inner surface. In 20 parallel measurements, the stop probability of PNIPAM microsphere after the phase transition is also about 57 % in rough hydrophilic microchannel.

The results indicate that rough inner surface of microchannel can also contribute to the site-specific targeting of PNIPAM microspheres during the phase transition. However, it is found that too rough inner surface (e.g., inner surface coated with multilayer of SiO_2 film) can result in a slack movement of the microsphere in the microchannel.

3.4.7 Flow Behaviors of PNIPAM Microspheres in Microchannel with Hydrophobic and Rough Surface During the Phase Transition

As discussed above, both roughness and hydrophobicity of microchannel inner surface can contribute to the site-specific targeting of PNIPAM microspheres during the phase transition. Therefore, it is valuable to investigate the flow behaviors of PNIPAM microspheres in the microchannel with both hydrophobic and rough inner surface. Figure 3.17 shows the variation of average velocity of a PNIPAM microsphere when one flows along microchannel with hydrophilic and smooth inner surface and another along the microchannel with hydrophobic and rough inner surface. In the microchannel with hydrophobic and rough inner surface, the microsphere easily stops at a certain position after the phase transition (Fig. 3.17). Likewise as mentioned above, not all samples can stop after the phase transition

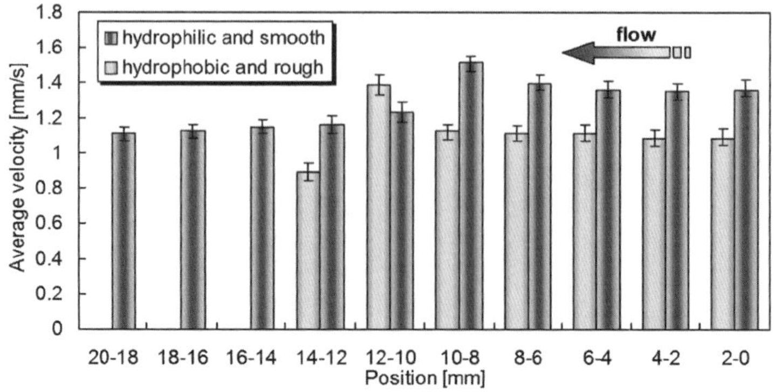

Fig. 3.17 The average velocity distribution of a PNIPAM microsphere varied with position in microchannels with different surface wettabilities and roughness. The fluid flow rate is 20 μl/min. Twenty independent measurements of the microsphere velocities in microchannels are carried out (Reproduced with permission from Ref. [5], Copyright (2009), Elsevier)

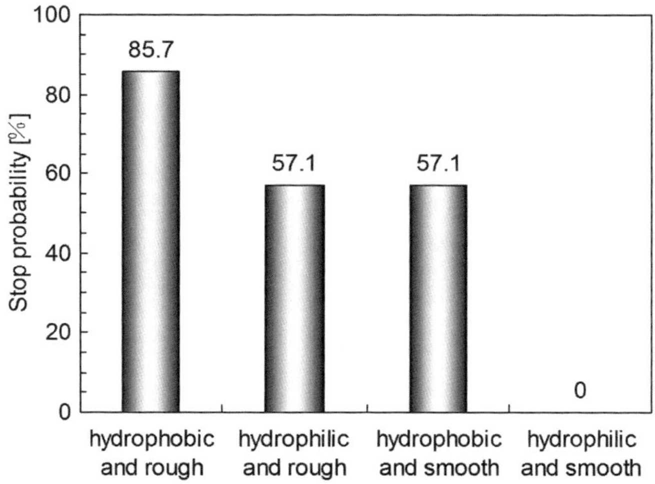

Fig. 3.18 The stop probabilities of a PNIPAM microsphere after the phase transition in microchannels with different surface wettabilities and roughness. The fluid flow rate is 20 μl/min. Twenty independent measurements of the stop probability of microsphere after the phase transition are carried out (Reproduced with permission from Ref. [5], Copyright (2009), Elsevier)

in the microchannel with hydrophobic and rough inner surface. However, the stop probability of PNIPAM microsphere after the phase transition is as high as about 86 % in 20 parallel measurements.

Figure 3.18 summarizes the post-phase-transition stop probabilities of an individual PNIPAM microsphere in microchannels with different surface wettability and roughness conditions. When flowing in the microchannel during the phase transition with the flow rate of liquid fixed at 20 μl/min, a post-phase-transition

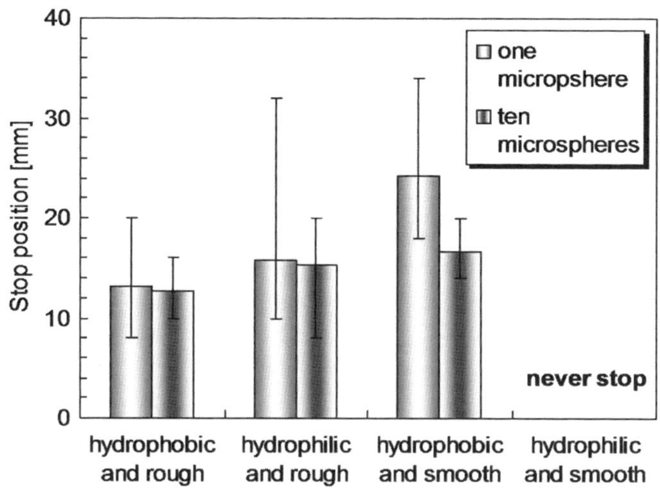

Fig. 3.19 Stop positions of a PNIPAM microsphere and ten PNIPAM microspheres in microchannels with different surface wettabilities and roughness. The fluid flow rate is 20 μl/min. Three independent measurements of the stop position of microsphere after the phase transition are carried out (Reproduced with permission from Ref. [5], Copyright (2009), Elsevier)

PNIPAM microsphere probably stops at a certain position in the microchannels with hydrophobic and rough inner surface, with hydrophobic and smooth inner surface, and with hydrophilic and rough inner surface. Among these three microchannels, the stop probability in microchannel with hydrophobic and rough inner surface is the highest. Such a high stop probability of microspheres in microchannel with hydrophobic and rough inner surface may come from a combination of both the hydrophobic association and significant friction force. High stop probability of microspheres after the phase transition can enhance the efficiency of site-specific targeting of thermo-responsive drug carriers.

If the drug carriers can stop as soon as they arrive at pathologically changed sites, the site-specific targeting delivery is more efficient. Therefore, not only the stop probability but also the stop position of drug carriers is an important evaluation criterion for the efficiency of targeting delivery. The stop positions of a PNIPAM microsphere and that of ten PNIPAM microspheres in the microchannel after the phase transition have been investigated (Fig. 3.19), where the stop position is defined as the distance from the start position scale "0" to the position where the microsphere stops. In this study, ten microspheres are chosen to investigate the stop probability and position mainly due to the following reasons. When less than ten microspheres aggregate in microchannels with different surface wettability and roughness conditions, the aggregated microspheres keep moving forward during the whole phase transition process. On the other hand, when the number of microspheres came to ten or more, the microspheres aggregate together and stop at a certain position after the phase transition. Furthermore, the experimental results prove that there are little distinct differences of stop position between ten microspheres and more than ten microspheres. Thus, ten PNIPAM microspheres are chosen to study the stop position of microspheres in microchannels with different surface wettability and roughness conditions. When a PNIPAM microsphere flows in the microchannels with different inner surfaces, it stops earlier in hydrophobic and rough microchannel than it does either in hydrophilic and rough microchannel or in hydrophobic and smooth microchannel. However, the microsphere never stops after the phase transition in hydrophilic and smooth microchannel. On the other hand, when ten microspheres flow in microchannels, they aggregate together during the phase transition. Compared with an individual PNIPAM microsphere, ten aggregated microspheres stop earlier in microchannels after the phase transition because the aggregation of microspheres enhances both the above-mentioned hydrophobic association and the friction force.

3.5 Summary

The flow and aggregation characteristics of thermo-responsive PNIPAM hydrogel spheres during the phase transition, the flow characteristics of thermo-responsive PNIPAM microspheres during the phase transition in microchannels, and the effects of microchannel surface wettability and roughness on the flow characteristics of thermo-responsive PNIPAM microspheres during the phase transition in microchannels are reported and discussed systematically. Many interesting flow behaviors of the PNIPAM microspheres during the phase transition are found in microchannels. The PNIPAM microspheres can respond to environment temperature quickly, and the response time is less than 10 s for the phase transition in the microchannel. The velocity of one microsphere with $d_s/D_{mc} = 0.913$ in horizontal microchannel increases first during the phase transition and then decreases after the phase transition. The PNIPAM microspheres are easy to aggregate together in the microchannel during the phase transition, even if the initial distance between two PNIPAM microspheres at entrance of the observation section is as long as 1,000 μm. The aggregation configuration of PNIPAM microspheres is determined by the value of d_s/D_{mc}, which ultimately plays a crucial part in the velocity variation of PNIPAM microspheres during the phase transition.

The wettability and roughness of microchannel have no influence on aggregation characteristics of PNIPAM microspheres during the phase transition. Because of boundary slip in hydrophobic microchannel, the average velocity of fluid without microspheres in hydrophobic microchannel is higher than that in hydrophilic microchannel. Hydrophobic surface of microchannel can enhance the post-phase-transition stop probability of PNIPAM microspheres. On the other hand, because rough inner surface enhances friction force, the average velocity of fluid in rough microchannel is lower than that in smooth microchannel. As a result, it may be easier for PNIPAM microspheres to stop at a certain position in rough microchannel than in smooth microchannel. Furthermore, both hydrophobic association and the friction force can contribute to the increase in stop probability of the microsphere during the phase transition. In addition, the wettability and roughness significantly affect the stop position of PNIPAM microsphere during the phase transition. The results indicate that hydrophobic or rough inner surface of microchannel, alone or in combination, is favorable for thermo-responsive site-specific targeting of PNIPAM microspheres in microchannels.

In a word, the results in this chapter provide valuable information for future applications of thermo-responsive PNIPAM microspheres in the therapeutic and biotechnological fields, especially in thermo-responsive site-specific targeting drug delivery systems.

References

1. Taniwaki H, Kawagishi T, Emoto M et al (1999) Correlation between the intima-media thickness of the carotid artery and aortic pulse-wave velocity in patients with type 2 diabetes – vessel wall properties in type 2 diabetes. Diabetes Care 22:1851–1857
2. Spinetti G, Kraenkel N, Emanueli C et al (2008) Diabetes and vessel wall remodelling: from mechanistic insights to regenerative therapies. Cardiovasc Res 78:265–273
3. Zhou MY, Chu LY, Chen WM et al (2006) Flow and aggregation characteristics of thermo-responsive poly(*N*-isopropylacrylamide) spheres during the phase transition. Chem Eng Sci 61:6337–6347
4. Zhou MY, Xie R, Ju XJ et al (2009) Flow characteristics of thermo-responsive microspheres in microchannel during the phase transition. AIChE J 55:1559–1568
5. Zhou MY, Xie R, Yu YL et al (2009) Effects of surface wettability and roughness of microchannel on flow behaviors of thermo-responsive microspheres therein during the phase transition. J Colloid Interface Sci 336:162–170
6. Ramadan A, Skalle P, Johansen ST (2003) A mechanistic model to determine the critical flow velocity required to initiate the movement of spherical bed particles in inclined channels. Chem Eng Sci 58:2153–2163
7. Mairey E, Genovesio A, Donnadieu E et al (2006) Cerebral microcirculation shear stress levels determine Neisseria meningitidis attachment sites along the blood-brain barrier. J Exp Med 203:1939–1950
8. Pham T, Jackson JB, Halas NJ et al (2002) Preparation and characterization of gold nanoshells coated with self-assembled monolayers. Langmuir 18:4915–4920
9. Idota N, Kikuchi A, Kobayashi J et al (2005) Microfluidic valves comprising nanolayered thermoresponsive polymer-grafted capillaries. Adv Mater 17:2723–2727
10. Tretheway DC, Meinhart CD (2002) Apparent fluid slip at hydrophobic microchannel walls. Phys Fluids 14:L9–L12
11. Tretheway DC, Meinhart CD (2004) A generating mechanism for apparent fluid slip in hydrophobic microchannels. Phys Fluids 16:1509–1515
12. Pit R, Hervet H, Leger L (2000) Direct experimental evidence of slip in hexadecane: solid interfaces. Phys Rev Lett 85:980–983
13. Zhu YX, Granick S (2002) Limits of the hydrodynamic no-slip boundary condition. Phys Rev Lett 88:106102

Chapter 4
Polyphenol-Induced Phase Transition of Thermo-responsive Hydrogels

Abstract Polyphenols such as tannic acid (TA) and ethyl gallate (EG) are important substances in food and medical, biological, and chemical fields and are used widely. In this chapter, polyphenol-induced phase transition behaviors of thermo-responsive PNIPAM hydrogels are introduced. Because microgels can respond much faster than macroscale hydrogels to environmental stimuli due to their small dimensions, PNIPAM hydrogel microgels rather than macroscale hydrogels are used to study the phase transition behaviors. Both equilibrium and dynamic phase transition behaviors of monodisperse PNIPAM microgels induced by TA and those of core-shell microcapsules with a PNIPAM shell and a colored oil core induced by EG are introduced.

4.1 Introduction

As mentioned in Chap. 1, poly(*N*-isopropylacrylamide) (PNIPAM) is a popular thermo-responsive polymer. In aqueous solutions, the phase transition of PNIPAM induced by environment temperature change could be controlled by the equilibrium of the hydrophilic and hydrophobic interactions, which leads to diverse phase transition behaviors of PNIPAM in different solvents owing to the interactions between PNIPAM and solvent molecules. Besides the solvent properties, the presence of certain guest molecules in solutions could also affect the volume-phase transition of PNIPAM and alter its transition temperature [1–4].

Polyphenols such as tannic acid (TA) and ethyl gallate (EG) are important substances in food and medical, biological, and chemical fields and are used widely. Tannic acid (TA), whose chemical structure is shown in Fig. 4.1, is a plant secondary metabolite and a popular polyphenol that can be found in approximately 80 % of woody and 15 % of herbaceous dicotyledonous species. It has been known that TA is an important protein precipitator, antioxidant, and curative for body injuries.

L.-Y. Chu et al., *Smart Hydrogel Functional Materials*,
DOI 10.1007/978-3-642-39538-3_4,

© Chemical Industry Press, Beijing and Springer Berlin Heidelberg 2013

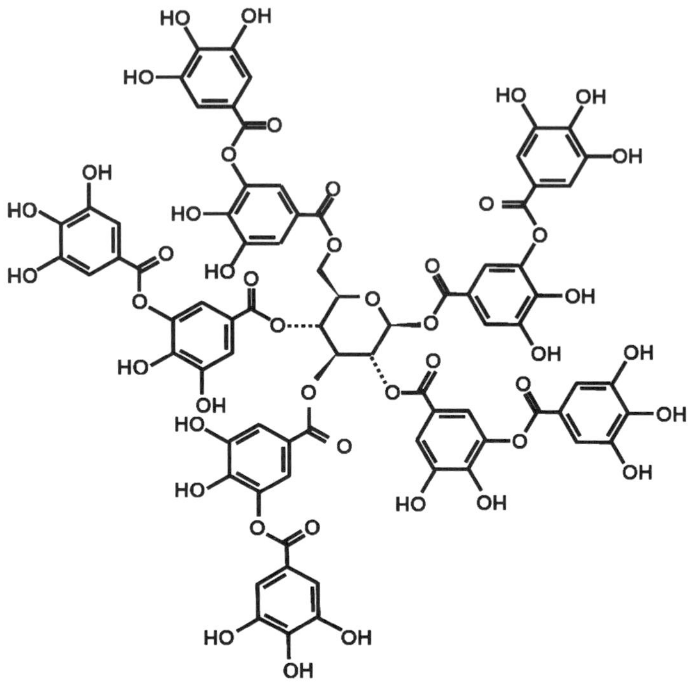

Fig. 4.1 Chemical structure of tannic acid (Reproduced with permission from Ref. [3], Copyright (2010), Elsevier)

The alkane esters of gallic acid are phenols and abundant in medicinal plants and fermented beverages, such as Lagerstroemia speciosa and red wine, and known to display a wide variety of biological functions in addition to their antioxidant activity. Ethyl gallate (EG), 3,4,5-trihydroxybenzoic acid ethyl ester, is one of typical gallic acid alkane esters and has a wide range of biological activities including antioxidant, antimicrobial, and anti-inflammatory functions [5, 6].

In this chapter, polyphenol-induced phase transition behaviors of thermo-responsive PNIPAM hydrogels are introduced. Because microgels can respond much faster than macroscale hydrogels to environmental stimuli because of their small dimensions, it should be more prompt and accurate to study the phase transition behavior of PNIPAM by using PNIPAM hydrogel microgels rather than macroscale hydrogels. In this chapter, a microfluidic technique [7] is applied to fabricate monodisperse PNIPAM microgels and core-shell microcapsules with a PNIPAM shell and a colored oil core, and both equilibrium and dynamic phase transition behaviors of PNIPAM microgels induced by TA [3] and those of core-shell PNIPAM microcapsules induced by EG [4] are introduced.

4.2 Phase Transition Behaviors of PNIPAM Microgels Induced by Tannic Acid

4.2.1 Preparation of Monodisperse PNIPAM Microgels

Monodisperse emulsions are prepared with a capillary microfluidic device, and monodisperse PNIPAM microgels are synthesized by free-radical polymerization with the emulsions as templates (Fig. 4.2). The disperse phase for droplets is aqueous solution containing monomer NIPAM (1 mol/L), cross-linker MBA (20 mmol/L), and initiator APS (3 g/L). The continuous phase is soybean oil containing 13 % (w/v) PGPR 90 as surfactant and 2 % (v/v) TEMED as reaction accelerator. The aqueous phase is broken into droplets at the tip of the injection tube, and resultant droplets are then carried away by continuous phase flow as monodisperse W/O emulsion drops. The W/O emulsion drops are then collected in a container with soybean oil phase containing 13 % (w/v) PGPR 90 and 4 % (v/v) TEMED. TEMED is both oil- and water-soluble. When TEMED in the continuous phase diffuses into the aqueous droplets and meets the initiator APS, a redox reaction starts to polymerize the monomers. The reaction is kept for 12 h at 20 °C.

4.2.2 Dynamic Isothermal Volume-Phase Transition of PNIPAM Microgels Induced by TA

Experiments on the dynamic volume-phase transition behavior of PNIPAM microgels are carried out by measuring the sizes of PNIPAM microgels at 25 °C and certain time intervals. A batch of PNIPAM microgels (more than 10) are immersed

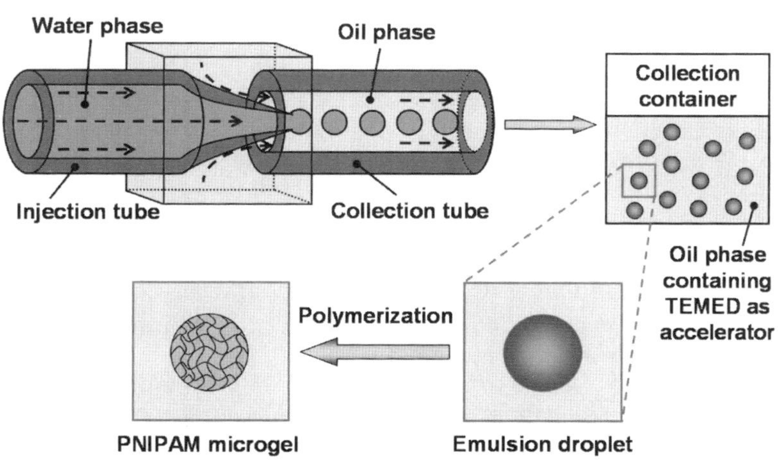

Fig. 4.2 Schematic illustration of preparation of monodisperse PNIPAM microgels by the microfluidic approach (Reproduced with permission from Ref. [3], Copyright (2010), Elsevier)

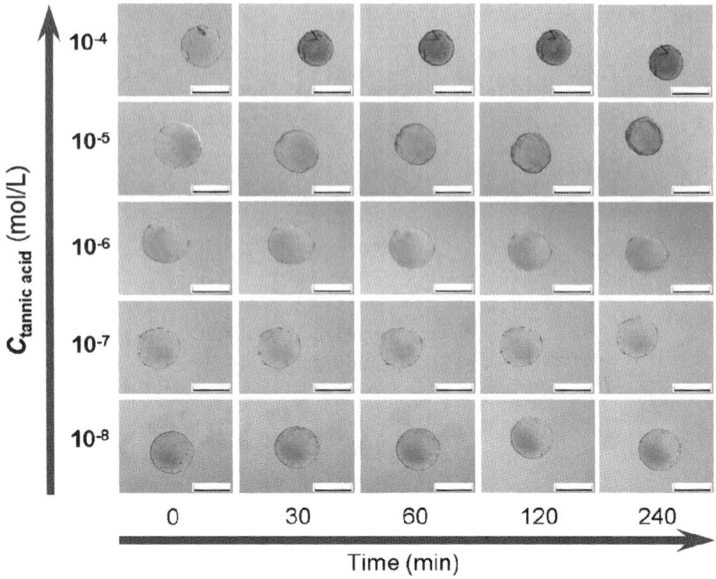

Fig. 4.3 Optical micrograms of PNIPAM microgels at different time intervals during the deswelling process in TA solutions with different concentrations at 25 °C. Scale bar is 200 μm (Reproduced with permission from Ref. [3], Copyright (2010), Elsevier)

in 2 ml aqueous TA solution for at least 4 h at 25 °C (below the LCST) that is controlled by a thermostatic stage system to ensure equilibration of isothermal phase transition that induced by TA. The dynamic isothermal deswelling degree of the PNIPAM microgels is calculated by Eq. 4.1:

$$R_t = \left(\frac{V_t}{V_0}\right) = \left(\frac{D_t}{D_0}\right)^3 \tag{4.1}$$

where R_t is the isothermal deswelling ratio of the PNIPAM microgel at time t during the phase transition process induced by TA, V_t and D_t are the volume (μm^3) and diameter (μm) of the PNIPAM microgel at time t, and V_0 and D_0 are the volume (μm^3) and diameter (μm) of the PNIPAM microgel initially in water before addition of TA.

Figures 4.3 and 4.4 show the dynamic isothermal volume-phase transition behaviors of PNIPAM microgels in solutions with different TA concentrations at 25 °C. The results demonstrate that the volumes of PNIPAM microgels do not change much at lower TA concentrations such as 10^{-8}, 10^{-7}, and 10^{-6} mol/L. As the TA concentration increases to 10^{-5} and 10^{-4} mol/L, the PNIPAM microgels exhibit more significant isothermal volume change. The higher the TA concentration, the faster the isothermal volume-phase transition of PNIPAM microgels induced by TA.

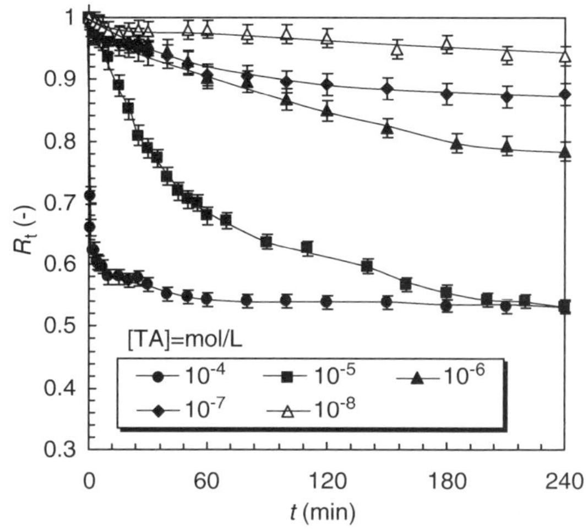

Fig. 4.4 Dynamic deswelling behaviors of PNIPAM microgels in TA solutions with different concentrations at 25 °C (Reproduced with permission from Ref. [3], Copyright (2010), Elsevier)

During the isothermal phase transition process of PNIPAM microgels induced by TA, the response rate of PNIPAM microgels to TA presence is mainly governed by diffusion-limited transport of TA molecules into PNIPAM polymeric networks. At the initial stage of the isothermal phase transition, the rate is mainly controlled by external diffusion in which TA molecules are transferred from the bulk solution to the interface of microgels/liquid. Then, the TA molecules further go through a slower internal diffusion process within PNIPAM polymeric networks in the microgels to approach the binding sites and finally reach adsorption equilibrium. Therefore, the more the TA molecules in the aqueous solution, the more the TA molecules diffusing into the PNIPAM microgels within the same time interval, and then the faster the deswelling rate of PNIPAM microgels. At TA concentrations of 10^{-4} and 10^{-5} mol/L, although the PNIPAM microgels reach equilibrium state finally with almost the same phase transition degrees, it takes less than 50 min for the microgels to reach the equilibrium state in the case of 10^{-4} mol/L while it takes more than 200 min in the case of 10^{-5} mol/L.

4.2.3 Equilibrium Isothermal Volume-Phase Transition of PNIPAM Microgels Induced by TA

With the same operation method as mentioned above for the dynamic study, the sizes of PNIPAM microgels before and after the isothermal phase transition are obtained. The equilibrium isothermal phase transition degree is calculated by the following equation:

$$R_C = \left(\frac{V_C}{V_0}\right) = \left(\frac{D_C}{D_0}\right)^3 \tag{4.2}$$

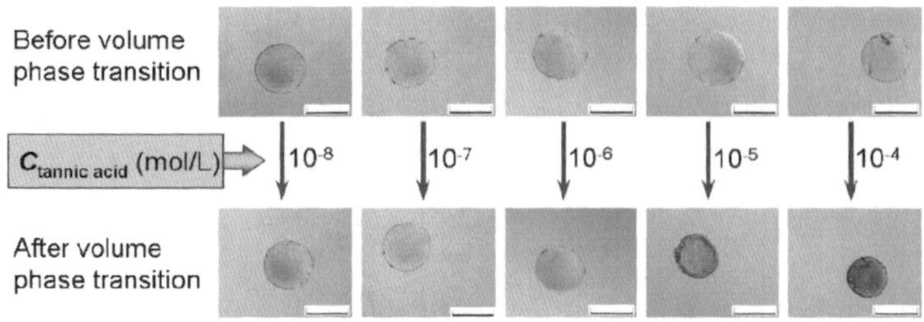

Fig. 4.5 Optical micrograms of PNIPAM microgels before and after volume-phase transition in TA solutions with different concentrations at 25 °C. Scale bar is 200 μm (Reproduced with permission from Ref. [3], Copyright (2010), Elsevier)

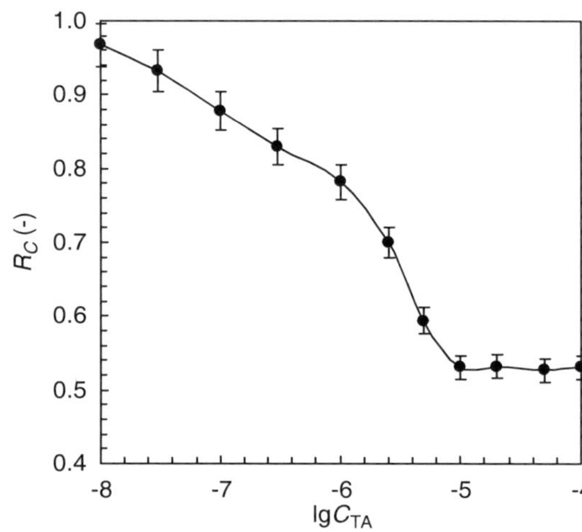

Fig. 4.6 Equilibrium isothermal phase transition degrees of PNIPAM microgels in TA solutions with different concentrations at 25 °C (Reproduced with permission from Ref. [3], Copyright (2010), Elsevier)

where R_C is the equilibrium isothermal deswelling ratio of PNIPAM microgels induced by TA at a concentration of C, V_C and D_C are the volume (μm^3) and diameter (μm) of PNIPAM microgels after the equilibrium of the isothermal phase transition induced TA at a concentration of C, and V_0 and D_0 are the same as those in Eq. 4.1.

At 25 °C, the equilibrium TA-induced isothermal volume-phase transitions of PNIPAM microgels in aqueous solutions with different TA concentrations are investigated. Figure 4.5 shows the optical micrograms of PNIPAM microgels before and after the equilibrium volume-phase transition in aqueous solutions with different TA concentrations at 25 °C, from which it is obviously found that the size of PNIPAM microgels decreases with increasing the TA concentration. Figure 4.6 further presents the equilibrium isothermal phase transition degrees of PNIPAM microgels as a function of TA concentration in aqueous solutions at 25 °C.

The equilibrium isothermal volume-phase transition degree of PNIPAM microgels is obviously concentration dependent. When the TA concentration is lower than 10^{-5} mol/L, the phase transition degree of PNIPAM microgels increases remarkably

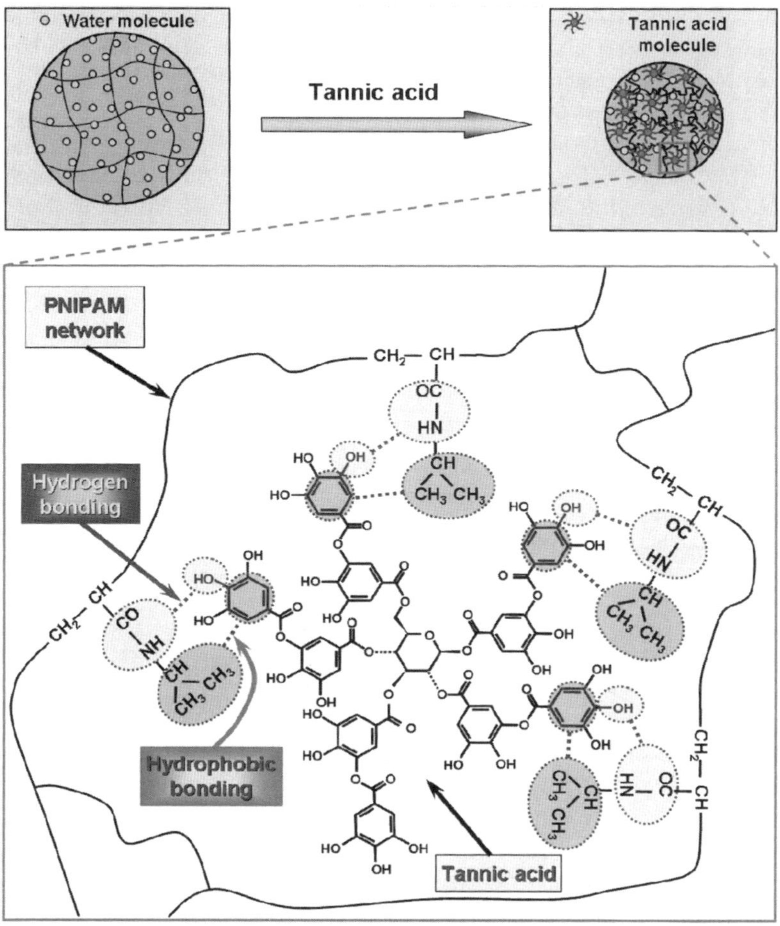

Fig. 4.7 Schematic illustration of volume-phase transition of PNIPAM microgel in aqueous solution at 25 °C induced by TA (Reproduced with permission from Ref. [3], Copyright (2010), Elsevier)

with increasing the TA concentration; however, when the TA concentration is higher than 10^{-5} mol/L, the phase transition degrees are almost the same. The results indicate that 10^{-5} mol/L is a critical TA concentration for PNIPAM microgels to reach the state of full isothermal phase transition. The critical concentration of TA is much lower than that of phenol (55 mM), resorcinol (50 mM), and phlorglucinol (25 mM) [2], because the number of hydroxyl groups in TA is much larger than those in small phenols molecules.

TA is a kind of polyphenol with plenty of hydroxyl groups. It is well known that the hydrogen bonding can be formed between the hydroxyl groups of phenols and amide groups of PNIPAM, and it has been reported that TA molecules can interact with PNIPAM polymeric chains through hydrogen-bonding and hydrophobic interaction in aqueous solutions [8]. Just as schematically illustrated in Fig. 4.7, due to a lot of hydroxyl groups in a TA molecule, it is possible for one TA molecule to occupy a number of PNIPAM binding sites (i.e., amide groups) and to act as physical cross-linker for the polymeric networks to some extent. During the adsorption process, the TA molecules bind to PNIPAM polymeric networks and expel the water molecules originally located on the adsorption cites. As a result, when enough TA molecules are adsorbed onto the polymeric networks, the PNIPAM microgels shrink. In addition, the aromatic groups in TA molecules strengthen the hydrophobic interaction in polymeric networks and break the balance between the hydrophilic and hydrophobic interaction, which also leads to the shrinking of PNIPAM microgels. As the TA concentration increases, more TA molecules diffuse into the PNIPAM microgels and bind to polymeric networks and enhance the equilibrium isothermal volume-phase transition of the PNIPAM microgels.

4.2.4 Thermosensitive Phase Transition of PNIPAM Microgels in TA Solutions

Before investigating the thermosensitive phase transition behaviors of PNIPAM microgels in aqueous solutions with different TA concentrations (10^{-8}, 10^{-7}, 10^{-6}, 10^{-5}, and 10^{-4} mol/L, respectively), the PNIPAM microgels have been immersed in the aqueous TA solution in a transparent sample holder at 25 °C for more than 4 h to ensure equilibration of isothermal phase transition. The temperature of TA solution containing PNIPAM microgels are increased step by step from 25 °C to 44.6 °C across the LCST of PNIPAM. At each predetermined temperature, the temperature is kept constant for more than 15 min to ensure the full equilibrium state of thermo-responsive phase transition. The equilibrium thermo-responsive phase transition degree of PNIPAM microgels is defined as

$$R_{T/25} = \left(\frac{V_T}{V_{25}}\right) = \left(\frac{D_T}{D_{25}}\right)^3 \tag{4.3}$$

where $R_{T/25}$ is the equilibrium thermo-responsive deswelling ratio of PNIPAM microgels at T °C as compared to the initial size of PNIPAM microgels at 25 °C (reference), V_T and D_T are the volume (μm^3) and diameter (μm) of PNIPAM microgels at T °C, and V_{25} and D_{25} are the volume (μm^3) and diameter (μm) of PNIPAM microgels at 25 °C.

Figure 4.8 shows the optical micrograms of PNIPAM microgels at different temperatures after being equilibrated in TA solutions with different concentrations, and Fig. 4.9 shows the thermo-responsive phase transition behaviors of PNIPAM

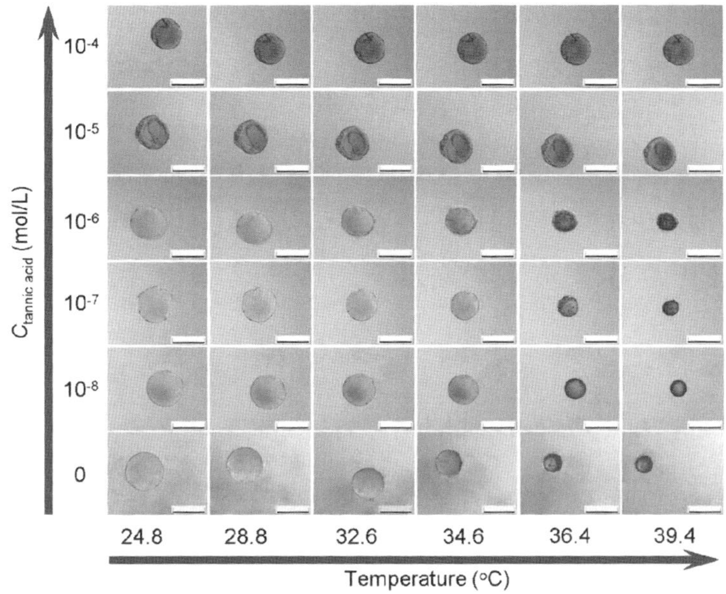

Fig. 4.8 Optical micrograms of PNIPAM microgels at different temperatures after being equilibrated in TA solutions with different concentrations. Scale bar is 200 μm (Reproduced with permission from Ref. [3], Copyright (2010), Elsevier)

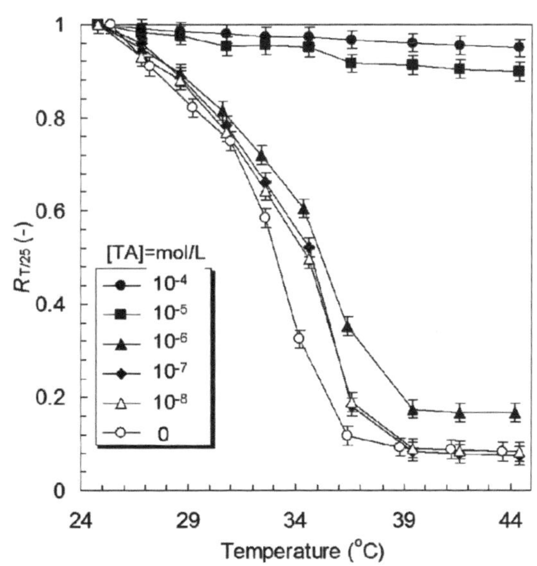

Fig. 4.9 Thermo-responsive phase transition behaviors of PNIPAM microgels after being equilibrated in TA solutions with different concentrations (Reproduced with permission from Ref. [3], Copyright (2010), Elsevier)

microgels. Obvious differences between the thermosensitive phase transition behaviors of PNIPAM microgels can be seen at different TA concentrations. When the TA concentration is higher than 10^{-5} mol/L, most of the hydrogen-bonding sites inside the PNIPAM microgels have been firmly occupied by TA molecules and most of water molecules inside the PNIPAM microgels have been repelled out from the polymeric networks at 25 °C. As a result, the PNIPAM microgels do not show significant thermo-responsive phase transition behavior anymore.

When the TA concentration is lower than 10^{-6} mol/L, the isothermal phase transition of PNIPAM microgels induced by TA at 25 °C is not so significant (Fig. 4.6), i.e., there are still plenty of water molecules inside the PNIPAM microgels at 25 °C. Therefore, the PNIPAM microgels still show remarkable thermo-responsive phase transition behaviors. The lower the TA concentration, the larger the thermo-induced deswelling ratio of the PNIPAM microgels. For example, the equilibrium thermo-responsive phase transition degree of PNIPAM microgels at 44.6 °C ($V_{44.6}/V_{25}$) in TA solutions with concentrations less than 10^{-7} mol/L is about 8 %, but it becomes about 17 % when the TA concentration increases to 10^{-6} mol/L. Because the TA molecules act as physical cross-linkers to some extent as mentioned above, more TA molecules result in a more serious physical cross-linkage between the polymeric networks inside the PNIPAM microgels; as a result, the thermo-induced shrinking degree decreases at a higher TA concentration.

Another interesting phenomenon is that, when the TA concentration is lower than 10^{-6} mol/L, the LCST of PNIPAM microgels slightly shifts to higher temperatures with increasing the TA concentration. For example, the LCST of PNIPAM microgel in pure water is about 33 °C, while it increases to about 35.5 °C when the TA concentration is 10^{-6} mol/L. This result is different from the previously reported result that the LCST of PNIPAM shifts to lower temperatures due to the adsorption of benzoic acid (BA) or phenol (Ph) molecules to the PNIPAM chains [9]. The reason is that, in the binding of BA and Ph, the hydrophobic interaction is considered to play a critical role because of the hydrophobic nature [9]. As a result, the LCST of PNIPAM shifts to lower temperatures due to the enhancement of hydrophobicity inside the PNIPAM microgels after the binding of BA and Ph [9]. In the case of TA, although part of hydroxyl groups in the TA molecule can bind to the amide groups on the PNIPAM polymeric networks through hydrogen bonding, there are still part of free hydroxyl groups that cannot bind to the amide groups on the PNIPAM polymeric networks because of steric hindrance (Fig. 4.7). These free hydroxyl groups of the TA molecule entrapped within the polymeric networks of PNIPAM microgels enhance the hydrophilicity of the PNIPAM microgels. As a result, the LCST of PNIPAM microgel shifts to higher temperatures due to the binding of TA onto PNIPAM polymeric networks.

4.3 Phase Transition Behaviors of PNIPAM Microgels Induced by Ethyl Gallate

4.3.1 Preparation of PNIPAM Microspheres and Core-Shell PNIPAM Microcapsules

A capillary microfluidic device is assembled for fabrication of emulsion droplets according to a well-established method [7, 10–12], as shown in Fig. 4.10. The outer diameters of all the cylindrical capillary tubes are 0.99 mm. The square capillary tubes have an inner dimension of 1.0 mm. The inner diameters of the injection tube, the transition tube, and the collection tube are 500, 200, and 450 μm, respectively. A micropuller (PN-30, Narishige) and a microforge (MF-830, Narishige) are used to fabricate the capillary microfluidic device. The inner diameters of the tapered end of the injection and transition tubes are 60 and 140 μm, respectively. To prepare PNIPAM microspheres, the second stage of the capillary microfluidic device is utilized to prepare single water/oil (W/O) emulsion droplets as templates. The water phase of W/O emulsion droplets is an aqueous solution (4 ml) containing NIPAM monomer (0.452 g), Pluronic F-127 (0.04 g, emulsifier), MBA (0.0062 g, cross-linker), APS (0.024 g, initiator), and glycerol (0.4 g, to adjust the viscosity). Soybean oil containing 5 % (w/v) PGPR 90 (emulsifier) is used as the oil phase. The flow rate of water phase and oil phase are 500 and 2,000 μL/h, respectively. To generate O/W/O emulsion droplets as templates for preparation of core-shell PNIPAM microcapsules, soybean oil containing 3 % (w/v) PGPR 90 and 0.1 % (w/v) LR300 is used as the inner oil phase. The middle water phase and outer oil phase are the same as mentioned above in preparation of single W/O emulsion droplets. The flow rate of inner oil phase, water phase, and oil phase are 500, 800, and 2,000 μL/h, respectively. The oil phase in a container for collecting the generated W/O and O/W/O emulsion droplets is soybean oil containing 5 % (w/v) PGPR 90 and 1 % (w/v) 2,2-dimethoxy-2-phenylacetophenone (BDK). The

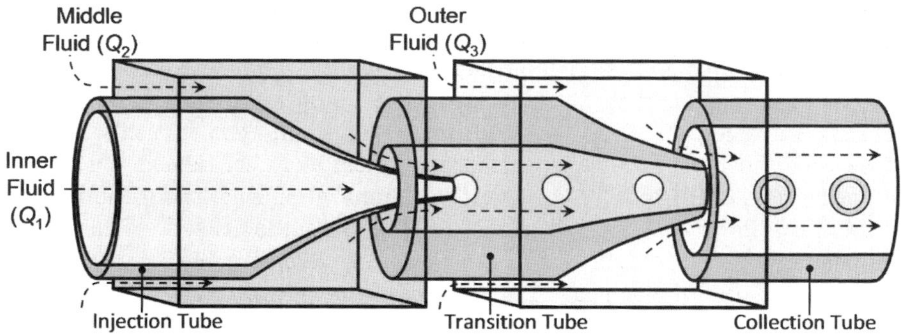

Fig. 4.10 Schematic illustration of the microfluidic device for fabricating monodisperse double emulsion templates (Reproduced with permission from Ref. [4], Copyright (2012), Elsevier)

formation process of O/W/O emulsion droplets in the capillary microfluidic device is recorded with a high-speed compact camera (Phantom Miro 3, Vision Research, USA). The polymerizations of the water phases in W/O and O/W/O emulsion droplets are initialized via UV irradiation in an ice bath for 12 min. A 250 W UV lamp with an illuminance spectrum of 250–450 nm is employed to produce UV light. The ice-water bath is used to ensure that the polymerization is carried out at a low temperature. After polymerization for 12 h, the PNIPAM microspheres are washed with isopropyl alcohol six times and subsequently washed with deionized (DI) water six times to remove the oil phase and isopropyl alcohol, and the PNIPAM microcapsules are washed with DI water for several times to remove the outer oil phase. Then, the PNIPAM microspheres and core-shell PNIPAM microcapsules are re-dispersed in DI water for further characterizations.

4.3.2 Thermo-responsive Phase Transition Behaviors of PNIPAM Microspheres in EG Solution

Thermo-responsive phase transition behaviors of PNIPAM microspheres in EG aqueous solution with varying concentration are investigated with an optical microscope (BX61, Olympus) equipped with a digital CCD camera and a thermostatic stage system (TS62, Instec). Prior to investigation of their thermo-responsive phase transition behaviors, the PNIPAM microspheres had been immersed in EG aqueous solution with varying concentration (0, 10, 20, 30 mmol/L, respectively) in a transparent sample holder at 5 °C for more than 2 h to ensure equilibrium of isothermal phase transition is reached. The temperature of each EG aqueous solution containing PNIPAM microspheres is increased step by step from 5 °C to 40 °C to across the VPTT of PNIPAM gels. At each predetermined temperature, it is kept constantly for more than 30 min to ensure the full equilibrium state of thermo-responsive phase transition. The equilibrium thermo-responsive phase transition deswelling ratio of PNIPAM microspheres ($R_{T/5}$) is determined, which is defined as

$$R_{T/5} = \left(\frac{V_T}{V_5}\right) = \left(\frac{d_T}{d_5}\right)^3 \tag{4.4}$$

where V_T and d_T are the volume (μm^3) and the diameter (μm) of PNIPAM microspheres at T °C and V_5 and d_5 are the volume (μm^3) and the diameter (μm) of PNIPAM microspheres at 5 °C. In this study, the VPTT value is defined as the temperature at which the absolute value of the slope of the $R_{T/5}$ versus T curve is the largest, which means the volume of the PNIPAM microspheres changes most rapidly at this temperature.

Figure 4.11 shows the optical micrographs of PNIPAM microspheres at different temperatures after being equilibrated in aqueous solution with varying EG concentration. The PNIPAM microspheres are transparent and swollen when the

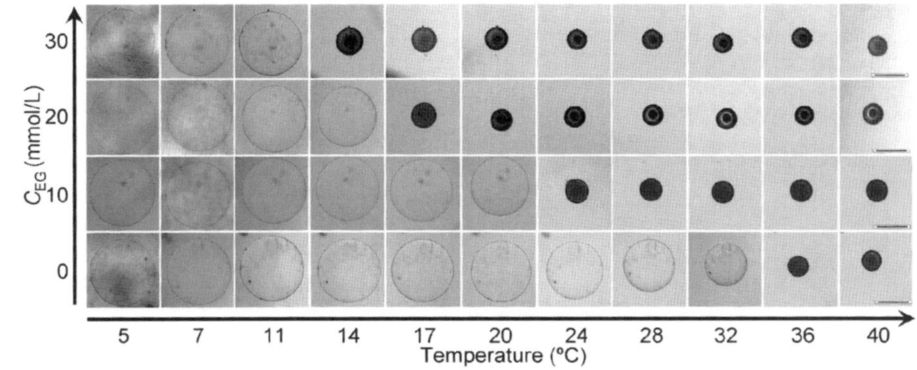

Fig. 4.11 Optical micrographs of the PNIPAM microspheres at different temperatures after being equilibrated in aqueous solution with varying EG concentration. Scale bars are 200 μm (Reproduced with permission from Ref. [4], Copyright (2012), Elsevier)

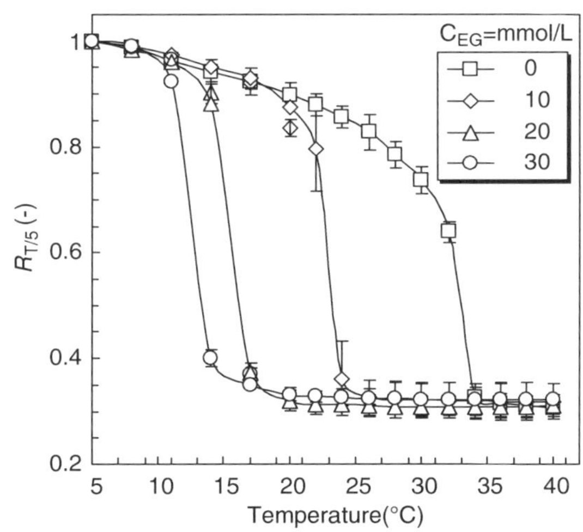

Fig. 4.12 Thermo-responsive phase transition behaviors of PNIPAM microspheres after being equilibrated in aqueous solution with varying EG concentration. The error bars represent the deviation of the mean from more than ten microspheres (Reproduced with permission from Ref. [4], Copyright (2012), Elsevier)

temperature of EG aqueous solution is lower than the corresponding VPTT value (e.g., ≤11 °C for an EG concentration of 30 mmol/L), but opaque and shrunken when the temperature of EG solution is higher than the corresponding VPTT value (≥14 °C in that case). Figure 4.12 shows the thermo-responsive phase transition behaviors of PNIPAM microspheres after being equilibrated in aqueous solution with varying EG concentration. With increasing the EG concentration, the VPTT value of PNIPAM microspheres shifts to lower temperature, as shown in Fig. 4.13.

The shift of the VPTT value of PNIPAM to lower temperature in EG aqueous solution can be explained by the hydrophilic and hydrophobic interaction of EG with PNIPAM. Hydrogen bonding could be formed between the hydroxyl groups of EG molecules and amide groups of PNIPAM, and benzene rings of EG molecules and isopropyl groups of PNIPAM form hydrophobic interaction with each other. As

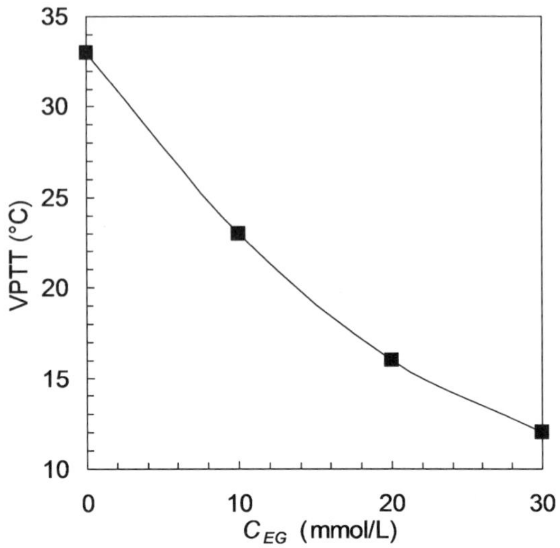

Fig. 4.13 Effect of EG concentration in aqueous solution on the VPTT of PNIPAM microspheres (Reproduced with permission from Ref. [4], Copyright (2012), Elsevier)

a result, EG disrupts the mechanism of water ordering around the PNIPAM polymer chains, which leads to the shift of VPTT of PNIPAM in EG aqueous solution to lower temperature.

4.3.3 The Intact-to-Broken Transformation Behaviors of Core-Shell PNIPAM Microcapsules in Aqueous Solution with Varying EG Concentration

The observation of change in size and structure of monodisperse core-shell PNIPAM microcapsules in EG aqueous solutions (10, 20, 30 mmol/L at 20 °C and 20 mmol/L at 25 °C) is carried out using a CLSM (TCS SP5-II, Leica) equipped with a thermostatic stage system (TSA02i, Instec). EG aqueous solutions are placed in water bath with a designed temperature for at least 2 h before measurement. Core-shell PNIPAM microcapsules are exposed to DI water at the same designed temperature controlled by the thermostatic stage for at least 30 min before measurement. The DI water around the core-shell PNIPAM microcapsules is replaced with EG aqueous solution with a preset concentration at the same temperature. The size and structure transformation behaviors of PNIPAM microcapsules are recorded by the CLSM. The dynamic deswelling ratio of the PNIPAM microcapsules at time t ($R_{d,t}$) is calculated by

$$R_{d,t} = \frac{d_t}{d_0} \tag{4.5}$$

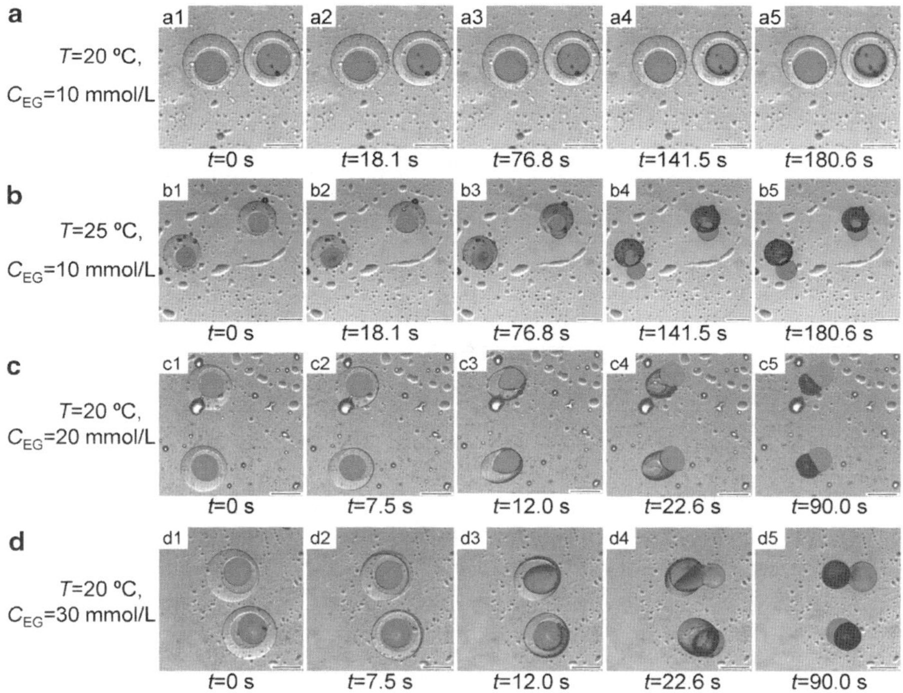

Fig. 4.14 CLSM micrographs of dynamic phase transition behaviors of PNIPAM microcapsules with colored oil core in aqueous solution with varying EG concentration and at different temperatures. The volume ratio between the inner core and the whole microcapsule is about 0.27. (**a**) $T = 20$ °C, $C_{EG} = 10$ mmol/L; (**b**) $T = 25$ °C, $C_{EG} = 10$ mmol/L; (**c**) $T = 20$ °C, $C_{EG} = 20$ mmol/L; (**d**) $T = 20$ °C, $C_{EG} = 30$ mmol/L. Scale bars are 200 μm (Reproduced with permission from Ref. [4], Copyright (2012), Elsevier)

where d_t is the diameter (μm) of the PNIPAM microcapsules at time t (s) and d_0 is the diameter (μm) of the PNIPAM microcapsules initially in DI water just before the water is replaced by the EG aqueous solution.

Figure 4.14 shows the CLSM micrographs of dynamic phase transition behaviors of core-shell PNIPAM microcapsules in aqueous solution with varying EG concentrations and at different temperatures. For the detailed phase transition behaviors corresponding to an EG concentration of 10 mmol/L at 20 °C, when the surrounding DI water is replaced by the EG solution, the PNIPAM microcapsules shrink slightly in the first few seconds and then keep unchanged afterwards. Figure 4.14b shows that the PNIPAM microcapsules shrink slowly and squirt out the colored oil core when the temperature is increased to 25 °C. At time $t = 0$, the diameters of the PNIPAM microcapsules in water at 25 °C are smaller than that at 20 °C, because the PNIPAM shells shrink slightly when the temperature rise from 20 °C to 25 °C [13]. When the environmental water is replaced by aqueous solution with an EG concentration of 20 or 30 mmol/L at 20 °C, the shell of PNIPAM microcapsules

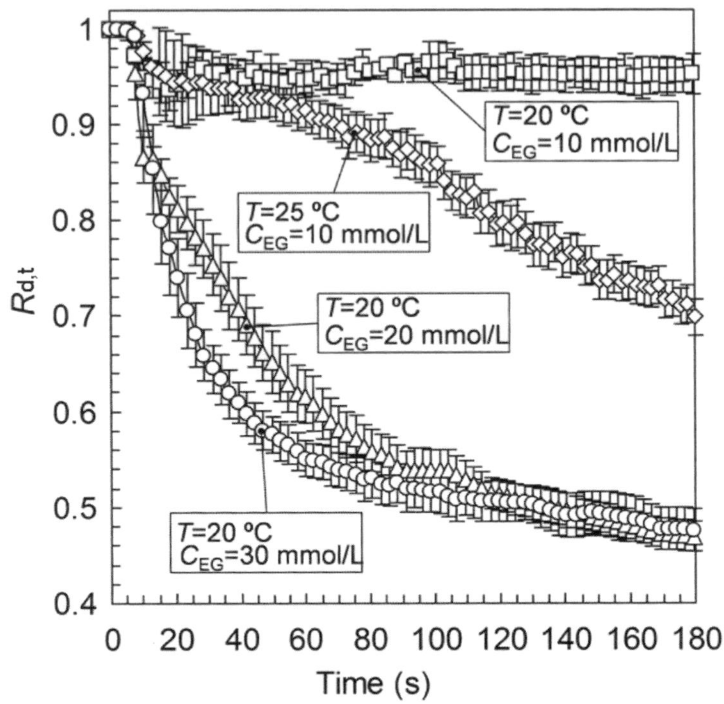

Fig. 4.15 Dynamic shrinking behaviors of PNIPAM microcapsules with a colored oil core in aqueous solution with varying EG concentration. The error bars represent the deviation from more than ten microcapsules (Reproduced with permission from Ref. [4], Copyright (2012), Elsevier)

shrinks rapidly within 12 s, and the core-shell microcapsules are transformed from the intact state to the broken state because the PNIPAM shell cannot contain the colored oil core anymore, as shown in Fig. 4.14c, d.

The dynamic shrinking behaviors of PNIPAM microcapsules with a colored oil core in aqueous solution with varying EG concentration are shown in Fig. 4.15. By comparing with the CLSM micrographs in Fig. 4.14, it can be seen that the PNIPAM microcapsules can burst and release their oil core when the $R_{d,t}$ value is reduced to a range between 0.88 and 0.7. When the concentration of EG aqueous solution is fixed at 10 mmol/L, the microcapsules can transform from the intact state to the broken state with increasing the temperature from 20 °C (below the VPTT, which is approximately 23 °C, as shown in Fig. 4.13) to 25 °C (above the VPTT). When the EG concentration is 20 or 30 mmol/L, the environment temperature 20 °C is higher than the corresponding VPTT values (approximately 16 or 12 °C, respectively, as shown in Fig. 4.13). Therefore, under these conditions the microcapsules can transform from the intact state to the broken state. The larger the difference between the operation temperature and the corresponding VPTT is, the faster the shrinkage of the core-shell microcapsules. For example, at 20 °C the PNIPAM microcapsules

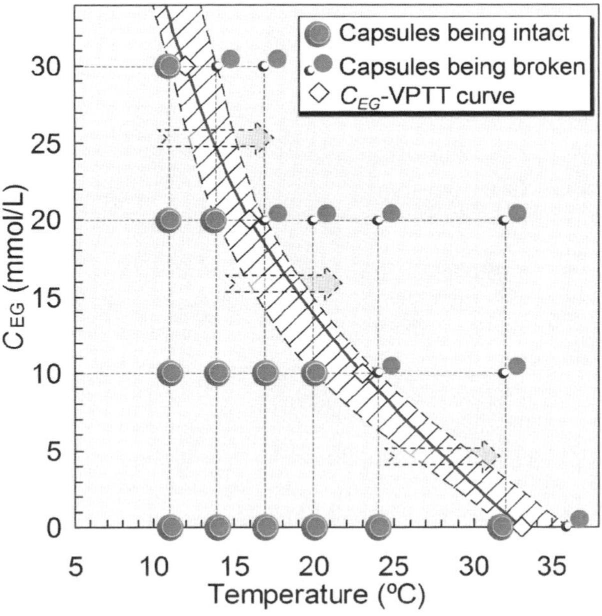

Fig. 4.16 State diagram of the intact-to-broken transformation of core-shell PNIPAM microcapsules in aqueous solution as a function of temperature and EG concentration (Reproduced with permission from Ref. [4], Copyright (2012), Elsevier)

in aqueous solution with EG concentration of 30 mmol/L shrink faster than that in aqueous solution with EG concentration of 20 mmol/L, because the difference between the operation temperature and the corresponding VPTT under the former condition is 8 °C while that under the later condition is just 4 °C.

Besides the EG concentration and temperature, the effect of the volume ratio between the inner core and the whole microcapsule on the intact-to-broken transformation behaviors as well as the $R_{d,t}$ value when microcapsule bursts is also investigated. The volume ratio between the inner core and the whole microcapsule of the above-mentioned microcapsules in Figs. 4.14 and 4.15 is about 0.27. Another batch of microcapsules is prepared with the volume ratio between the inner core and the whole microcapsule being about 0.36. The experimental results show that, in the experimental range, the intact-to-broken transformation behaviors and the $R_{d,t}$ value when microcapsule bursts of two batches of microcapsules are almost the same.

Figure 4.16 illustrates the state diagram of the intact-to-broken transformation of core-shell PNIPAM microcapsules in aqueous solutions as a function of temperature and EG concentration. The intact-state region is located in the lower left region and the broken-state region is located in the upper right region, which are separated by a narrow region near the C_{EG}-VPTT curve. From the state diagram, the state of PNIPAM microcapsules is related to the temperature and the EG concentration. The results indicate that the core-shell PNIPAM microcapsules may be highly potential to be used as sensors and/or indicators for some simple detection of EG concentration if the EG concentration falls in the range from 0 to 35 mmol/L. When the core-shell PNIPAM microcapsules are put in EG aqueous solution and are heated up gradually, their size and the structure will change and transform from an intact state to a burst state. Therefore, the EG concentration can be determined simply by measuring the corresponding critical temperature for the microcapsules to squirt out their colored oil cores.

4.4 Summary

Significant isothermal phase transitions of PNIPAM microgels induced by TA and EG molecules in aqueous solutions at temperature below the LCST are observed. Both the dynamic phase transition rate and equilibrium shrinking degree of PNIPAM microgels in TA and EG solutions are dependent on the TA or EG concentrations. Such isothermal phase transitions of PNIPAM microgels resulted from the adsorption of TA or EG molecules to PNIPAM polymeric networks through hydrogen-bonding and hydrophobic interaction. The higher the TA or EG concentration in the aqueous solution, the faster the isothermal phase transition rate of PNIPAM microgels induced by TA or EG molecules, and the more significant the equilibrium phase transition degree. There exists a critical TA or EG concentration for the isothermal phase transition behavior, below which the equilibrium shrinking degree of PNIPAM microgels becomes larger with increasing the TA or EG concentration, while above which the equilibrium shrinking degrees of PNIPAM microgels remain almost the same. A state diagram of the intact-to-broken transformation of core-shell PNIPAM microcapsules in aqueous solution as a function of temperature and EG concentration is constructed, which might be used to determine the EG concentration in aqueous solution via measuring the critical temperature to cause the microcapsules to squirt out their colored oil cores. Therefore, besides the applications as thermo-responsive materials, PNIPAM hydrogels or microgels can also be used as material candidates for some new applications, such as TA or EG sensors, indicators, and/or adsorbents.

References

1. Laszlo K, Kosik K, Rochas C et al (2003) Phase transition in PNIPA hydrogels induced by phenols. Macromolecules 36:7771–7776
2. Kosik K, Wilk E, Geissler E et al (2007) Distribution of phenols in thermoresponsive hydrogels. Macromolecules 40:2141–2147
3. Chen G, Niu CH, Zhou MY et al (2010) Phase transition behaviors of poly(*N*-isopropylacrylamide) microgels induced by tannic acid. J Colloid Interface Sci 343:168–175
4. Mou CL, He XH, Ju XJ et al (2012) Change in size and structure of monodisperse poly(*N*-isopropylacrylamide) microcapsules in response to varying temperature and ethyl gallate concentration. Chem Eng J 210:212–219
5. Wu J, Sugiyama H, Zeng LH et al (1998) Evidence of Trolox and some gallates as synergistic protectors of erythrocytes against peroxyl radicals. Biochem Cell Biol 76:661–664
6. Soe WM, Myint NL, Sing LC et al (2011) Ethyl gallate as a combination drug can overcome resistance in MRSA. Lett Drug Des Discov 8:65–68

7. Chu LY, Utada AS, Shah RK et al (2007) Controllable monodisperse multiple emulsions. Angew Chem Int Ed 46:8970–8974
8. Erel-Unal I, Sukhishvili SA (2008) Hydrogen-bonded multilayers of a neutral polymer and a polyphenol. Macromolecules 41:3962–3970
9. Koga S, Sasaki S, Maeda H (2001) Effect of hydrophobic substances on the volume-phase transition of *N*-isopropylacrylamide gels. J Phys Chem B 105:4105–4110
10. Shah RK, Shum HC, Rowat AC et al (2008) Designer emulsions using microfluidics. Mater Today 11:18–27
11. Wang W, Liu L, Ju XJ et al (2009) A novel thermo-induced self-bursting microcapsule with magnetic-targeting property. Chem Phys Chem 10:2405–2409
12. Liu L, Yang JP, Ju XJ et al (2011) Monodisperse core-shell chitosan microcapsules for pH-responsive burst release of hydrophobic drugs. Soft Matter 7:4821–4827
13. Wu C, Zhou S (1997) Light scattering study of spherical poly(*N*-isopropylacrylamide) microgels. J Macromol Sci B Phys 36:345–355

Chapter 5
Functional Membranes with Thermo-responsive Hydrogel Gates

Abstract In this chapter, the design, fabrication, and performance of functional membranes/systems with thermo-responsive hydrogel gates are introduced. The thermo-responsive polymeric hydrogel gates are either grafted or filled into the membrane pores. The response temperature of the prepared membranes can be regulated by the introduction of hydrophilic or hydrophobic monomers in the comonomer solution. The thermo-responsive gating characteristics of membranes are affected by the grafting degree, length, and density of polymeric gates as well as preparation temperature, etc. In addition, the membranes exhibit a satisfactory thermo-responsive controlled-release characteristics and affinity adsorption/desorption behavior. The results in this study provide valuable guidance for designing, fabricating, and operating thermo-responsive gating membranes with desirable performances.

5.1 Introduction

Because environmental temperature fluctuations is one of the most important factors in the natural environment and can be easily designed and artificially controlled, much attention has been focused on the thermo-responsive membranes. The thermo-responsive membranes are usually prepared by grafting the thermo-responsive polymeric chains onto both the external membrane surface and inner surface of membrane pores of the existing substrate membranes. These prepared membranes combine the desirable mechanical robustness of substrate membranes and good thermo-responsive characteristics of polymers. The grafting methods include photo-initiated grafting, radiation-induced grafting, plasma-induced grafting, and controlled radical grafting polymerization such as atom-transfer radical polymerization (ATRP) and reversible addition-fragmentation chain transfer (RAFT) polymerization [1–6]. There are inorganic porous substrate membranes, such as anodic

L.-Y. Chu et al., *Smart Hydrogel Functional Materials*,
DOI 10.1007/978-3-642-39538-3_5,
© Chemical Industry Press, Beijing and Springer Berlin Heidelberg 2013

aluminum oxide (AAO) membrane and Shirasu porous glass (SPG) membrane, and organic ones, such as Nylon-6 (N6), polyethylene (PE), poly(vinylidene fluoride) (PVDF), and polycarbonate (PC).

Here, the thermo-responsive membranes are fabricated by grafting poly(*N*-isopropylacrylamide) (PNIPAM) chains onto the substrate membrane via plasma-induced pore-filling graft polymerization and ATRP or by filling PNIPAM into membrane pores via free-radical polymerization. The influential factors on the thermo-responsive gating characteristics of the prepared membranes as well as their applications such as controlled release and affinity adsorption are investigated systematically.

5.2 Functional Membranes with Thermo-responsive Hydrogel Gates Fabricated by Plasma-Induced Pore-Filling Graft Polymerization

5.2.1 Regulation of Response Temperature of Thermo-responsive Membranes

The response temperature of the thermo-responsive membrane is defined as the temperature at which the water flux or hydraulic permeability across the membrane changes dramatically. The response temperature of PNIPAM-based thermo-responsive membranes is always around 32 °C, which is consistent to the lower critical solution temperature (LCST) of PNIPAM polymer. However, in most applications higher or lower response temperatures of the thermo-responsive membranes are preferred. Therefore, the investigation on how to adjust the response temperature of the thermo-responsive membranes will provide valuable guidance for the design and preparation of the thermo-responsive membranes with desired response temperatures for different applications.

The adjustment or manipulation of the response temperature of thermo-responsive membranes is systematically investigated by adding hydrophilic or hydrophobic monomers into the *N*-isopropylacrylamide (NIPAM) comonomer solution in the fabrication of the thermo-responsive gates [7]. The thermo-responsive PNIPAM-based polymeric gates are grafted onto porous PVDF or N6 membrane substrates by virtue of plasma-induced pore-filling graft polymerization method. Experimental results of water flux during pressure-driven filtration show that the response temperature of the thermo-responsive membrane could be effectively controlled by simply adjusting the content of hydrophilic or hydrophobic monomers. The response temperatures of gating membranes are found to be linearly increased with increasing the molar ratio of hydrophilic monomer acrylamide (AAM) in the NIPAM comonomer solution, but linearly decreased with increasing the molar ratio of hydrophobic monomer butyl methacrylate (BMA), respectively (Fig. 5.1). The response temperature of

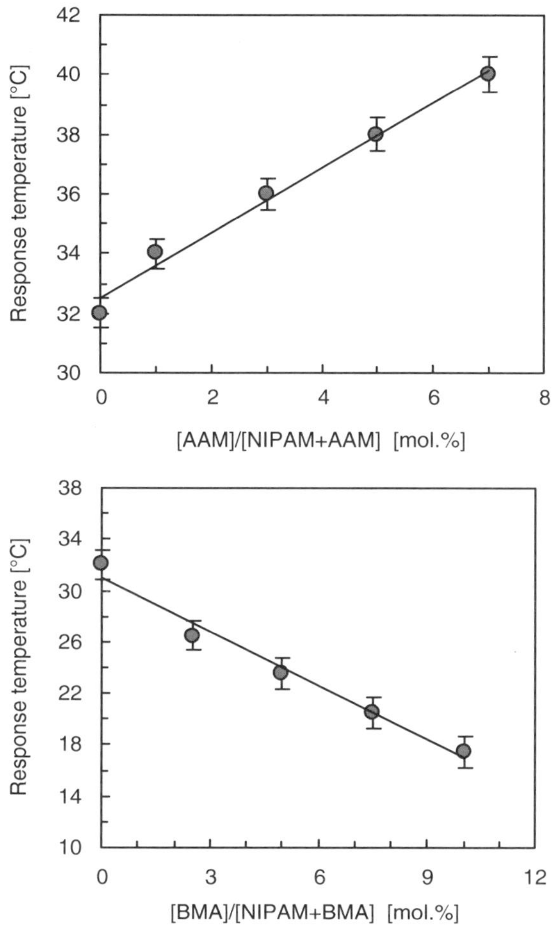

Fig. 5.1 Effects of the molar ratio of AAM or BMA in NIPAM comonomer solutions on the response temperature of PNA-grafted (**a**) and PNB-grafted (**b**) membranes (Reproduced with permission from Ref. [7], Copyright (2007), Elsevier)

poly(*N*-isopropylacrylamide-*co*-acrylamide) (PNA)-grafted PVDF membrane is raised to 40 °C when 7 mol% of AAM is added into the NIPAM comonomer solution. However, that of poly(*N*-isopropylacrylamide-*co*-butyl methacrylate) (PNB)-grafted N6 membrane is reduced to 17.5 °C when 10 mol% of BMA is introduced into the NIPAM comonomer solution.

5.2.2 Effect of Grafting Degree on the Thermo-responsive Gating Characteristics

The grafting degree of thermo-responsive membranes could be represented by either grafting yield or pore-filling ratio. The grafting yield of the membrane is defined as the weight increase of the membrane after the grafting per unit mass or per unit area and could be calculated according to the Eqs. 5.1 and 5.2, respectively. However, the pore-filling ratio of the membrane is defined as the volume ratio of grafted polymeric gates to membrane pores (Eq. 5.3):

$$Y_W = \frac{W_g - W_0}{W_0} \times 100\ \% \tag{5.1}$$

$$Y_A = \frac{W_g - W_0}{A} \times 100\ \% \tag{5.2}$$

where Y_W and Y_A are the grafting yields calculated by unit mass and unit area of substrate membrane [wt% or mg·cm^{-2}], respectively. W_g and W_0 represent the mass [mg] of the membrane after and before grafting, respectively. A stands for the area of the substrate membrane [cm^2];

$$Y_F = \frac{V_g}{V_m} \times 100\ \% \tag{5.3}$$

where Y_F is the pore-filling ratio [%] and V_g and V_m are the total volume [cm^3] of the grafted PNIPAM polymeric gates in the membrane and the membrane pores before grafting, respectively.

The thermo-responsive gating characteristics of the membranes are always investigated by conducting pressure-driven filtration experiment and concentration-driven diffusion experiment at different operation temperatures. To quantitatively describe the thermo-responsive gating characteristics of the membranes, a parameter known as the thermo-responsive gating coefficient is defined. It is the ratio of hydraulic permeability, diffusional permeability, or membrane pore sizes at temperature above the LCST of PNIPAM (usually 40 °C) to that below the LCST (usually 25 °C). These thermo-responsive gating coefficients are called the flux thermo-responsive coefficient R_J (Eq. 5.4), diffusion thermo-responsive coefficient R_D (Eq. 5.5), and thermo-responsive coefficient of membrane pore size R_d (Eq. 5.6), respectively. The higher the thermo-responsive gating coefficients are, the better the thermo-responsive gating characteristics of the membranes are.

$$R_J = \frac{J_{40}}{J_{25}} \tag{5.4}$$

$$R_D = \frac{D_{40}}{D_{25}} \tag{5.5}$$

$$R_d = \frac{d_{40}}{d_{25}} = \sqrt[4]{\frac{J_{40}\eta_{40}}{J_{40}\eta_{25}}} \tag{5.6}$$

where J_{40} and J_{25} are the water fluxes or hydraulic permeabilities [ml·cm^{-2}·s^{-1}] of thermo-responsive membrane at 40 °C and 25 °C, respectively. D_{40} and D_{25}

are the diffusional coefficients or permeability coefficients [$cm^2 \cdot s^{-1}$] of model drug molecules across membrane at 40 °C and 25 °C, respectively. d_{40} and d_{25} are the effective pore diameters [cm] of grafted membranes at 40 °C and 25 °C under the same operation pressures, respectively. The effective pore diameters could be calculated by Hagen-Poiseuille's law (Eq. 5.7):

$$J = \frac{n\pi d^4 P}{128\eta l} \tag{5.7}$$

where J is the water flux or hydraulic permeability [$ml \cdot cm^{-2} \cdot s^{-1}$] of membrane, n for the number of pores per unit membrane area [cm^{-2}], d for the membrane pore mean diameter [cm], P the transmembrane pressure [Pa], η for the liquid viscosity [Pa·s], and l the thickness of membrane [cm].

The architectures and relatively physicochemical properties of the grafted PNIPAM linear chains include the length and density of the PNIPAM chains. They will change with the variation of the grafting degree and have remarkable effect on the thermo-responsive gating characteristics of thermo-responsive membranes [8]. The effect of grafting degree on the thermo-responsive gating characteristics of membranes is investigated systematically by thermo-responsive filtration experiment and diffusion experiment.

***Thermo-responsive hydraulic permeability*:** A series of thermo-responsive membranes, with a wide range of grafting yields and pore-filling ratios, are prepared by grafting PNIPAM via plasma-induced grafting polymerization method [8–12]. The membrane substrates are either hydrophilic N6 and PE membranes or hydrophobic PVDF and PC track-etched substrate membranes. For all of the investigated thermo-responsive membranes, the grafted PNIPAM polymers are found homogeneously distributed on the inner surfaces of the pores throughout the entire membrane thickness. The hydraulic permeability and diffusional permeability across these thermo-responsive membranes are dependent on the grafting degree which in turn heavily affects the thermo-responsive gating characteristics.

For PNIPAM-grafted PVDF membranes with moderate grafting yields (e.g., 0.5–3.0 %), the water flux across membranes changes dramatically at 32 °C (around the LCST of PNIPAM) [8]. With an increase in the grafting yield, the hydraulic permeability decreases rapidly at both 25 °C and 40 °C because of the decrease in the pore size of membranes. When the grafting yield is smaller than 2.81 %, both the R_J and R_d values increase with an increase in the grafting yield; however, when the grafting yield is higher than 6.38 %, both coefficients are always equal to 1. When the grafting yield exceeds 6.38 %, the length of grafted PNIPAM chains is long enough to "choke" the membrane pores and thus there is no obvious change of the pore size at temperatures across the LCST (Range II in Fig. 5.2). As a result, the thermo-responsive coefficients of grafted membranes with the grafting yields higher than 6.38 % are equal to unity which indicates there are no thermo-responsive gating characteristics anymore. However, when the grafting yield is smaller than 6.38 %, the length of PNIPAM chains grafted in the membrane pores

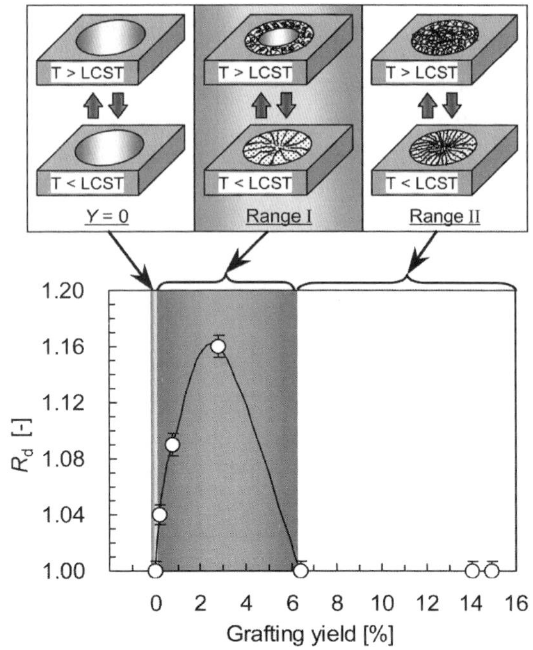

Fig. 5.2 Effect of the grafting yield on the thermo-responsive gating characteristics and schematic illustration of the thermo-responsive pore size (Reproduced with permission from Ref. [8], Copyright (2004), American Chemical Society)

is too short to "choke" the membrane pores (Range I in Fig. 5.2). There is a distinct difference between the lengths of PNIPAM chains at temperature below and above the LCST. As a result, the obvious change of membrane pore and hydraulic permeability at temperatures across the LCST is obtained. That is, the good thermo-responsive gating characteristics are achieved at the grafting yields of 0.5–3.0 %. The maximum flux thermo-responsive coefficient R_J and thermo-responsive coefficient of membrane pore size R_d are 2.54 and 1.16 when the grafting yield is 2.81 %.

Similar to the trend of the PNIPAM-grafted PVDF membranes, the water flux through the PNIPAM-grafted hydrophilic N6 membranes with proper grafting yields also has a dramatic change at 32 °C. For the PNIPAM-grafted N6 membranes, the flux thermo-responsive coefficient R_J reaches the maximum value of 15.41 when the grafting yield is 7.47 % [9]. When the grafting yield of PNIPAM-grafted N6 membranes is less than 7.47 %, the R_J value increases with increasing grafting yield. However, the R_J value tends to be 1.0 as the grafting yield approaches 12.84 %. When the grafting yield is larger than 12.84 %, the length and density of the PNIPAM-grafted chains are large enough to choke the membrane pores. As a result, no water flux could permeate through the membranes at both 25 °C and 40 °C and therefore no thermo-responsive gating characteristics exist anymore. Interestingly, the PNIPAM-grafted N6 membranes exhibit much larger maximum thermo-responsive gating coefficient R_J than that of the PNIPAM-grafted PVDF membranes (i.e., 15.41 and 2.54). Moreover, the optimum grafting yield of PNIPAM

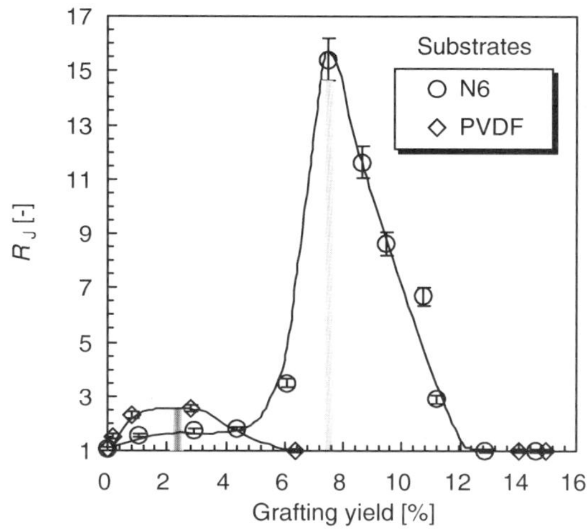

Fig. 5.3 Effect of grafting yield on the thermo-responsive gating characteristics of PNIPAM-grafted membranes with different substrates (Reproduced with permission from Ref. [9], Copyright (2006), Wiley-VCH Verlag GmbH & Co. KGaA)

(7.47 %) for the PNIPAM-grafted N6 membranes corresponding to the maximum R_J value is also higher than that (2.81 %) of the PNIPAM-grafted PVDF membranes (Fig. 5.3). These phenomena should result from the differences in the physical and chemical properties of the porous substrates, such as the hydrophilicity and microstructure. During the design of thermo-responsive gating membranes, it is quite important to choose appropriate porous membrane substrates, with different physical and chemical properties to achieve different gating membrane responses.

In addition, PNIPAM is also successfully grafted on the surfaces and in the pores of PC track-etched membranes by using plasma-induced pore-filling graft polymerization method. Because of the cylindrical and straight pores with narrow pore size distribution, the track-etched membranes are considered as excellent substrates to study the microstructures of gating membranes. The PNIPAM chains are grafted inside the pores throughout the entire membrane thickness, and there is no dense PNIPAM layer formed on the membrane surface even when the pore-filling ratio is as high as 76.1 % [10]. For the grafted membranes with pore-filling ratio smaller than 44.2 %, the water flux across the PNIPAM-grafted PC membranes at 40 °C is always larger than that at 25 °C, and the water fluxes at both 25 °C and 40 °C decrease with increasing the pore-filling ratio (Fig. 5.4). With the pore-filling ratio increasing, the pore diameters of PNIPAM-grafted membranes become smaller. When the pore-filling ratio is smaller than 44.2 %, the pores of PNIPAM-grafted PC membranes show thermo-responsive gating characteristics because of the conformational change of grafted PNIPAM in the pores. However, when the pore-filling ratio is larger than 44.2 %, the pores of membranes immersed in water are choked by the volume expansion of the grafted PNIPAM polymers, and the membranes does not show thermo-responsive gating characteristics any longer. The critical pore-filling ratio for choking the membrane pores is in the range from 30 % to 40 %. The pore diameter of the PNIPAM-grafted PC membrane with pore-filling ratio of 23.9 % increases dramatically when the temperature changes from 28 °C

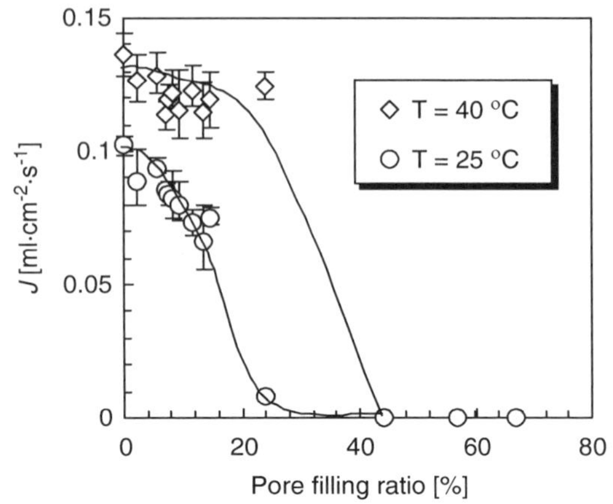

Fig. 5.4 Effect of pore-filling ratios on water flux across the PNIPAM-grafted PC membranes (Reproduced with permission from Ref. [10], Copyright (2005), Elsevier)

to 34 °C but keep unvaried at the temperatures lower than 28 °C and/or higher than 34 °C. The pore diameter of the grafted membrane at 40 °C is almost twofold of that at 25 °C according to Hagen-Poiseuille's law. The contact angle of PNIPAM-grafted PC membrane increases from 58.5° to 87.9° when the temperature changes from 25 °C to 40 °C. The thermo-responsive gating characteristics of the water flux of PNIPAM-grafted PC membranes are mainly dependent on the pore size change rather than the variation of membrane/pore surface hydrophilicity/hydrophobicity.

At last, the effect of grafting yield on the thermo-responsive water flux through porous PE membranes with PNIPAM gates is investigated [11]. At 25 °C, the grafted PNIPAM chains in the membrane pores are in the swollen state, and therefore the pore size decrease rapidly with increasing the grafting yield of PNIPAM. As a result, the water flux decreases rapidly at 25 °C with increasing the grafting yield. At environmental temperature of 40 °C, however, the water flux across the grafted membrane increases firstly to a peak, and then decreases and finally tends to zero because of the pore size becoming smaller and smaller (Fig. 5.5). It is suggested to set the grafting yield in the range 0.4∼0.8 mg·cm^{-2} for the membrane to get an effective thermo-responsive gating water flux.

To draw a conclusion, the PNIPAM-grafted membranes must be designed and prepared with a proper grafting yield to obtain a desired or satisfactory thermo-responsive gating performance.

***Thermo-responsive diffusional permeability*:** During the diffusional permeation experiments, the thermo-responsive membranes show positive and negative thermo-responsive models depending on the grafting yield. In detail, the diffusional coefficient of solute molecules across membranes with low grafting yields increase with temperature, while that across membranes with high grafting yields decrease with temperature [8]. Figure 5.6 shows the thermo-responsive diffusional permeability through PNIPAM-grafted PVDF membranes with different grafting yields. The diffusional coefficient of sodium chloride (NaCl) across the grafted membrane changes dramatically at temperatures around the LCST of PNIPAM, which is due

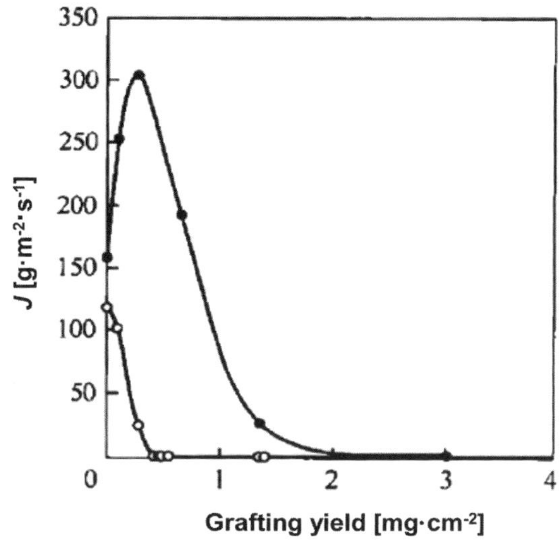

Fig. 5.5 Effect of grafting yield on the thermo-responsive water flux through the PNIPAM-grafted PE membranes

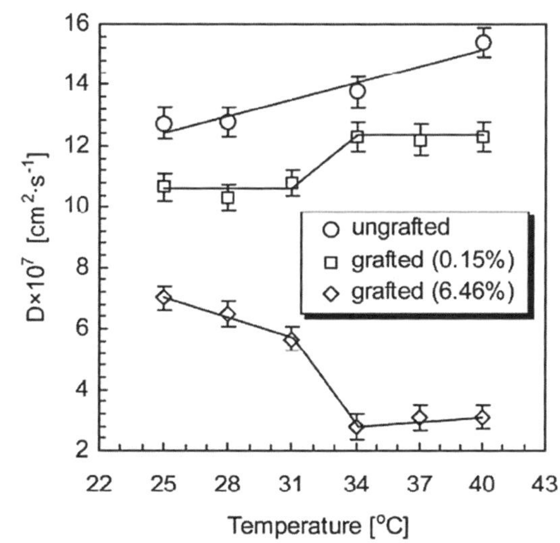

Fig. 5.6 Effect of the grafting yield on thermo-responsive diffusional permeation through PNIPAM-grafted PVDF membranes (Reproduced with permission from Ref. [8], Copyright (2004), American Chemical Society)

to the conformational change of the PNIPAM chains grafted in membrane pores. When the grafting yield of PNIPAM is low, the diffusional coefficient *D* of NaCl across the membrane is higher at temperature above the LCST than that below the LCST. It is attributable to the pores of the membrane being controlled open/closed by the shrinking/swelling mechanism of the grafted PNIPAM gates. However, when the grafting yield is high, the diffusional coefficient is lower at temperature above the LCST than that below the LCST owing to the hydrophilic/hydrophobic phase transition of the grafted PNIPAM gates. Because the solute used in the experiments is water soluble, any solute diffusion within the membranes occurs primarily within the water-filled regions in the spaces delineated by the grafted PNIPAM chains.

Therefore, it is easier for the solute to find water-filled regions in the membranes with hydrophilic PNIPAM gates (below the LCST) rather than in the membranes with hydrophobic PNIPAM gates (above the LCST) under the high grafting yield.

The grafting degree of PNIPAM-grafted PE membranes also has an effect on the thermo-responsive models [12]. When the pore-filling ratio of PNIPAM-grafted PE membrane is below 30 %, the diffusional coefficients of the solutes across the PNIPAM-grafted membranes are higher at temperatures above the LCST than those below the LCST. In contrast, when the pore-filling ratio is higher than 30 %, the diffusional coefficients are lower at temperatures above the LCST than those below the LCST. That is, the PNIPAM-grafted membranes change from positive thermo-responsive to negative thermo-responsive types with increasing pore-filling ratios at around 30 %. Phenomenological models for predicting the diffusion coefficient of a solute across PNIPAM-grafted membranes at temperatures both above and below the LCST are developed. The predicted diffusional coefficients of solutes across PNIPAM-grafted flat membranes are shown to fit with experimental values. To obtain ideal results for diffusional thermo-responsive controlled release through PNIPAM-grafted membranes, those substrates strong enough to prevent any conformation changing should be used in the preparation of the thermo-responsive membranes rather than weak substrates.

These results verify that it is also very important to choose or design a proper grafting yield of PNIPAM for obtaining a desired thermo-responsive diffusional permeability.

5.2.3 Gating Characteristics of Thermo-responsive Membranes with Grafted Linear and Cross-linked Hydrogel Gates

Both linear and cross-linked PNIPAM-grafted membranes with the same grafting yield (0.3 mg·cm^{-2}) are successfully prepared at 40 °C (above the LCST of PNIPAM) and 25 °C (below the LCST of PNIPAM) by using a plasma-induced grafting polymerization method [13]. To generate a cross-linked PNIPAM gate, cross-linker N,N′-methylenebisacrylamide (MBA) is added into the NIPAM monomer solution (the molar ratio of NIPAM to MBA is 100:1). The linear PNIPAM-grafted membranes prepared at 25 °C and 40 °C are respectively coded as LM-LT and LM-HT for short, and the cross-linked PNIPAM-grafted membranes prepared at 25 °C and 40 °C are respectively coded as CM-LT and CM-HT.

The effects of operation pressure and grafting temperature on the thermo-responsive gating characteristics of the prepared membranes are investigated systematically. For all the PNIPAM-grafted membranes, the water fluxes increase simply with increasing the operation pressure no matter whether the environmental temperature is 40 °C or 25 °C. Compared with the linear PNIPAM gates grafted in the LM-LT membrane, the cross-linked PNIPAM gates grafted in the CM-LT

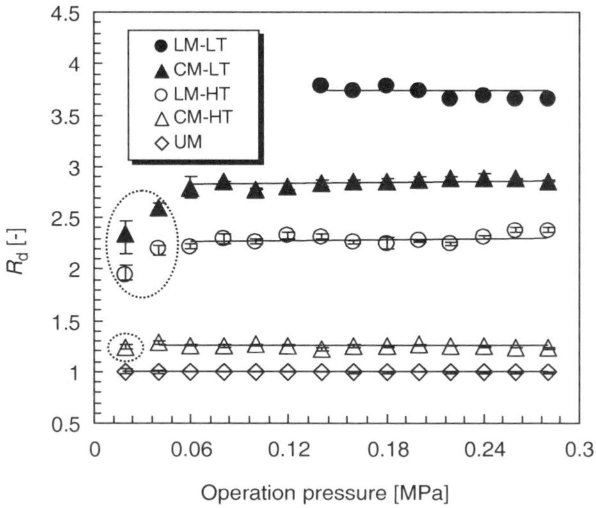

Fig. 5.7 Thermo-responsive gating coefficients of different membranes operated under different pressures (Reproduced with permission from Ref. [13], Copyright (2009), Wiley-VCH Verlag GmbH & Co. KGaA)

membrane are more stable at an environmental temperature of 25 °C and under high operation pressure. When the operation pressure increases to above 0.12 MPa, the effective pore size of the CM-LT membrane does not change. However, the pores of the LM-LT membrane change from the "fully closed" state at pressures lower than 0.12 MPa to the "slightly opened" state at operation pressures higher than 0.12 MPa. It results from the collapse of the linear-grafted PNIPAM chains under high operation pressure.

For all the PNIPAM-grafted membranes, the thermo-responsive gating coefficients under different operation pressures remain unchanged when each operation pressure is higher than a certain critical pressure value (Fig. 5.7). The critical pressures for the LM-LT, LM-HT, CM-LT, and CM-HT membranes are 0.14 MPa, 0.06 MPa, 0.06 MPa, and 0.04 MPa, respectively. However, at operation pressures lower than the critical pressure, the thermo-responsive gating coefficients are heavily affected by the thermo-responsive hydrophilicity/hydrophobicity change of pore surface. In order to get a stable thermo-responsive gating property of the PNIPAM-grafted membrane, the operation pressure should be higher than a critical value.

When the operation pressure is above 0.12 MPa, the thermo-responsive coefficient of membrane pore size R_d of the LM-LT membrane under a fixed operation pressure is the highest among the four kinds of grafted membranes, which reaches 3.7 (Fig. 5.7). It means that the effective pore diameter of the LM-LT membrane at 40 °C is nearly 4 times as large as that at 25 °C. The thermo-responsive coefficient of membrane pore size of the LM-HT and CM-LT membranes are 2.3 and 2.8, respectively. It also can be seen that the R_d value of the CM-HT membrane is the lowest among the four PNIPAM-grafted membranes, which is only 1.3.

For both membranes with grafted linear and cross-linked PNIPAM gates, the membranes prepared at 25 °C show larger thermo-responsive gating coefficients

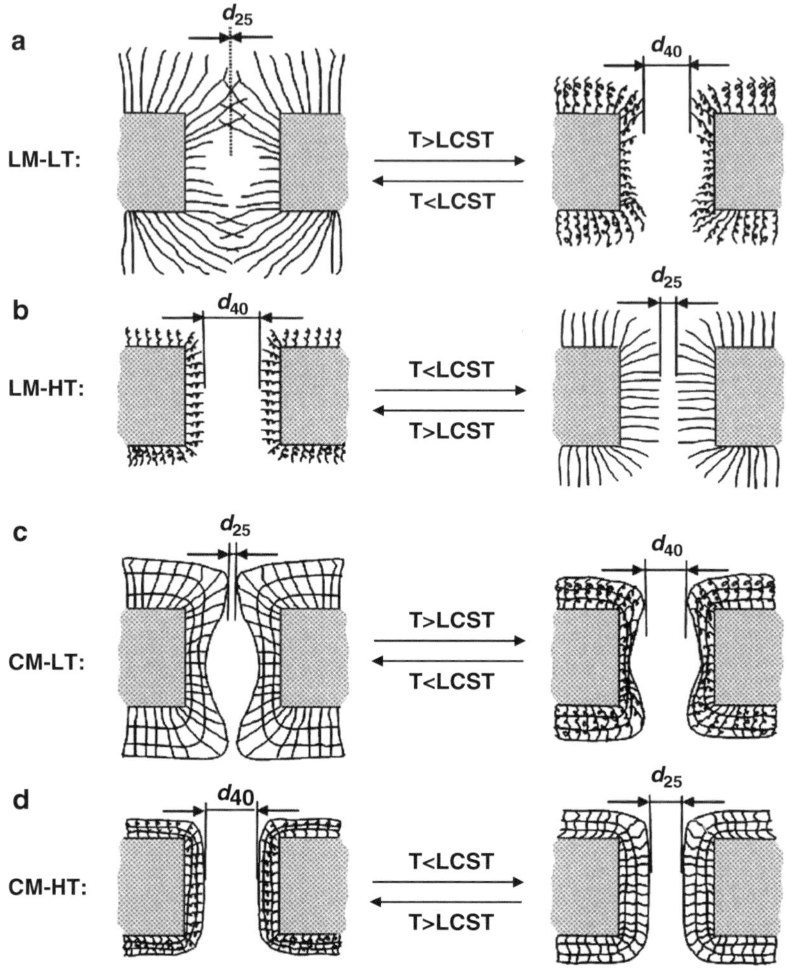

Fig. 5.8 Schematic illustrations of thermo-responsive microstructural changes of linear and cross-linked PNIPAM-grafted N6 membranes prepared at 25 °C and 40 °C, respectively. (**a**) LM-LT, (**b**) LM-HT, (**c**) CM-LT, and (**d**) CM-HT (Reproduced with permission from Ref. [13], Copyright (2009), Wiley-VCH Verlag GmbH & Co. KGaA)

than those prepared at 40 °C. It is attributable to the effect of grafting temperature on the microstructures of the grafted polymeric gates in the membrane pores (Fig. 5.8). The distributions of grafted polymeric gates in the PNIPAM-grafted membranes with the same grafting yield but prepared at different temperatures are also verified by SEM images. At the grafting temperature of 25 °C, the grafted layer on the membrane surface and at the pore entrance near the membrane surface is thicker than that in the middle of the membrane pores. On the other hand, at the grafting temperature of 40 °C, more homogeneous grafted layer is fabricated throughout the whole membrane thickness. Therefore, with the same grafting yield, the effective pore sizes of PNIPAM-grafted membranes, with either linear or cross-linked gates prepared at 25 °C, are smaller than those prepared at 40 °C. In addition, the cross-linked network structures of grafted PNIPAM layers of CM-LT and CM-HT membranes should be more compact than the linear PNIPAM chains with free ends of LM-LT and LM-HT membranes. Consequently, the effective pore sizes of cross-linked PNIPAM-grafted membranes are slightly larger than those of linear PNIPAM-grafted membranes when the grafting yields and grafting temperature are the same (CM-LT vs. LM-LT, and CM-HT vs. LM-HT). Therefore, the effective pore sizes of the LM-LT and CM-HT membranes at 25 °C are the smallest and largest among the four grafted membranes, respectively. Because the effective pore sizes of the membranes at 40 °C are almost the same, the thermo-responsive gating coefficients of membranes is mainly determined by the effective pore sizes at 25 °C. That's why the thermo-responsive gating coefficients of the LM-LT and CM-HT membranes are the highest and lowest one, respectively.

Both linear and cross-linked PNIPAM gates in the grafted membranes exhibit stable and repeatable thermo-responsive "open-close" switch performance under operation pressure of 0.26 MPa even the membranes have been tested for 20 runs. To get desired and satisfactory thermo-responsive gating characteristics of PNIPAM-grafted membranes, it is quite important and essential to design the grafted gates with proper structures (linear chains or cross-linked networks), to fabricate the PNIPAM-grafted gates at a proper temperature (higher or lower than the LCST of PNIPAM), and to operate the membrane under a proper pressure (should be higher than a critical value). The results in this study provide valuable guidance for designing, fabricating, and operating thermo-responsive gating membranes with desirable performances.

5.2.4 Membranes with Negatively Thermo-responsive Hydrogel Gates

All the membranes with thermo-responsive gates mentioned above exhibit the transition from "closed" to "open" when the environmental temperature increases from below the LCST of PNIPAM to above the LCST during the pressure-driven filtration experiment. It is attributable to the swelling/shrinking configuration change of PNIPAM at temperatures below and above the LCST of PNIPAM. In certain applications, however, an inverse mode of the thermo-responsive gating behavior of the thermo-responsive membranes is preferred. A novel family of thermo-responsive gating membranes with negatively thermo-responsive gating characteristics is developed [14]. That is, the "opening" of membrane pores is induced by a decrease rather than an increase in the environmental temperature. The proposed membrane is prepared by grafting cross-linked polyacrylamide (PAAM) gates on N6 membranes via plasma-induced pore-filling graft polymerization, and then polymerizing poly(acrylic acid) (PAAC) by sequential interpenetrating polymer networks (IPN) synthesis.

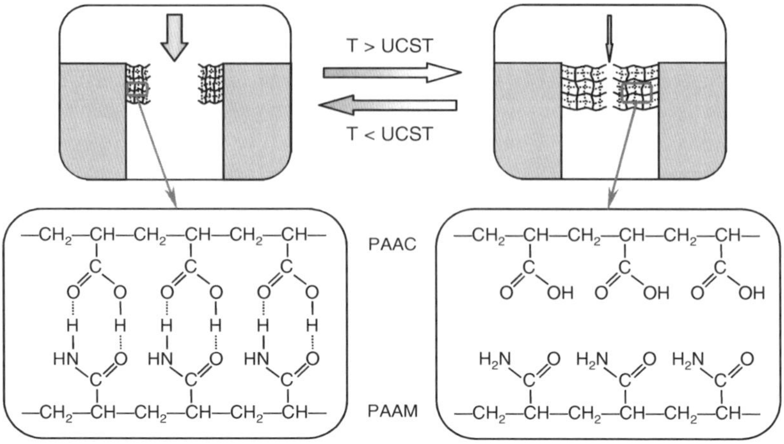

Fig. 5.9 Schematic illustration of the concept of the proposed negatively thermo-responsive membrane (Reproduced with permission from Ref. [14], Copyright (2005), Wiley-VCH Verlag GmbH & Co. KGaA)

The functional gates of membrane are thermo-responsive IPN composed of PAAM and PAAC, in which the volume-phase transition is driven by the cooperative "zipping" interactions between PAAM and PAAC (Fig. 5.9). PAAC forms intermolecular hydrogen bonds with PAAM when the environmental temperature is lower than the upper critical solution temperature (UCST) of the PAAM/PAAC-based IPN hydrogel. The IPN hydrogels have a shrunken state as a result of the interaction between two polymer chains, which is the so-called chain-chain zipper effect. However, when the environmental temperature is higher than the UCST of the IPN hydrogel, intermolecular PAAM/PAAC hydrogen bonds are disrupted and the IPN hydrogels become swollen. Consequently, the membrane pores change from an "open" gate to a "closed" gate when the temperature increases from below the UCST to above the UCST.

The water flux of the membranes exhibits significant negative gating characteristics. The water flux across the membrane with PAAM/PAAC-based IPN gates in the temperature range from 10 °C to 15 °C is much larger than that in the range from 30 °C to 40 °C. A sharp transition of the hydraulic permeability occurs on going from 20 °C to 25 °C, which corresponded to the UCST of PAAM/PAAC-based IPN hydrogels [15, 16]. The membranes with PAAM/PAAC-based IPN gates show satisfactorily reversible and reproducible thermo-responsive permeation characteristics. These gating membranes provide a new mode of behavior for thermo-responsive "smart" or "intelligent" membrane actuators, which are highly attractive for targeting drug delivery systems, chemical separations, and sensors.

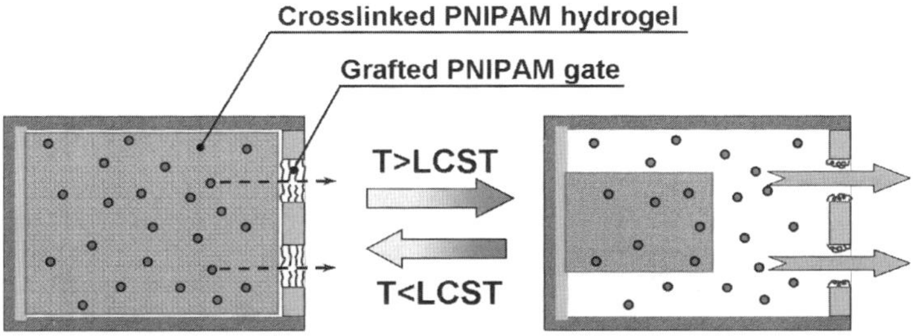

Fig. 5.10 Schematic illustration of the thermo-responsive controlled-release behavior of solutes from the proposed composite system (Reproduced with permission from Ref. [17], Copyright (2007), Wiley-VCH Verlag GmbH & Co. KGaA)

5.2.5 Composite Thermo-responsive Membrane System

The composite thermo-responsive membrane system is composed of a porous PVDF membrane with linearly grafted PNIPAM gates acting as functional valves, and a cross-linked PNIPAM hydrogel inside the reservoir acting as the solute carrier (Fig. 5.10) [17]. The thermo-responsive controlled-release characteristics of the proposed system are studied when the ambient temperature is continuously increased from 20 °C to 45 °C (across the LCST of PNIPAM) at a constant rate of $1.5\ °C \cdot min^{-1}$. To quantitatively describe the thermo-responsive controlled-release characteristics, a parameter known as the controlled factor (CF) is defined as

$$CF = \frac{\nu_2}{\nu_1} \tag{5.8}$$

where ν_1 and ν_2 stand for the release rates of vitamin B12 (VB12) $[mol \cdot s^{-1}]$ at temperatures below the LCST and above the LCST, respectively.

For the proposed composite system with both cross-linked PNIPAM hydrogel inside the reservoir and grafted PNIPAM gates in the membrane ($Y_{PNIPAM} = 7.02\ \%$), the CF for the VB12 release is increased to as high as 2.25 (Fig. 5.11). However, the CF values for the system with substrate PVDF membrane and no PNIPAM hydrogels inside the reservoir, the system with PNIPAM-grafted PVDF membrane ($Y_{PNIPAM} = 7.02\ \%$) but no PNIPAM hydrogel inside the reservoir, and the system with substrate PVDF membrane and cross-linked PNIPAM hydrogel inside the reservoir are 1.00, 1.40 and 1.28 within the temperature range tested, respectively. Due to the dual function of both the grafted PNIPAM gates in the membrane pores and the cross-linked PNIPAM hydrogel in the reservoir, the CF is higher than for both of the systems with thermo-responsive membrane only and with cross-linked PNIPAM hydrogel only. That is, with the cooperative action of the gating membrane and the cross-linked hydrogel, the proposed composite system exhibits better performance of thermo-responsive controlled release than those single-functional systems currently in existence.

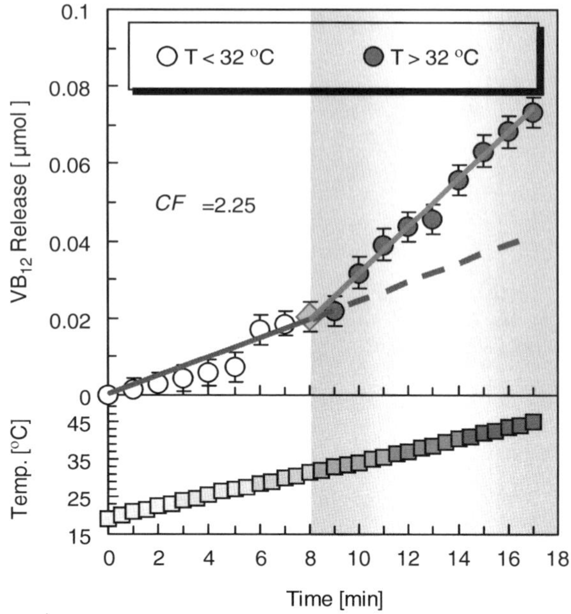

Fig. 5.11 Thermo-responsive gating characteristics of VB12 release from the proposed composite system with both PNIPAM hydrogel inside the reservoir and PNIPAM-grafted PVDF membrane ($Y_{PNIPAM} = 7.02$ %) (Reproduced with permission from Ref. [17], Copyright (2007), Wiley-VCH Verlag GmbH & Co. KGaA)

Due to the existence of cross-linked hydrogel inside the reservoir, the drug leakage problem of the reservoir-type systems could be effectively avoided. Also, because of the protection by the reservoir shell and the gating membrane, the mechanical strength problem of the hydrogel matrix-type systems could also be effectively solved. The proposed system provide a new mode for thermo-responsive "smart" or "intelligent" controlled release, which is highly attractive for targeting drug delivery systems, chemical carriers, and sensors, etc. Furthermore, the proposed new mode could be simply applied to micron- or nanometer-scale systems by incorporating a porous microcapsule or nanocapsule with functional gates as the reservoir with gating membranes and fabricating microgels or hydrogel nanoparticles inside the microcapsule or nanocapsule.

5.2.6 Thermo-responsive Affinity Membrane

The thermo-responsive affinity membranes with nanostructured pore surfaces and grafted PNIPAM surface layer for hydrophobic adsorption have been successfully developed [18]. SPG membranes with mean pore size of 1.8 μm are used as substrate membranes. The nanostructured pore surfaces are homogeneously formed by depositing 125 nm SiO_2 nanoparticles onto the SPG membrane pore surfaces. Subsequently, PNIPAM brushes are then grafted on the nanostructured pore surfaces of membranes by plasma-induced grafting polymerization method. The nanostructures

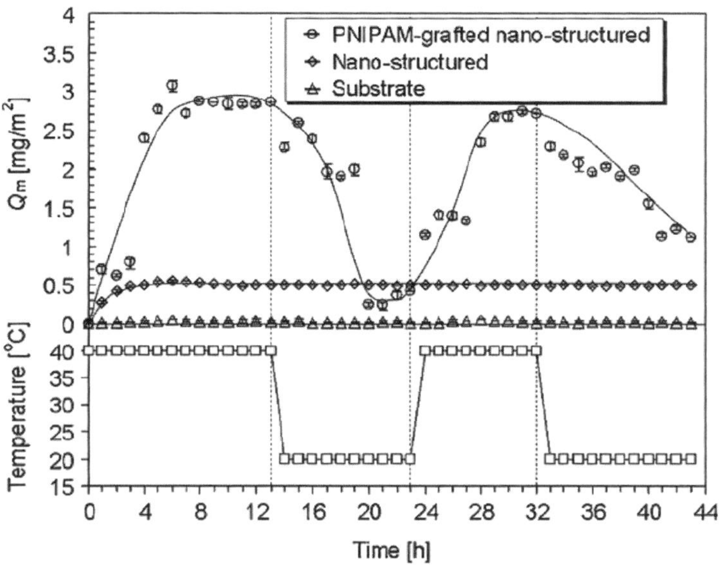

Fig. 5.12 Temperature-dependent dynamic adsorption and desorption of BSA on the PNIPAM-grafted nanostructured SPG membrane ($Y_{\mathrm{PNIPAM}} = 0.1$ %) (Reproduced with permission from Ref. [18], Copyright (2010), Elsevier)

on the membrane pore surfaces are verified to be beneficial for improving not only the hydrophobicity at temperature above the LCST but also the hydrophilicity at temperature below the LCST of PNIPAM-grafted membranes. The as-prepared PNIPAM-grafted nanostructured membranes present very hydrophobic surfaces with water contact angle of 130° at 40 °C but very hydrophilic surfaces with water contact angle of 0° at 20 °C. Such thermo-responsive hydrophilic/hydrophobic surface wettability change of the prepared membranes is perfectly reversible and reproducible.

The temperature-dependent dynamic adsorption and desorption of model protein Bovine serum albumin (BSA) on the PNIPAM-grafted nanostructured SPG membrane ($Y_{\mathrm{PNIPAM}} = 0.1$ %) are shown in Fig. 5.12. The substrate SPG membrane nearly does not adsorb BSA molecules whether the environmental temperature is 20 °C or 40 °C. As for the ungrafted nanostructured SPG membrane, the unit adsorption amount Q_{m} value increases to about 0.5 mg·m^{-2} at the first 3 h and then stays the same no matter how the environmental temperature changes. As for the PNIPAM-grafted nanostructured SPG membrane, the Q_{m} value at 40 °C increases sharply to about 2.95 mg·m^{-2} within the first 6 h and stays almost the same in the next 7 h. With decreasing the environmental temperature to 20 °C, the Q_{m} value goes down rapidly to about 0.25 mg·m^{-2} within 6 h, and then stays almost the same in next 4 h. When the environmental temperature is heated back to 40 °C, the Q_{m} value goes up to about 2.95 mg·m^{-2} again. The as-prepared PNIPAM-grafted nanostructured SPG membrane exhibits a satisfactory "adsorbing at temperature above the LCST-desorbing at temperature below the LCST" performance for BSA molecules. This is a simple and efficient mode for the adsorption/desorption of hydrophobic solutes. For the PNIPAM-grafted nanostructured SPG membrane with grafting yield of $Y_{PNIPAM} = 0.1$ %, the difference between the equilibrium Q_m value at 40 °C and the equilibrium Q_m value at 20 °C is about 2.7 mg·m^{-2}. That is, more than 90 % of BSA proteins adsorbed on the as-prepared membrane at 40 °C can be desorbed just by cooling the environmental temperature down to 20 °C, which means a convenient and efficient approach for the adsorption/desorption of BSA proteins.

5.3 Functional Membranes with Thermo-responsive Hydrogel Gates Fabricated by Atom-Transfer Radical Polymerization

A series of thermo-responsive membranes with controllable length and density of grafted polymer chains is prepared by grafting PNIPAM chains in the pores of AAO porous membranes with a silanization, acylation, and ATRP method (Table 5.1) [19]. The length and density of the grafted PNIPAM chains which form a uniform layer in the membrane pores are independently controlled. The density of grafted PNIPAM chains can be regulated by adjusting the reaction time during silanization (to generate $-NH_2$ groups on membrane surface). The length of grafted PNIPAM chains can be regulated by changing the grafting temperature, grafting time, and NIPAM concentration in monomer solution. Both the length and density of grafted PNIPAM chains are key factors for obtaining PNIPAM-grafted membranes with desired thermo-responsive gating characteristics. The thermo-responsive gating characteristics of the proposed PNIPAM-grafted AAO membranes are investigated by tracking the diffusional permeation of VB12 at temperatures below and above the LCST of PNIPAM.

Table 5.1 The information of PNIPAM-grafted AAO membranes[a] (Reproduced with permission from Ref. [19], Copyright (2009), Elsevier)

			PNIPAM-grafted AAO		
No.	$NH_2\%$ [wt%]	$Br\%$ [wt%]	C_{NIPAM} [g ml^{-1}]	Y_{PNIPAM} [wt%]	R_D [−]
M-1	0.16	0.06	0.05	5.55	1.58
M-2	0.15	0.06	0.08	11.9	8.1
M-3	0.24	0.1	0.04	12.1	4.4

[a]$NH_2\%$ and $Br\%$ are the densities of $-NH_2$ and $-Br$ groups on the membranes, respectively. C_{NIPAM} is the feed concentration of NIPAM monomer, Y_{PNIPAM} is the grafting yield of PNIPAM on AAO membrane, and R_D is the diffusion thermo-responsive coefficient of PNIPAM-grafted AAO membrane

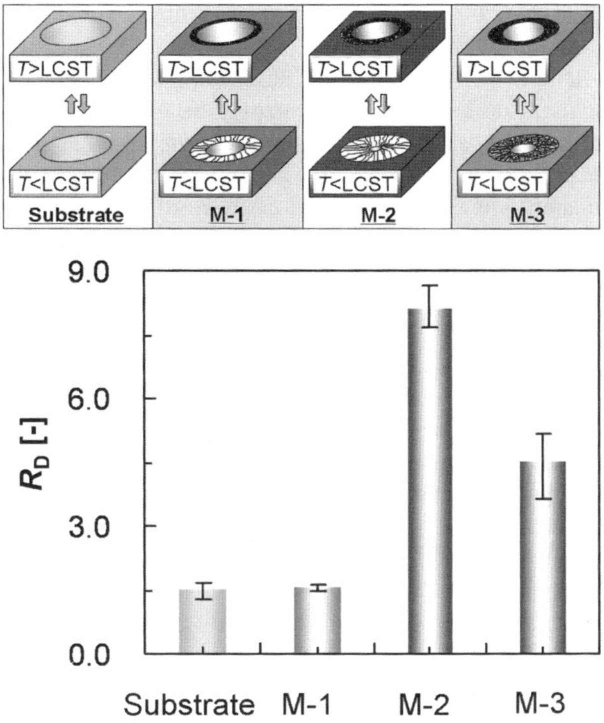

Fig. 5.13 Thermo-responsive gating characteristics of PNIPAM-grafted AAO membranes with different lengths and densities of grafted chains (Reproduced with permission from Ref. [19], Copyright (2009), Elsevier)

Figure 5.13 shows thermo-responsive gating factors of PNIPAM-grafted AAO membranes, and the schematic illustration of the thermo-responsive change of membrane pores at temperatures above and below the LCST of PNIPAM. The diffusional coefficient of the solute across the substrate membrane increases slightly with increasing the environmental temperature. It is because the decreased viscosity of water according to the Stokes-Einstein equation [12]. Therefore, the R_D value of AAO substrate membrane is 1.47.

If the density of –Br groups (*Br%*) on the membranes is identical, the grafted membranes with higher grafting yields will have longer PNIPAM chains grafted in the membrane pores. From Table 5.1, the length of grafted PNIPAM chains of membrane M-1 ($Y_{PNIPAM} = 5.55$ %) is shorter than that of membrane M-2 ($Y_{PNIPAM} = 11.9$ %) while the density of grafted chains on both membranes are identical. The diffusional coefficients at 40 °C (D_{40}) of membrane M-1 is twice as that of membrane M-2 while the diffusional coefficients at 25 °C (D_{25}) of membrane M-1 is 10 times as high as that of membrane M-2. Therefore, the diffusion thermo-responsive coefficient of membrane M-2 ($R_D = 8.1$) is much larger than that of membrane M-1 ($R_D = 1.58$) (Fig. 5.13). To study the effect of the density of grafted PNIPAM chains on the thermo-responsive performance of PNIPAM-grafted membranes, the R_D values of membrane M-1 and M-3 are compared. The density of grafted PNIPAM chains on the membrane M-3 is 1.67 times higher than that on membrane M-1, and the grafting yield Y_{PNIPAM} of membrane M-3 is 2.18 times more than that of membrane M-1 (Table 5.1). Therefore, the length of grafted PNIPAM chains on membrane M-3 is 1.3 times longer than that on membrane M-1. The diffusion thermo-responsive coefficient of membrane M-3 ($R_D = 4.4$) is approximately three times as that of membrane M-1 ($R_D = 1.58$) (Fig. 5.13). Compared with membrane M-1, the larger R_D value of membrane M-3 resulted from both the higher density and the longer length of grafted PNIPAM chains. Moreover, for membranes M-2 and M-3, the grafting yields of PNIPAM are almost the same while the density of –Br on membrane M-2 is smaller than that on membrane M-3. Therefore, the length of grafted PNIPAM chains in the pores of membrane M-2 is longer than that in the pores of membrane M-3 (Table 5.1 and Fig. 5.13). The diffusion thermo-responsive coefficient of membrane M-2 ($R_D = 8.1$) is larger than that of membrane M-3 ($R_D = 4.4$). It indicates the length of grafted PNIPAM chains in the membrane pores has more significant effect on the thermo-responsive gating characteristics of PNIPAM-grafted membranes than the density of grafted PNIPAM chains. The PNIPAM chains grafted by ATRP also show satisfactorily reversible and reproducible thermo-responsive behaviors in the PNIPAM-grafted membrane pores.

5.4 Functional Membranes with Thermo-responsive Hydrogel Gates Fabricated by Free-Radical Polymerization

Thermo-responsive membranes are prepared by fabricating cross-linked PNIPAM hydrogel gates inside the pores of porous N6 membranes by the free-radical polymerization initiated at different preparation temperatures [20]. Both PNIPAM-filled N6 membranes prepared at 60 °C (above the volume-phase transition temperature (VPTT) of PNIPAM) and at 25 °C (below the VPTT of PNIPAM) exhibit significant reversible and reproducible thermo-responsive diffusional permeability (Fig. 5.14). It is because the 3-dimensional interpenetrating network structure of the cross-linked PNIPAM hydrogels inside the N6 membranes is robust. A dramatic change in the diffusion coefficient of VB12 solute across the PNIPAM-filled N6 membranes prepared at both 25 °C and 60 °C occurs on going from 31 °C up to 34 °C, which corresponds to the VPTT of PNIPAM (~32 °C).

The values of diffusion thermo-responsive coefficient for the PNIPAM-filled N6 membrane prepared at 25 °C and that prepared at 60 °C are 7.84 and 6.68, respectively. When the PNIPAM hydrogel is prepared at 25 °C, the equilibrium volume deswelling ratio is much larger than that prepared at 60 °C, resulting in a larger diffusion thermo-responsive coefficient. If the diffusion thermo-responsive coefficient of the thermo-responsive membrane is the main target index, the PNIPAM-filled membrane should be prepared at 25 °C.

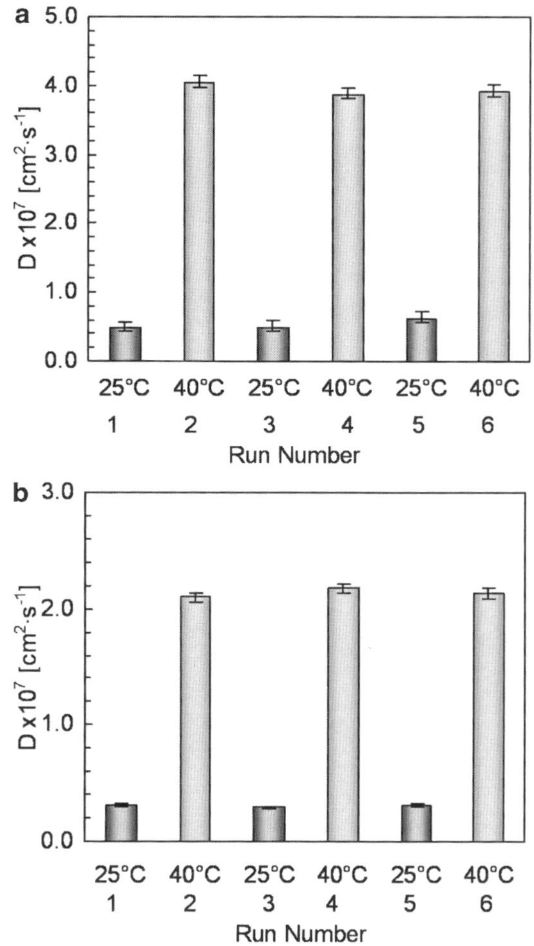

Fig. 5.14 Reversible thermo-responsive release characteristics of PNIPAM-filled N6 membranes prepared at 25 °C (**a**) and 60 °C (**b**) (Reproduced with permission from Ref. [20], Copyright (2006), Wiley-VCH Verlag GmbH & Co. KGaA)

5.5 Summary

In summary, the thermo-responsive membranes with PNIPAM gates are successfully fabricated by free-radical polymerization, plasma-induced pore-filling graft polymerization, and ATRP. The as-prepared membranes can be either positive or negative thermo-responsive models. The grafting degree, length and density of PNIPAM chains, the microstructures of substrate membrane, and preparation temperature have great effect on the thermo-responsive gating characteristics of membranes. The response temperatures of thermo-responsive membranes are found to be linearly increased with increasing the molar ratio of hydrophilic monomer, but linearly decreased with increasing the molar ratio of hydrophobic monomer. For each type of substrate membranes, there is an optimum grafting degree to achieve desired or satisfactory thermo-responsive gating performance.

The composite thermo-responsive membrane system exhibits better performance of thermo-responsive controlled release than those single-functional systems currently in existence by virtue of the cooperation of the gating membrane and the cross-linked hydrogel. Moreover, the PNIPAM-grafted nanostructured SPG membrane exhibits a satisfactory "adsorbing at temperature above the LCST-desorbing at temperature below the LCST" performance for BSA molecules. The results in this study provide valuable guidance for designing, fabricating, and operating thermo-responsive gating membranes with desirable performances.

References

1. Yang B, Yang WT (2003) Thermo-sensitive switching membranes regulated by pore-covering polymer brushes. J Membr Sci 218:247–255
2. Geismann C, Ulbricht M (2005) Photoreactive functionalization of poly(ethylene terephthalate) track-etched pore surfaces with "smart" polymer systems. Macromol Chem Phys 206:268–281
3. Kuroki H, Ohashi H, Ito T et al (2010) Isolation and analysis of a grafted polymer onto a straight cylindrical pore in a thermal-responsive gating membrane and elucidation of its permeation behavior. J Membr Sci 352:22–31
4. Alem H, Duwez AS, Lussis P et al (2008) Microstructure and thermo-responsive behavior of poly(*N*-isopropylacrylamide) brushes grafted in nanopores of track-etched membranes. J Membr Sci 308:75–86
5. Friebe A, Ulbricht M (2007) Controlled pore functionalization of poly(ethylene terephthalate) track-etched membranes via surface-initiated atom transfer radical polymerization. Langmuir 23:10316–10322
6. Yu HY, Li W, Zhou J et al (2009) Thermo- and pH-responsive polypropylene microporous membrane prepared by the photoinduced RAFT-mediated graft copolymerization. J Membr Sci 343:82–89
7. Xie R, Li Y, Chu LY (2007) Preparation of thermo-responsive gating membranes with controllable response temperature. J Membr Sci 289:76–85
8. Li Y, Chu LY, Zhu JH et al (2004) Thermoresponsive gating characteristics of poly(*N*-isopropylacrylamide)-grafted porous polyvinylidene fluoride membranes. Ind Eng Chem Res 43:2643–2649
9. Yang M, Chu LY, Li Y et al (2006) Thermo-responsive gating characteristics of poly(*N*-isopropylacrylamide)-grafted membranes. Chem Eng Technol 29:631–636
10. Xie R, Chu LY, Chen WM et al (2005) Characterization of microstructure of poly(*N*-isopropylacrylamide)-grafted polycarbonate track-etched membranes prepared by plasma-graft pore-filling polymerization. J Membr Sci 258:157–166
11. Chu LY, Zhu JH, Chen WM et al (2003) Effect of graft yield on the thermo-responsive permeability through porous membranes with plasma-grafted poly(*N*-isopropylacrylamide) gates. Chin J Chem Eng 11:269–275
12. Chu LY, Niitsuma T, Yamaguchi T et al (2003) Thermo-responsive transport through porous membranes with grafted PNIPAM gates. AIChE J 49:896–909
13. Chen YC, Xie R, Yang M et al (2009) Gating characteristics of thermo-responsive membranes with grafted linear and crosslinked poly(*N*-isopropylacrylamide) gates. Chem Eng Technol 32:622–631
14. Chu LY, Li Y, Zhu JH et al (2005) Negatively thermoresponsive membranes with functional gates driven by zipper-type hydrogen-bonding interactions. Angew Chem Int Ed 44:2124–2127

15. Katono H, Maruyama A, Sanui K et al (1991) Thermo-responsive swelling and drug release switching of interpenetrating networks composed of poly(acrylamide-co-butyl methacrylate) and poly(acrylic acid). J Control Release 16:215–227
16. Xiao XC, Chu LY, Chen WM et al (2003) Positively thermo-sensitive monodisperse core-shell microspheres. Adv Funct Mater 13:847–852
17. Hu L, Chu LY, Yang M et al (2007) A composite thermo-responsive membrane system for improved controlled-release. Chem Eng Technol 30:523–529
18. Meng T, Xie R, Chen YC et al (2010) A thermo-responsive affinity membrane with nano-structured pores and grafted poly(*N*-isopropylacrylamide) surface layer for hydrophobic adsorption. J Membr Sci 349:258–267
19. Li PF, Xie R, Jiang JC et al (2009) Thermo-responsive gating membranes with controllable length and density of poly(*N*-isopropylacrylamide) chains grafted by ATRP method. J Membr Sci 337:310–317
20. Li PF, Ju XJ, Chu LY et al (2006) Thermo-responsive membranes with crosslinked poly(*N*-isopropylacrylamide) hydrogels inside porous substrates. Chem Eng Technol 29:1333–1339

Chapter 6
Functional Microcapsules with Thermo-responsive Hydrogel Shells

Abstract In this chapter, the design, fabrication, and property of functional microcapsules with thermo-responsive hydrogel shells are introduced. The as-designed thermo-responsive microcapsules can be microcapsules with membranes grafted by poly(*N*-isopropylacrylamide) (PNIPAM) chains or interspersed with PNIPAM microgels as gates, or microcapsules with whole thermo-responsive PNIPAM membranes. Both Shirasu porous glass (SPG) membrane emulsification and capillary microfluidic emulsification are employed to generate the monodisperse emulsions with controllable size, and subsequently the emulsions as template are polymerized into microcapsules. The prepared thermo-responsive microcapsules with membranes grafted or embedded PNIPAM gates display satisfactory reversible and reproducible thermo-responsive controlled-release characteristics, while those microcapsules with whole PNIPAM membranes exhibit prompt and complete temperature-triggered bursting-release characteristics. These thermo-responsive microcapsules are highly attractive for various promising applications, such as site-targeting drug delivery system, controlled release of chemicals, microreactors, biomedical and/or chemical sensors, immobilization of cells and enzymes, and encapsulation of foods and cosmetics.

6.1 Introduction

Thermo-responsive microcapsules have attracted widespread interest in the last decade due to their potential applications in numerous fields, including controlled drug delivery systems (DDS), sensors, and vesicles for enzymes and chemicals [1, 2]. For DDS, the monodisperse small-sized microcapsules are preferred because the distribution of the microcapsules within the body is greatly affected by the particle size and their drug release kinetics can be manipulated [3, 4]. In addition, the prepared microcapsules are expected to have a faster response time compared with the microgels or microspheres due to the decreased diffusion resistant across the thin microcapsule membranes [5].

L.-Y. Chu et al., *Smart Hydrogel Functional Materials*,
DOI 10.1007/978-3-642-39538-3_6,

© Chemical Industry Press, Beijing and Springer Berlin Heidelberg 2013

These thermo-responsive microcapsules can be classified as two main types: microcapsules with membranes modified by poly(*N*-isopropylacrylamide) (PNIPAM) [6–9] and microcapsules with whole PNIPAM membranes [10–15]. The microcapsules with PNIPAM modified membranes are usually prepared by grafting PNIPAM chains onto the pores of porous membrane or embedding the PNIPAM microgels during the fabrication of membrane. However, the microcapsules with whole PNIPAM membranes are formed by emulsion template method, that is, polymerizing the emulsion containing *N*-isopropylacrylamide (NIPAM) monomers into hydrogel membranes. The methods adopted to generate the emulsion can be Shirasu porous glass (SPG) membrane emulsification and scalable capillary microfluidic technique. SPG membrane emulsification is based on the injection of a disperse phase through a porous membrane, with the resulting droplets forming at the end of pores on the membrane surface [4, 16–18]. By the use of this technique, it is easier to control the droplet size and size distribution. The capillary microfluidic technique can independently control both the size and the number of inner droplets to fabricate highly monodisperse multiple emulsion and is easily scalable to higher-order multiple emulsions [19]. Therefore, the emulsion generated by the SPG membrane emulsification and capillary microfluidic technique is monodispersed with controllable particle size.

In this chapter, the thermo-responsive microcapsules with different gates (grafted PNIPAM chains and embedded PNIPAM microgels) and microcapsules with whole PNIPAM membranes) are designed and developed. Their temperature-triggered, controlled-release, or bursting-release characteristics are investigated systematically.

6.2 Functional Microcapsules with Grafted Thermo-responsive Hydrogel Chains in the Porous Membranes as Gates

As mentioned above, to make the thermo-responsive microcapsules more suitable for DDS, it is essential to develop monodispersed small-sized thermo-responsive microcapsules. Recently, the monodispersed thermo-responsive microcapsules with a mean diameter of about 4 μm have been successfully prepared, which are composed of porous polyamide membranes and PNIPAM-grafted gates [4]. By virtue of SPG membrane emulsification technique, the small-sized monodispersed oil-in-water (O/W) emulsions are generated and then form core-shell porous polyamide microcapsules via interfacial polymerization. The PNIPAM chains are grafted into the pores of the microcapsule membranes by plasma-induced pore-filling graft polymerization.

Figure 6.1 shows the thermo-responsive release of NaCl from PNIPAM-grafted microcapsules with a mean diameter of about 4 μm. The NaCl concentration of the bulk solution increases slowly at 25 °C and rapidly at 40 °C; that means the release of NaCl from the PNIPAM-grafted microcapsules is slow at 25 °C and fast

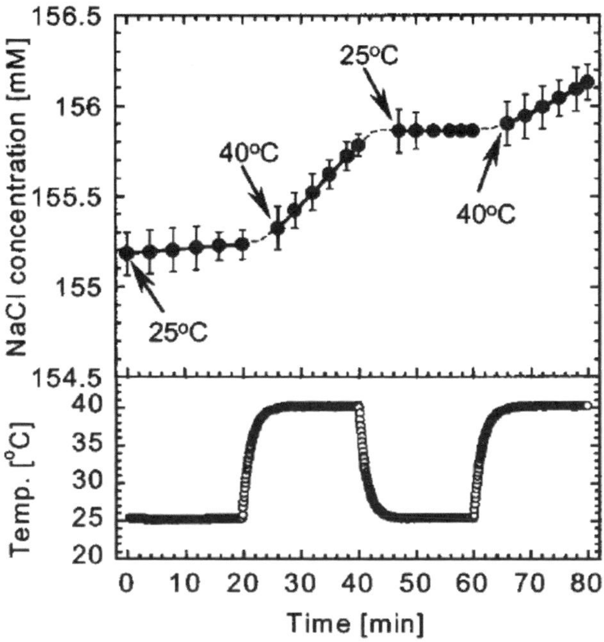

Fig. 6.1 Thermo-responsive release of NaCl from the PNIPAM-grafted microcapsules with a mean diameter of about 4 μm (Reproduced with permission from Ref. [4], Copyright (2002), American Chemical Society)

at 40 °C. At temperatures below the lower critical solution temperature (LCST), the PNIPAM-grafted chains in the pores of the microcapsules are swollen and the membrane pores are closed by the PNIPAM gates. As a result, the release of NaCl molecules across the microcapsule membranes is slow. In contrast, at temperatures above the LCST, the PNIPAM-grafted chains in the membrane pores are shrunk, and therefore, the pores of the microcapsule membranes are open and resulted in a faster release rate of the NaCl molecules across the microcapsules. Therefore, the release rate of the solute molecules from this kind of PNIPAM-grafted microcapsules at higher temperatures (above the LCST) is greater than that at lower temperatures (below the LCST). The prepared PNIPAM-grafted microcapsules with a mean diameter of about 4 μm show satisfactory reversible and reproducible thermo-responsive release characteristics. Similar to that of NaCl, the release of vitamin B12 (VB12) from the PNIPAM-grafted microcapsules is slow at 25 °C and fast at 40 °C, which is also due to the closed-open state of the grafted "gates". The diffusion thermo-responsive coefficient $R_D = P_{40}/P_{25}$ of the PNIPAM-grafted microcapsules toward VB12 is much larger than that of NaCl.

Similar to the results of flat membranes in Chap. 5 [9, 20–22], the proposed thermo-responsive microcapsules could exhibit a positive thermo-responsive controlled-release mode or a negative thermo-responsive one by changing the PNIPAM grafting yield [5]. The thermo-responsive controlled-release characteristics of VB12 from PNIPAM-grafted microcapsules with different grafting yields are illustrated in Fig. 6.2. At low grafting yields, the PNIPAM-grafted microcapsules show a positive thermo-responsive controlled-release mode, while at high grafting yields,

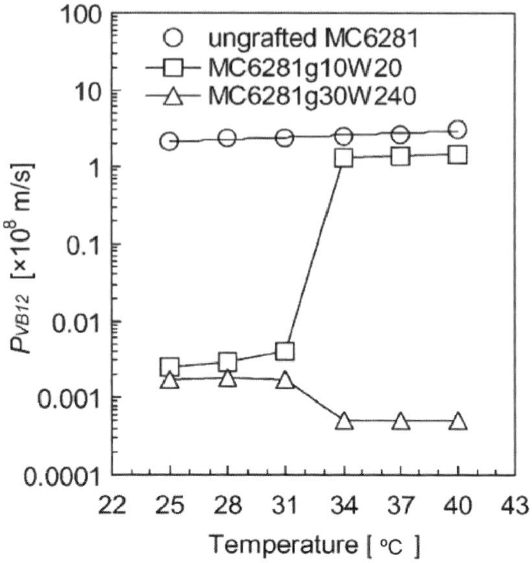

Fig. 6.2 Thermo-responsive controlled release of VB12 from PNIPAM-grafted microcapsules with different grafting yields (Reproduced with permission from Ref. [5], Copyright (2001), Elsevier)

they show a negative thermo-responsive controlled-release mode. At temperature above the LCST, the pores of PNIPAM-grafted microcapsules with low grafting yield are open, and solute diffusion occurs within the pores with the openings larger than the solute size. However, the pores are closed due to swollen PNIPAM gate at temperature below the LCST, and solute diffusion occurs within the PNIPAM hydrogels. Therefore, the permeability coefficient of VB12 from the microcapsules is higher at temperatures above the LCST than that below the LCST, due to the "open-closed" pores in the microcapsule membranes controlled by the PNIPAM gates. The prepared PNIPAM-grafted microcapsules show a satisfactorily reversible and reproducible thermo-responsive controlled release.

In contrast, when the grafting yield of PNIPAM-grafted microcapsules is high, the permeability coefficient of VB12 at temperatures above the LCST is lower than that below the LCST, due to the hydrophilic/hydrophobic phase transition of the PNIPAM gates. At a high grafting yield, because there is too much grafted polymer in the pores of microcapsules, the pores cannot reopen even at high temperatures (above the LCST), i.e., the pore is choked. However, the grafted PNIPAM chains are still highly hydrophilic and water soluble at temperature below the LCST and dramatically become hydrophobic and insoluble in water at temperature above the LCST, with a phase transition. As the solute is water soluble, any solute diffusion within the membrane occurs primarily within the water-filled regions in the spaces delineated by the polymer chains. It is easier for the solute to find water-filled regions in the membrane with hydrophilic PNIPAM gates rather than in the membrane with hydrophobic PNIPAM gates. Therefore, the permeability coefficient of the solute molecules from PNIPAM-grafted microcapsules at low temperatures (below the LCST) is higher than that at high temperatures (above the LCST).

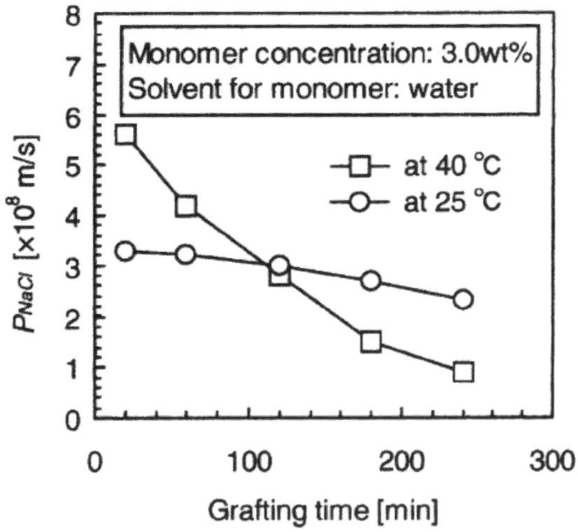

Fig. 6.3 Thermo-responsive permeability coefficients of NaCl from PNIPAM-grafted microcapsules with different grafting yields (Reproduced with permission from Ref. [5], Copyright (2001), Elsevier)

The thermo-responsive permeability coefficient of NaCl through the PNIPAM-grafted microcapsules with different grafting time is shown in Fig. 6.3. In a previous study [23], it was shown that the grafting yield was directly proportional to the grafting time in the plasma-induced pore-filling graft polymerization, with all other experimental conditions being the same. Therefore, the result in Fig. 6.3 reflects the effect of the grafting yield on the thermo-responsive permeability coefficient of the PNIPAM-grafted microcapsules. It is interesting to note that the grafting yields have an opposite effect on the permeability coefficients of the microcapsules. It also indicates that two distinct modes of gating functions exist as positive thermo-responsive mode and negative thermo-responsive mode, depending on the grafting yield. It is seen that at shorter grafting times (or lower grafting yield), the permeability at 40 °C is higher than that at 25 °C; while at longer grafting times (or higher grafting yield), the permeability at 40 °C is lower than that at 25 °C.

Based on our previous work [5], a superparamagnetic property is introduced into the porous microcapsule membrane with thermo-responsive gates [24]. Before preparing the microcapsule, oleic acid (OA)-modified Fe_3O_4 nanoparticles are synthesized using a chemical coprecipitation route followed by coating with OA. Subsequently, the modified Fe_3O_4 nanoparticles are introduced to prepare the polyamide microcapsules with magnetic porous membranes during interfacial polymerization process. Later, the microcapsule membranes are grafted with PNIPAM chains by employing plasma-induced pore-filling graft polymerization. When the temperature is below the LCST, the grafted PNIPAM chains in the magnetic thermo-responsive microcapsule membranes are in the swollen state and the gates of membrane pores are "close"; on the other hand, when the temperature is above the LCST, the PNIPAM chains are in the shrunken state and the gates of membrane pores are "open." Thus, the release of substance form the microcapsules is controlled by changing the environmental temperature.

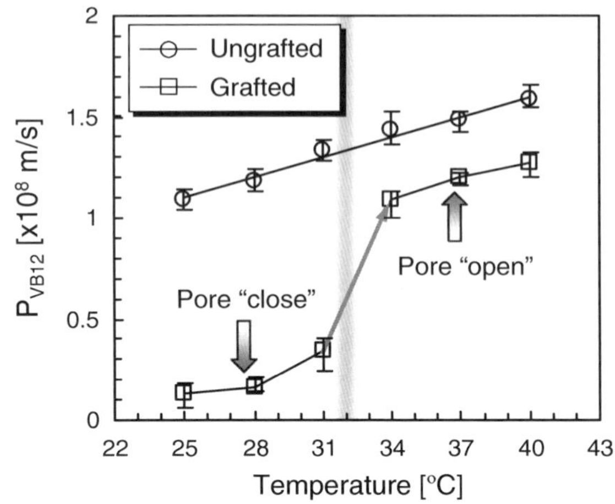

Fig. 6.4 Thermo-responsive release of VB12 from ungrafted and PNIPAM-grafted microcapsules with magnetic property (Reproduced with permission from Ref. [24], Copyright (2008), Elsevier)

The permeability coefficients *P* of VB12 releasing from ungrafted and PNIPAM-grafted magnetic microcapsules are shown as a function of temperature (Fig. 6.4). For the PNIPAM-grafted microcapsules, the *P* values are low when the environmental temperature is below 31 °C, and they increase little between 25 °C and 31 °C. On the other hand, the *P* values are much higher when the environmental temperature is above 34 °C, while the *P* values also increase little with temperature increasing from 34 °C to 40 °C. A sharp transition of the permeability coefficient occurs on going from 31 °C to 34 °C, which corresponds to the LCST of PNIPAM (around 32 °C). However, the *P* value of ungrafted microcapsules does not show such a sharp transition between 31 °C and 34 °C under the same experimental conditions. Consequently, the release rate of the solute molecules VB12 from the PNIPAM-grafted microcapsules is much larger at temperatures above the LCST than that below the LCST. The prepared microcapsule membranes exhibit time-independent superparamagnetic property with good magnetic-responsive ability, and satisfactory thermo-responsive controlled-release property. The combined properties of such dual stimuli-responsive microcapsules make them highly attractive for various promising applications, such as site-targeting drug delivery, controlled release of chemicals, microreactors, biomedical and/or chemical sensors, and separations.

6.3 Functional Microcapsules with Thermo-responsive Microgels in the Membranes as Gates

As the thermo-responsive membrane gates, PNIPAM microgels are successfully embedded in the Ca-alginate capsule wall by a co-extrusion minifluidic approach [25]. The prepared Ca-alginate capsules are highly monodisperse (coefficient of

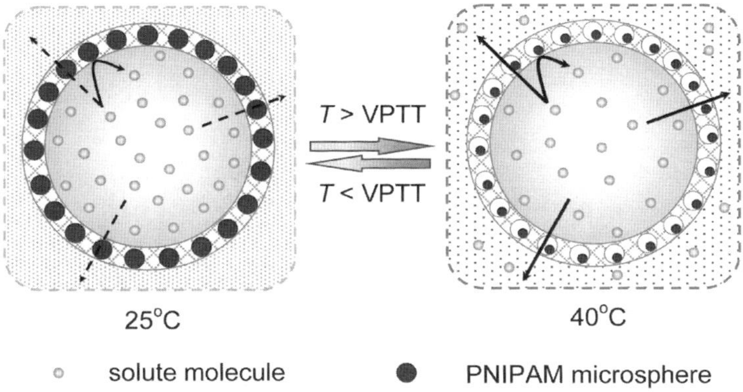

Fig. 6.5 Schematic illustration of the thermo-responsive characteristic of Ca-alginate capsule with thermo-responsive microgels as gates (Reproduced with permission from Ref. [25], Copyright (2011), Elsevier)

variation $CV = 3.12$ %, calculated by Eq. 6.1). The average diameter and membrane thickness of Ca-alginate capsules are about 2.96 and 0.11 mm, respectively. When the environmental temperature is 25 °C (below the volume phase transition temperature (VPTT)), it is hard for the solute molecules to permeate across the Ca-alginate capsule membrane due to the "closed" gates (PNIPAM microgels in swollen state); however, the solute molecules can permeate easily across the membrane as a result of the "open" gates at 40 °C (PNIPAM microgels in shrunken state) (Fig. 6.5):

$$CV = 100\,\% \times \left(\sum_{i=1}^{N} \frac{\left(D_{\mathrm{i}} - \overline{D_{\mathrm{n}}}\right)^2}{N-1} \right)^{\frac{1}{2}} \Big/ \overline{D_{\mathrm{n}}} \qquad (6.1)$$

where D_{i} is the diameter of the ith microsphere [cm], N is the total number of the microgels counted [–], and $\overline{D_{\mathrm{n}}}$ is the arithmetic average diameter [cm].

The diffusional permeability coefficients of solute molecules (i.e., VB12, PEG2000 and PEG20000) across Ca-alginate capsules with different contents of PNIPAM microgels are investigated at temperatures below and above the VPTT of PNIPAM microgels. With the content of PNIPAM microgels increasing, the permeability coefficients of solute molecules across the Ca-alginate capsules at 25 °C are almost the same. The solute molecules inside the capsules diffuse into the surrounding medium merely through the solution filled in the space between polymer networks of PNIPAM microgels and Ca-alginate capsule membrane. However, the permeability coefficients across the capsules at 40 °C gradually increase as the content of PNIPAM microgels in the capsule membranes increases. Since the volumes of PNIPAM microgels in water at 40 °C decrease by 80 % than those at 25 °C, the embedded PNIPAM microgels act as "open" gates (Fig. 6.5). The "open" gates form much larger diffusion channels, which contribute to faster diffusion of solute molecules to the surrounding medium. The more the PNIPAM

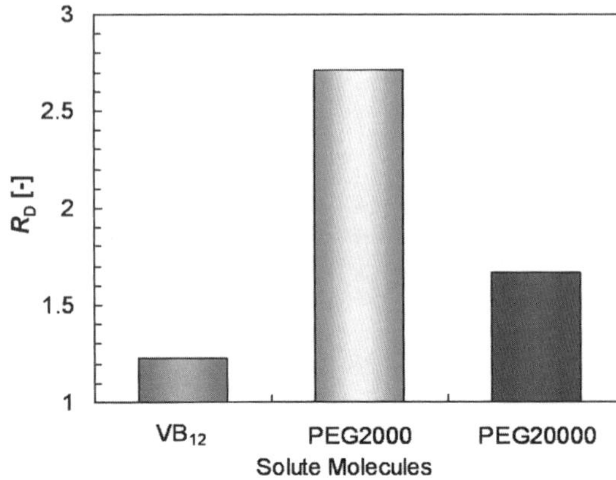

Fig. 6.6 Comparison of thermo-responsive gating characteristics of Ca-alginate capsules with the same components ($C_{PNIPAM}/C_{alginate} = 5.0/15$ g/L) for different solute molecules (Reproduced with permission from Ref. [25], Copyright (2011), Elsevier)

microgels embedded in the capsules, the larger the permeability coefficients of solute molecules in the investigated range. The results show that the diffusional permeability of solute across the Ca-alginate capsules with the thermo-responsive "smart" gates can be adjusted.

The thermo-responsive gating characteristics of Ca-alginate capsules are valuated by the diffusion thermo-responsive coefficient R_D. It is defined as the permeability coefficient of solute molecules across the membranes at temperatures higher than the VPTT to that at temperatures lower than the VPTT of PNIPAM microgels. The larger the R_D value is, the better the thermo-responsive gating characteristics of Ca-alginate capsules. For the Ca-alginate capsules with the same content of PNIPAM microgels ($C_{PNIPAM}/C_{alginate} = 5/15$ g/L), the R_D values for different solute molecules VB12, PEG2000, and PEG20000 are 1.23, 2.71, and 1.67, respectively (Fig. 6.6). Such a phenomenon attributes to the match between the molecule size of solute and the gate size of capsule membrane during the trans-membrane diffusion. The thermo-responsive property of Ca-alginate capsules make them highly attractive for various promising applications, such as microreactors, immobilization of cells and enzymes in bioreactors, controlled release of chemicals, and encapsulation of foods and cosmetics.

6.4 Functional Microcapsules with Thermo-responsive Cross-linked Hydrogels as Membranes

A novel and simple method for preparation of monodisperse hollow PNIPAM microcapsules via SPG membrane emulsification and UV-initiated polymerization at the interface of water-in-oil (W/O) single emulsions is successfully developed [26]. Because the polymerization is carried out at a low temperature (20 °C) which is below the VPTT of PNIPAM, the emulsions are stable during the reaction process. Therefore, the monodispersity of the hollow PNIPAM microcapsules is almost the same as that of the emulsions template. Before and after the polymerization, the particle size dispersal coefficient δ (Eq. 6.2) is maintained at about 0.27:

$$\delta = \frac{D_{90} - D_{10}}{D_{50}} \quad (6.2)$$

where D_n ($n = 10, 50, 90$) denotes the cumulative number percentage of particles with diameter up to D_n and n is the percentage. The smaller the value of δ is, the narrower the size distribution.

The prepared PNIPAM hollow microcapsules exhibit a reversibly excellent thermo-responsive characteristic and fast response to environmental temperature. When the temperature is changed from 15 °C to 45 °C and from 45 °C to 15 °C repeatedly, the prepared hollow PNIPAM microcapsules show good reversibility in the shrinking/swelling volume changes. Figure 6.7a shows the temperature-dependent equilibrium diameter change of resultant hollow PNIPAM microcapsules. The diameter of microcapsules decreases significantly with increasing the temperature near the VPTT, but no significant diameter change occurs anymore when the temperature is above 38 °C. Figure 6.7b shows the time-dependent volume change of the PNIPAM microcapsules when the temperature increases from 20 °C to 40 °C within 30 s. Microcapsules shrink rapidly within 60 s and then reach their equilibrium swelling/shrinking states. The equilibrium volume of microcapsules at 20 °C is about 17 times as large as that at 40 °C. The results indicate that the prepared microcapsules are featured with fast response to environmental temperature, which is just needed for most potential applications of microgels.

This approach exhibits great advantages in preparing monodisperse thermo-responsive microcapsules for encapsulating bioactive materials or drugs which require the mild encapsulation conditions, because of the flexibility in choosing substances being dissolved in the water phase. The fabrication methodology demonstrated in this study provided a unique approach for preparing monodisperse hollow polymeric microcapsules with W/O single emulsions, which can be easily obtained from some advanced emulsification methods, such as membrane emulsification, microchannel emulsification, microfluidic emulsification, and so on. Furthermore, by using the preparation methodology proposed in this study, the preparation of monodisperse hollow polymeric microcapsules can also be easily carried out continuously in microfluidic devices.

The microfluidic emulsification followed by free radical polymerization is employed to fabricate the thermo-responsive PNIPAM microcapsules [19]. The multiple emulsion produced by microfluidic technique can perform a polymerization reaction in a specific layer since each fluid layer is sandwiched between two immiscible fluids. A water-in-oil-in-water-in-oil (W/O/W/O) triple emulsion is formed and the monomer, cross-linker, and initiator are added in the outer aqueous layer. An accelerator is added to the inner oil phase, where it diffuses into the outer aqueous shell and speeds polymerization of the microcapsule. An as-prepared microcapsule consisting of a shell of thermo-responsive hydrogel encapsulates an oil drop containing several water droplets (Fig. 6.8a). Upon heating from 25 °C

Fig. 6.7 Temperature-dependent equilibrium diameter change (**a**) and time-dependent volume change (**b**) of hollow PNIPAM microcapsules (Reproduced with permission from Ref. [26], Copyright (2007), Elsevier)

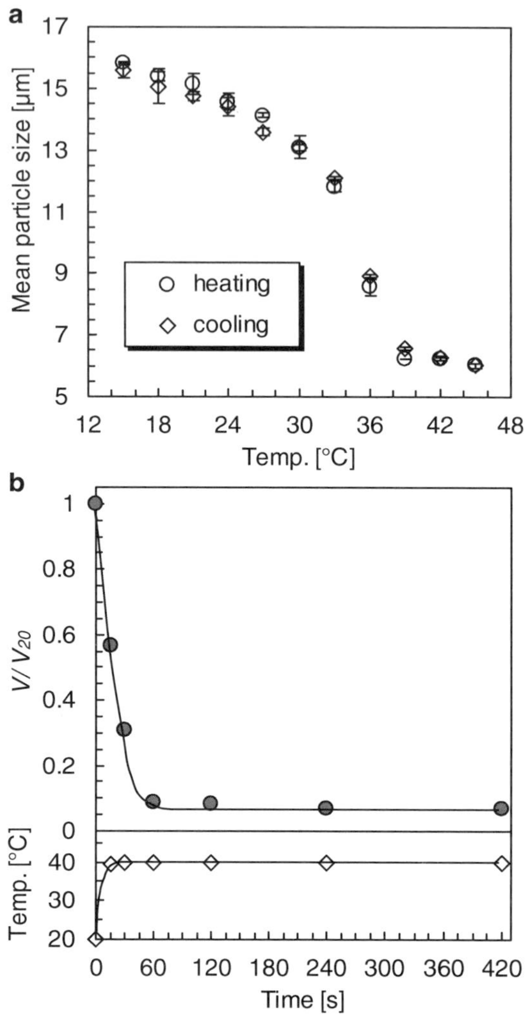

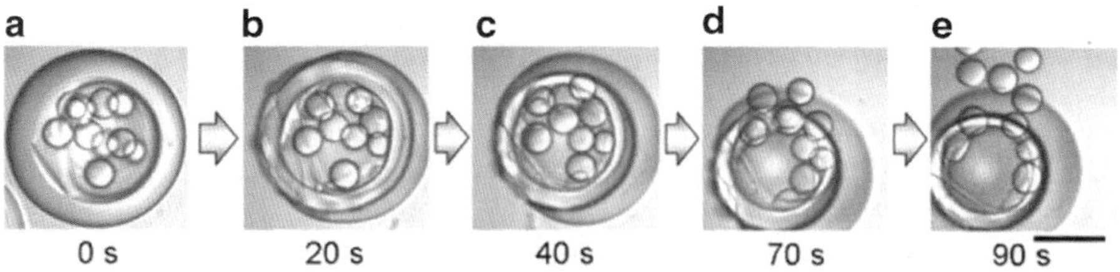

Fig. 6.8 Thermo-responsive microcapsule for pulsed release. The scale bar is 200 mm (Reproduced with permission from Ref. [19], Copyright (2007), Wiley-VCH Verlag GmbH & Co. KGaA)

to 50 °C, the thermo-responsive microcapsule rapidly shrinks by expelling water; however, because of the incompressibility of the inner oil, the PNIPAM shell breaks, providing spontaneous, pulsed release of the innermost water droplets into the continuous oil phase, as shown in Fig. 6.8b–e. This structure has a Trojan horse-like behavior, protecting the innermost water droplets in the PNIPAM shell until their thermo-triggered release. This experiment demonstrates the utility of our technique to generate highly controlled capsules with multiple internal volumes that remain separate from each other; it also highlights the potential of this microfluidic device to create highly engineered structures for controlled release of active substances. Further refinements could adjust the thickness of the layers and the number of droplets, thus enabling fine control over diffusion of compounds contained within the innermost droplets, which would facilitate their highly controlled release.

The high degree of control and scalability afforded by microfluidic technique makes it a flexible and promising route for engineering highly controlled multiple emulsions and microcapsules. It can be used to optimize multiple emulsion systems for a broad range of applications in pharmaceutics, foods, cosmetics, and separations. Moreover, its generality will enable fabrication of novel materials containing complex internal structures. Future work must focus on refining methods proposed herein to allow the full potential of the technique to be realized.

A thermo-triggered squirting microcapsule is designed to deliver nanoparticles inspired by the squirting cucumber ejecting its seeds [27]. The ripe fruit of *Ecballium elaterium* (Fig. 6.9a), also called squirting cucumber or exploding cucumber, is highly turgid. Due to its own ripeness or being disturbed by sniffing animals or whatsoever, the ripe fruit squirts a stream of mucilaginous liquid containing its seeds into air for a considerable long distance by sudden contraction of the wall of the fruit (Fig. 6.9b). The proposed microcapsule is composed of a cross-linked PNIPAM hydrogel shell and encapsulates nanoparticles by emulsifying the nanoparticle aqueous suspension in the water-in-oil (W/O) emulsion core (Fig. 6.9c). Because the encapsulated nanoparticles exist in the water phase of the W/O emulsion core inside the microcapsule, the swollen and hydrophilic PNIPAM hydrogel membrane of the microcapsule can protect the encapsulated nanoparticles when the temperature is below the VPTT (Fig. 6.9c). Upon heating, the PNIPAM hydrogel shell rapidly shrinks, which results in a sudden increase of the liquid pressure inside the microcapsule because both the continuous oil phase and the dispersed water phase in the capsule are incompressible. When the internal pressure increases to a critical value, the PNIPAM hydrogel shell ruptures suddenly, due to its limited mechanical strength. At the same time, the encapsulated nanoparticles are squirted from the microcapsule together with the oil phase stream into the environment with a high momentum (Fig. 6.9d), just like the seed-ejecting of ripe squirting cucumber.

The squirting of nanoparticles from the prepared microcapsules upon heating is observed. When the environmental temperature increases from 20 °C to 50 °C, the hydrogel shell of the microcapsule shrinks rapidly. The inner oil phase cannot permeate through the shrinking hydrogel membrane, leading to deformation of the microcapsule. During the deformation in the thermo-triggered squirting process,

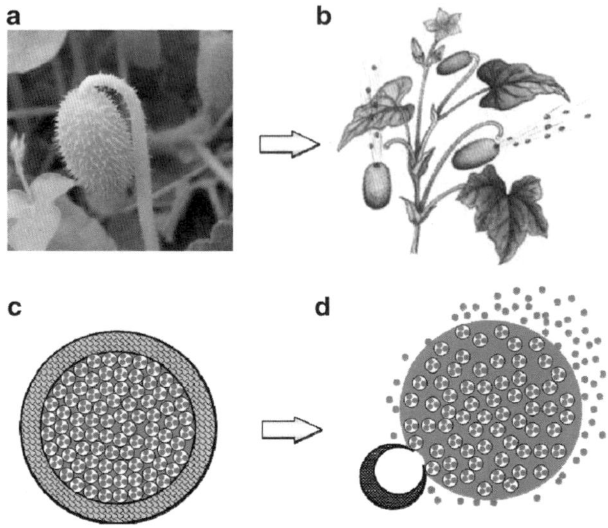

Fig. 6.9 (**a**) A picture of squirting cucumber, (**b**) a schematic illustration of squirting cucumbers ejecting seeds together with a stream of mucilaginous liquid, (**c**) a microcapsule with cross-linked PNIPAM hydrogel shell containing nanoparticles in the inner water phase of W/O emulsion core at a temperature below the VPTT, (**d**) nanoparticles being squirted out from the microcapsule together with the oil phase stream due to the dramatic shrinkage and sudden rupture of the PNIPAM hydrogel shell triggered by increasing the environmental temperature above the VPTT (Reproduced with permission from Ref. [27], Copyright (2010), Royal Society of Chemistry)

the encapsulated W/O phase tends to breach the thinner side of the hydrogel membrane, which is stretched by the incompressible inner oil phase. When the shrinkage reaches a high degree, the hydrogel membrane turns into an "8" shape, of which one head (the side with thinner membrane) is full of the encapsulated W/O primary emulsion and the hydrogel shell becomes extremely thin. When the inner pressure reaches a critical value, the hydrogel membrane ruptures and the contained oil phase, together with the encapsulated nanoparticles, is squirted out to the surrounding water (Fig. 6.10a). Figure 6.10b shows the snapshots of the thermo-triggered squirting process in dark field. Because the squirting direction is upwards, the "8" deformation is not as obvious as that shown in Fig. 6.10a. During the squirting process, the large bright area indicates the considerable wide distribution of the squirted substance. The squirting capsule can completely squirt out the encapsulated nanoparticles with a high momentum, just like a nanoparticle bomb. It provides a novel mode for drug and diagnostic reagent delivery systems. As a temperature stimulus is convenient to manipulate and can be generated remotely, the site-specific targeted delivery of nanoparticles can be achieved with such microcapsules.

To make these squirting microcapsules more suitable for practical use, the VPTT of the PNIPAM hydrogel shell can be easily adjusted by simply copolymerizing the NIPAM monomer with hydrophilic or hydrophobic monomers. If

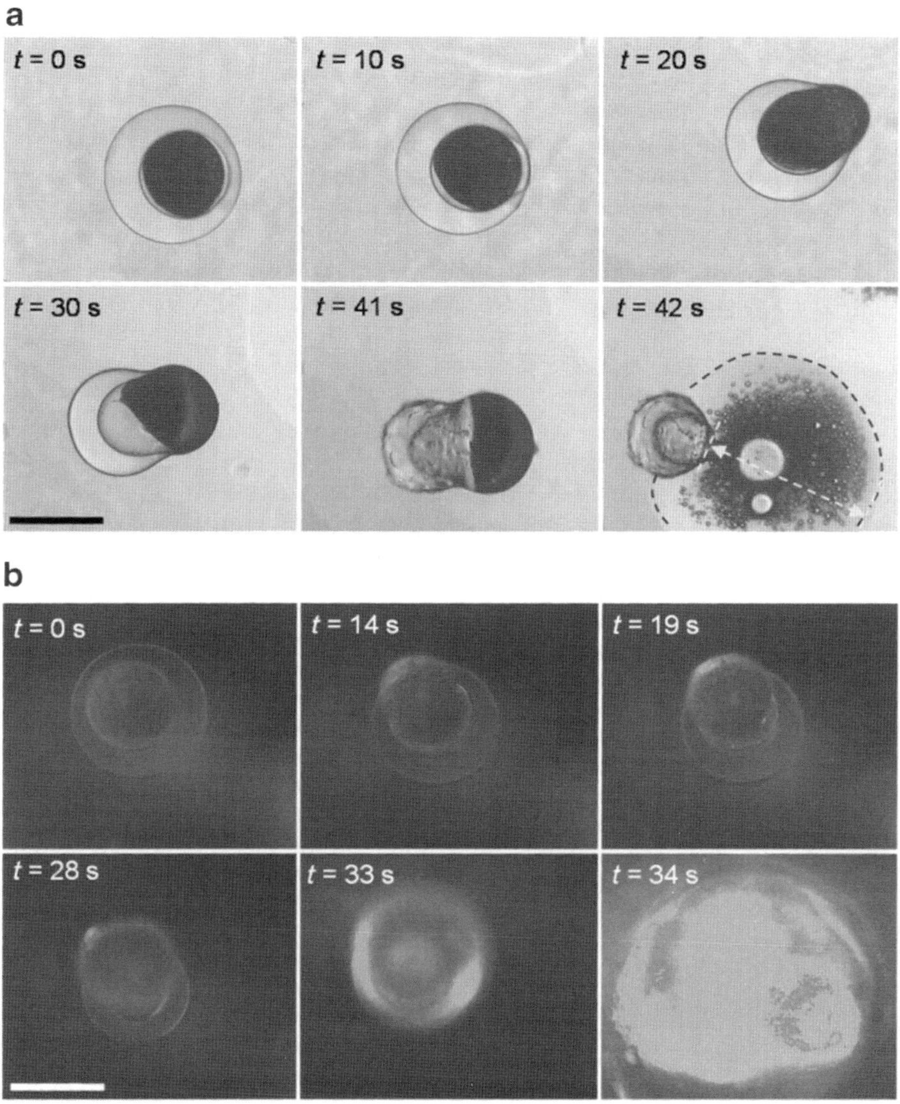

Fig. 6.10 Bright-field (**a**) and dark-field (**b**) microscope snapshots of thermo-triggered squirting of nanoparticles from microcapsules by increasing the environmental temperature from 20 °C to 50 °C. The scale bars are 200 mm (Reproduced with permission from Ref. [27], Copyright (2010), Royal Society of Chemistry)

the VPTT is several degrees higher than the normal physiological temperature, therefore our proposed microcapsules will remain in a swollen state and protect the inner encapsulated substances from leaking under physiological temperature. The slight temperature fluctuation within the physiological range will never cause microcapsule rupture. Upon site-specific local thermotherapy, for instance, local treatment with infrared irradiation, microwave or ultrasound, the site-specific local temperature is increased to be higher than the VPTT of the microcapsule shell materials. As a result, the hydrogel capsule shell undergoes volume phase transition, which causes the squirting delivery of nanoparticles from the microcapsule on demand. The applications of these systems to biomedical fields would require a smaller capsule size, which can be achieved with developing the microfluidic and nanofluidic techniques.

Stimuli-responsive microcapsules functionalized with magnetic nanoparticles have been designed for magnetic-guided drug delivery and subsequent controlled drug release by an external trigger such as temperature, pH, ultrasonic, and high frequency magnetic field. Most of the stimuli-responsive carriers functionalized with magnetic nanoparticles are designed for drug delivery of hydrophilic drugs [11, 24, 28, 29], while it is worth noting that currently available drugs such as anticancer drugs are usually lipophilic molecules. Therefore, design of carriers for lipophilic drugs is of great importance and necessity. A novel type of monodisperse thermo-triggered self-bursting microcapsules with oil cores for encapsulating lipophilic substances is developed by microfluidic technique [30]. The thermo-responsive PNIPAM polymeric shell embedded with superparamagnetic Fe_3O_4 nanoparticles. It enables not only magnetic-guided targeting but also thermo-triggered rapid and complete burst release of encapsulated lipophilic chemicals, and there is no leakage of encapsulated substances at all before the thermo-triggering.

Figure 6.11 is a series of snapshots showing the magnetic-guided targeting performance of Sudan III-loaded microcapsules from site A to site B. The microcapsules are randomly dispersed in deionized water at 20 °C at the beginning (Fig. 6.11a). When a magnet is placed under the Petri dish, the microcapsules are attracted together in site A (Fig. 6.11b). After that, aggregated microcapsules as a whole are moved quickly following the arrows from site A to site B under the magnetic guide and finally trapped in the targeted site B (Fig. 6.11b–f). Such a magnetic-responsive property can make the microcapsules achieve the purpose of site and/or route-specific targeting drug delivery.

Once the as-prepared microcapsules are specifically delivered to the desired site, the release of the encapsulated chemicals can be triggered by local heating. Figure 6.12a illustrates the burst release of the inner oil core from the microcapsule when temperature is increased from 20 °C to 60 °C. With increase in temperature, thermo-responsive PNIPAM shell of the microcapsule shrinks dramatically. Since the inner oil core is incompressible, the internal pressure in oil core keeps increasing due to the shell shrinkage. The PNIPAM shell finally ruptures because of the limited mechanical strength, which results in burst releasing of the inner oil core (Fig. 6.12(a3), (a4)). With the shrinkage and rupture of PNIPAM shell, the inner oil phase is squeezed out of the microcapsule within a very short time, and spreads fast into the surrounding environment. As a result, the release from such a microcapsule is complete, leaving just a hollow cavity without any leftover inside the microcapsule (Fig. 6.12b). Such a rapid and complete burst release of the encapsulated oil phase and lipophilic chemicals enables that a high local drug

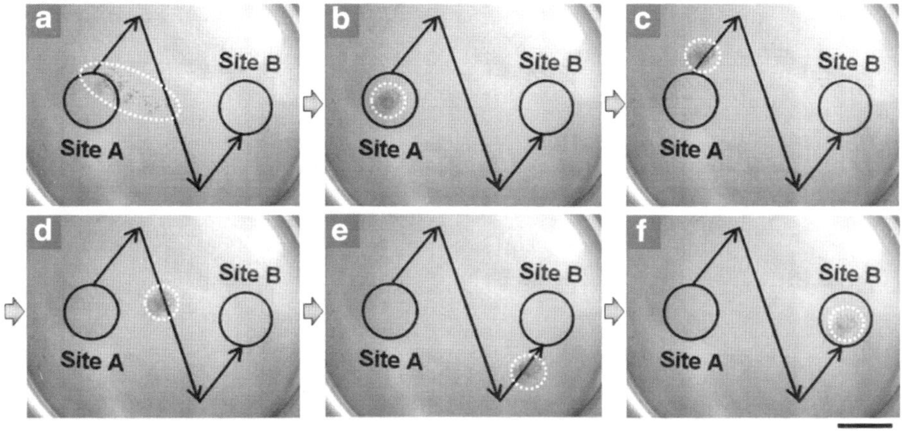

Fig. 6.11 Snapshots of the magnetic-guided targeting behavior of microcapsules loaded with Sudan III in water at 20 °C. A cylindrate NdFeB magnet with size of Φ12 mm × 8 mm is placed under the Petri dish to guide the microcapsules. The magnetic field strength of the magnet is 0.3 T (Reproduced with permission from Ref. [30], Copyright (2009), Wiley-VCH Verlag GmbH & Co. KGaA)

concentration can be rapidly achieved. To determine the release rate of the inner oil core, the burst-release behavior during the first 3.2 s after the PNIPAM shell ruptures is investigated (Fig. 6.12c). The optical microscope snapshots show that the encapsulated oil phase shoots out very quickly due to the strong boost that resulted from shrinkage and squeeze function of microcapsule shell. Besides, radius of the circular edge of released oil phase increases by ~250 μm within 3.2 s. This spread speed is much faster than that in diffusion-driven release systems. Such a quick release and spread rate may make our microcapsules be of specific interest and significance especially in certain cases where released substances need to cross some media with high viscosity or low permeability.

The as-proposed microcapsules can be easily converted into other stimuli-triggered self-bursting ones by simply changing the thermo-responsive shells into other stimuli-responsive ones, such as pH-induced, molecular recognition-induced, and glucose-induced. Besides the application in the field of targeted delivery and controlled release of drugs, the proposed multifunctional microcapsules can find myriad applications in various fields. For example, these microcapsules can be applied to site- and/or route-specifically transport and release lipophilic functional substances such as corrosion inhibitors, self-healing agents, and lubricants and other chemicals to certain sites, even to some hand-unreachable micro-spaces. Different reagents in a chemical reaction, which could react with each other once they meet, can also be targetedly transported by such microcapsules to a predefined spot where reaction starts after stimuli-triggered burst release. Moreover, development in microfluidic technology gives us the opportunity to prepare emulsions with tunable sizes, so microcapsules with different sizes and thicknesses can be obtained

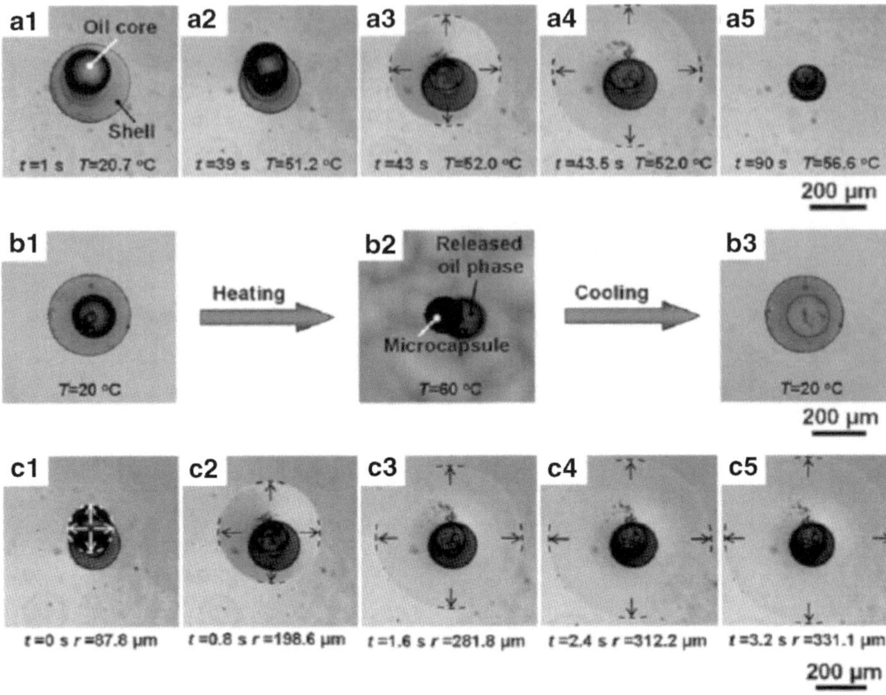

Fig. 6.12 Optical microscope snapshots of the thermo-triggered self-bursting behavior (**a**) and complete release performance (**b**) of the as-prepared microcapsules. (**c**) Thermo-triggered burst-release process of inner oil core from the microcapsule [shown in (**a**)] during the first 3.2 s of release, in which time-dependent increase of the radius (*r*) of the circle edge of released oil phase is used to valuate the release rate (Reproduced with permission from Ref. [30], Copyright (2009), Wiley-VCH Verlag GmbH & Co. KGaA)

by further adjusting the microchannel dimensions of the microfluidic devices and the fluid flow rates. In a nutshell, the proposed thermo-triggered self-bursting microcapsules with magnetic-guided targeting property provide a novel mode for controlled release and can be applied in a broad range of applications.

6.5 Summary

The thermo-responsive microcapsules with grafted PNIPAM chains or embedded PNIPAM microgels as gates as well as those with whole PNIPAM membranes are successfully fabricated. As for the microcapsules with grafted PNIPAM chains or embedded PNIPAM microgels as gates, the thermo-responsive controlled-release characteristics of solute molecules are satisfactorily reversible and reproducible. Similar to the results of flat membranes, the proposed PNIPAM-grafted microcapsules show a positive thermo-responsive controlled-release mode at low grafting yields of PNIPAM; however, they display a negative thermo-responsive one at high grafting yields. For Ca-alginate capsules with PNIPAM microgels as smart gates, the permeability coefficients of solute can be adjusted by the content of the PNIPAM microgels embedded in the capsules. The optimum thermo-responsive gating coefficient is achieved by the solute molecules whose molecule size is matched with the gate size of capsule membrane during the transmembrane diffusion. As far as the microcapsules with whole PNIPAM membranes are concerned, they exhibit a rapid and complete burst-release mode of encapsulated chemicals which leads to a high local drug concentration within a short time at the target site. These thermo-responsive microcapsules are highly attractive for various promising application, such as site-targeting drug delivery system, controlled release of chemicals, microreactors, biomedical and/or chemical sensors, immobilization of cells and enzymes, and encapsulation of foods and cosmetics.

References

1. Sukhorukov G, Fery A, Möhwald H (2005) Intelligent micro- and nanocapsules. Prog Polym Sci 30:885–897
2. Chu LY, Yamaguchi T, Nakao S (2002) A molecular-recognition microcapsule for environmental stimuli-responsive controlled release. Adv Mater 14:386–389
3. Shiga K, Muramatsu N, Kondo T (1996) Preparation of poly(D, L-lactide) and copoly(lactide-glucolide) microspheres of uniform size. J Pharm Pharmacol 48:891–895
4. Chu LY, Park SH, Yamaguchi T et al (2002) Preparation of small-sized monodispersed thermo-responsive core-shell microcapsules. Langmuir 18:1856–1864
5. Chu LY, Park SH, Yamaguchi T et al (2001) Preparation of thermo-responsive core-shell microcapsules with a porous membrane and poly(N-isopropylacrylamide) gates. J Membr Sci 192:27–39
6. Okahata Y, Noguchi H, Seki T (1986) Thermoselective permeation from a polymer-grafted capsule membrane. Macromolecules 19:493–494
7. Ichikawa H, Fukumori Y (2000) A novel positively thermosensitive controlled-release microcapsule with membrane of nano-sized poly(N-isopropylacrylamide) gel dispersed in ethylcellulose matrix. J Control Release 63:107–119
8. Kono K, Okabe H, Morimoto K et al (2000) Temperature-dependent permeability of polyelectrolyte complex capsule membranes having *N*-isopropylacrylamide domains. J Appl Polym Sci 77:2703–2710
9. Chu LY, Niitsuma T, Yamaguchi T et al (2003) Thermo-responsive transport through porous membranes with grafted PNIPAM gates. AIChE J 49:896–909
10. Zha LS, Zhang Y, Yang WL et al (2002) Monodisperse temperature-sensitive microcontainers. Adv Mater 14:1090–1092
11. Gao HF, Yang WL, Min K et al (2005) Thermosensitive poly(*N*-isopropylacrylamide) nanocapsules with controlled permeability. Polymer 46:1087–1093
12. Glinel K, Sukhorukov GB, Möhwald H et al (2003) Thermosensitive hollow capsules based on thermoresponsive polyelectrolytes. Macromol Chem Phys 204:1784–1790
13. Zhang YW, Jiang M, Zhao JG et al (2005) A novel route to thermosensitive polymeric core-shell aggregates and hollow spheres in aqueous media. Adv Funct Mater 15:695–699
14. Nayak S, Gan DJ, Serpe MJ et al (2005) Hollow thermoresponsive microgels. Small 1:416–421

15. Sun QH, Deng YL (2005) In situ synthesis of temperature-sensitive hollow microspheres via interfacial polymerization. J Am Chem Soc 127:8274–8275
16. Kandori K, Kishi K, Ishikawa T (1991) Preparation of monodispersed W/O emulsions by Shirasu-porous-glass filter emulsification technique. Colloids Surf 55:73–78
17. Chu LY, Xie R, Zhu JH et al (2003) Study of SPG membrane emulsification processes for the preparation of monodisperse core-shell microcapsules. J Colloid Interface Sci 265:187–196
18. Cheng CJ, Chu LY, Xie R (2006) Preparation of highly monodisperse W/O emulsions with hydrophobically modified SPG membranes. J Colloid Interface Sci 300:375–382
19. Chu LY, Utada AS, Shah RK et al (2007) Controllable monodisperse multiple emulsions. Angew Chem Int Ed 46:8970–8974
20. Li Y, Chu LY, Zhu JH et al (2004) Thermoresponsive gating characteristics of poly(N-isopropylacrylamide)-grafted porous polyvinylidene fluoride membranes. Ind Eng Chem Res 43:2643–2649
21. Chu LY, Zhu JH, Chen WM et al (2003) Effect of graft yield on the thermo-responsive permeability through porous membranes with plasma-grafted poly(*N*-isopropylacrylamide) gates. Chin J Chem Eng 11:269–275
22. Peng T, Cheng YL (1998) Temperature-responsive permeability of porous PNIPAAm-g-PE membranes. J Appl Polym Sci 70:2133–2142
23. Yamaguchi T, Nakao S, Kimura S (1991) Plasma-graft filling polymerization: preparation of a new type of pervaporation membrane for organic liquid mixtures. Macromolecules 24:5522–5527
24. Yang WC, Xie R, Pang XQ et al (2008) Preparation and characterization of dual stimuli-responsive microcapsules with a superparamagnetic porous membrane and thermo-responsive gates. J Membr Sci 321:324–330
25. Wang JY, Jin Y, Xie R et al (2011) Novel calcium-alginate capsules with aqueous core and thermo-responsive membrane. J Colloid Interface Sci 353:61–68
26. Cheng CJ, Chu LY, Ren PW et al (2007) Preparation of monodisperse thermo-sensitive poly(N-isopropylacrylamide) hollow microcapsules. J Colloid Interface Sci 313:383–388
27. Liu L, Wang W, Ju XJ et al (2010) Smart thermo-triggered squirting capsules for nanoparticle delivery. Soft Matter 6:3759–3763
28. Soppimath KS, Liu LH, Seow WY et al (2007) Multifunctional core/shell nanoparticles self-assembled from pH-induced thermosensitive polymers for targeted intracellular anticancer drug delivery. Adv Funct Mater 17:355–362
29. Satarkar NS, Hilt JZ (2008) Magnetic hydrogel nanocomposites for remote controlled pulsatile drug release. J Control Release 130:246–251
30. Wang W, Liu L, Ju XJ et al (2009) A novel thermo-induced self-bursting microcapsule with magnetic-targeting property. Chem Phys Chem 10:2405–2409

Part II
pH-Responsive Hydrogel Functional Materials

Chapter 7
Preparation and Properties of Monodisperse pH-Responsive Microgels

Abstract In this chapter, the design, preparation, and performance of various monodisperse pH-responsive microgels with pH-responsive swelling/shrinking property or pH-responsive stability are introduced. Monodisperse pH-responsive chitosan microspheres and microcapsules with acid-triggered swelling and decomposition properties are developed using uniform-sized water-in-oil (W/O) emulsions as preparation templates via microfluidic approaches. Monodisperse cationic pH-responsive poly(*N*,*N*-dimethylaminoethyl methacrylate) (PDM) microgels are successfully prepared by dispersion polymerization in ethanol/water mixture using poly(vinyl pyrrolidone) (PVP) as the steric stabilizer. Monodisperse cationic pH-responsive PDM microcapsules are prepared via UV-initiated polymerization based on a double initiation system. These monodisperse pH-responsive microgels with good biocompatibility are of great potential as smart drug delivery carriers.

7.1 Introduction

pH change occurs at many specific and/or pathological body sites, such as the stomach, intestine, endosome, lysosome, blood vessels, vagina, and tumor extracellular sites. So, pH variation is an important chemical stimulus for stimuli-responsive "intelligent" materials in biomedical applications. For example, there is an obvious pH change along the gastrointestinal tract from the stomach ($pH = 1 \sim 3$) to intestine ($pH = 5 \sim 8$). Moreover, there are also more subtle pH changes within different tissues. Certain tumors as well as inflamed or wound tissues exhibit a pH value different from 7.4 as it is in circulation. For example, chronic wounds have been reported to have pH values between 7.4 and 5.4 [1], and tumor tissue is also reported to be acidic extracellularly [2, 3]. Therefore, pH-responsive hydrogel functional materials have attracted considerable interest in biomedical fields, especially for controlled release of drug at desired sites for greater efficacy.

L.-Y. Chu et al., *Smart Hydrogel Functional Materials*,
DOI 10.1007/978-3-642-39538-3_7,

© Chemical Industry Press, Beijing and Springer Berlin Heidelberg 2013

It is well known that all pH-responsive materials with pH-dependent swelling properties contain either acidic or basic groups, which respond to changes in environmental pH by gaining or losing protons. For example, hydrogels with carboxylic groups are known to have deswelling properties corresponding to the acidic pH sensitivity of the isoelectric point of the carboxylic acid.

Recently, the authors' group has designed and developed several kinds of monodisperse pH-responsive microgels, such as chitosan microspheres, chitosan microcapsules, and cationic poly(*N*,*N*-dimethylaminoethyl methacrylate) (PDM) microgels and microcapsules. In this chapter, the design, preparation, and pH-responsive properties of these monodisperse chitosan and PDM microgels will be introduced.

7.2 Monodisperse pH-Responsive Chitosan Microgels

Up to now, a lot of researches have been carried out on anionic pH-responsive microgels based on poly(acrylic acid) and poly(methacrylic acid) [4, 5]. Such microgels are capable of swelling at high pH and shrinking at low pH because of the carboxyl groups being ionized at high pH and unionized at low pH. Very little research on cationic pH-responsive microgels has been reported. Whereas, the cationic pH-responsive microgels have pH-responsive swelling property in acidic condition due to the protonation, which is preferred in many circumstances [6, 7].

Chitosan is an amino-polysaccharide obtained by alkaline deacetylation of chitin, a natural component of shrimp or crab shells. Due to its excellent biological activity, good biocompatibility and biodegradability, and antiulcer and antimicrobial properties, chitosan is receiving great interest for biomedical and pharmaceutical applications [8–11]. As a cationic polysaccharide with a p*K*a value around $6.2 \sim 7.0$, chitosan has the ability to form hydrogels with a pH-responsive volume change [12]. When ambient pH is lower than the p*K*a value, the chitosan hydrogel swells due to the protonation of its free amino groups. As shown in Fig. 7.1, the cross-linked chitosan hydrogels can be prepared using terephthalaldehyde as cross-linker via formation of a Schiff base bonding between amino groups of chitosan and aldehyde groups of terephthalaldehyde in neutral medium. The Schiff base bondings between chitosan and terephthalaldehyde also possess an interesting pH-responsive stability.

The microfluidic technique provides a facile method to produce monodisperse single or double emulsions which could be utilized as formation templates to prepare microspheres and microcapsules [13–17]. The authors' group prepared monodisperse pH-responsive chitosan microspheres using uniform-sized water-in-oil (W/O) emulsions as the formation templates. As illustrated in Fig. 7.2, the highly monodisperse W/O emulsion templates are fabricated by capillary microfluidic technique according to the published method [17], which guarantees good reproducibility and precise controllability of the emulsion size. In the W/O emulsion templates, chitosan is dissolved in the inner water phase, and the outer oil phase contains oil-soluble terephthalaldehyde as cross-linker. The chitosan microsphere is

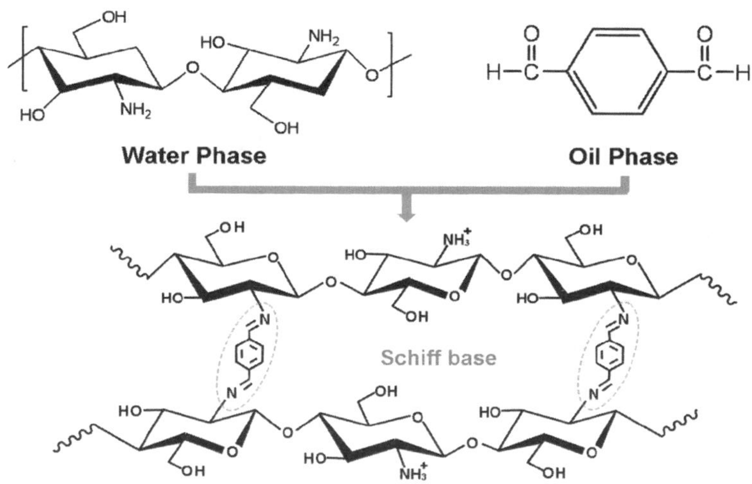

Fig. 7.1 Schematic illustration of chitosan hydrogel with terephthalaldehyde as cross-linker

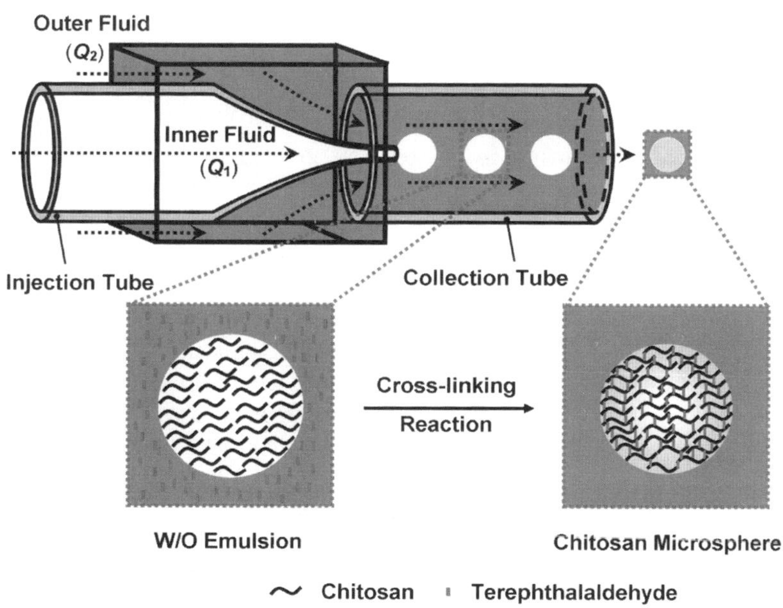

Fig. 7.2 Schematic illustration of capillary microfluidic device for preparation of monodisperse W/O emulsions and the formation process of the chitosan microsphere

formed via the cross-linking reaction at the W/O interface as soon as the inner water fluid contacts the outer oil fluid in the collection tube of the microfluidic device. In the formation process, 2.0 wt % water-soluble chitosan is dissolved in water as the inner water phase using 1 mol/L NaOH to adjust its pH value to 6.7. A mixture of

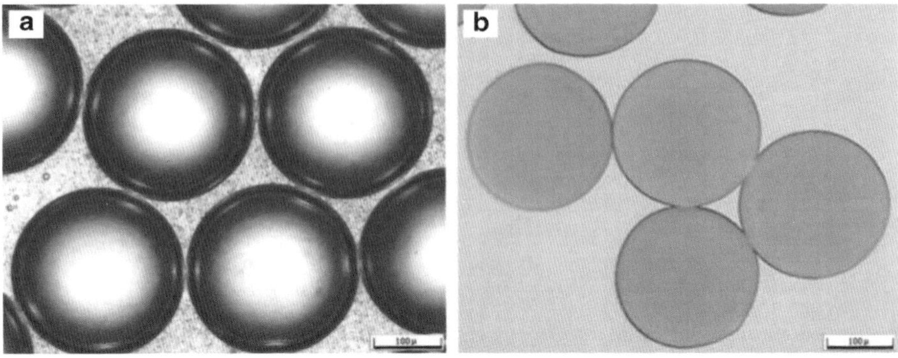

Fig. 7.3 Optical photos of the W/O emulsions (**a**) and the prepared chitosan microspheres (**b**) dispersed in water. The scale bars are 100 μm

soybean oil and benzyl benzoate (1:1, v/v) containing 2.0 wt % terephthalaldehyde and 8.0 wt % polyglycerol polyricinoleate (PGPR) is used as the outer oil phase. The flow rates of the inner and outer fluids are $Q_1 = 600$ μL/h and $Q_2 = 2{,}000$ μL/h, respectively. The production of W/O emulsions and subsequent interfacial cross-linking reaction are all performed at 20 °C, and the cross-linking reaction is carried out for 30 min.

Figure 7.3 shows the optical photos of the highly monodisperse W/O emulsions and the prepared chitosan microspheres. The cross-linking stability of prepared chitosan microspheres closely depends on environmental pH. In neutral medium, the microspheres maintain good spherical shape and structural integrity, as shown in Fig. 7.3b. Whereas in acidic environment with low pH value, the amino groups of chitosan are protonated and positively charged, so the intramolecular electrostatic repulsion and enhanced hydrophilicity make the chitosan microspheres swell. With the protonation of the amino groups of chitosan going on, the Schiff base bondings become instable, and finally the cross-linked chitosan microspheres are decomposed in acidic environment as shown in Fig. 7.4. The green fluorescence of chitosan microspheres comes from the Schiff base bondings with autofluorescent properties [18, 19].

The size of the chitosan microspheres mentioned above is almost above 100 μm in diameter, which is too big to apply as drug delivery system transporting in blood vessels. Chitosan microcapsules with small size (<50 μm) are fabricated using W/O emulsions as templates by a simple microfluidic technique. The microfluidic device simply based on cover slips and microscope glass slides according to the authors' reported method [20] is designed with an expanding nozzle to generate monodisperse picoliter-sized W/O emulsions. Figure 7.5a illustrates the fabricating method of the simple microfluidic device, and the optical image and schematic of the shear focusing flow-focus device for O/W emulsions formation are shown in Fig. 7.5b, c. The width of the expanding nozzle is about 60 μm and the surface wettability of microchannels in the devices is modified to be hydrophobic by coating self-assembled monolayer (SAM) with chlorotrimethylsilane on the channel

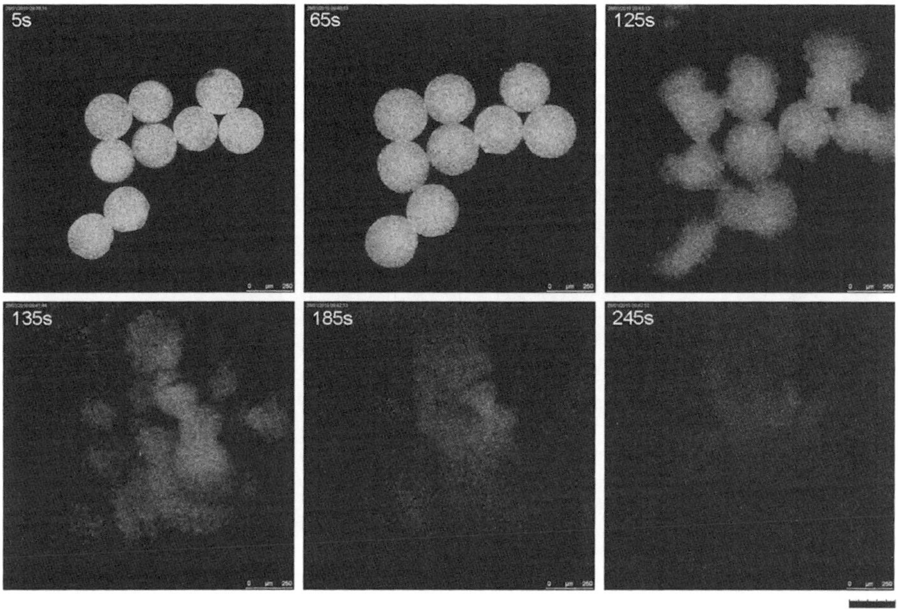

Fig. 7.4 The confocal laser scanning microscopy (CLSM) images of the dissolving process of chitosan microspheres in an acidic buffer solution with pH = 2.9 at 25 °C. The scale bars are 250 μm

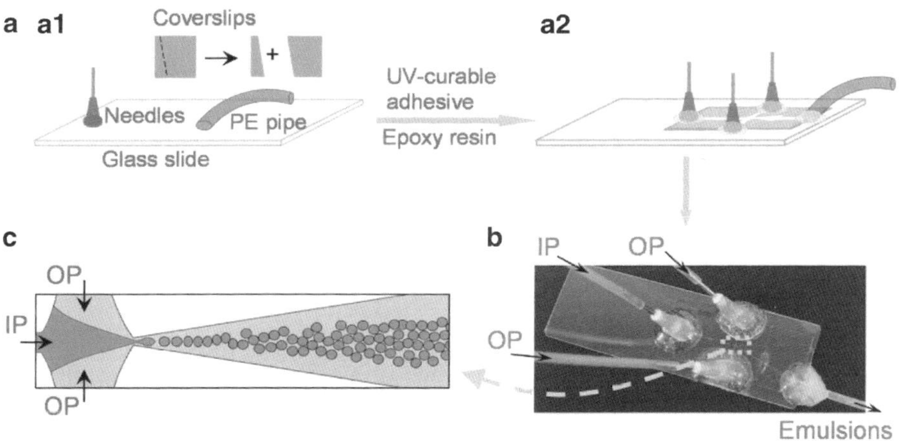

Fig. 7.5 (**a**) Schematic illustration for fabricating the simple microfluidic device based on microscope glass slides and coverslips; (**b**) optical image and schematic illustration (**c**) of the shear focusing flow-focus device for W/O emulsions formation

surface. For the preparation of W/O emulsions, 2.0 wt % water-soluble chitosan and 0.5 wt % hydroxyethylcellulose are dissolved into deionized water as the inner water phase, soybean oil containing 0.05 ~ 1.0 wt % terephthalaldehyde as cross-linker and 4.0 wt % PGPR is used as the oil phase. Because terephthalaldehyde has

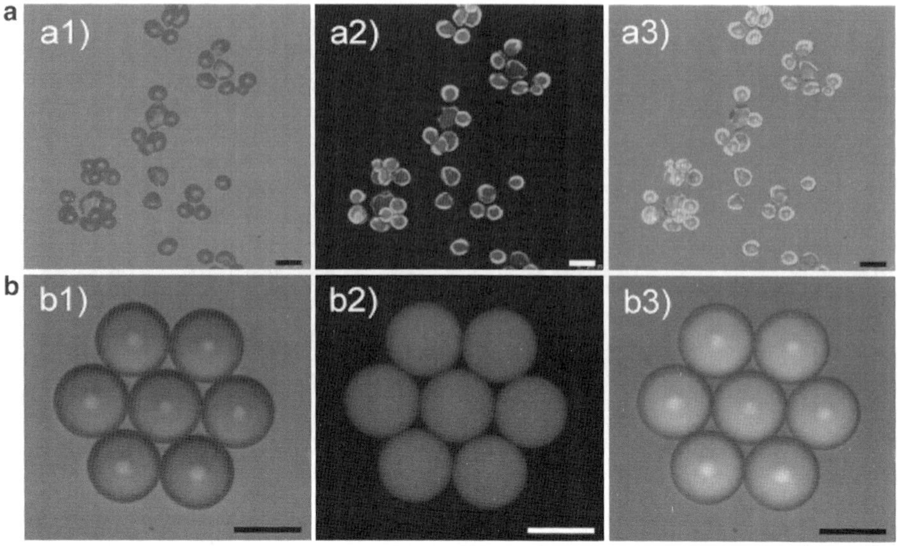

Fig. 7.6 CLSM images of the chitosan microgels with different cross-linking degrees: (**a**) $C_{PA} = 0.1$ wt %, $Q_{IP} = 100$ μL/h, $Q_{OP} = 1{,}000$ μL/h; (**b**) $C_{PA} = 0.5$ wt %, $Q_{IP} = 100$ μL/h, $Q_{OP} = 1{,}000$ μL/h. (*a1*) and (*b1*): transmission channel, (*a2*) and (*b2*): green fluorescent channel, (*a3*) and (*b3*): overlay channel. All the bars are 25 μm

a finite solubility in inner water phase, the chitosan microcapsules could be formed via diffusion of terephthalaldehyde from outer oil phase to inner chitosan droplets in neutral medium to form the Schiff base bondings.

Different chitosan microgels are prepared by controlling the cross-linker concentration (C_{PA}) in the outer oil phase at same gelation time. The morphology of the chitosan microgel changes from hollow capsule to solid sphere as C_{PA} increases. Figure 7.6 shows the CLSM images of the chitosan microgels with different cross-linking degree after water evaporation. Obviously, the microgels with low cross-linking degree ($C_{PA} = 0.1$ wt %) have a hollow capsule structure (Fig. 7.6a), while the microgels with high cross-linking degree ($C_{PA} = 0.5$ wt %) are solid spheres (Fig. 7.6b). Furthermore, it can be clearly seen that these chitosan microgels have small sizes less than 50 μm. These monodisperse chitosan microgels with small size also have good pH-responsive properties in acidic medium like the chitosan microspheres mentioned above, which have a great potential in smart drug delivery systems and high-throughput screening of enzymes and cells.

7.3 Monodisperse Cationic pH-Responsive Microgels

PDM is also a typical cationic pH-responsive material with good biocompatibility. Besides its traditional applications such as water treatment, rubber, and paint, PDM has been used for developing controlled drug delivery systems in recent years owing to its acid-induced swelling characteristics [21–23].

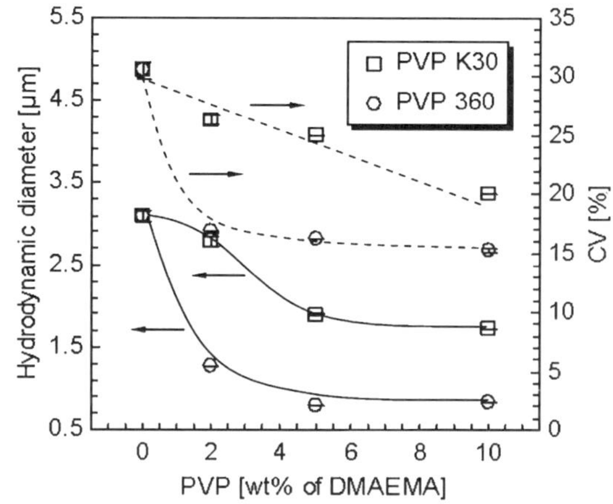

Fig. 7.7 Effects of PVP concentration and molecular weight on the hydrodynamic diameter (*solid line*) and size distribution (*dotted line*) of PDM microgels. The dosages of other ingredients are 0.5 mol/L DM, ethanol/water ratio of 1:9 (v/v), 0.5 wt % MBA (Reproduced with permission from Ref. [24], Copyright (2007), Elsevier)

Monodisperse cationic pH-responsive PDM microgels are successfully prepared by dispersion polymerization in ethanol/water mixture using poly(vinyl pyrrolidone) (PVP) as the steric stabilizer and *N*,*N*′-methylenebisacrylamide (MBA) as the cross-linker [24]. The control of microgel size and size distribution is achieved by properly varying the polymerization parameters. The research results show that increase of the polarity of the reaction system by increasing the volume ratio of ethanol to water could increase the size and size distribution of microgels. In addition, at a high concentration of the cross-linker, 2 wt %, macrogels instead of microgels may be formed. Increasing the concentration of *N*,*N*-dimethylaminoethyl methacrylate (DM) monomer could also increase the microgel size; however, the microgels become unstable when the monomer concentration reaches 1.0 mol/L.

Two kinds of PVP (PVP K30 and PVP 360) with different concentrations are used as the steric stabilizer to prepare PDM microgels. As shown in Fig. 7.7, it is demonstrated that increase of the concentration and the molecular weight of the PVP stabilizer could decrease the size and size distribution of the microgels [24]. The results can be explained as follows. The higher the concentration of stabilizer is, the faster the stabilizer is adsorbed, and consequently the larger the amount of the adsorbed stabilizer is. Hence, for a given duration, a greater number of smaller particles are stabilized during the primary stabilization process. On the other hand, increasing PVP concentration can increase the viscosity of the medium, which leads to a retardation of particles coagulation. Higher molecular weight of PVP is meant to increase the viscosity of the medium and improve the stabilization by longer chains of PVP [25]. Therefore, the size of microgels decreases and the size distribution becomes more monodisperse with an increase in the molecular weight of PVP.

In order to examine the pH-dependent stability of PDM microgels, zeta potentials of these PDM microgels are measured at different pH values as shown in Fig. 7.8 [24]. It is determined that the isoelectric point (IEP) of PDM microgels is at around pH 6. At pH lower than 5, the zeta potential of microgels is positive, as expected, due to protonation of the amine groups of PDM. The negative zeta potentials are determined at pH values of 7–11. This may be caused by deprotonation of

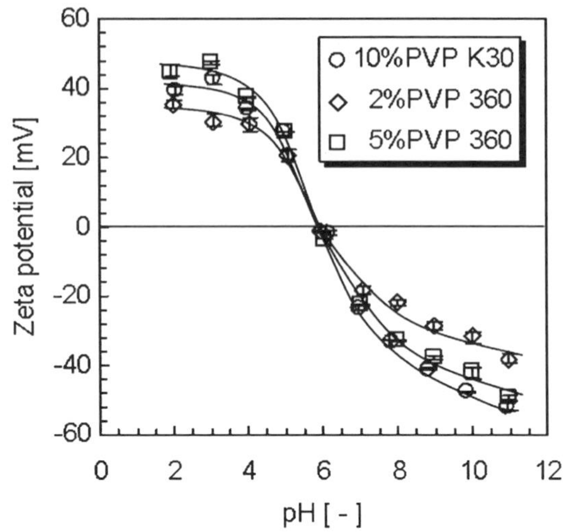

Fig. 7.8 Zeta potentials of PDM microgels prepared with 10 wt % PVP K30 (○), 2 wt % PVP 360 (◇), and 5 wt % PVP 360 (□) at different pH values at 25 °C. The dosages of other ingredients are 0.5 mol/L DM, ethanol/water ratio of 1:9 (v/v), and 0.5 wt % MBA. All the samples are carried out on purified dilute dispersions (Reproduced with permission from Ref. [24], Copyright (2007), Elsevier)

amine groups in the microgels that leads to overall negative charge. The results demonstrate that the prepared PDM microgels are able to maintain electrostatic stabilization at both low and high pHs. The microgels are partially or totally aggregated at pH values in the range of 5 ~ 7, which spans the IEP of the PDM microgels.

The PDM microgels exhibit excellent pH responsivity and significantly swell at low pH values. Figure 7.9 shows the optical micrographs of PDM microgels with different stabilizer PVP (10 wt % PVP K30, 2 wt % PVP 360, and 5 wt % PVP 360) at pH 2.5 and pH 11 [24]. It can be seen that all the microgels significantly swell at pH 2.5 compared to that at pH 11.

Figure 7.10 shows the hydrodynamic diameters of these PDM microgels at different pH values, and the degree of pH responsivity is determined by the maximum ratio of pH-dependent volume change of PDM microgels (V_{max}/V_{min}) [24]. The selected three types of microgels with different original hydrodynamic diameters have different maximum ratios of volume change at the tested pH range. The microgels prepared with 2 wt % PVP 360 have the largest ratio of 11.7. PDM bearing tertiary amine groups in the side chains can be protonated in acidic solution. The protonation of the amine groups would introduce positive charge to the polymer side chains. As a result, PDM microgels swell in acidic solution by electrostatic repulsion of positively charged amine groups within the gel and hydration of such functional groups. But on the other hand, the conformational entropy elasticity of the cross-linked polymer chains counteracts this swelling. Since the cross-linker is kept constant in all experiments, the large volume tends to consume a larger amount of cross-linker inside the microgels network. Therefore, for large microgels, such as the one prepared with 10 wt % PVP K30, the cross-linked polymer chains might restrict the gel swelling more. Therefore, the microgels prepared with 10 wt % PVP K30 have the lowest degree of pH responsivity among the selected three types of microgels. On the other hand, the small microgels prepared with 5 wt % PVP 360

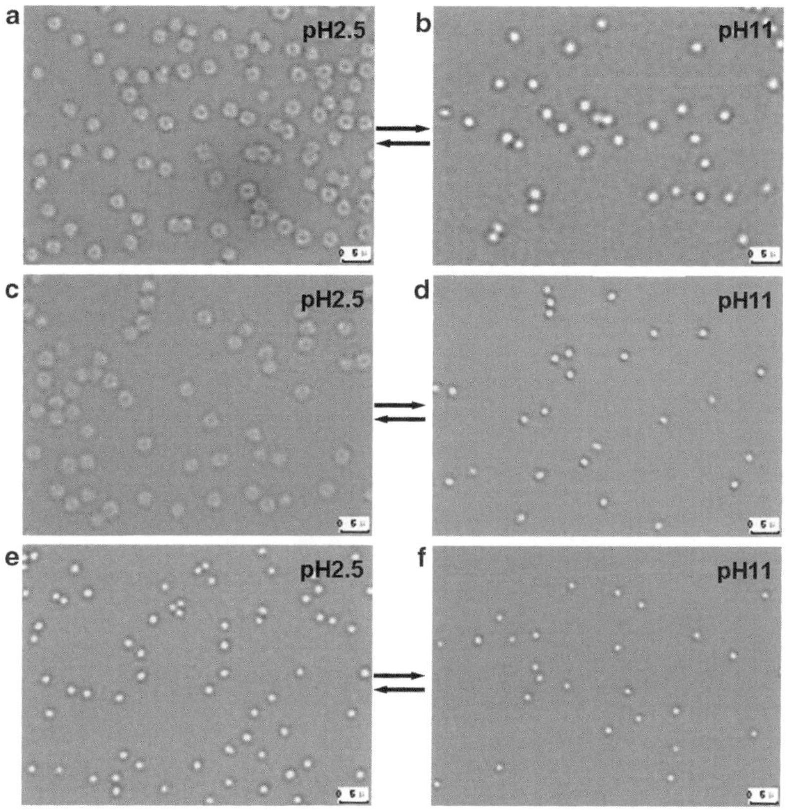

Fig. 7.9 Optical micrographs of PDM microgels prepared with 10 wt % of PVP K30 (**a**, **b**), 2 wt % of PVP 360 (**c**, **d**), and 5 wt % of PVP 360 (**e**, **f**) at pH 2.5 and pH 11. The dosages of other ingredients are 0.5 mol/L DM, ethanol/water ratio of 1:9 (v/v), and 0.5 wt % MBA. All the samples are carried out on purified dilute dispersions. Scale bar = 5 μm (Reproduced with permission from Ref. [24], Copyright (2007), Elsevier)

have relatively low degree of pH responsivity ($V_{max}/V_{min} = 6.6$) because of their low content of amine groups resulting in relatively weak electrostatic repulsive forces. Consequently, the microgels prepared with 2 wt % PVP 360 with moderate size has the highest degree of pH responsivity ($V_{max}/V_{min} = 11.7$).

7.4 Monodisperse Cationic pH-Responsive Hydrogel Capsules

As mentioned above, microfluidic techniques, which have been developed for generating highly monodisperse emulsions in recent years, provide a promising route for preparing microparticles with uniform size [13–17].

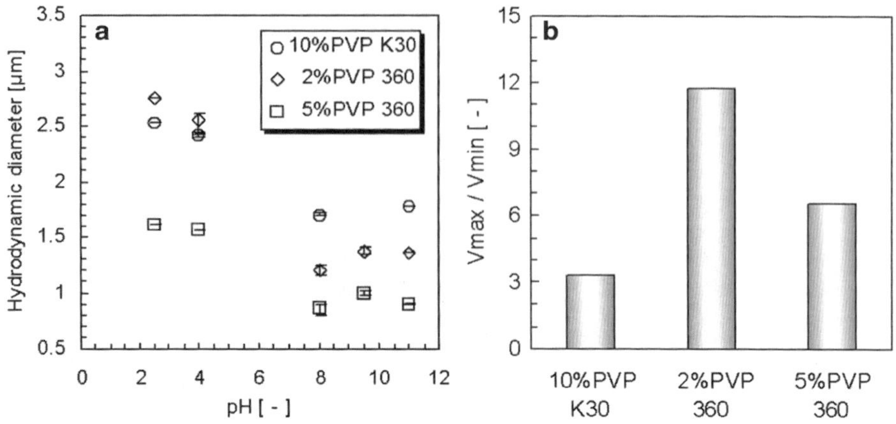

Fig. 7.10 (**a**) Hydrodynamic diameters of PDM microgels prepared with 10 wt % of PVP K30 (○), 2 wt % of PVP 360 (◇), and 5 wt % of PVP 360 (□) at different pH values and (**b**) the maximum pH-dependent volume-change ratios of PDM microgels prepared with different PVP. The dosages of other ingredients are 0.5 mol/L DM, ethanol/water ratio of 1:9 (v/v), and 0.5 wt % MBA. All the samples are carried out on purified dilute dispersions (Reproduced with permission from Ref. [24], Copyright (2007), Elsevier)

Table 7.1 The compositions of different monomer aqueous fluids (Reproduced with permission from Ref. [26], Copyright (2011), Elsevier)

Code	DM (M)	AAm (M)	MBA (M)	pH
1#	1.0	/	0.050	4.3
2#	1.0	/	0.050	7.8
3#	1.0	/	0.025	4.3
4#	1.0	/	0.100	4.3
5#	1.5	/	0.075	7.8
6#	1.0	0.1	0.050	4.3

Note: In the monomer fluids, deionized water is used as solvent and concentrated HCl is used to modulate pH of the solution

Monodisperse PDM hollow microcapsules with cationic pH-responsive property are successfully prepared using oil-in-water-in-oil (O/W/O) double emulsions as the polymerization templates [26]. The compositions of monomer aqueous fluids are listed in Table 7.1. As illustrated in Fig. 7.11a, the emulsion templates are generated in a capillary microfluidic device according to the previous work [17]. Soybean oil containing 3 % (w/v) PGPR and 0.1 % (w/v) Sudan III is used as the inner fluid, and soybean oil containing 5 % (w/v) PGPR is employed as the outer fluid. The middle fluid is monomer aqueous solution containing monomer DM, cross-linker MBA, surfactant Pluronic F127 (1 %, w/v), initiator 2,2'-azobis(2-amidinopropane dihydrochloride) (V50) (0.05 %, w/v), and glycerin (5 %, w/v). The generated O/W/O emulsions are collected in a beaker containing excess soybean oil which contains 5 % (w/v) PGPR and 1 % (w/v) photo-initiator 2,2-dimethoxy-2-phenylacetophenone (BDK). The cross-linked PDM microcapsules are prepared via

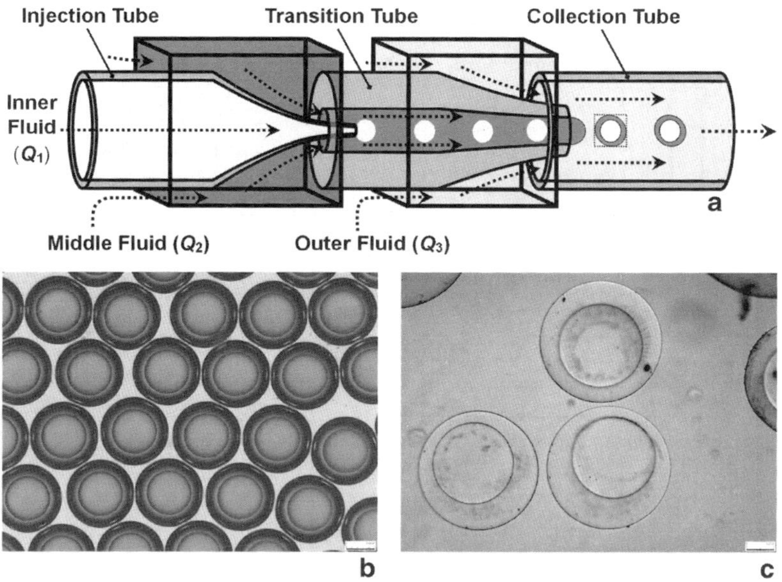

Fig. 7.11 Schematic illustration of the capillary microfluidic device for generating O/W/O emulsions (**a**), and optical microscope images of sample 2# O/W/O emulsions (**b**) and PDM microcapsules in buffer solution of pH 7.4 at 37 °C (**c**). The scale bars are 100 μm (Reproduced with permission from Ref. [26], Copyright (2011), Elsevier)

UV-initiated polymerization in an ice bath for 30 min. A double initiation system composed of water-soluble V50 and oil-soluble BDK photo-initiators is applied to ensure the successful synthesis of PDM microcapsules. Under UV irradiation, the oil-soluble photo-initiator dissociates to generate a great deal of active free radicals that diffuse across the oil-water interface to the water phase of O/W/O emulsions to start the polymerization at the oil-water interface. Such interface initiation would ensure the obtained microcapsules being of a good sphericity. On the other hand, the water-soluble photo-initiator in water phase can initiate the monomers to polymerize sufficiently.

The optical microscope image of O/W/O emulsions prepared by the microfluidic technique is shown in Fig. 7.11b, and the optical image of PDM microcapsules polymerized from these double emulsions is shown in Fig. 7.11c [26]. It can be clearly seen that the obtained O/W/O emulsions and PDM microcapsules exhibit good spherical shape and monodispersity and the PDM microcapsules present obvious hollow cavity structures.

The membrane of the prepared microcapsule is composed of cross-linked PDM, which can swell in acidic environment due to the protonation of $-N(CH_3)_2$ groups in the polymeric network. The effects of various preparation parameters, such as pH value of the monomer solution, concentrations of MBA cross-linker, concentration of DM monomer, and addition of copolymeric monomer acrylamide (AAm), on the

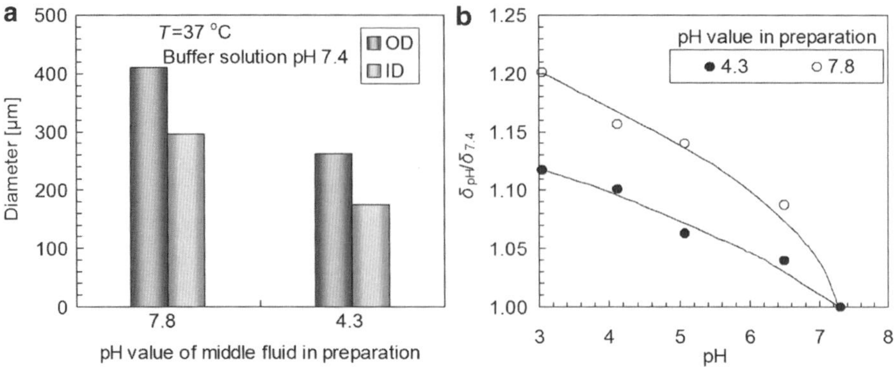

Fig. 7.12 Diameters (**a**) and pH-responsive swelling ratios (**b**) of PDM microcapsules prepared at different pH. The test temperature is 37 °C (Reproduced with permission from Ref. [26], Copyright (2011), Elsevier)

pH-responsive characteristics of the PDM microcapsules are systematically studied [26]. The swelling ratios of membrane thickness ($\delta_{pH}/\delta_{7.4}$) induced by external pH changing from 7.4 to certain pH are defined to characterize the pH-responsive behaviors of prepared PDM microcapsules.

Figure 7.12 shows the diameters and pH-responsive swelling ratios ($\delta_{pH}/\delta_{7.4}$) of PDM microcapsules prepared at different pH (1# and 2#) [26]. The microcapsule prepared at pH 4.3 is smaller than the microcapsule prepared at pH 7.8 when they are immersed in the same buffer solution (Fig. 7.12a). Both PDM microcapsules exhibit obvious pH-responsive characteristics that the membrane thickness increases with decreasing external pH value, as shown in Fig. 7.12b. Interestingly, the swelling ratios of PDM microcapsule prepared at pH 4.3 are lower than that prepared at pH 7.8. Figure 7.13 illustrates the explanation of these pH-responsive phenomena. When the pH value of monomer aqueous fluid is 4.3, the network of PDM microcapsules has already swollen to some extent due to the partial protonation of $-N(CH_3)_2$ groups during the preparation process. Therefore, the length of the polymeric chain between two cross-linking points is shorter due to the electrostatic repulsion of protonated $-N(CH_3)_2$ groups, which leads to lower pH-responsive swelling ratios, as shown in Fig. 7.13a. Moreover, in the network of PDM microcapsules prepared at pH 4.3, the number of $-N(CH_3)_2$ groups with protonation ability decreases, which would also lead to a lower swelling ratio when these microcapsules are put in acidic solution. On the contrary, the PDM microcapsules prepared at 7.8 have a longer polymeric chain between two cross-linking points because the PDM network is in a shrunken state during the preparation process. Therefore, the PDM microcapsules prepared at pH 7.8 exhibit larger swelling ratio than those prepared at pH 4.3, as shown in Fig. 7.13b.

Figure 7.14 shows the diameters and pH-responsive swelling ratios of PDM microcapsules prepared with different concentrations of cross-linker MBA [26]. It can be observed that, with the increase of cross-linking degree, both the outer and

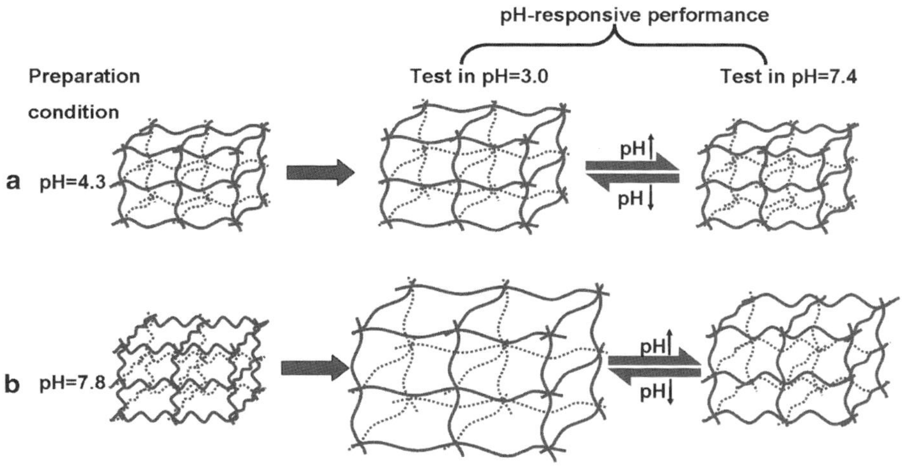

Fig. 7.13 Schematic illustration of pH-responsive mechanism of cross-linked PDM polymeric network prepared at different pH: (**a**) pH = 4.3 and (**b**) pH = 7.8 (Reproduced with permission from Ref. [26], Copyright (2011), Elsevier)

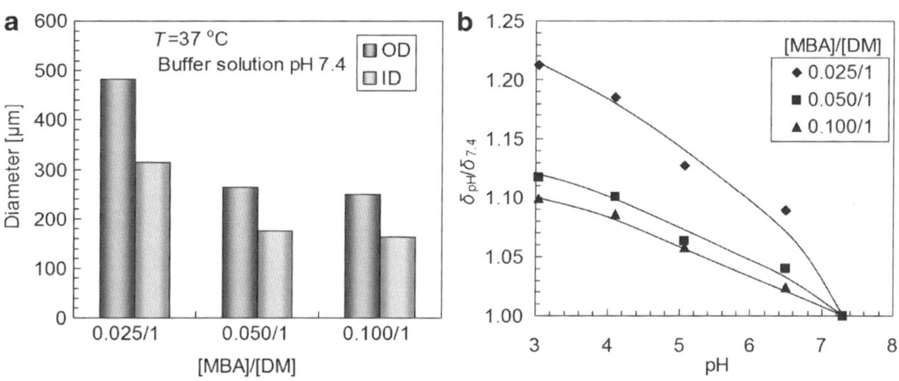

Fig. 7.14 Diameters (**a**) and pH-responsive swelling ratios (**b**) of PDM microcapsules prepared with different MBA concentrations. The test temperature is 37 °C (Reproduced with permission from Ref. [26], Copyright (2011), Elsevier)

inner diameters of PDM microcapsules decrease in the same buffer solution. As shown in Fig. 7.14b, all prepared PDM microcapsules exhibit good pH-sensitivities that the swelling ratios of membrane thickness increase with the decrease of pH in the external solution. Specially, the swelling ratios of PDM microcapsules also decrease with increasing the cross-linking degree of polymeric network. The increase of cross-linking degree would result in a decreased elasticity of the network and then a decreased swelling ratio.

The effect of monomer DM concentration on the pH-responsive swelling ratios of microcapsules is also investigated by keeping the same molar ratio of [MBA]/[DM]

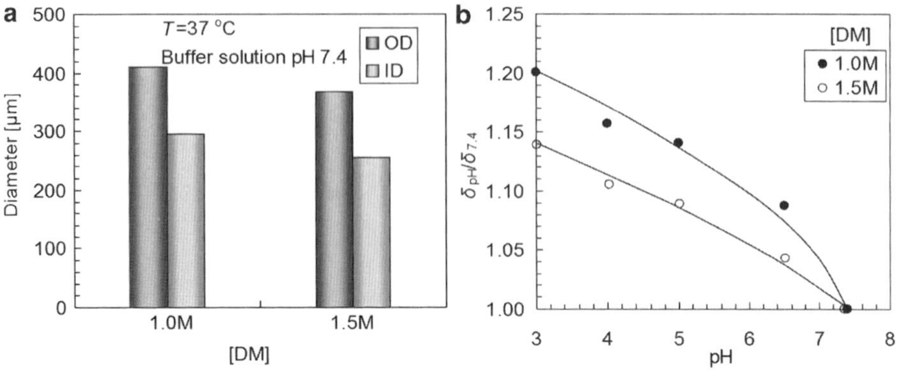

Fig. 7.15 Diameters (**a**) and pH-responsive swelling ratios (**b**) of PDM microcapsules prepared with different DM concentrations. The test temperature is 37 °C (Reproduced with permission from Ref. [26], Copyright (2011), Elsevier)

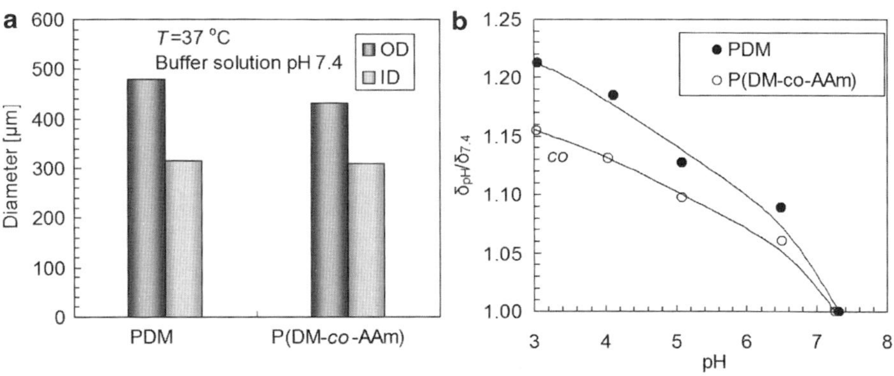

Fig. 7.16 Diameters (**a**) and pH-responsive swelling ratios (**b**) of PDM microcapsules prepared with the addition of AAm. The test temperature is 37 °C (Reproduced with permission from Ref. [26], Copyright (2011), Elsevier)

as 0.050/1. The size of PDM microcapsules in buffer solution of pH 7.4 decreases with increasing the DM content, as shown in Fig. 7.15a [26]. PDM microcapsules with lower DM content also show larger pH-responsive swelling ratios than those with higher DM content. To keep the same molar ratio of [MBA]/[DM], the cross-linker concentration also increases with increasing the DM concentration, which results in a larger density of the polymeric network and a decreased aperture of the PDM network.

The PDM microcapsules are also prepared by adding another comonomer AAm with no pH sensitivity. The PDM-based copolymeric microcapsules also have pH sensitivity, but the swelling ratios decrease, as shown in Fig. 7.16 [26]. The reason is that the formation of hydrogen bonds between amide groups in AAm and $-N(CH_3)_2$

groups in DM would protect $-N(CH_3)_2$ groups from exposing to the outside [27]. Therefore, the protonation ability of $-N(CH_3)_2$ groups decreases, and then a decrease of pH-responsive swelling ratio results.

The above results show that, when PDM microcapsules are prepared at high pH and with low cross-linking density and low DM monomer concentration, they exhibit high pH-responsive swelling ratios. The addition of copolymeric monomer AAm decreases the swelling ratios of PDM microcapsules. The prepared PDM microcapsules with both biocompatibility and cationic pH-responsive properties are of great potential as drug delivery carriers for tumor therapy. Moreover, the fabrication method and research results provide valuable guidance for preparation of core-shell microcapsules via free-radical polymerization based on synergistic effects of interfacial initiation and initiation in confined space.

7.5 Summary

Cationic pH-responsive microgels, which have pH-responsive swelling property in acidic condition due to protonation, are preferred in many circumstances. Several kinds of monodisperse cationic pH-responsive microgels, such as chitosan microspheres, chitosan microcapsules, and cationic PDM microgels, are introduced in this chapter. Monodisperse chitosan microspheres and microcapsules with acid-triggered swelling and decomposition properties can be achieved with the microfluidic approach. Monodisperse PDM microgels with acid-induced swelling property can be prepared by dispersion polymerization in ethanol/water mixture using PVP as the steric stabilizer. Monodisperse cationic pH-responsive PDM microcapsules can be prepared using O/W/O emulsions as templates via UV-initiated polymerization based on a double initiation system. Due to their cationic pH-responsive properties, these monodisperse microgels with good biocompatibility are of great potential as smart drug delivery carriers in response to external acidic condition.

References

1. Dissemond J, Witthoff M, Brauns TC et al (2003) pH values in chronic wounds. Evaluation during modern wound therapy. Hautarzt 54:959–965
2. Vaupel P, Kallinowski F, Okunieff P (1989) Blood flow, oxygen and nutrient supply, and metabolic microenvironment of human tumors: a review. Cancer Res 49:6449–6465
3. Rofstad EK, Mathiesen B, Kindem K et al (2006) Acidic extracellular pH promotes experimental metastasis of human melanoma cells in athymic nude mice. Cancer Res 66:6699–6707
4. Ballauff M, Lu Y (2007) "Smart" nanoparticles: preparation, characterization and applications. Polymer 48:1815–1823
5. Dai S, Ravi P, Tam KC (2008) pH-responsive polymers: synthesis, properties and applications. Soft Matter 4:435–449

6. Freiberg S, Zhu XX (2004) Polymer microspheres for controlled drug release. Int J Pharm 282:1–18
7. Esposito E, Sebben S, Cortesi R et al (1999) Preparation and characterization of cationic microspheres for gene delivery. Int J Pharm 189:29–41
8. Rabea EI, Badawy MET, Stevens CV et al (2003) Chitosan as antimicrobial agent: applications and mode of action. Biomacromolecules 4:1457–1465
9. Gulbake A, Jain SK (2012) Chitosan: a potential polymer for colon-specific drug delivery system. Expert Opin Drug Deliv 9:713–729
10. Jayakumar R, Prabaharan M, Nair SV et al (2010) Novel carboxymethyl derivatives of chitin and chitosan materials and their biomedical applications. Prog Mater Sci 55:675–709
11. Bhattarai N, Gunn J, Zhang MQ (2010) Chitosan-based hydrogels for controlled, localized drug delivery. Adv Drug Deliver Rev 62:83–99
12. Wu J, Sailor M (2009) Chitosan hydrogel-capped porous SiO_2 as a pH responsive nano-valve for triggered release of insulin. Adv Funct Mater 19:733–741
13. Joanicot M, Ajdari A (2005) Droplet control for microfluidics. Science 309:887–888
14. Dendukuri D, Doyle PS (2009) The synthesis and assembly of polymeric microparticles using microfluidics. Adv Mater 21:4071–4086
15. Tumarkin E, Kumacheva E (2009) Microfluidic generation of microgels from synthetic and natural polymers. Chem Soc Rev 38:2161–2168
16. Marre S, Jensen KF (2010) Synthesis of micro and nanostructures in microfluidic systems. Chem Soc Rev 39:1183–1202
17. Chu LY, Utada AS, Shah RK et al (2007) Controllable monodisperse multiple emulsions. Angew Chem Int Edit 46:8970–8974
18. Wei W, Yuan L, Hu G et al (2008) Monodisperse chitosan microspheres with interesting structures for protein drug delivery. Adv Mater 20:2292–2296
19. Wei W, Wang LY, Yuan L et al (2007) Preparation and application of novel microspheres possessing autofluorescent properties. Adv Funct Mater 17:3153–3158
20. Deng NN, Meng ZJ, Xie R et al (2011) Simple and cheap microfluidic devices for preparation of monodisperse emulsions. Lab Chip 11:3963–3969
21. Zhu CH, Jung S, Luo SB et al (2010) Co-delivery of siRNA and paclitaxel into cancer cells by biodegradable cationic micelles based on PDMAEMA-PCL-PDMAEMA triblock copolymers. Biomaterials 31:2408–2416
22. Liu YH, Gao XH, Luo MB et al (2009) Self-assembled micellar nanoparticles of a novel star copolymer for thermo and pH dual-responsive drug release. J Colloid Interface Sci 329:244–252
23. Brahio S, Narinesingh D, Guiseppi-Elie A (2003) Release characteristics of novel pH-sensitive p(HEMA-DMAEMA) hydrogels containing 3-(trimethoxy-silyl) propyl methacrylate. Biomacromolecules 4:1224–1231
24. Hu L, Chu LY, Yang M et al (2007) Preparation and characterization of novel cationic pH-responsive poly(N, N′-dimethylamino ethyl methacrylate) microgels. J Colloid Interface Sci 311:110–117
25. Shen S, Sudol ED, El-Aasser MS (1994) Dispersion polymerization of methyl methacrylate: mechanism of particle formation. J Polym Sci Pol Chem 32:1087–1100
26. Wei J, Ju XJ, Xie R et al (2011) Novel cationic pH-responsive poly(*N*,*N*-dimethylaminoethyl methacrylate) microcapsules prepared by a microfluidic technique. J Colloid Interface Sci 357:101–108
27. Cho SH, Jhon MS, Yuk SH et al (1997) Temperature-induced phase transition of poly(*N*,*N*-dimethylaminoethyl methacrylate-co-acrylamide). J Polym Sci Pol Phys 35:595–598

Chapter 8
pH-Responsive Membranes and Microcapsules for Controlled Release

Abstract In this chapter, the design, fabrication, and performance of the pH-responsive composite membrane system and core-shell microcapsule systems with improved controlled-release properties are introduced. The pH-responsive composite membrane system is featured with a high responsive release rate, which is composed of a porous membrane with linear-grafted positively pH-responsive polymer gates acting as functional valves, and a cross-linked negatively pH-responsive hydrogel inside the reservoir working as a functional pump. Two kinds of monodisperse core-shell microcapsule systems with acid-triggered burst release properties are developed by microfluidic approach. These pH-responsive systems provide new modes for smart controlled-release systems, which are highly attractive for drug delivery systems, chemical carriers, sensors, and so on.

8.1 Introduction

Many investigations have been carried out with the efforts to improve drug treatment through rate- and time-programmed and site-specific controlled-release drug delivery [1, 2]. As mentioned in Chap. 7, pH variation is an important chemical stimulus to design stimuli-responsive materials in biomedical applications. The pH-responsive controlled-release delivery systems enable drugs to be targeted at specific areas in the body such as tumors, sites of inflammation and inflection, or the colonic region [3–6]. Therefore, the development of pH-responsive controlled-release delivery systems is of both scientific and technological interests.

For the environmental stimuli-responsive controlled-release systems, a fast response is the key for their successful applications. Furthermore, it is equally important that the delivery systems release drugs as quickly as possible upon meeting environmental stimuli. To increase the response dynamics of stimuli-responsive controlled-release systems, several strategies have been explored, such as improving the internal architecture or structure of the hydrogels [7–9], developing microgels or hydrogel particles with micro- or nano-dimensions [10–13], and introducing

L.-Y. Chu et al., *Smart Hydrogel Functional Materials*,
DOI 10.1007/978-3-642-39538-3_8,

© Chemical Industry Press, Beijing and Springer Berlin Heidelberg 2013

linear-grafted chain with freely mobile ends into hydrogel configurations [14]. However, most of the investigations are concentrated on the improvement of the response time of phase transitions of the stimuli-responsive materials themselves, but not on the improvement of the stimuli-responsive release rate. Because the driving force for drug release in those systems is the concentration gradient of drug, the release rate is limited by the solute diffusion. The limitation of the release rate restrains the development of rate-programmed fast-response drug delivery systems.

Recently, the authors' group has designed and developed several kinds of pH-responsive systems, such as composite membrane system and core-shell microcapsule systems, with improved controlled-release performances. The pH-responsive composite membrane system is composed of a porous membrane with positively pH-responsive linear-grafted polymer gates acting as functional valves and a negatively pH-responsive cross-linked hydrogel inside the reservoir working as a functional pumping element. This composite membrane system is featured with a high responsive release rate due to the pumping effects of the negatively responsive hydrogel that goes effectively beyond the limit of concentration-driven diffusion [15]. The core-shell microcapsule system based on cross-linked chitosan membrane with acid-triggered burst and complete release characteristics is designed and successfully prepared by a microfluidic approach [16]. Another core-shell microcapsule system with chitosan microsphere as core and PNIPAM hydrogel as shell membrane is also developed with acid-triggered self-bursting performance. These proposed core-shell microcapsule systems provide a promising drug carrier candidate for pH-responsive quick release. In this chapter, the design, fabrication, and performance of these pH-responsive composite membrane system and core-shell microcapsule systems will be introduced.

8.2 pH-Responsive Gating Membrane System with Pumping Effect for Improved Controlled Release

The concept of the developed pH-responsive gating membrane system with pumping effects is schematically illustrated in Fig. 8.1 [15]. The system is composed of a gating membrane with linear-grafted poly(methacrylic acid) (PMAA) gates and a cross-linked poly(*N,N*-dimethylaminoethyl methacrylate) (PDM) hydrogel inside the reservoir. The pH-responsive phase transition characteristics of PMAA and PDM have been verified to be opposite [17]. Linear-grafted PMAA chains are featured with a positively pH-responsive volume-phase transition characteristic, i.e., the swelling of polymer chains is induced by an increase in the environmental pH value. On the contrary, PDM hydrogel shows a negatively pH-responsive volume-phase transition characteristic, i.e., the hydrogel swelling is induced by a decrease in the environmental pH value. So, linear-grafted positively pH-responsive PMAA chains in the membrane pores act as pH-sensitive "gates" and cross-linked negatively pH-responsive PDM hydrogel in the reservoir acts as a pH-sensitive "pumping element."

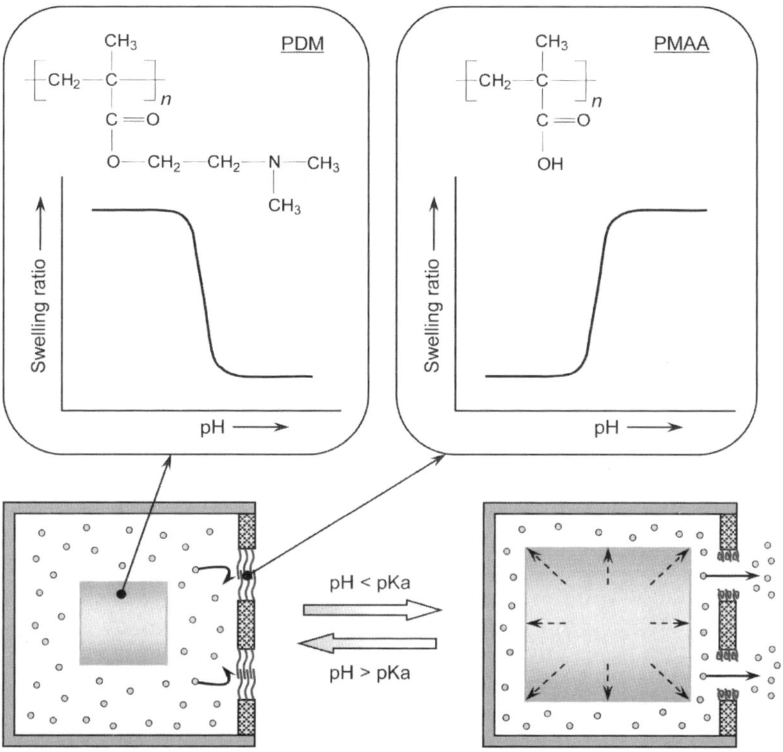

Fig. 8.1 Schematic representation of the proposed pH-responsive controlled-release system with functional gating and pumping effects (Reproduced with permission from Ref. [15], Copyright (2006), Wiley-VCH Verlag GmbH & Co. KGaA)

When the environmental pH is higher than the value of max(pK_{PMAA}, pK_{PDM}), the inner PDM hydrogel shrinks. At the same time, the grafted PMAA chains swell and then close the membrane pores, and as a result, the release rate is slow. On the other hand, when the ambient pH is decreased to be lower than the value of min(pK_{PMAA}, pK_{PDM}), the grafted PMAA chains shrink and consequently the membrane pores open. At the same time, the PDM hydrogel swells and then accelerates the release rate. Therefore, the maximum release rate of drugs from the proposed membrane system responding to environmental pH changes can be effectively improved, because the limitation of the release rate restricted by the concentration-driven diffusion can be broken by the pumping effect of the negatively pH-responsive hydrogel inside the reservoir.

The proposed pH-responsive gating membranes are prepared by grafting linear PMAA chains onto porous polyvinylidene fluoride (PVDF) membrane substrates using plasma-graft pore-filling polymerization, and cross-linked PDM hydrogels are synthesized by free-radical cross-linking polymerization [15]. In the preparation of PMAA-grafted PVDF (PMAA-*g*-PVDF) gating membranes, the argon plasma treatment power is varied from 30 to 50 W, and the plasma treatment time

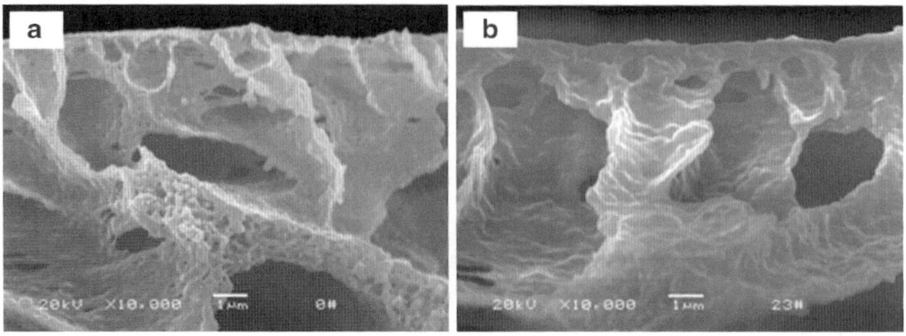

Fig. 8.2 SEM images of the cross sections of (**a**) virgin PVDF substrate membrane and (**b**) PMAA-*g*-PVDF membrane ($Y = 8.58$ %). Scale bars are 1 μm (Reproduced with permission from Ref. [15], Copyright (2006), Wiley-VCH Verlag GmbH & Co. KGaA)

is 60 s. The MAA monomer concentration is varied from 3.0 to 10.0 vol.%, and the graft polymerization is carried out at 30 °C for a fixed period (from 3 to 10 h). In the preparation of cross-linked PDM hydrogels, the reaction solution is composed of monomer DM ([DM] = 1.4 or 1.8 mol/L), cross-linker *N'N'*-methylenebisacrylamide ([MBA] = 40.0, 37.5 or 30.0 mmol/L), and initiator potassium persulfate ([KPS] = 3.0 g/L) dissolved in deionized water. The oxygen dissolved in the reaction solution is removed by nitrogen gas before polymerization, and then the solution immediately transfers into small glass tubes and is sealed. The polymerization is carried out at 60 °C for 24 h. After the gelation is completed, the prepared cylindrical hydrogels are pushed out from the glass tubes and washed with excess deionized water. The grafting yield of the membrane is defined as the mass increase in the membrane after the grafting, which is given by

$$Y = \frac{W_g - W_0}{W_0} \times 100\,\% \tag{8.1}$$

where Y stands for the grafting yield of PMAA onto the membrane substrate [%] and W_g and W_0 stand for the mass of the membrane after and before grafting, respectively [g].

Scanning electron microscopy (SEM) is used to ascertain the microstructures of the membranes before and after grafting PMAA polymer chains. Figure 8.2 shows the SEM micrographs of the cross sections of virgin PVDF membrane substrate and PMAA-*g*-PVDF membrane with $Y = 8.58$ % [15]. It can be clearly seen that the microstructures of the cross sections of virgin membrane (Fig. 8.2a) and PMAA-grafted (Fig. 8.2b) membrane are quite different. After grafting PMAA onto the inner pore surfaces of the porous PVDF membrane substrate by plasma-graft pore-filling polymerization, a significant grafted PMAA layer is homogeneously formed across the cross section of the membrane. The microstructure change in the membrane indicated that PMAA had been successfully grafted onto the porous membrane substrate.

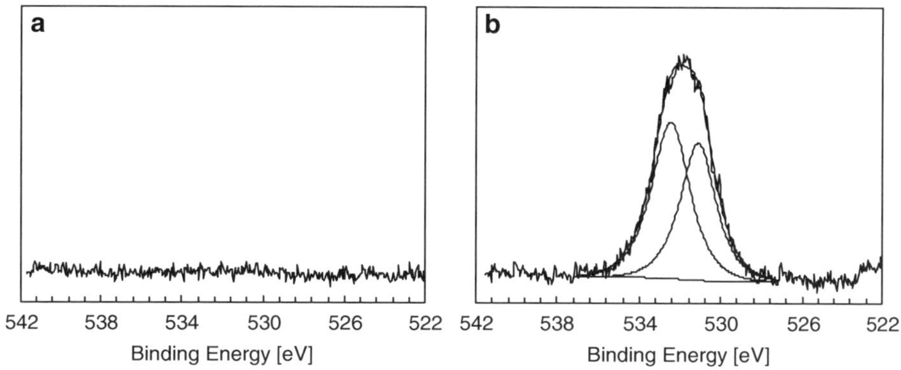

Fig. 8.3 XPS O_{1s} core-level spectra of (**a**) virgin PVDF substrate membrane and (**b**) PMAA-*g*-PVDF membrane ($Y = 5.80$ %) (Reproduced with permission from Ref. [15], Copyright (2006), Wiley-VCH Verlag GmbH & Co. KGaA)

To confirm the fabrication of membranes with grafted PMAA gates, the virgin PVDF membrane substrate and PMAA-*g*-PVDF membrane are analyzed by X-ray photoelectron spectroscopy (XPS) [15]. Figure 8.3 shows the XPS O_{1s} core-level spectra of the virgin PVDF membrane substrate and PMAA-*g*-PVDF membrane with $Y = 5.80$ %. There is no peak in the O_{1s} core-level spectrum of PVDF membrane substrate (Fig. 8.3a), i.e., there is no oxygen (O) element in the composition of the membrane substrate. While the O_{1s} core-level spectrum of the PMAA-*g*-PVDF membrane is curve-fitted with two peak components (Fig. 8.3b), in which the binding energy of 531.1 eV stands for the C═O species and 532.5 eV for the O—H species. The new peak components are assigned to the COOH groups of the grafted PMAA polymers in the membrane. These experimental observations confirm the PMAA grafting in the fabrication of the PMAA-*g*-PVDF gating membranes.

The pH-responsive changes in the pore size of the PMAA-*g*-PVDF membranes are estimated by filtration experiments with various pH buffer solutions (pH is in the range from 2 to 7) [15]. Ionic strengths of the buffer solutions are all adjusted to 0.1 mol/L by adding certain amounts of NaCl, and transmembrane pressures of filtration experiments are all carried out under 90 kPa. Temperatures of the membrane system and pH buffer solutions are all controlled at 37 °C. The pH-responsive change in the pore size of PMAA-*g*-PVDF membranes could be calculated according to Hagen-Poiseuille's law [18, 19]. The ratio of the effective pore diameter of PMAA-*g*-PVDF membrane at $\mathrm{pH} = x$ (x is in the range from 2 to 7) to that at $\mathrm{pH} = 7$, which is defined as the pH-responsive gating factor of the membrane pore size, could be evaluated using the measured water fluxes with the following equation:

$$R_{d,\mathrm{pH}x/\mathrm{pH7}} = \frac{d_{\mathrm{pH}=x}}{d_{\mathrm{pH}=7}} = \left(\frac{J_{\mathrm{pH}=x}}{J_{\mathrm{pH}=7}}\right)^{\frac{1}{4}} \tag{8.2}$$

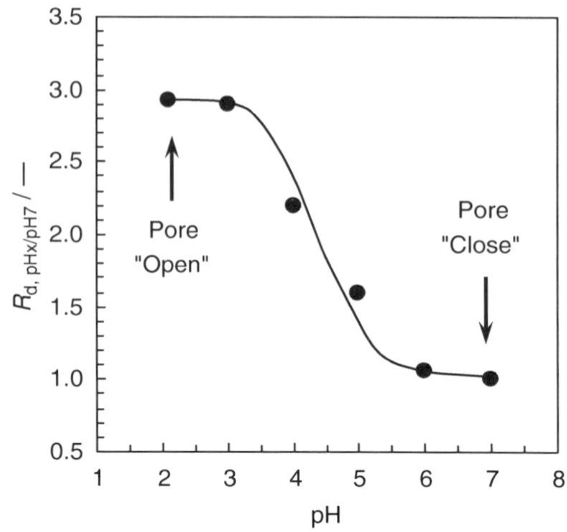

Fig. 8.4 pH-responsive change of the effective pore size of PMAA-*g*-PVDF membrane ($Y = 5.98$ %) (Reproduced with permission from Ref. [15], Copyright (2006), Wiley-VCH Verlag GmbH & Co. KGaA)

where $R_{d,pHx/pH7}$ stands for the pH-responsive gating factor of the membrane pore size [−]; $d_{pH=x}$ and $d_{pH=7}$ are the effective pore diameters of the PMAA-grafted membranes at pH = x and pH = 7, respectively [cm]; and $J_{pH=x}$ and $J_{pH=7}$ are the measured water fluxes at pH = x and pH = 7, respectively [mL·cm^{-2}·s^{-1}].

Figure 8.4 shows the pH-responsive change in the effective pore size of the PMAA-*g*-PVDF membrane with $Y = 5.98$ % [15]. As expected, the effective pore size of the PMAA-grafted membrane changes dramatically at pH values around the pK_{PMAA} as a result of conformational change of the grafted PMAA chains. When the ambient pH value is less than 3 or larger than 6, the effective pore size remains nearly unchanged. It is due to the fact that the conformation of the grafted PMAA chains presented a steady state at these pH ranges.

The pH-responsive swelling/deswelling properties of cross-linked PDM hydrogels are determined by measuring the pH-responsive swelling ratio of PDM hydrogels [15]. To determine their pH-responsive volume-change properties, pre-weighed freeze-dried hydrogel samples are immersed in buffer solutions with different pH values. The buffer solutions are kept at 37 °C and their ionic strengths are all adjusted to 0.1 mol/L by adding certain amounts of NaCl. When the equilibrium swelling/deswelling state has been reached (for measuring the equilibrium swelling/deswelling characteristics) or after a fixed period of time (for measuring the swelling kinetics), the hydrogel samples are taken out from the buffer solutions. After being wiped off the excess solutions on their surfaces, the hydrogel samples are immediately weighed. The swelling ratio of hydrogel is defined as W_s/W_d, where W_s and W_d show the weights of the hydrogel in buffer solution at a certain pH and in a dry state, respectively. To determine the pH-responsive dynamic volume-change rate of PDM hydrogels, the pH value of the ambient buffer solution is designed to change from 7 to 2 at 37 °C. At predetermined time intervals, the PDM hydrogels are weighed and the swelling ratios are calculated as described above.

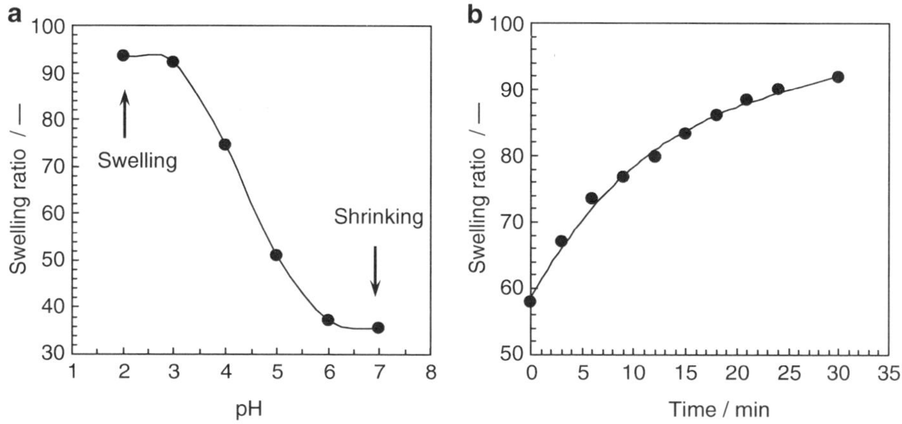

Fig. 8.5 pH-responsive (**a**) equilibrium volume-change and (**b**) dynamic volume-change behaviors of PDM hydrogel prepared with [DM] = 1.4 mol/L and [MBA] = 37.5 mmol/L (Reproduced with permission from Ref. [15], Copyright (2006), Wiley-VCH Verlag GmbH & Co. KGaA)

Figure 8.5a shows the pH-responsive equilibrium volume change of PDM hydrogel prepared with [DM] = 1.4 mol/L and [MBA] = 37.5 mmol/L [15]. When the ambient pH changes across the pK_{PDM}, the volume of PDM hydrogel changes dramatically. When the ambient pH is lower than the pK_{PDM}, the amine groups are protonated and the electrostatic repulsion initiates a swelling of the cross-linked PDM polymer network. When the ambient pH is higher than the pK_{PDM}, the PDM hydrogel shrinks due to the deprotonation of the amine groups and the swelling ratio of hydrogels decreases abruptly. Figure 8.5b shows the pH-responsive dynamic volume-change property of the PDM hydrogel when the pH value of the ambient buffer solution is changed from 7 to 2 [15]. It can be seen that the PDM hydrogel swells quickly when the environmental pH value changes from 7 to 2. The swelling ratio increases from 57.8 to 73.4 within 6 min and to 90.1 within 24 min. The significant volume-phase transition and the fast-response rate of the PDM hydrogels are satisfactory, as required for developing the proposed composite membrane system illustrated in Fig. 8.1.

The pH-responsive release rate of the composite membrane system is experimentally investigated using vitamin B12 (VB_{12}) as a model drug [15]. After immersing the prepared system (as the donor cell) into the pure buffer solution (without VB_{12} at first) inside a receptor cell, the release rate of VB_{12} from the system is measured by determining the increase in VB_{12} concentration of the surrounding medium over time. The PMAA-*g*-PVDF gating membrane and the cross-linked PDM hydrogels are coupled in the prepared system. Before being put into the reservoir of the prepared system, the PDM hydrogels are cut into segments with the same length at 25 °C, and then the hydrogel segments are immersed in 0.2 mmol/L VB_{12} buffer solution with pH = 7 at room temperature for at least 72 h. The spare space inside the reservoir of the prepared system is filled with 0.2 mmol/L VB_{12} buffer solution with pH = 7. Each test membrane is immersed in 0.2 mmol/L VB_{12} buffer solution with pH = 7 for 24 h before starting the release experiments. The prepared composite system is then immersed in pure buffer solution (100 mL) with pH = 7 inside the receptor cell. The temperature of the whole release system is always maintained at 37 °C during the experiments. To investigate the pH-responsive controlled release of VB_{12} from the proposed system, the pH values of the solutions in both donor and receptor cells are both changed from pH = 7 (higher than the value of max(pK_{PMAA}, pK_{PDM})) to pH = 2 (lower than the value of min(pK_{PMAA}, pK_{PDM})) at the same time after a period of release. Ionic strengths of pure buffer solutions and VB_{12} buffer solutions are both adjusted to 0.1 mol/L. To quantitatively describe the pH-responsive controlled-release characteristic, a parameter called the controlled factor (*CF*) is defined as [15]

$$CF = \frac{v_{pH=2}}{v_{pH=7}} \tag{8.3}$$

where $v_{pH=2}$ and $v_{pH=7}$ stand for the release rates of VB_{12} at pH = 2 and pH = 7, respectively [mol/m^2·s].

Figure 8.6 shows the pH-responsive controlled release of VB_{12} from different systems [15]. Detailedly, Fig. 8.6a shows the release results from a system without pH-responsive gates in the membrane and no PDM hydrogel inside the reservoir either. Figure 8.6b shows the release results from a system with grafted PMAA gates in the membrane but no PDM hydrogel inside the reservoir, and Fig. 8.6c shows the release results from a system with only PDM hydrogel inside the reservoir but without pH-responsive gates in the membrane. Figure 8.6d shows the release results from the proposed composite system with both cross-linked PDM hydrogel inside the reservoir and grafted PMAA gates in the membrane.

For the system with a virgin PVDF membrane and no PDM hydrogels inside the reservoir (Fig. 8.6a), the controlled factor (*CF*) for the VB_{12} release is 1.00 when the environmental pH is changed from 7 to 2. In other words, there is no change in the release rate of VB_{12} at all when the ambient pH changes from 7 to 2. That means the VB_{12} release from this system does not show any pH-responsive characteristic.

For the system with the PMAA-*g*-PVDF membrane ($Y = 15.65$ %) but no PDM hydrogel inside the reservoir (Fig. 8.6b), *CF* value for the VB_{12} release is increased to 1.86 when the environmental pH is changed from 7 to 2. When the environmental pH changes from 7 to 2, the pores of the PMAA-*g*-PVDF membrane change from a "closed" situation into an "open" situation due to the conformational change of grafted PMAA chains in the membrane pores. Therefore, the diffusion channels for VB_{12} became wider, and as a result, the release rate of VB_{12} became faster. The results verify that the grafted PMAA chains in the membrane pores act successfully as pH-responsive functional valves for VB_{12} release.

For the system with a virgin PVDF membrane and cross-linked PDM hydrogel inside the reservoir (Fig. 8.6c), *CF* value for the VB_{12} release reaches 2.84 when the environmental pH is changed from 7 to 2. The pumping effects of PDM hydrogel on the release rate are significant. When the ambient pH is higher than the value of

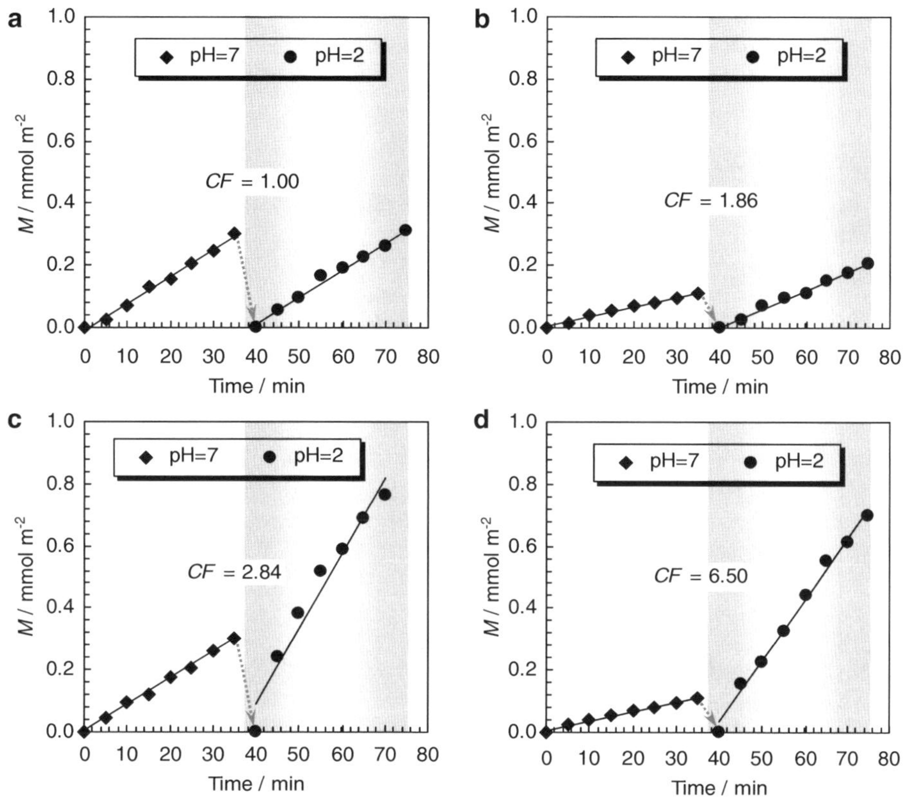

Fig. 8.6 pH-responsive controlled-release characteristics of VB_{12} from different systems. (**a**) With virgin PVDF membrane substrate and without PDM hydrogel inside the reservoir; (**b**) with PMAA-*g*-PVDF gating membrane ($Y = 15.65$ %) but without PDM hydrogel inside the reservoir; (**c**) with virgin PVDF membrane substrate but with PDM hydrogel ([DM] = 1.4 mol/L and [MBA] = 30.0 mmol/L) inside the reservoir; and d) with PMAA-*g*-PVDF gating membrane ($Y = 15.65$ %) and with PDM hydrogel ([DM] = 1.4 mol/L and [MBA] = 30.0 mmol/L) inside the reservoir (Reproduced with permission from Ref. [15], Copyright (2006), Wiley-VCH Verlag GmbH & Co. KGaA)

max(pK_{PMAA}, pK_{PDM}), the PDM hydrogel is in a shrinking state; when the pH is decreased to be lower than the value of min(pK_{PMAA}, pK_{PDM}), the PDM hydrogel becomes swollen and acts as a micro-pumping element for VB_{12} release from the system. The results verify the pH-responsive pumping function for driving the solute release by the inner PDM hydrogel.

For the proposed composite system with both cross-linked PDM hydrogel inside the reservoir and grafted PMAA gates in the membrane (Fig. 8.6d), *CF* value for the VB_{12} release is increased as large as 6.50. Due to the dual functions of the grafted PMAA gates in the porous membrane and the cross-linked PDM hydrogel in the reservoir, *CF* value is higher than both systems with the pH-responsive gating membrane only (Fig. 8.6b) and with the PDM hydrogel only (Fig. 8.6c). Specially, compared to that of the single PMAA-grafted membrane system shown in Fig. 8.6b, *CF* value of the proposed composite system is improved 3.5 times. That means that with the cooperative action of the membrane "gating" function and the cross-linked hydrogel "pumping" effect, the proposed composite system exhibits a much better performance of pH-responsive controlled release than those currently existing gating membrane systems, and the limitation of the release rate restricted by the concentration-driven diffusion has been effectively broken by the pumping effects of the negatively responsive hydrogel.

In summary, the developed pH-responsive composite system, which is composed of a porous membrane with linear-grafted PMAA gates acting as functional valves and a cross-linked PDM hydrogel inside the reservoir working as a functional pumping element, demonstrates an improved controlled-release performance. Because of the cooperative action of the PMAA-*g*-PVDF gating membrane and the cross-linked PDM hydrogel, the composite system exhibits a large responsive release rate that goes effectively beyond the limit of concentration-driven diffusion. This kind of composite system provides a new mode for pH-responsive smart or intelligent controlled-release systems.

8.3 pH-Responsive Microcapsules for Burst Release of Hydrophobic Drugs

Monodisperse core-shell microcapsules based on cross-linked chitosan membrane with acid-triggered burst release properties are also successfully developed recently by a microfluidic approach [16]. As illustrated in Fig. 8.7, the chitosan microcapsules are fabricated using uniform-sized oil-in-water-in-oil (O/W/O) double emulsions as templates, which are generated by capillary microfluidic device and then converted into core-shell microcapsules via interfacial cross-linking reaction. In the O/W/O emulsion templates, chitosan is in the middle water layer and the inner oil phase contains oil-soluble terephthalaldehyde acting as cross-linker. The cross-linking reaction occurs at the inner O/W interface in the double emulsion template to form the cross-linked chitosan membrane.

As described in Chap. 7, based on the pH-responsive stability of the Schiff bases between chitosan and terephthalaldehyde, an interesting acid-triggered burst release from the proposed chitosan microcapsule can be achieved. On the other hand, lipophilic drug can be easily encapsulated into the proposed core-shell chitosan microcapsules within oil cores through microfluidic approach. As illustrated in Fig. 8.8, in neutral medium, the microcapsules maintain good spherical shape and structural integrity, which promises that lipophilic drug would not be released [16]. While in acidic medium at low pH value, amino groups of chitosan are protonated and positively charged, so the intramolecular electrostatic repulsion and enhanced hydrophilicity make the chitosan membranes swell dramatically. With the

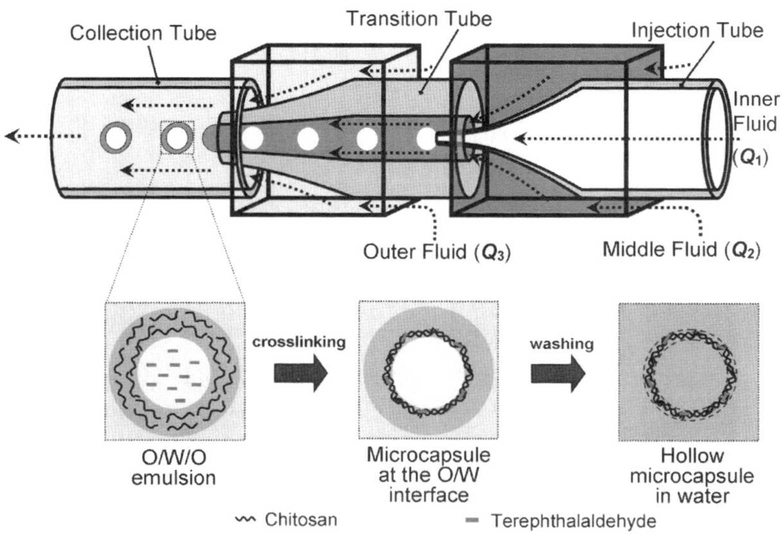

Fig. 8.7 Schematic illustration of the microfluidic preparation process of cross-linked chitosan microcapsule (Reproduced with permission from Ref. [16], Copyright (2011), Royal Society of Chemistry)

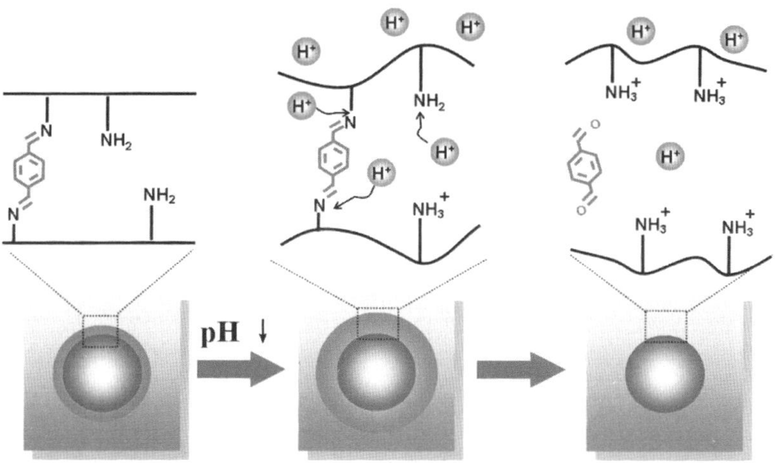

Fig. 8.8 Schematic illustration of the process of the acid-induced burst release behavior of the terephthalaldehyde cross-linked chitosan microcapsule (Reproduced with permission from Ref. [16], Copyright (2011), Royal Society of Chemistry)

protonation of the amino groups going on, the Schiff bases become instable and decomposition of the cross-linked chitosan membrane happens, which finally causes an acid-triggered burst and complete release of lipophilic drug from the proposed microcapsule.

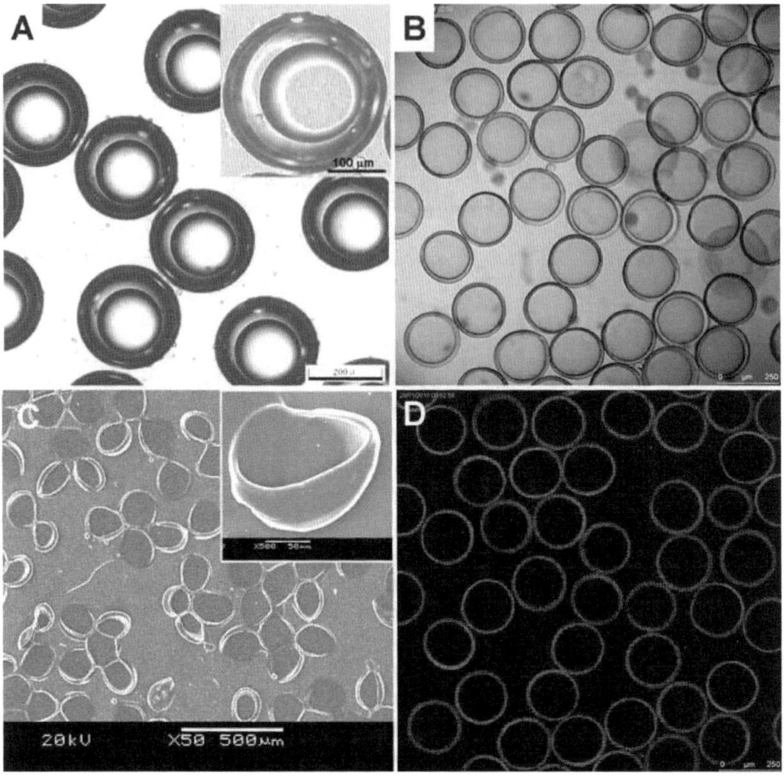

Fig. 8.9 (**a**) Optical micrograph of O/W/O double emulsions (scale bar = 200 μm) and the insert picture shows a double emulsion after complete cross-linking (scale bar = 100 μm). (**b**) CLSM image of the chitosan microcapsules on the transmission channel (scale bar = 250 μm). (**c**) SEM image of air-dried chitosan microcapsules (scale bar = 500 μm) and the insert picture shows a collapsed chitosan microcapsule at high magnification (scale bar = 50 μm). (**d**) CLSM image of the chitosan microcapsules on the green fluorescent channel (scale bar = 250 μm) (Reproduced with permission from Ref. [16], Copyright (2011), Royal Society of Chemistry)

The optical micrographs of the prepared O/W/O double emulsions and the resultant hollow chitosan microcapsules dispersed in water are shown in Fig. 8.9a, b [16]. Both the emulsions and the microcapsules are highly monodisperse, and the number of inner drops in each emulsion template is strictly limited to one. The insert picture in Fig. 8.9a shows a double emulsion after 24 h cross-linking reaction, which indicates the formation mechanism of the microcapsule. From the insert picture, a thin hydrogel membrane can be clearly seen at the inner W/O interface. As terephthalaldehyde molecules diffuse from the inner oil phase to the inner O/W interface and reacts with chitosan from the middle water phase, a thin chitosan membrane forms at the inner O/W interface. Because of the steric hindrance from the chitosan membrane, it becomes harder and harder for terephthalaldehyde to diffuse from the inner oil phase to the middle water phase as the microcapsule membrane grows

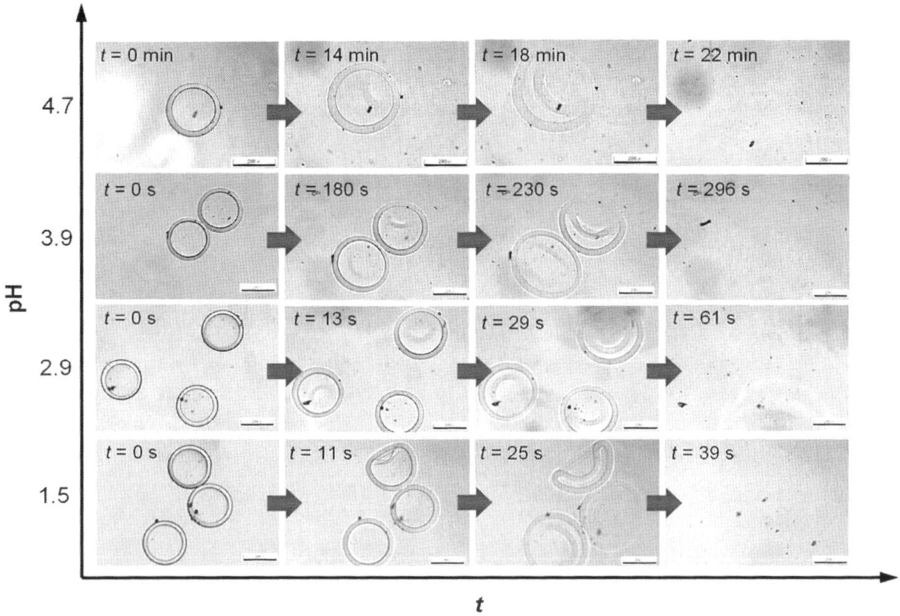

Fig. 8.10 Optical microscope snapshots of the swelling, collapsing, and decomposing of the chitosan microcapsules in various pH mediums at 37 °C. The scale bars are 200 μm (Reproduced with permission from Ref. [16], Copyright (2011), Royal Society of Chemistry)

thicker. So the average outer diameter of the resultant microcapsules is smaller than that of the double emulsion precursors.

Since the prepared chitosan microcapsule membrane is very thin and soft, the dried microcapsules can hardly maintain spherical shape as those in water. Figure 8.9c clearly shows SEM image of the collapsed microcapsules after drying in air. The collapse of microcapsules is caused by volatilization of the inner water during drying, which confirms the hollow structure of the microcapsules [16]. The inserted high-magnification picture in Fig. 8.9c clearly shows a dried microcapsule with hollow structure. Figure 8.9d shows the confocal laser scanning microscope (CLSM) image of chitosan microcapsules cross-linked by terephthalaldehyde on the green fluorescent channel [16]. The green fluorescence from the membrane layer of the microcapsules reveals that the chitosan microcapsule membrane is completely cross-linked [20, 21].

To estimate the capability of acid-triggered burst release from microcapsules, the decomposition processes of chitosan microcapsule membranes in the pH range of 1.5 ~ 4.7 are studied systematically [16]. Microcapsules are firstly immersed in deionized water, and then a sudden change in the pH value of their ambient solution is caused by quickly adding HCl or phosphate buffer solution with different pH values. Figure 8.10 shows the acid-triggered decomposition process of the chitosan microcapsules in different pH buffer solutions at 37 °C [16]. All the microcapsules swell first and then gradually collapse and finally decompose. The lower the environmental pH value is, the faster the microcapsule membranes swell and decompose. When the environmental pH value is 4.7, it takes 22 min for the microcapsules to completely decompose, whereas when the environmental pH value decreases to 1.5, the microcapsules decompose rapidly in 39 s.

The acid-triggered morphological change behavior of chitosan microcapsule membranes can be explained by the mechanism as follows. In acidic medium with a pH value in the range of 1.5 ∼ 4.7, the free amino groups of chitosan are protonated first. As a result, the electrostatic repulsion among the positively charged polymer chains and the increased hydrophilicity make the microcapsule membranes swell. Because the microcapsule membrane swells so fast that the water molecules outside the microcapsules cannot diffuse into the inner space fast enough in time, the microcapsules are not able to expand fully and start to collapse due to the lack of enough liquid to fill the inner spherical space of the microcapsule. The microcapsule membranes decompose and finally dissolve in acidic solutions due to the acidic hydrolysis of Schiff bases between chitosan and terephthalaldehyde.

To demonstrate the feasibility of acid-triggered burst release of lipophilic drugs using the prepared microcapsules, LR300, a red fluorescent dye, is encapsulated into the cross-linked chitosan microcapsules as a lipophilic model drug. Figure 8.11 presents the CLSM microscope snapshots of the acid-triggered burst release process [16]. In the CLSM images, the chitosan membranes exhibit green fluorescence and the oil cores containing LR300 exhibit red fluorescence. When the LR300-encapsulated chitosan microcapsules are subjected to acidic buffer solution with pH of 3.1, the microcapsules decompose rapidly, leaving the LR300-loaded oil cores exposed in 122 s. Since the microcapsule membrane is already in a considerable swollen state caused by detergent OP-10 in wash process before the pH changes, the microcapsule membrane does not swell much during the whole decomposition process. The proposed microcapsules are highly attractive for achieving acid-triggered stomach-targeted delivery carriers with prompt onset and complete release characteristics.

8.4 Monodisperse Core/Shell Microcapsules for pH-Responsive Controlled Release

Another novel monodisperse core-shell microcapsule system with acid-triggered self-bursting property is also developed recently. As illustrated in Fig. 8.12, monodisperse chitosan microspheres prepared according to the microfluidic technique described above are further modified by acylation reaction with 2-bromoisobutyryl bromide (BIBB) to produce chitosan-Br microspheres. And then the core-shell microcapsule system is prepared by grafting cross-linked poly(*N*-isopropylacrylamide) (PNIPAM) hydrogel membrane onto the chitosan-Br microsphere by atom transfer radical polymerization (ATRP) method. Figure 8.13 shows the illustration of the acid-triggered self-bursting performance of the

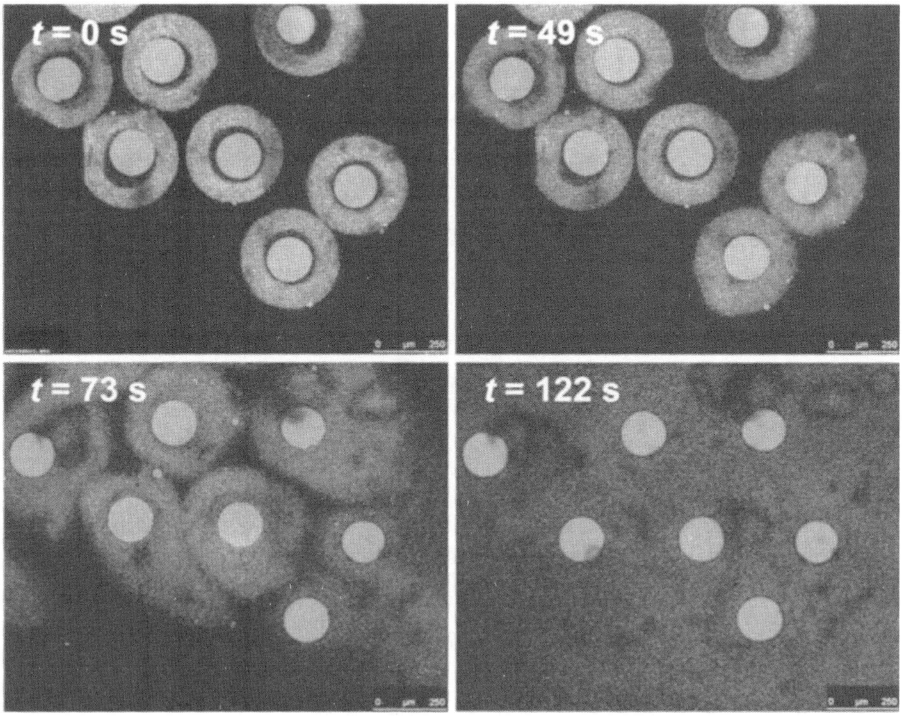

Fig. 8.11 CLSM microscope snapshots of the acid-triggered burst release process of chitosan microcapsules. The acidic buffer solution with pH of 3.1 is added at $t = 0$ s. The scale bars are 250 μm (Reproduced with permission from Ref. [16], Copyright (2011), Royal Society of Chemistry)

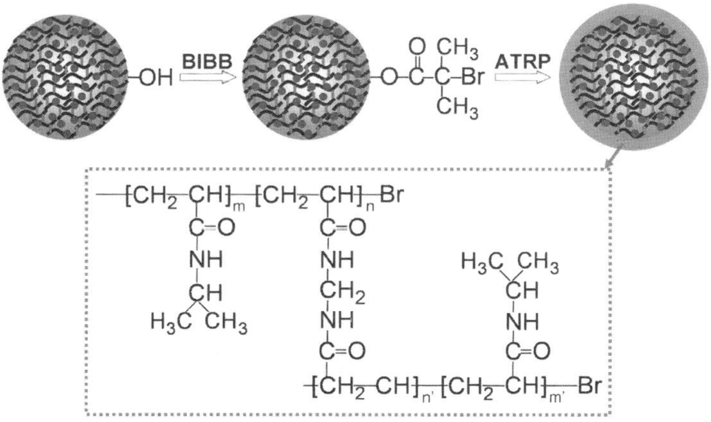

Fig. 8.12 The preparation route of core-shell microcapsules with chitosan microsphere as the core and PNIPAM hydrogel as the shell membrane

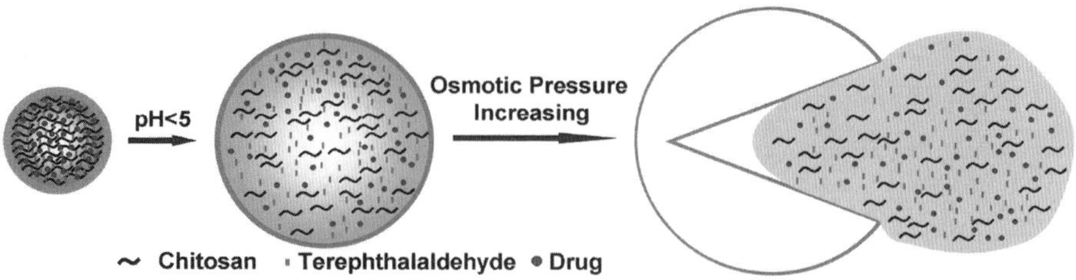

Fig. 8.13 Self-bursting performance of the core-shell microcapsules with chitosan microsphere core and PNIPAM hydrogel shell membrane

core-shell microcapsule system. As mentioned in Sect. 8.3, the chitosan microspheres cross-linked by terephthalaldehyde would also decompose and finally dissolve in acidic solution due to the acidic hydrolysis of Schiff bases. So when the prepared core-shell microcapsule system is in an acidic environment ($pH < 5$), the acidic solution penetrates across the membrane and into the interior of the microcapsule. The osmotic pressure inside the microcapsule would increase due to the dissolution of chitosan microsphere by acidic solution, which could make the PNIPAM capsule membrane swell. The PNIPAM membrane becomes thicker and thicker with the gradual increase of the inner osmotic pressure, and finally when the internal pressure reaches a certain critical value, the membrane ruptures due to its limited mechanical strength.

Before coated with PNIPAM membrane, the chitosan-Br microspheres should be dehydrated in order to decrease the microsphere size and then obtain higher osmotic pressure to make the microcapsules rupture. The research results show that the core-shell microcapsule system prepared with chitosan-Br microsphere dehydrated under vacuum condition at normal temperature with ethanol as the solvent shows the best self-bursting performances under acidic environment. Figure 8.14 shows the dissolution process of dehydrated chitosan-Br microspheres in the acidic buffer solution with pH of 2.9 at 25 °C. When placed into the acidic buffer solution, the chitosan-Br microspheres swell first and finally decompose completely within 14 min. The result also indicates that the chitosan microspheres after modified by acylation reaction still have acid-soluble characteristics. The cross-linked degree for the PNIPAM shell should not be too high, and the optimum cross-linked degree in this study is NIPAM/MBA = 20/1 (mol/mol) and the optimum NIPAM monomer concentration is 0.05 mol/L.

Figure 8.15 presents the self-bursting process of the core-shell microcapsule with chitosan microsphere as core and PNIPAM hydrogel as shell membrane in the acidic buffer solution with pH of 2.9 at 25 °C. Within first 5 min, the chitosan microspheres exhibit green fluorescence due to the Schiff bases. With chitosan microspheres being decomposed by the acidic buffer solution, their fluorescence intensity gradually decreases and the PNIPAM membranes swell due to the increased inner osmotic pressure. From the 9th min, the PNIPAM membranes begin to rupture when the internal pressure reaches the certain critical value.

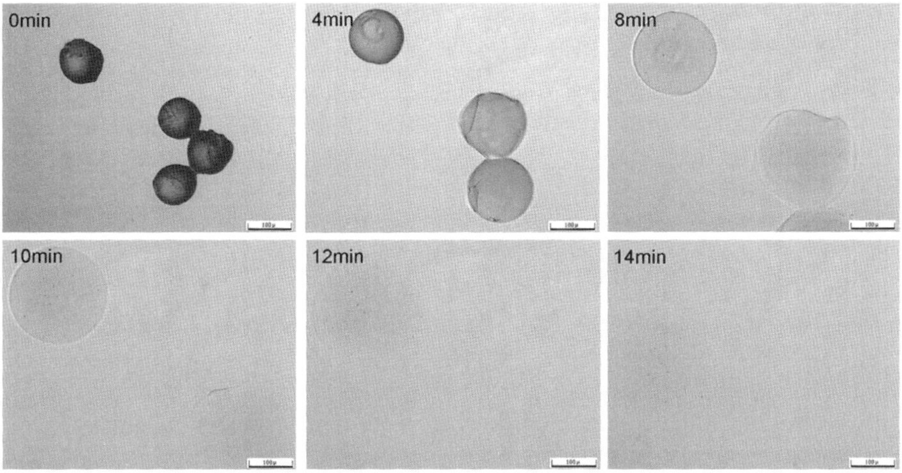

Fig. 8.14 The dissolution characteristic of dehydrated chitosan-Br microspheres in an acidic buffer solution with pH of 2.9 at 25 °C. The chitosan-Br microspheres are dehydrated under vacuum condition at normal temperature with ethanol as solvent. The scale bars are 100 μm

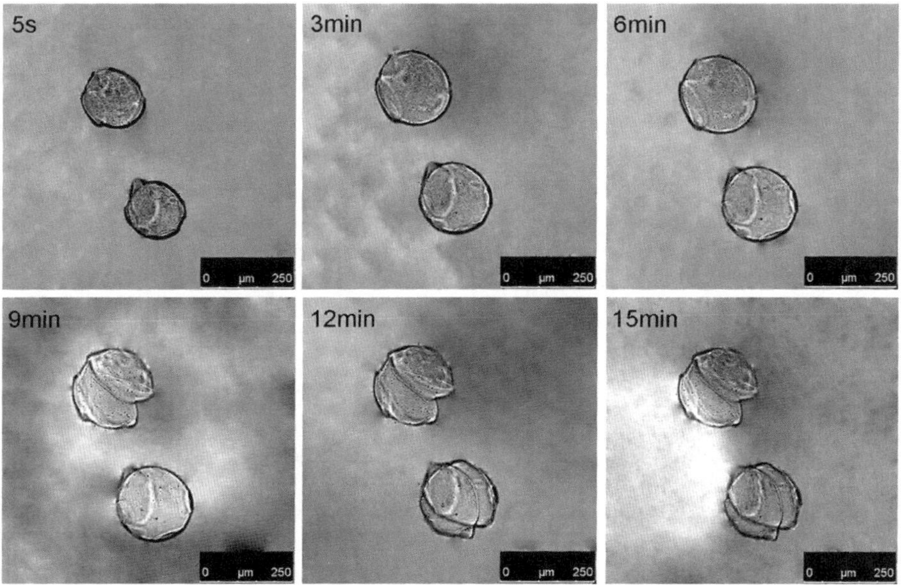

Fig. 8.15 The self-bursting process of the core-shell microcapsule with chitosan microsphere core and PNIPAM shell membrane in the acidic buffer solution with pH of 2.9 at 25 °C. The scale bars are 250 μm

The proposed core-shell microcapsule system with acid-triggered self-bursting performance provides a novel mode for the pH-responsive self-bursting controlled-release systems.

8.5 Summary

In summary, the developed pH-responsive composite membrane system and core-shell microcapsule systems demonstrate the improved controlled-release performance. Because of the cooperative action of the PMAA-*g*-PVDF gating membrane and the cross-linked PDM hydrogel, the pH-responsive composite membrane system exhibits a high responsive release rate that goes effectively beyond the limit of concentration-driven diffusion. The core-shell microcapsule system based on cross-linked chitosan membrane exhibits acid-triggered burst and complete release of lipophilic drugs due to the pH-responsive stability of the Schiff bases in the membrane network. The core-shell microcapsule system with chitosan microsphere as core and PNIPAM hydrogel as shell membrane is also developed with acid-triggered self-bursting performance. These pH-responsive systems would provide new modes for pH-responsive smart or intelligent controlled release.

References

1. Breimer DD (1999) Future challenges for drug delivery. J Control Release 62:3–6
2. Allen TM, Cullis PR (2004) Drug delivery systems: entering the mainstream. Science 303:1818–1822
3. Yatvin MB, Kreutz W, Horwitz BA et al (1980) Shinitzky M. pH-sensitive liposomes: possible clinical implications. Science 210:1253–1255
4. Na K, Lee ES, Bae YH (2003) Adriamycin loaded pullulan acetate/sulfonamide conjugate nanoparticles responding to tumor pH: pH-dependent cell interaction, internalization and cytotoxicity in vitro. J Control Release 87:3–13
5. Rodriguez M, Vila-Jato JL, Torres D (1998) Design of a new multiparticulate system for potential site-specific and controlled drug delivery to the colonic region. J Control Release 55:67–77
6. Hu Z, Kimura G, Ito Y et al (1999) Technology to obtain sustained release characteristics of drugs after delivered to the colon. J Drug Target 6:439–448
7. Yoshida R, Uchida K, Kaneko Y et al (1995) Comb-type grafted hydrogels with rapid deswelling response to temperature changes. Nature 374:240–242
8. Wu XS, Hoffman AS, Yager PJ (1992) Synthesis and characterization of thermally reversible macroporous poly(*N*-isopropylacrylamide) hydrogels. J Polym Sci Pol Chem 30:2121–2129
9. Kwon IC, Bae YH, Kim SW (1991) Electrically credible polymer gel for controlled release of drugs. Nature 354:291–293
10. Pelton R (2000) Temperature-sensitive aqueous microgels. Adv Colloid Interface 85:1–33
11. Gan D, Lyon LA (2001) Interfacial nonradiative energy transfer in responsive core-shell hydrogel nanoparticles. J Am Chem Soc 123:8203–8209
12. Xiao XC, Chu LY, Chen WM et al (2003) Positively thermo-sensitive monodisperse core-shell microspheres. Adv Funct Mater 13:847–852

13. Xiao XC, Chu LY, Chen WM et al (2004) Preparation of submicron-sized monodispersed thermoresponsive core-shell hydrogel microspheres. Langmuir 20:5247–5253
14. Chu LY, Park SH, Yamaguchi T et al (2002) Preparation of small-sized monodispersed thermo-responsive core-shell microcapsules. Langmuir 18:1856–1864
15. Qu JB, Chu LY, Yang M et al (2006) A pH-responsive gating membrane system with pumping effects for improved controlled-release. Adv Funct Mater 16:1865–1872
16. Liu L, Yang JP, Ju XJ et al (2011) Monodisperse core-shell chitosan microcapsules for pH-responsive burst release of hydrophobic drugs. Soft Matter 7:4821–4827
17. Salamone JC (ed) (1996) Polymeric materials encyclopedia, vol 10. CRC Press, Florida
18. Li Y, Chu LY, Zhu JH et al (2004) Thermo-responsive gating characteristics of poly(*N*-isopropylacrylamide)-grafted porous polyvinylidene fluoride membranes. Ind Eng Chem Res 43:2643–2649
19. Xie R, Chu LY, Chen WM et al (2005) Characterization of microstructure of poly(*N*-isopropylacrylamide)-grafted polycarbonate track-etched membranes prepared by plasma-graft pore-filling polymerization. J Membrane Sci 258:157–166
20. Wei W, Yuan L, Hu G et al (2008) Monodisperse chitosan microspheres with interesting structures for protein drug delivery. Adv Mater 20:2292–2296
21. Wei W, Wang LY, Yuan L et al (2007) Preparation and application of novel microspheres possessing autofluorescent properties. Adv Funct Mater 17:3153–3158

Part III
Thermo-/pH-Dual-Responsive Hydrogel Functional Materials

Chapter 9
Thermo-/pH-Dual-Responsive Hydrogels with Rapid Response Properties

Abstract In this chapter, strategies for fabricating thermo-/pH-dual-responsive hydrogels with rapid response properties are introduced systematically. The types of hydrogels include thermo-/pH-dual-responsive anionic poly(*N*-isopropylacrylamide-*co*-acrylic acid) (P(NIPAM-*co*-AAc)) hydrogels and microgels and cationic poly(*N*-isopropylacrylamide-*co*-*N,N'*-dimethylamino ethyl methacrylate) (poly(NIPAM-*co*-DMAEMA)) hydrogels.

9.1 Introduction

A hydrogel system that swells and shrinks in response to environmental stimuli such as temperature [1–3], pH [4, 5], ionic strength [6], and certain chemicals [7] has attracted much attention in the past decades. It has potential applications in numerous fields including sensors, actuators, chemical separation, and drug delivery systems [8–15]. Poly(*N*-isopropylacrylamide) (PNIPAM) gel is a typical temperature-responsive gel exhibiting volume-phase transition at approximately 33 °C. The temperature sensitivity of PNIPAM has been extensively studied [16–19]. In many cases, multiple environmental stimuli may occur at the same time. Therefore, from an application point of view, it is much more favorable that hydrogels could respond to more than one stimulus simultaneously. Temperature and pH are two important environmental factors in typical physiological, biological, and/or chemical systems and can be manipulated easily in many applications. Therefore, dual temperature- and pH-responsive hydrogels have attracted much attention [4, 20–26].

For a lot of potential applications of hydrogels, a fast response is necessary for their practical usage. Some target drug delivery systems need hydrogels to release drugs immediately in a specific time and location, and an acting actuator requires an instantaneous feedback after receiving signals. However, the swelling and deswelling behaviors of conventional hydrogels are dominated by

L.-Y. Chu et al., *Smart Hydrogel Functional Materials*,
DOI 10.1007/978-3-642-39538-3_9,

© Chemical Industry Press, Beijing and Springer Berlin Heidelberg 2013

diffusion-controlled transport through the polymer networks, and the response rate of hydrogels is inversely proportional to the square of the smallest dimension of the gel [27, 28]. A conventional PNIPAM hydrogel shrinks very slowly after the temperature is increased from 10 °C to 40 °C due to the formation of dense skin layers, requiring more than a month to reach equilibrium [12]. As previously reported, for thermo-/pH-dual-responsive hydrogels, the combination of pH-responsive materials might reduce or even eliminate the thermosensitivity of the resulting hydrogels [29–31].

In this chapter, strategies for fabricating thermo-/pH-dual-responsive hydrogels with rapid response properties will be introduced systematically. The type of hydrogels includes anionic poly(*N*-isopropylacrylamide-*co*-acrylic acid) (P(NIPAM-*co*-AAc)) hydrogels [32] and microgels [33], as well as cationic poly(*N*-isopropylacrylamide-*co*-*N,N'*-dimethylamino ethyl methacrylate) (poly(NIPAM-*co*-DMAEMA)) hydrogels [34].

9.2 Thermo-/pH-Dual-Responsive Hydrogels with Rapid Response

Considering that a rapid deswelling single thermo-responsive comb-type grafted PNIPAM hydrogel has been reported by Yoshida et al. [12], a novel strategy is developed to improve the response rate of thermo-/pH-dual-responsive P(NIPAM-*co*-AAc) hydrogels by introducing comb-type grafted chains through a modification of the molecular structure [32]. The comb-type grafted P(NIPAM-*co*-AAc) hydrogels are synthesized by radical copolymerization of PNIPAM macromonomer with NIPAM and AAc, as illustrated in Fig. 9.1.

*9.2.1 Fabrication of Comb-Type Grafted P(NIPAM-*co*-AAc) Hydrogels*

9.2.1.1 Synthesis of Macromonomers

The *N*-isopropylacrylamide (NIPAM) macromonomers are synthesized as follows [35]. First, a PNIPAM polymer with a terminal hydroxyl end group (PNIPAM-OH) is prepared. NIPAM (16.95 g, 0.15 mol), 2-hydroxyethanethiol (0.117 g, 1.5 mmol), and benzoyl peroxide (0.0242 g, 0.1 mmol) are dissolved in THF (50 mL). The sample containing the monomer solution is degassed by a freeze-thaw cycle and sealed in vacuum. Polymerization is carried out by heating the reaction at 70 °C for 15 h. To precipitate PNIPAM-OH, the reactant is poured into diethyl ether. PNIPAM-OH is collected by filtration and is purified by repeated precipitation in diethyl ether from acetone. The polymer is isolated by freeze-drying

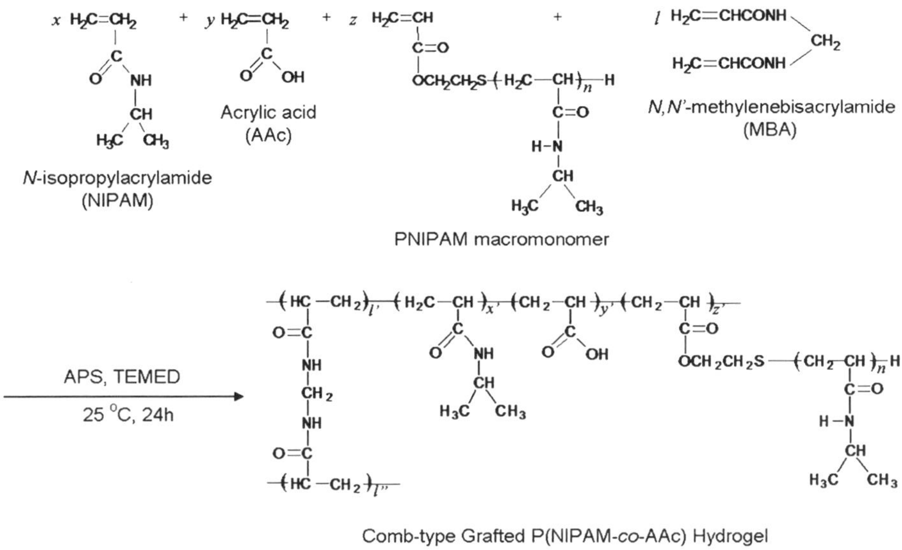

Fig. 9.1 Synthetic scheme for the preparation of comb-type grafted P(NIPAM-*co*-AAc) hydrogels by radical copolymerization (Reproduced with permission from Ref. [32], Copyright (2007), Elsevier)

the aqueous solution. The purified PNIPAM-OH is dissolved in chloroform, acryloyl chloride is instilled, and the reaction is stirred at 40 °C for 2 h under nitrogen atmosphere. NIPAM macromonomer is isolated using the procedure described above for PNIPAM-OH. ^{1}H NMR spectra is recorded on a Bruker-300 spectroscopy using D_2O as the solvent. The molecular weight of semitelechelic PNIPAM-OH is estimated by gel permeation chromatography (GPC, Waters 515 pump with Waters 2,410 refractive-index detector) using THF as the mobile phase and polystyrene as the standard. The weight-average and number-average molecular weights of NIPAM macromonomers are respectively determined to be 8,778 and 4,579 by gel permeation chromatography.

9.2.1.2 Synthesis of Comb-Type Grafted P(NIPAM-*co*-AAc) Hydrogels

To synthesize the comb-type grafted P(NIPAM-*co*-AAc) hydrogels, the NIPAM and AAc monomers, NIPAM macromonomer, cross-linker MBA, and accelerator TEMED are dissolved in 20 mL of ultrapure water (18.2 MΩ cm) and bubbled with nitrogen gas for 10 min, at which time APS is added as an initiator. The solution is then injected between two glass plates covered with PE films and separated by a glass gasket (3.6 mm). The feed compositions of the monomers and other reactants are listed in Table 9.1. Polymerization is carried out at room temperature for 1 day, and the resulting hydrogels are cut into disks (7 mm in diameter) with a cork borer.

Table 9.1 Feed compositions for the preparation of P(NIPAM-*co*-AAc) hydrogels

	Sample ID					
Component	NNA20	GNA20-1	GNA20-2	NNA30	GNA30-1	GNA30-2
NIPAM (g)	1.0	0.6	0.5	1.0	0.6	0.5
Macromonomer (g)	0	0.4	0.5	0	0.4	0.5
AAc (μL)	20	20	20	30	30	30

Note: MBA as a cross-linker = 0.02 g; TEMED as an accelerator = 50 μL; 5 wt % APS as an initiator = 0.2 mL; solution (ultrapure water) = 20 mL

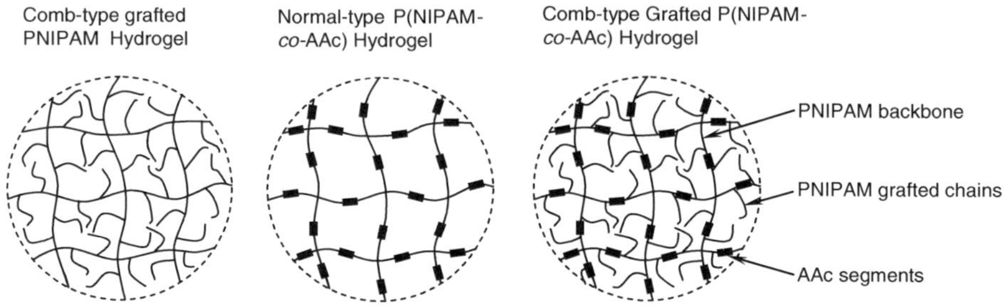

Fig. 9.2 Schematic illustration of the structures of comb-type grafted PNIPAM and normal-type P(NIPAM-*co*-AAc) hydrogels, as well as a comb-type grafted P(NIPAM-*co*-AAc) hydrogel (Reproduced with permission from Ref. [32], Copyright (2007), Elsevier)

To leach out unreacted compounds and allow the hydrogels to equilibrium, the gel disks are immersed in ultrapure water, which is changed twice every day, for at least 6 days at room temperature. The P(NIPAM-*co*-AAc) hydrogels with grafted chains and normal-type P(NIPAM-*co*-AAc) hydrogels are denoted as GNA and NNA, respectively.

The backbone networks of the prepared hydrogels are made up of the NIPAM and AAc components, and the linear PNIPAM polymers serve as the freely mobile chains and are grafted onto the backbone by fixing one end structurally. Within the hydrogel, the grafted chains have freely mobile ends, distinct from the typical network structure in which both ends of chains are cross-linked and relatively immobile. Schematic structures of comb-type PNIPAM, normal-type P(NIPAM-*co*-AAc), and comb-type P(NIPAM-*co*-AAc) hydrogels are provided in Fig. 9.2. As listed in Table 9.1, the grafted P(NIPAM-*co*-AAc) gels constructed with different contents of AAc and macromonomer are designated as GNA20-1, GNA30-1, GNA20-2, and GNA30-2, respectively; normal-type P(NIPAM-*co*-AAc) gels without grafted chains are designated as NNA20 and NNA30 with a different content of AAc. The total weight of NIPAM and PNIPAM macromonomer is kept constant, although their ratio is varied, ensuring that the quantity of thermo-responsive components is constant.

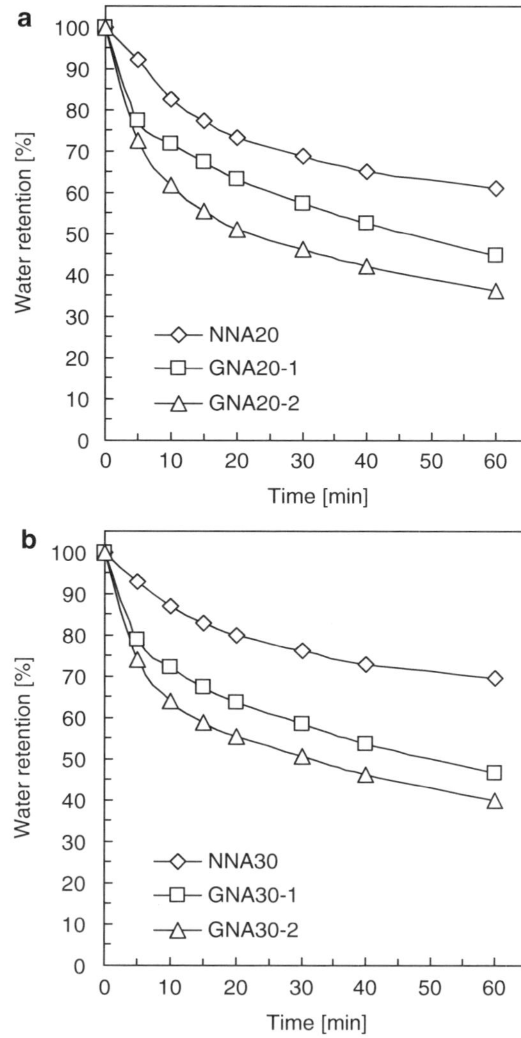

Fig. 9.3 Deswelling kinetics of hydrogels at 60 °C as measured from an equilibrium swelling condition at 25 °C in water (Reproduced with permission from Ref. [32], Copyright (2007), Elsevier)

9.2.2 Deswelling Kinetics of Hydrogels in Various Conditions

9.2.2.1 Deswelling Behavior of Hydrogels in Pure Water with Temperature Changes

Figure 9.3 shows the deswelling kinetics of NNA and GNA gels after a temperature jump from the equilibrium state in ultrapure water at 25 °C to above the LCST at 60 °C. The graft-type hydrogels shrink rapidly on the time scale and entrapped water is rapidly squeezed out from the gels interior for the experiment, and the quick response is ranked in the order of the feed weight of macromonomers in response to temperature changes. In contrast, shrinking of the normal-type P(NIPAM-*co*-AAc) hydrogels is much slower, and it is hardly shrunken at all after 60 min. In this process, such rapid shrinking of the GNA gels is due to the immediate dehydration of freely mobile grafted chains in the gel matrixes, followed by subsequent hydrophobic interactions between dehydrated grafted chains preceding shrinkage of the whole network [12]. On the other hand, surface stable dense shrunken layers within polymer networks containing the hydrophilic AAc comonomer are not formed due to decreased hydrophobic aggregation forces [29]. Therefore, a rapid expulsion of water from the GNA gel matrixes is observed, while this is not the case with pure PNIPAM gel [36]. When NIPAM units aggregate at higher temperatures, incorporate segments which maintain hydration also restrict the shrinkage of hydrogels synchronously. Even though the two forces are contradictory, the hydrophobic forces overwhelm the hydrophilic forces in the matrixes, especially in the graft-type hydrogels with freely mobile ends. Clearly, greater hydrophobic aggregation forces are engendered within the GNA gels having a greater number of grafted chains. In addition, the hydrogels with less AAc exhibit a faster rate of deswelling than those with more AAc.

9.2.2.2 Deswelling Behavior of Hydrogels in pH Buffers at Room Temperature with pH Changes

Figure 9.4 shows the deswelling behavior of hydrogels from a swollen equilibrium in a pH 7.4 buffer solution at room temperature to a pH 2.0 buffer solution. In this process, only a change in pH, with no change of temperature, is measured. The deswelling rate of all gels is relatively slow due to the mild surroundings and lack of a rapid dehydration phenomenon responded to temperature change. The deswelling occurs because of the pH sensitivity of the AAc incorporated in the backbone of hydrogels. With the same AAc content, GNA deswells faster than NNA, and the deswelling rate increases with an increasing number of grafted chains. For the grafted comb-type gels, the AAc units in the backbone may become closer, owing to the higher density of pH-responsive monomers. Therefore, the ability of the hydrogel to respond to pH changes is enhanced, thereby reducing the response time. As expected, an increased amount of AAc units in the hydrogels also results in a faster deswelling rate.

9.2.2.3 Deswelling Behavior of Hydrogels in pH 7.4 Buffer Undergoing Temperature Changes

Large differences are observed in the deswelling process of hydrogels that had been pre-equilibrated in pH 7.4 buffer solutions at 25 °C and are subsequently

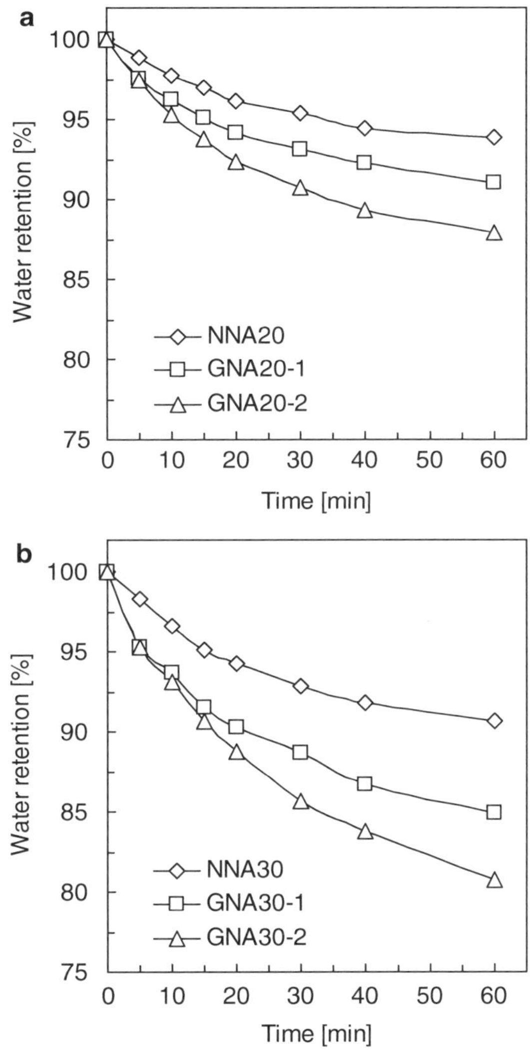

Fig. 9.4 Deswelling kinetics of hydrogels in buffer at pH 2.0 as measured from an equilibrium swelling condition in buffer at pH 7.4 at room temperature (25 °C) (Reproduced with permission from Ref. [32], Copyright (2007), Elsevier)

elevated to a temperature of 60 °C. The deswelling kinetics results are illustrated in Fig. 9.5. Unlike the deswelling behaviors of hydrogels mentioned above, the order of response rates is ranked the other way round. The abnormal phenomenon described above might occur for the following reasons. According to their pH-responsive nature, the polymer chains (backbone) expand because of strong electrostatic repulsions among the charged carboxyl groups of AAc in alkaline solution. Upon an external temperature increase, the freely mobile grafted chains would collapse and the backbone of gels would shrink at the same time due to their NIPAM component. For the NNA gels, the holistic shrinking forces on the backbone aroused by NIPAM would be greater than the expanding forces aroused by AAc. On the other hand,

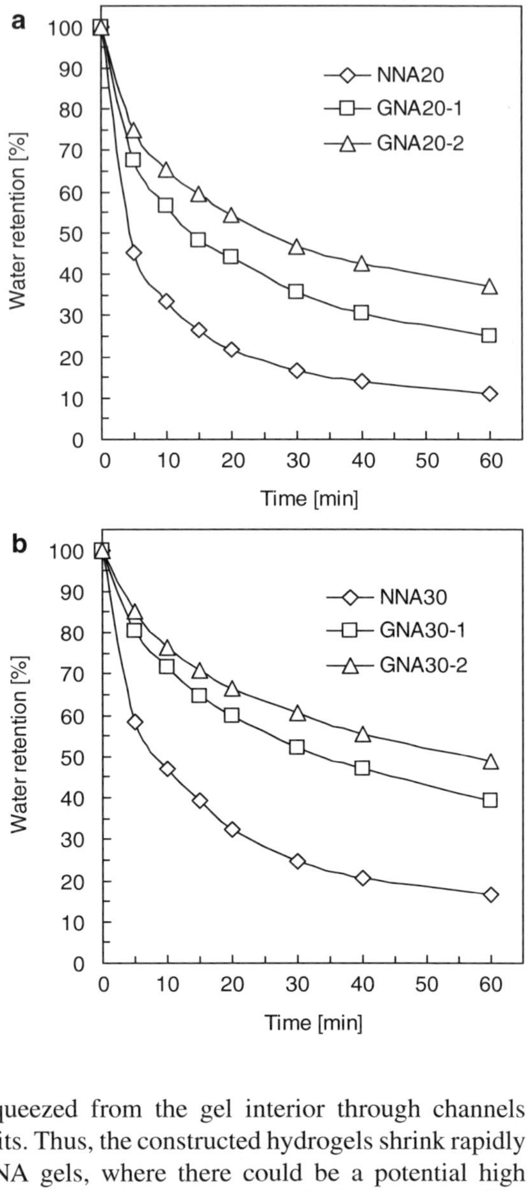

Fig. 9.5 Deswelling kinetics of hydrogels at 60 °C as measured from an equilibrium swelling condition at 25 °C in buffer solution (pH = 7.4) (Reproduced with permission from Ref. [32], Copyright (2007), Elsevier)

trapped water could be easily squeezed from the gel interior through channels formed by the hydrophilic AAc units. Thus, the constructed hydrogels shrink rapidly under such conditions. In the GNA gels, where there could be a potential high density of pH-responsive AAc units in the backbone networks, the largest expanding forces would counteract the shrinking forces to a certain extent, despite the presence of strong aggregated forces among the grafted chains. As a result, the backbone networks are unable to shrink to a large extent. Obviously, the density of AAc in the backbone networks increases either indirectly with the increase of macromonomers or directly with the increase of AAc, and thus the shrinkage of the hydrogels with more grafted chains or AAc is much slower.

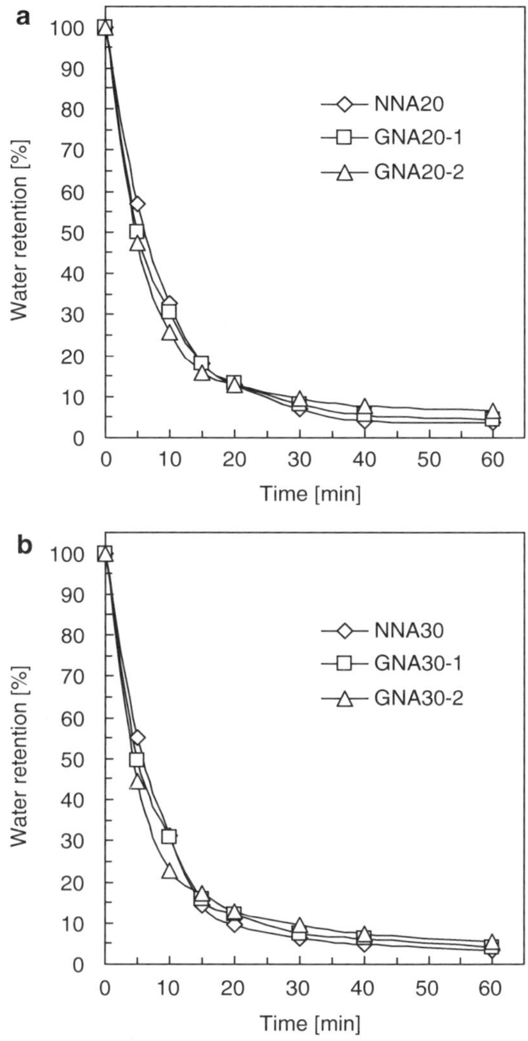

Fig. 9.6 Deswelling kinetics of hydrogels at 60 °C as measured from an equilibrium swelling condition at 25 °C in buffer solution (pH = 2.0) (Reproduced with permission from Ref. [32], Copyright (2007), Elsevier)

9.2.2.4 Deswelling Behavior of Hydrogels in pH 2.0 Buffer with Temperature Changes

Figure 9.6 shows the shrinking kinetics of the hydrogels in a pH 2.0 buffer solution after a temperature jump from the equilibrated state at 25 °C to 60 °C. It can be clearly seen that the shrinking rates of all the gels are much higher than those described above. One interesting trend is that the GNA gels deswell slightly faster than the NNA gels in the beginning but are slower to reach a new equilibrium. Upon reaching a new equilibrium, the water retention of GNA is greater than NNA. The hydrogels are pre-equilibrated in an acidic environment at 25 °C and are thus in a contracted state. Upon experiencing an external temperature shift above the LCST, for the GNA gels, the immediate response to temperature occurs and can be attributed to the freely mobile grafted chains, which would cause a much faster contraction in the absence of the counteraction caused by AAc. Likewise, AAc would produce contracting forces in an acidic solution, but the forces would be milder than those caused by the temperature response. Thus, the rapid shrinking would destroy the layers formed by AAc units for packing water, leading to a rapid release of water from the gels' interior. Subsequently, the rate of deswelling slows because of weakened thermo-responsive effects when the hydrophobic aggregation reaches an utmost, and the ceaselessly formed layers could not be destroyed, resulting in the formation of some uneven bubbles containing aqueous solution. The NNA hydrogels shrink slower than GNA hydrogels in the early phase of the deswelling process because of a lack of grafted chains, and then more aqueous solution releases through the not-well-compacted skin layers due to the lower density of AAc in the backbone networks, whereas they reach a smaller contracted state finally owing to the aggregation effects of the NIPAM main chains. On the other hand, an obvious effect of AAc quantity in the hydrogels is not detected, as the rate of deswelling of all hydrogels occurs too rapidly for such measurements.

9.2.2.5 Deswelling Behavior of Hydrogel in Buffer with Dual Thermo-change and pH Change

In the above-mentioned experiments and results, only one of the environmental elements (temperature or pH) is changed, and the other is fixed to measure the deswelling kinetics of hydrogels in different conditions. In fact, environmental elements always change together in potential applications. Figure 9.7 shows the time course of deswelling for hydrogels undergoing shrinking in pH 2.0 buffer at 60 °C after an abrupt change from pH 7.4 and 25 °C. The NNA hydrogels shrink slowly like their behaviors in ultrapure water at 60 °C. On the other hand, the GNA hydrogels shrink rapidly, exceeding the shrink rates of them in response to either temperature or pH alone, on the minute time scale. The actual shrinking processes of disk-shaped NNA30, GNA30-1, and GNA30-2 gels are demonstrated in the series of photographs in Fig. 9.8. The hydrogels swell in pH 7.4 buffer solutions at 25 °C and the volume is ranked in order of the number of grafted chains, as forenamed. Transparent blister formation is immediately observed on the surfaces of the hydrogels after an abrupt environmental change, and then, these bubbles break continuously during the collapse. This temporal formation of surface layers is probably engendered by the firstly occurred hydrophobic aggregation and constriction of the hydrogels in response to the abrupt pH change to 2.0 at the interface.

The GNA hydrogels, GNA30-1 and GNA30-2, have smaller bubbles than those of NNA30. The strong aggregation accumulates large internal pressure to create the bubble formation on the surface structure [37], and the power of these strong aggregation forces surpasses the influences of the flimsy surface layers on the gel's

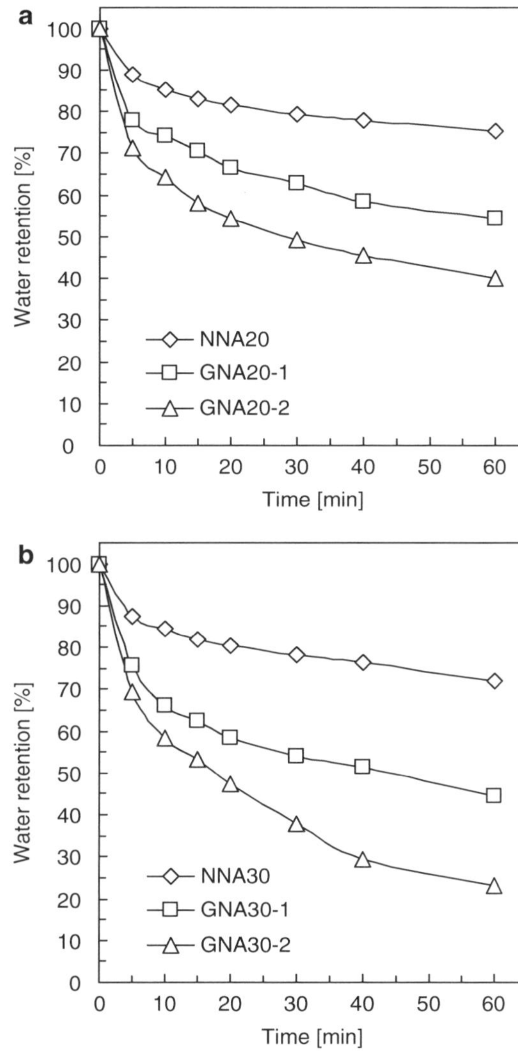

Fig. 9.7 Deswelling kinetics of hydrogels in buffer at pH 2.0 and 60 °C as measured from an equilibrium swelling condition in buffer at 7.4 and 25 °C (Reproduced with permission from Ref. [32], Copyright (2007), Elsevier)

collapse. Meanwhile, contracting forces bring about by the changes in pH further enhances the hydrostatic pressures within the hydrogels. Thus, in a few barrages, the entrapped water is rapidly released through the broken bubbles. In particular, GNA30-2 gel does exhibit mechanical bucking. After a brief time, the hydrogels exhibit a dramatically decreased volume. The deswelling mechanism is close to that of the comb-type grafted PNIPAM gels [37]. However, there are no strong aggregations within the normal-type hydrogels due to the bubbles that are not broken and the water which is prevented from entering the bulk polymer networks. In fact, to some extent, all of hydrogels shrink faster in the beginning (several minutes) of the change in conditions, but this rate becomes slower subsequently. One possible

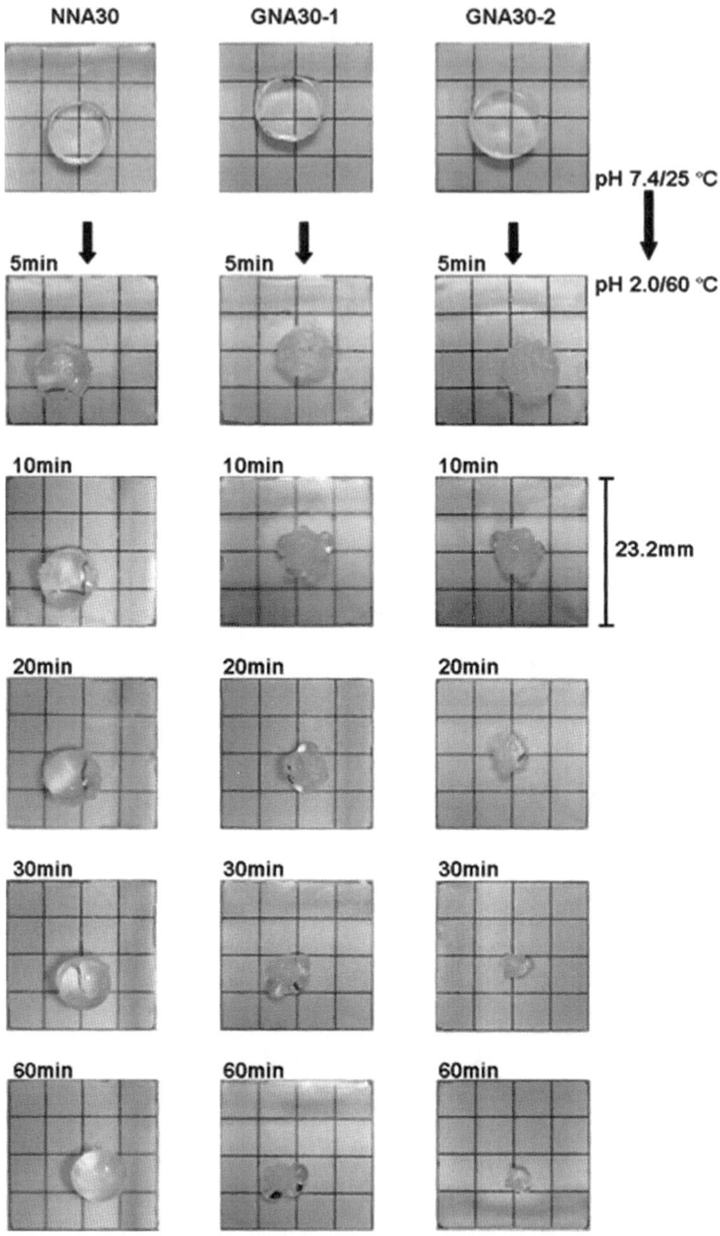

Fig. 9.8 Photographs of the deswelling process of a disk-shaped normal-type hydrogel (NNA30) and comb-type grafted hydrogels (GNA30-1 and GNA30-2) undergoing shrinking at pH 2.0 and 60 °C after being removed from an equilibrium condition of pH 7.4 at 25 °C (Reproduced with permission from Ref. [32], Copyright (2007), Elsevier)

explanation for this phenomenon is that the bubbles are formed incompletely in the beginning of shrinkage so that the water is repulsed rapidly with less delay; however, after the formation of intact bubbles, water is restricted in the bubbles resulting in slower shrinkage of the hydrogels. As mentioned above, the pH sensitivity of hydrogels is also one of the factors responsible for shrinking of the hydrogels. Therefore, faster deswelling rates are observed for hydrogels that have a higher AAc content.

9.3 Graft-Type Microgels with Rapid Thermo-responsive and pH-Responsive Properties

The size of the hydrogel should be small enough to exert effects in many particular regions, especially in the drug delivery systems. Thereupon, microgels with much faster volume change than macroscopic hydrogels with the same chemical structure have attracted increasing attention. However, to broaden the functions of microgels in the aforementioned applications, it is desirable to incorporate other functional groups within the hydrogel matrix. Along these lines, microgels with temperature and pH sensitivities have been commonly investigated in many recent works, because both parameters are important environmental factors in biomedical and other systems. In this section, thermo-/pH-dual-sensitive macromonomers are prepared first and then are incorporated into thermo-/pH-dual-sensitive polymer backbone networks as grafted chains with freely mobile ends via free-radical copolymerization to form graft-type microgels with rapid thermo- and pH-responsive properties [33].

9.3.1 Fabrication of Graft-Type Microgels

9.3.1.1 Synthesis of Poly(NIPAM-*co*-AAc) Macromonomers

A poly(NIPAM-*co*-AAc) macromonomer with a terminal hydroxyl end group poly(NIPAM-*co*-AAc)-OH is prepared by radical telomerization of NIPAM monomer and AAc monomer using HESH as a chain transfer agent (Fig. 9.9). NIPAM (6.78 g, 0.06 mol), AAc (1.03 mL, 0.015 mol), HESH (0.105 mL, 1.5 mmol), and BPO (0.0363 g, 0.15 mmol) are dissolved in THF (25 mL). The monomer solution is degassed by a freeze-thaw cycle and sealed in vacuum. Polymerization is carried out by heating the reaction at 70 °C for 12 h. To precipitate poly(NIPAM-*co*-AAc)-OH, the reactant is poured into diethyl ether. Poly(NIPAM-*co*-AAc)-OH is collected by filtration and is purified by repeated precipitation in diethyl ether from acetone. The polymer is isolated by freeze-drying from aqueous solution. The purified poly(NIPAM-*co*-AAc)-OH is dissolved in chloroform, and

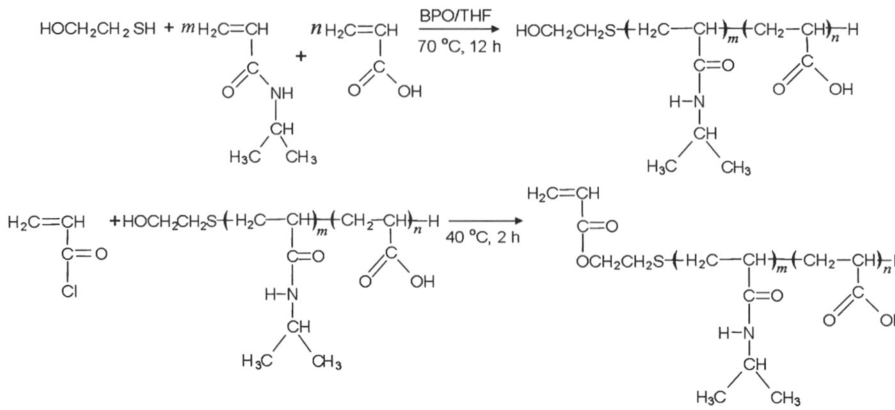

Fig. 9.9 Preparation of poly(NIPAM-*co*-AAc) macromonomer (Reproduced with permission from Ref. [33], Copyright (2008), Elsevier)

acryloyl chloride (excess) is instilled, and the reaction is stirred at 40 °C for 2 h under nitrogen atmosphere, and finally poly(NIPAM-*co*-AAc) macromonomer is isolated using the procedure described above for poly(NIPAM-*co*-AAc)-OH.

^{1}H NMR and FT-IR spectra are recorded on a Bruker-400 spectroscopy using D_2O as the solvent and on a NICOLET-560 spectroscopy, respectively. The molecular weight of semitelechelic poly(NIPAM-*co*-AAc)-OH is estimated by gel permeation chromatography (GPC, Waters 515 pump with Waters 2,410 refractive-index detector) using THF as the mobile phase and polystyrene as the standard. Molar component ratio of NIPAM and AAc in the macromonomer is, respectively, determined to be 79.84 mol % and 20.16 mol % by ^{1}H NMR spectroscopy.

9.3.1.2 Synthesis of Graft-Type Poly(NIPAM-*co*-AAc) Microgels

Graft-type poly(NIPAM-*co*-AAc) microgel is prepared by free-radical copolymerization of poly(NIPAM-*co*-AAc) macromonomer with NIPAM and AAc in the presence of MBA as a cross-linker. Normal-type microgel is also prepared without adding macromonomer, as illustrated in Fig. 9.10. The backbone networks are made up of the NIPAM and AAc components, and the linear poly(NIPAM-*co*-AAc) polymers served as the freely mobile chains and are grafted onto the backbone by fixing one end structurally. Within the microgel, the grafted chains had freely mobile ends, distinct from the typical network structure in which both ends of chains are cross-linked and relatively immobile. Schematic structures of normal-type and graft-type microgels are provided in Fig. 9.11. Poly(NIPAM-*co*-AAc) microgels are prepared by inverse suspension polymerization using kerosene as the continuous phase and PGPR90 as oil-soluble surfactant. Certain NIPAM, AAc, poly(NIPAM-*co*-AAc) macromonomer, and 0.015 g MBA are dissolved in 5 mL ultrapure water containing 10 mg APS, and the mixture is bubbled with

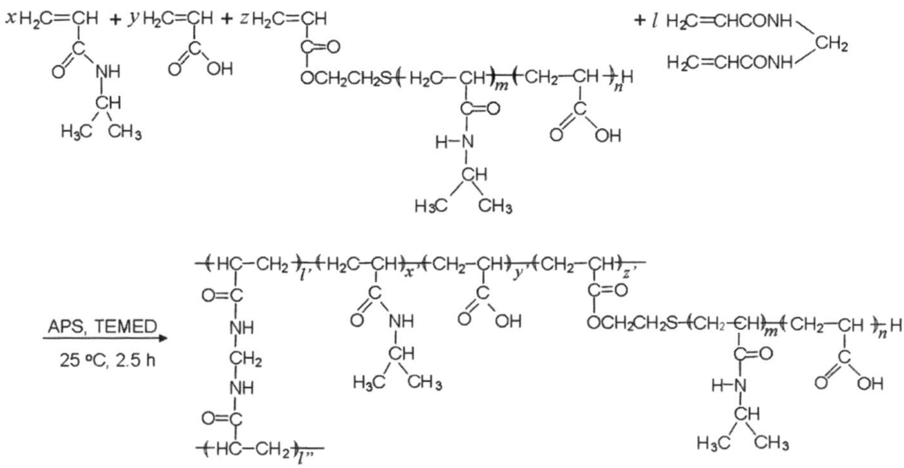

Fig. 9.10 Synthetic scheme for the preparation of the poly(NIPAM-*co*-AAc) microgel with dual sensitive grafted chains (Reproduced with permission from Ref. [33], Copyright (2008), Elsevier)

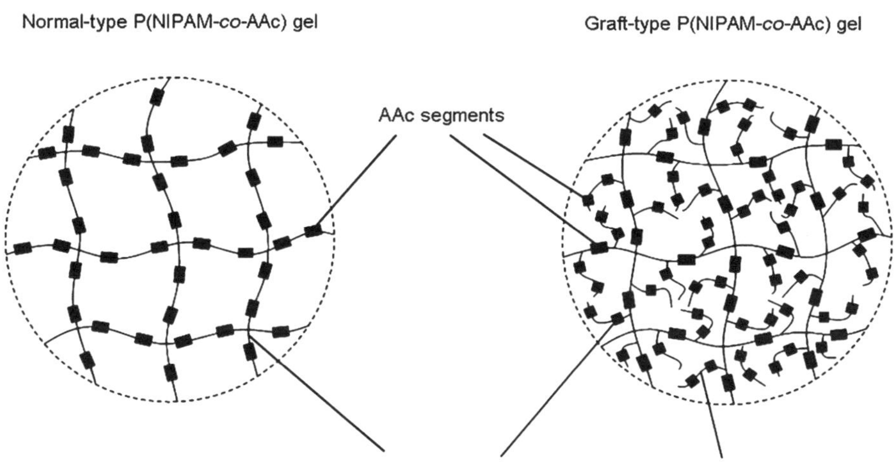

Fig. 9.11 Schematic illustration of the microstructures of normal-type and graft-type poly(NIPAM-*co*-AAc) microgels (Reproduced with permission from Ref. [33], Copyright (2008), Elsevier)

nitrogen for 15 min to remove dissolved oxygen. This solution is immediately poured into 80 mL kerosene containing 3.368 g PGPR90, which is previously purged with nitrogen. The reaction mixture is stirred at 1,450 rpm under nitrogen atmosphere for 25 min. After formation of relatively uniform aqueous emulsion droplets in continuous phase is confirmed, 50 μL of TEMED is added to the continuous phase to initiate polymerization. The polymerization is allowed to proceed for 2.5 h at 25 °C (Fig. 9.10). After polymerization, the microgels

Table 9.2 Feed compositions for the preparation of poly(NIPAM-*co*-AAc) microgels

	Sample code		
Component	PNA[a]	PNAm-1[a, b]	PNAm-2[a, b]
NIPAM monomer (mg)	528.2	431.0	397.7
AAc monomer (μL)	35	20	15
PNA macromonomer (mg)	0	114.0	151.5

Note: Cross-linker MBA = 15 mg; accelerator TEMED = 50 μL; APS initiator = 10 mg; solution (ultrapure water) = 5 mL; emulsifier PGPR = 3.368 g; and continuous phase kerosene = 80 mL
[a]The total molar ratio of NIPAM to AAc in the whole sensitive monomers is kept to about 9:1
[b]The molar ratio of NIPAM to AAc in macromonomer is about 4:1 according to the result of ^{1}H NMR spectroscopy

are separated from the oil phase and are washed with mixture of abluent and ultrapure water one time and then washed with ultrapure water five times. The feed compositions of monomers and other chemicals are listed in Table 9.2. As listed in Table 9.2, the graft-type microgels constructed with different contents of macromonomer and AAc are designated as PNAm-1 and PNAm-2, respectively; normal-type microgels without grafted chains are designated as PNA. The total weight of NIPAM, AAc, and poly(NIPAM-*co*-AAc) macromonomer in PNAm-1, PNAm-2, and PNA microgels is kept constant. Furthermore, the respective ratio of thermosensitive and pH-sensitive components in each kind of microgels is ensured constant.

9.3.2 *Temperature Dependence of Swelling/Deswelling Degree of Microgels in Water*

Equilibrium swelling/deswelling degrees of graft-type and normal-type poly(NIPAM-*co*-AAc) microgels in ultrapure water are plotted as a function of temperature in Fig. 9.12. The introduction of hydrophilic AAc into the polymer framework causes the LCST of microgels to increase, as a higher temperature is needed to drive the disruption of strengthened hydrogen bonds. Nevertheless, the phase transition behaviors indicate that the graft-type microgels show the same phase transition temperature (~53 °C) as that of the normal-type microgels. That means the AAc content in every type of microgels is kept constant, and the prescription to prepare both PNA and PNAm microgels ensures that the quantity of thermosensitive components is constant as well as pH-sensitive components. PNAm microgels exhibit higher swelling degree in swollen state bellow the LCST and lower deswelling degree in collapsed state above the LCST than PNA microgels. As the grafted chains are structurally separated from the backbone cross-linked network,

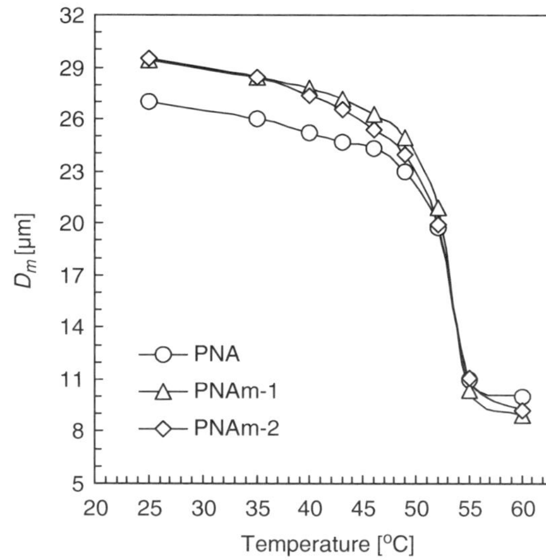

Fig. 9.12 Temperature dependence of the equilibrium mean diameters of normal-type and graft-type poly(NIPAM-*co*-AAc) microgels in water in the temperature range from 25 °C to 60 °C (Reproduced with permission from Ref. [33], Copyright (2008), Elsevier)

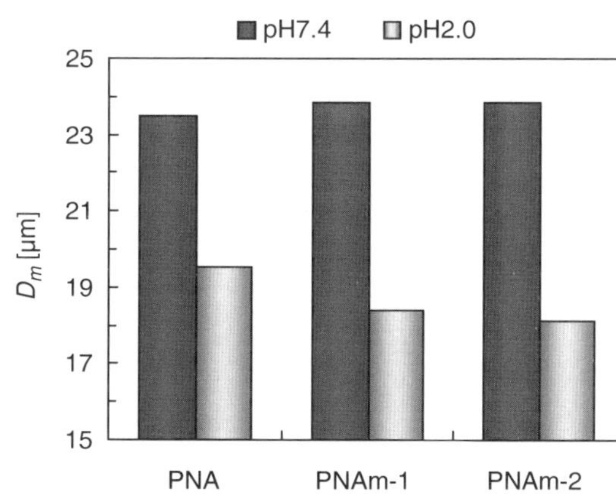

Fig. 9.13 Equilibrium mean diameters of normal-type and graft-type poly(NIPAM-*co*-AAc) microgels in buffer solutions at pH 7.4 and pH 2.0, respectively, at room temperature (25 °C) (Reproduced with permission from Ref. [33], Copyright (2008), Elsevier)

stronger hydration may be possible. This chain expansion is considered to result in increased hydration in PNAm microgels. Controlling the amount of grafted chains can regulate the equilibrium swelling properties of microgels.

9.3.3 Equilibrium Swelling/Deswelling Degree of Microgels in pH Buffers

Figure 9.13 shows the equilibrium swelling/deswelling degrees of graft-type and normal-type poly(NIPAM-*co*-AAc) microgels in pH buffers at 25 °C. Due to the effects of ionic strength, the equilibrium swelling degrees of the microgels in pH buffers are universally smaller than those in ultrapure water [38]. With increasing pH, more and more carboxylic groups (—COOH) within the microgels are deprotonated to the anionic —COO— form so that the internal electrostatic repulsion within the microgels is enhanced [39, 40]. On the other hand, there is a probably intrachain hydrogen-bonding association between NIPAM and AAc units at low pH surrounding [41]. Thus, all of microgels are swollen in pH 7.4 buffer and are shrunk in pH 2.0 buffer. Compared with PNA microgels, PNAm ones are a little more swollen in pH 7.4 buffer, but more shrunk in pH 2.0 buffer to a certain extent. That is, the incorporation of the grafted chains into microgel makes the pH-sensitive swelling/deswelling degree improved. Comparing with PNAm-1 microgels, the PNAm-2 microgels with more side chains show more prominent phenomenon. On the condition at pH 7.4, the hydration of mobile grafted chains are stronger and the electrostatic repulsive forces operating between the charged carboxyl groups of acrylic acid make the grafted chains more extended which tow the whole networks to a more swollen state. On the other hand, at pH 2.0, the grafted chains shrink because the electrostatic repulsive force is vanished between the uncharged carboxyl groups that make their hydration decreased and then tow the whole networks to a more collapsed state. Obviously, the pH sensitivity brought from the AAc on grafted chains is the decisive effect to the response of microgels to pH changes, but the effect of the AAc on rigid main chains is relatively puny.

9.3.4 Deswelling Kinetics of Microgels in Ultrapure Water

Figure 9.14 shows the deswelling kinetics of PNA and PNAm microgels after a temperature jump from the equilibrium state in ultrapure water at 25 °C to above the LCST (at 63 °C). The microgels swell at 25 °C and the volume is ranked in order of the number of grafted chains. The normal-type microgels shrink slowly in response to temperature changes with a nearly changeless stage in the beginning 20 s and reach the minimal shrinkage state after 120 s. In contrast, shrinking of the graft-type microgels is much rapid with a sharp deswelling and reaches the minimal shrinkage state after about 60 s, and the quick response is ranked in the order of the content of macromonomer. For instance, at the point of 40 s, PNAm-1 and PNAm-2 microgels have deswelled about 70 % and 95 % by volume, respectively, but the normal-type microgels have deswelled only about 17 % by volume. On the other hand, the mean diameters of all microgels are almost the same after 120 s.

The actual shrinking processes of PNA, PNAm-1, and PNAm-2 microgels are demonstrated in the series of photographs in Fig. 9.15. Comparing with that of the normal-type PNA microgels with rigid frame architecture, such rapid shrinkage of the graft-type PNAm microgels is due to the immediate dehydration of NIPAM moieties on freely mobile grafted chains in the gel matrixes, followed by subsequent

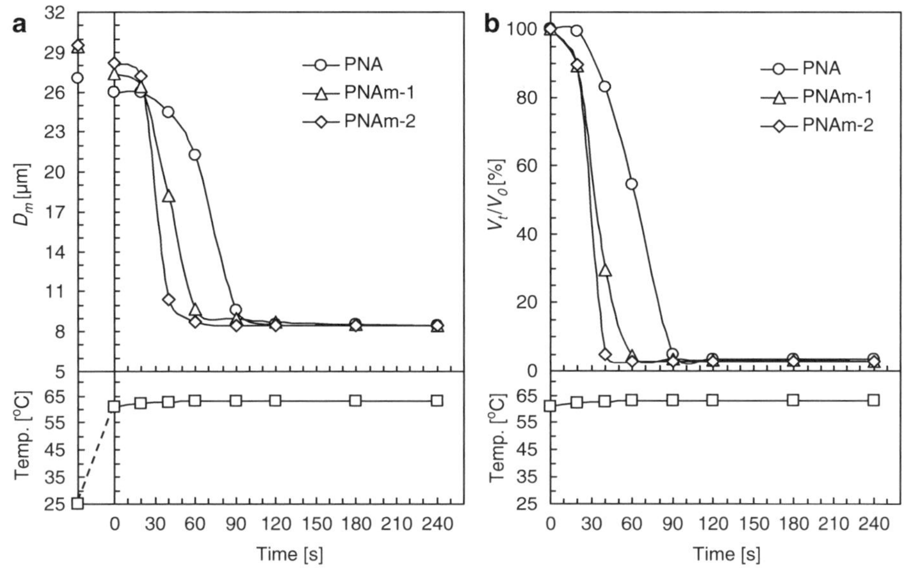

Fig. 9.14 Deswelling kinetics of microgels at about 63 °C that are measured from an equilibrium swelling condition at 25 °C in ultrapure water (Reproduced with permission from Ref. [33], Copyright (2008), Elsevier)

hydrophobic interactions between dehydrated grafted chains preceding shrinkage of the whole network and resulting in rapid expulsion of water from the gel matrix [42]. On the other hand, when NIPAM units aggregate at higher temperatures, the incorporated hydrophilic AAc segments, which bring counterforce to hydrophobic aggregation forces, maintain hydration and also restrict the shrinkage of microgels synchronously. Even though the two forces are contradictory, the hydrophobic forces overwhelm the hydrophilic forces in the matrixes. Nevertheless, there is the same content of AAc in each type of microgels as forenamed, but their distribution is different. The hydrophilic AAc segments located on the main chains of the normal-type microgels restrict the shrinkage of microgels; as a result, PNA microgels shrink slowly to a certain extent. For PNAm microgels, the AAc moieties are not only located on the main chains but also distributed on the grafted chains with freely mobile ends; therefore, the hydrophilic effects might be extremely delicate compared with the strong hydrophobic aggregation effects. In short, the resistances to shrinkage are smaller in PNAm microgels than that in PNA microgels. Clearly, greater hydrophobic aggregation forces are engendered within PNAm microgels having a greater number of branched grafted chains according to the difference between the PNAm-1 and PNAm-2 microgels. Nevertheless, all microgels show almost the same sizes at the equilibrium shrunken state, because they are small enough to squeeze out the interior water in short order.

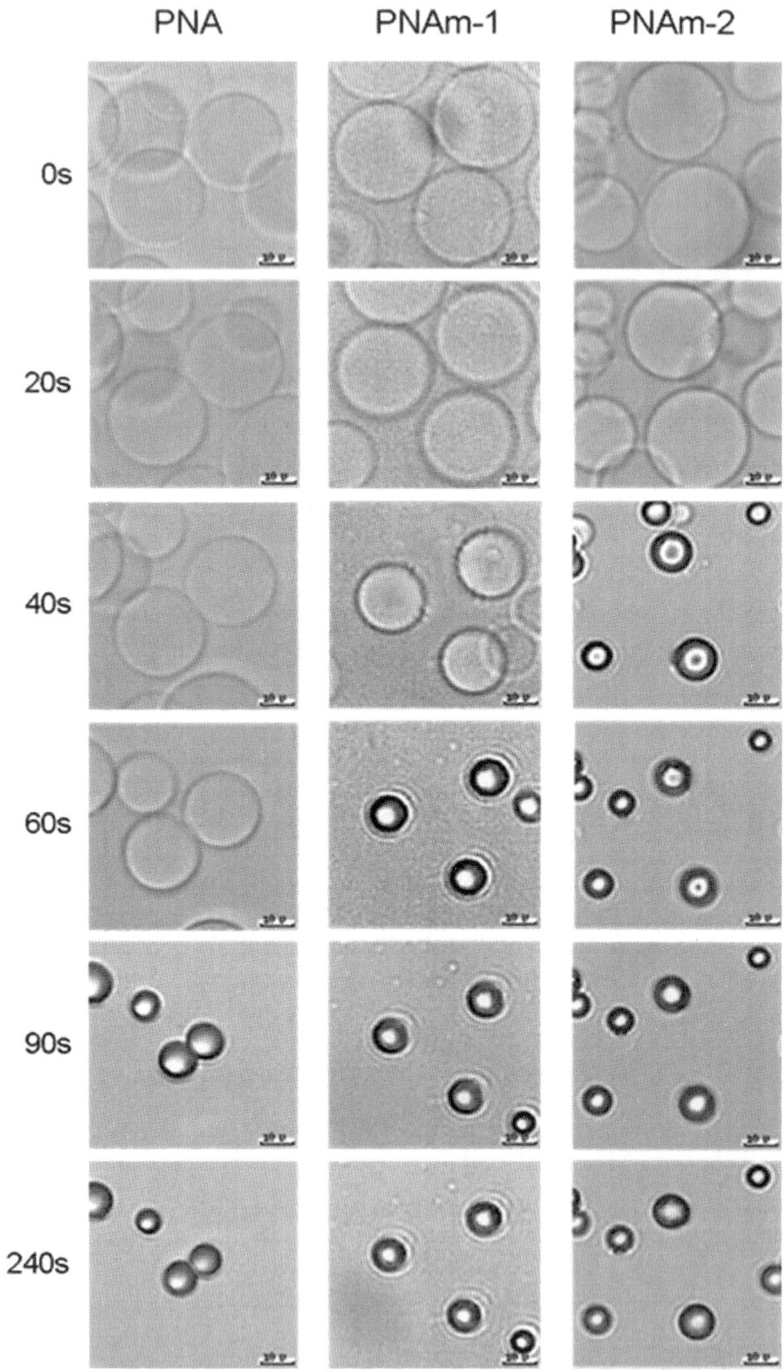

Fig. 9.15 Photographs of the deswelling process of normal-type (PNA) and graft-type microgels (PNAm-1 and PNAm-2) undergoing shrinking at about 63 °C after being suddenly heated from an equilibrium condition at 25 °C. Scale bar is 10 μm (Reproduced with permission from Ref. [33], Copyright (2008), Elsevier)

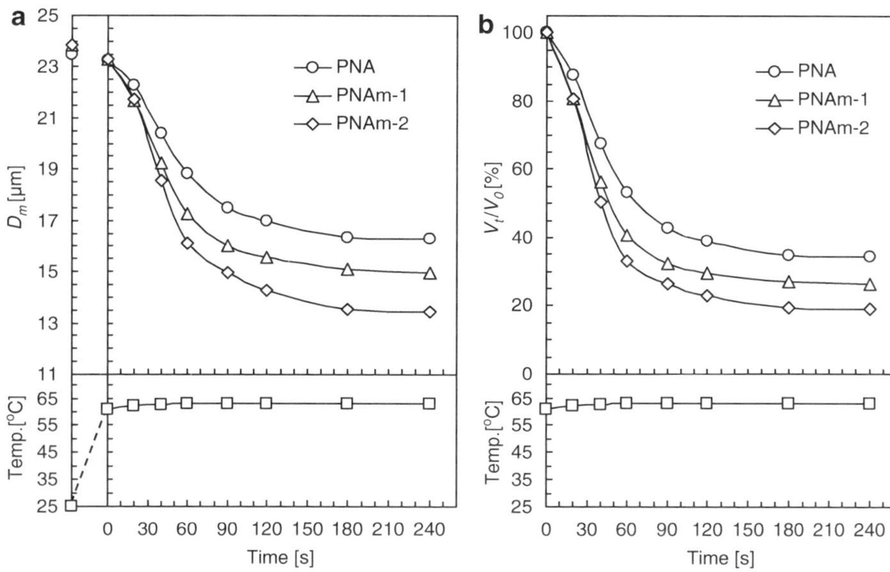

Fig. 9.16 Deswelling kinetics of microgels at about 63 °C that are measured from an equilibrium swelling condition at 25 °C in buffered solution with pH = 7.4 (Reproduced with permission from Ref. [33], Copyright (2008), Elsevier)

9.3.5 Deswelling Kinetics of Microgels in pH Buffers

In order to probe into the dual stimuli sensitivity of the microgels, the deswelling kinetics of the microgels is studied under the conditions with both temperature and pH adjusted. In the experiments, the microgels have been pre-equilibrated in pH 7.4 buffered solutions at 25 °C before measurement, and then the environment temperature is suddenly increased to 63 °C. Similar phenomena to those described above are observed in the deswelling process, as illustrated in Fig. 9.16. Again, the PNAm-1 and PNAm-2 microgels with grafted chains possessing freely mobile ends in the polymeric networks show more rapid deswelling than normal-type PNA microgels. On the other hand, the volume change degrees of all microgels are not as large as those in ultrapure water. According to their pH-sensitive nature, the polymer chains expand because of electrostatic repulsions among the charged carboxyl groups of AAc in alkaline solution. Upon an external temperature increase, the freely mobile grafted chains would collapse and the backbone of networks would also shrink due to their NIPAM component at the same time. For PNA microgels, the electrostatic repulsions of AAc moieties on the backbone would impair the shrinking forces, saying nothing of the absence of strong collapses of grafted chains. In PNAm microgels, where the AAc moieties are partly on the main chains and the others are on the grafted chains with freely mobile ends, the repulsive forces would be too small to counteract with shrinking forces brought from grafted chains and

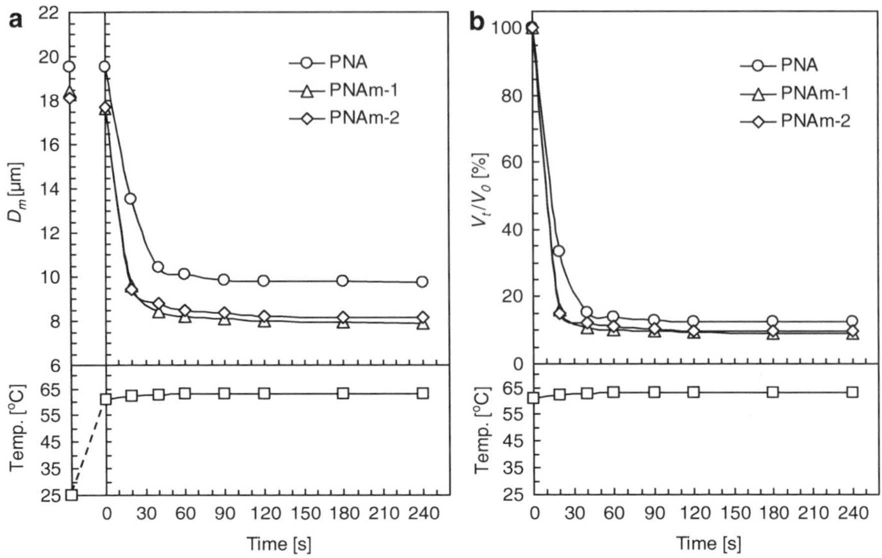

Fig. 9.17 Deswelling kinetics of microgels at about 63 °C that are measured from an equilibrium swelling condition at 25 °C in buffered solution with pH = 2.0 (Reproduced with permission from Ref. [33], Copyright (2008), Elsevier)

the backbone of networks together. Comparing PNAm-2 microgels with PNAm-1 ones, the result shows that the more grafted chains the microgels have, the larger the shrinking force creates and the more rapid deswelling the microgels exhibit. Besides the hydrophilic effects of AAc moieties, just as those in microgels during the deswelling process in ultrapure water, there are still the electrostatic repulsions of AAc moieties in all microgels under this condition. These two effects restrain the shrinkage of microgels to a greater extent; as a result, the deswelling degrees of microgels are smaller than those in ultrapure water.

Figure 9.17 shows the shrinking kinetics of microgels in a pH 2.0 buffer solution after a temperature jump from the equilibrated state at 25 °C to 63 °C. It can be clearly seen that the shrinking rates of all the microgels are much larger than those described above. They have achieved the shrunken equilibrium in about 40 s. All the PNAm microgels deswell more quickly than PNA microgels. One interesting trend is that the PNAm-2 microgels deswell slightly faster than the PNAm-1 microgels in the beginning but become slightly slower to reach a new equilibrium, and the other is that the last mean diameters of the PNA microgels are slightly larger than that in ultrapure water. Upon experiencing an external temperature increase to above the LCST, for PNAm microgels, the immediate response to temperature occurred, which would cause a much faster shrinkage in the lack of counteraction of electrostatic repulsions. Likewise, AAc moieties and the intrachain hydrogen-bonding association would produce contracting layers in an acidic solution, but the forces would be milder than those caused by the temperature response. Thus, the

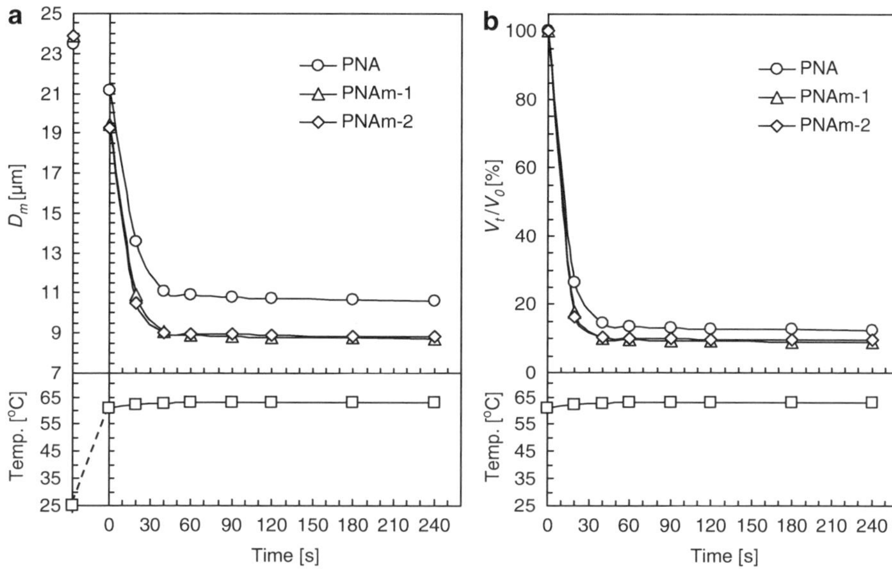

Fig. 9.18 Deswelling kinetics of microgels in buffer at pH 2.0 and 63 °C that are measured from an equilibrium swelling condition in buffer at pH 7.4 and 25 °C (Reproduced with permission from Ref. [33], Copyright (2008), Elsevier)

rapid shrinking would destroy the layers for packing solution, leading to a rapid release of solution from the microgels' interior in a short time. Subsequently, when the hydrophobic aggregation reaches an utmost, the deswelling rate of the microgels with more grafted chains slows because of weakened thermosensitive effects and less content of NIPAM moieties on the main chains according to the recipe. The PNA microgels shrunk slower than PNAm microgels in the early phase of the deswelling process because of a lack of grafted chains, and then more aqueous solution might be preserved by the not-well-destroyed skin layers due to higher density of AAc in the backbone networks. Thus, they have larger sizes than that in ultrapure water at last.

Investigations on microgels in response to dual temperature and pH changes are also carried out. Figure 9.18 shows the time course of deswelling for microgels undergoing shrinking in pH 2.0 buffer at 63 °C after an abrupt change from pH 7.4 and 25 °C. The results are almost the same as the trends in the foregoing condition. These microgels are so small that the pH buffered solution could soak into them in short order. Thus, the actual environment is similar to the condition at pH 2.0 after changing. It is obvious to note that these microgels have shrunk quite a bit in a few seconds after the pH transfer before heating from 25 °C, especially those microgels with grafted chains possessing freely mobile ends inside the polymeric networks. That is, the deswelling rate of the microgels responding to single pH change is also rapid. The grafted chains do enhance the pH responsibility of the dual stimuli-sensitive microgels, just as they enhance the temperature responsibility.

9.4 Rapid pH-/Temperature-Responsive Cationic Hydrogels with Grafted Side Chains

Up to now, most thermo-/pH-dual-sensitive hydrogels are prepared by incorporating pH-responsive anionic components bearing carboxyl groups into PNIPAM-based networks. Such anionic dual stimuli-responsive hydrogels swell in alkaline pH surrounding but deswell in acidic pH environment. However, hydrogels that can swell in acidic pH surrounding and deswell in alkaline pH are necessary in certain cases, such as drug release and dye adsorption [43–46]. For example, the drug (chloramphenicol) should be released more rapidly from hydrogel in a pH 1.4 (close to the pH of the stomach) buffer solution than in a pH 7.4 (close to the pH of the intestine) one [43], in which the drug release is controlled by the swelling-deswelling behavior of the hydrogel. To achieve such functions, cationic hydrogels are needed. Since the dynamics property of environmental-stimuli-responsive hydrogels is vital to their applications, their response rates are usually expected to be as fast as possible. Here the preparation of a novel type of thermo-/pH-dual-sensitive comb-type grafted cationic hydrogels with rapid response rate will be introduced. Both cross-linked network backbones and grafted chains with freely mobile ends are composed of both NIPAM and *N,N*-dimethylamino ethyl methacrylate (DMAEMA) segments, i.e., the chemical compositions of them are poly(*N*-isopropylacrylamide-*co*-*N,N'*-dimethylamino ethyl methacrylate) (poly(NIPAM-*co*-DMAEMA)).

9.4.1 Fabrication of Cationic Hydrogels with Grafted Side Chains

9.4.1.1 Synthesis of Macromonomers

Before fabrication of comb-type poly(NIPAM-*co*-DMAEMA) hydrogels, a poly(NIPAM-*co*-DMAEMA) macromonomer is synthesized. The synthesis route is shown in Fig. 9.19. At first, poly(NIPAM-*co*-DMAEMA) polymer with a terminal hydroxyl group (poly(NIPAM-*co*-DMAEMA)-OH) is prepared by radical telomerization of NIPAM monomer and DMAEMA monomer using HESH as a chain transfer agent. NIPAM (0.054 mol), DMAEMA (0.0088 mol), HESH, and AIBN are dissolved in THF (20 mL). The molar percentages of HESH and AIBN in the total monomers are 3 and 0.8 mol %, respectively. The monomer solution is degassed by a freeze-thaw cycle and sealed in vacuum. Polymerization is carried out at 70 °C for 15 h. After the reaction finishes, the mixture solution is poured into diethyl ether to precipitate semitelechelic poly(NIPAM-*co*-DMAEMA)-OH. Then the precipitation is collected by filtration and is purified by repeated precipitation in diethyl ether from acetone. Secondly, after isolation by freeze-drying from aqueous solution, the purified powder is dissolved in chloroform

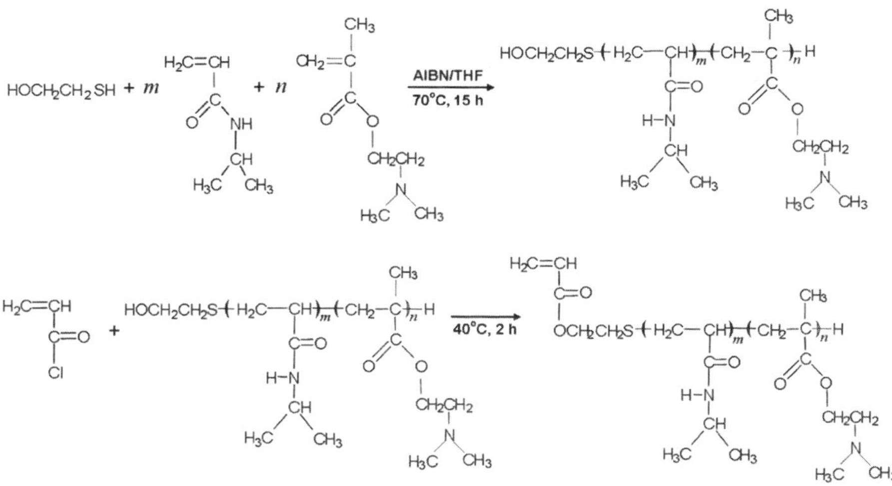

Fig. 9.19 Synthetic scheme for the preparation of the P(NIPAM-*co*-DMAEMA) macromonomer by radical telomerization (Reproduced with permission from Ref. [34], Copyright (2009), Elsevier)

and acryloyl chloride (excess) under stirring and nitrogen atmosphere at 40 °C for 2 h. Finally, the poly(NIPAM-*co*-DMAEMA) macromonomer is collected and purified using the same procedure described above. The molecular weight of macromonomer is estimated by gel permeation chromatography (GPC, Waters 515 pump with Waters 2,410 refractive-index detector) using THF as the mobile phase and polystyrene as the standard. FT-IR and ^{1}H NMR spectra are recorded on a NICOLET-560 spectroscopy and on a Bruker-400 spectroscopy using D_2O as the solvent, respectively.

9.4.1.2 Synthesis of Comb-Type Grafted Poly(NIPAM-*co*-DMAEMA) Hydrogels

Unlike normal-type copolymer hydrogel prepared by direct radical copolymerization of monomers NIPAM and DMAEMA, comb-type grafted hydrogel is synthesized by radical copolymerization of poly(NIPAM-*co*-DMAEMA) macromonomer with monomers NIPAM and DMAEMA. Monomers NIPAM and DMAEMA make up the cross-linked network backbones of the hydrogel, and poly(NIPAM-*co*-DMAEMA) chains with freely mobile ends are grafted onto the backbones. Within comb-type hydrogels, the grafted chains have freely mobile ends, which are distinct from typical network structures of normal-type cross-linked hydrogels. Schematic structures of normal-type and comb-type poly(NIPAM-*co*-DMAEMA) hydrogels are illustrated in Fig. 9.20. The feed compositions of the monomers and other reactants are listed in Table 9.3. The total weight of NIPAM and DMAEMA monomers and poly(NIPAM-*co*-DMAEMA) macromonomer is kept constant. To synthesize comb-type grafted poly(NIPAM-*co*-DMAEMA) hydrogels (PND-50), NIPAM and

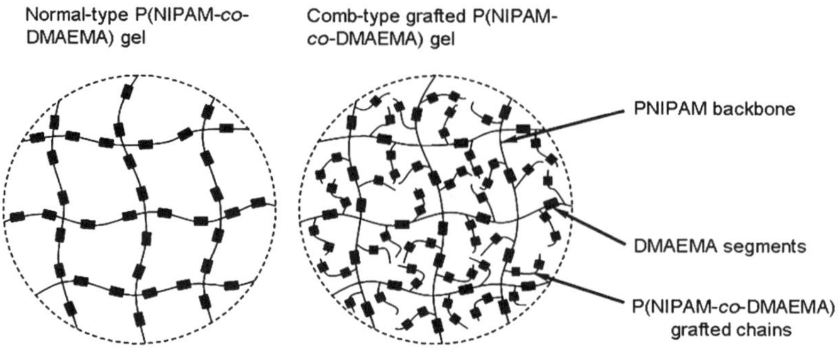

Fig. 9.20 Schematic illustration of the structures of the normal-type and comb-type grafted P(NIPAM-*co*-DMAEMA) hydrogels (Reproduced with permission from Ref. [34], Copyright (2009), Elsevier)

Table 9.3 Recipe for the preparation of poly(NIPAM-*co*-DMAEMA) hydrogels

	Sample code	
Component	PND-00	PND-50
NIPAM monomer (g)	1.8306	0.9153
DMAEMA monomer (g)	0.2826	0.1413
macromonomer (g)	0	1.0566

Note: Cross-linker MBA = 40 mg; accelerator TEMED = 100 μL; initiator APS = 0.015 g; solution (ultrapure water) = 18 mL. The molar ratio of NIPAM to DMAEMA in macromonomer is about 90:10 according to the result of ^{1}H NMR spectroscopy, and the total molar ratio of NIPAM to DMAEMA in the whole monomers is also kept to 90:10

DMAEMA monomers (50 wt %), prepared macromonomer (50 wt %), cross-linker MBA (1.9 wt % to monomers), and accelerator TEMED are dissolved in ultrapure water, which has been bubbled with nitrogen gas for 20 min, and then APS is added as an initiator. The solution is then injected between two glass plates covered with PE films and separated by a Teflon gasket (2 mm in thickness). Polymerization is carried out at room temperature for 24 h, and the resulting hydrogels are cut into disks (8 mm in diameter) with a cork borer. Then, the gel disk samples are immersed in ultrapure water to leach out unreacted chemical residues for at least 3 days at room temperature, and the water is changed twice every day. Swollen gel disks are initially dried under ambient condition for 3 days followed by thorough drying under vacuum at room temperature for fear that the hydrogel structure is destroyed by directly being exposed to vacuum. Normal-type hydrogels (PND-00), which serve as a reference, are also prepared using the protocol described above, except that macromonomer is not included in the recipe. To focus on the response properties of hydrogels due to different physicochemical structures, the ratio of MBA to total monomers (monomer and macromonomer) is fixed at the same value for preparing both comb-type and normal-type hydrogels.

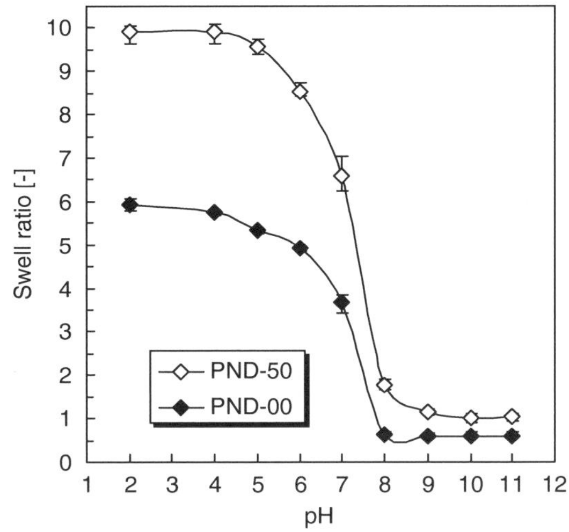

Fig. 9.21 Equilibrium swelling behaviors of hydrogels at 37 °C as a function of pH (Reproduced with permission from Ref. [34], Copyright (2009), Elsevier)

9.4.2 Effects of pH and Temperature on the Equilibrium Swelling Ratio (SR)

The equilibrium swelling ratio is defined as the weight of water absorbed in the equilibrium swollen hydrogel (W_s) divided by the weight of dried hydrogel (W_d) (Eq. 9.1).

$$SR = \frac{W_s}{W_d} \tag{9.1}$$

Due to the natural body temperature and the physiological pH which are about 37 °C and pH 7.4, respectively, the equilibrium swelling ratio (*SR*) of cationic hydrogel disks is measured gravimetrically after wiping excess water from the gel surface with moistened filter paper as a function of pH values at constant temperature of 37 °C or as a function of temperatures at constant pH of 7.4. The ionic strength of all pH buffers is adjusted to 0.1 M beforehand. In the case of constant temperature, dried gel disks have been immersed in a series of pH buffers for 24 h at 37 °C before the measurement. In the case of constant pH, dried gel disks are first equilibrated in pH 7.4 buffer solutions at 20 °C for 12 h and, after being weighted, equilibrated again at another temperature until 48 °C. The transition temperature and pH values of the hydrogels are determined as the corresponding temperature and pH values at which the swelling ratio decreases to half of the original values.

To investigate the effect of pH on the equilibrium swelling ratio, the hydrogels are equilibrated in buffer solutions at pH ranging from 2.0 to 11.0 at 37 °C. Figure 9.21 shows the dependence of equilibrium *SR* of synthesized normal-type and comb-type poly(NIPAM-*co*-DMAEMA) hydrogels on pH values at 37 °C. Both hydrogels

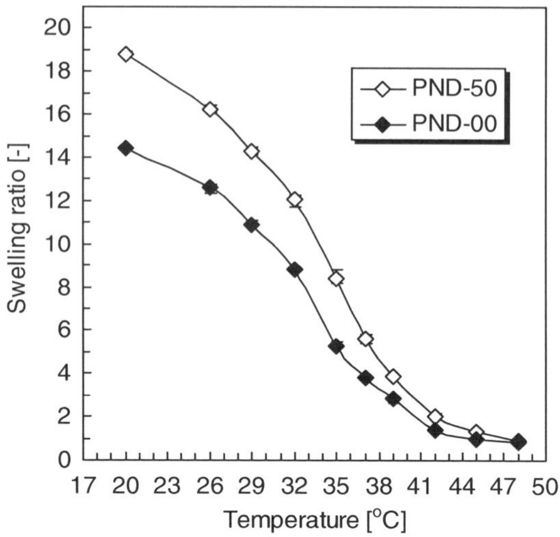

Fig. 9.22 Equilibrium swelling behaviors of hydrogels at pH 7.4 as a function of temperature (Reproduced with permission from Ref. [34], Copyright (2009), Elsevier)

have higher *SR* values at an acidic pH than at a basic pH, and the critical pH values of two hydrogels for the transition are approximately 7.3. Amino groups in cationic poly(NIPAM-*co*-DMAEMA) hydrogels are ionized in the lower pH region and positively charged. As a result, poly(NIPAM-*co*-DMAEMA) hydrogels become more extended because of the increased osmotic pressure among hydrogels. On the other hand, it is difficult for poly(NIPAM-*co*-DMAEMA) hydrogels to produce enough ionized amino groups in the higher pH condition so that the hydrogels take a contracted form [43]. The accordant critical pH values for the transition of PND-00 and PND-50 hydrogels confirm that the ratio of NIPAM to DMAEMA is the same in both hydrogels. Although the equilibrium *SR* difference between two types of hydrogels at alkaline pH is not so significant, PND-50 hydrogel has much larger *SR* value at acidic pH than PND-00 hydrogel. As far as PND-50 hydrogel is concerned, the acidic pH makes the dual sensitive grafted linear chains having freely mobile ends more extended because of the electrostatic repulsion between ionizable groups, which tows the whole networks to a more swollen state. On the other hand, PND-00 hydrogel has no additional aid for swelling. Therefore, comb-type poly(NIPAM-*co*-DMAEMA) hydrogels have higher equilibrium *SR* values at acidic pH than normal-type hydrogels.

The effect of temperature on the equilibrium *SR* of hydrogels in a buffer solution (pH 7.4) is shown in Fig. 9.22. The comb-type poly(NIPAM-*co*-DMAEMA) grafted hydrogel has the same phase transition temperature (about 34 °C) as the normal-type poly(NIPAM-*co*-DMAEMA) hydrogel. This result confirms again that two types of hydrogels are composed of copolymers with the same ratio of NIPAM to DMAEMA. Because amino groups of NIPAM and DMAEMA in hydrogels form intermolecular hydrogen bond with surrounding water at low temperature, hydrogels extend and obtain large *SR*, while hydrogen bonds are overwhelmed by hydrophobic interactions among hydrophobic groups over the LCST, which cause phase separation and shrinkage of hydrogel matrix. However, PND-50 hydrogel shows more swollen state than PND-00 below the LCST. As the grafted chains are structurally separated from the backbone cross-linked network, stronger hydration may be possible [47]. The inherent mobile nature of dual sensitive grafted chains in PND-50 hydrogel makes them readily exposed to water [12]. This chain expansion is considered to result in increased hydration in the comb-type grafted polymeric hydrogels.

9.4.3 Dynamic Swelling/Deswelling Behaviors of Hydrogels in pH Buffer Solutions at Fixed Temperature

The dynamic swelling behaviors of hydrogels are investigated by measuring *SR* variation along with time at room temperature (18 °C) after dried gels are suddenly immersed in the pH solutions. At regular time intervals, the gel sample is retrieved, wiped, and weighed. The dynamic deswelling behaviors of hydrogels are investigated by measuring the water retention of hydrogel in different designed conditions at different time intervals, which is calculated using the following equation:

$$\text{Water retention} = \frac{(W_t - W_d) \times 100\,\%}{W_s} \tag{9.2}$$

where W_t is the weight of hydrogel in certain pH buffer solution at time t, and the other symbols are the same as those described above. Firstly, the gel sample is allowed to reach equilibrium in given buffer solutions at a given temperature; then, the equilibrated gel sample is quickly transferred to another condition with pH and/or temperature stimuli. At regular time intervals, the weight datum is measured. The *SR* and water retention are determined by the weight datum with an average of three samples.

Figure 9.23 shows the *SR* variation of dried poly(NIPAM-*co*-DMAEMA) gels with time after being immersed in different pH buffer solutions at room temperature (18 °C). For the same type of hydrogels, they have trends to swell faster and to higher degree at an acidic solution than at an alkaline solution. In the dual thermo-/pH-sensitive poly(NIPAM-*co*-DMAEMA) hydrogels, NIPAM segments respond to temperature and DMAEMA segments respond to pH. According to the above-mentioned results in Figs. 9.21 and 9.22, poly(NIPAM-*co*-DMAEMA) hydrogels are swollen at lower temperature (below 34 °C) and acidic pH (lower than 7.3) and become shrunken at higher temperature (above 34 °C) and alkaline pH (higher than 7.3). At lower temperature (18 °C) and acidic pH (pH 2.0), both swelling effects derived from thermo- and pH-sensitive components make hydrogels expand sufficiently, while at lower temperature (18 °C) and alkaline pH (pH 10.0), swelling effects derived from the hydration of thermosensitive components and shrinking effects from decreased electrostatic repulsion attributing to pH-sensitive components counteract mutually to some extent, which leads to insufficient swollen

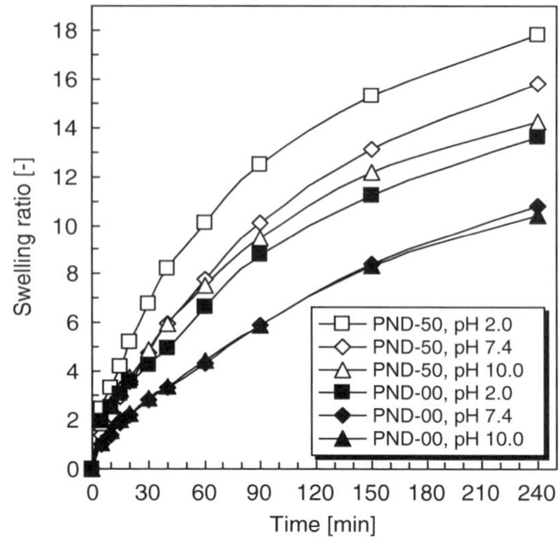

Fig. 9.23 Dynamic swelling behaviors of hydrogels at 18 °C in different pH buffer solutions (Reproduced with permission from Ref. [34], Copyright (2009), Elsevier)

state. As a result, the swelling rate and degree of each type of hydrogels are larger at acidic pH solution than at alkaline pH solution at the same temperature of 18 °C. Moreover, although hydrogels receive adverse interactions at 18 °C and pH 10.0, they still exhibit swelling tendency, which shows that the effect of environment temperature overwhelms that of pH. The molar ratio of NIPAM to DMAEMA is 90–10 for both types of hydrogels, i.e., the amount of pH-sensitive component in hydrogels is much smaller than that of thermosensitive component. Therefore, the pH-response ability of hydrogels is milder than that of thermo-responsibility. For the same type of hydrogels, however, the higher the pH is, the slower the swelling rate. This confirms that the shrinking effect at alkaline pH restricts the swelling behavior of hydrogels to a certain extent. In addition, PND-50 hydrogel swells much more rapidly and to a higher degree at all pH investigated (i.e., pH = 2.0, 7.4, and 10.0) than PND-00 hydrogel. In PND-50 hydrogels, the dual sensitive grafted chains, which could hydrate with little restriction, could tow the whole networks to swell quickly. On the other hand, the shrinking effect of the dual sensitive grafted chains at alkaline pH may be concealed by the expanded effect owing to the strong hydration. Since the interactions in PND-50 hydrogels are larger than that in PND-00 hydrogels, PND-50 hydrogels have higher swelling rates and swollen degree than PND-00 hydrogels.

Figure 9.24 shows the dynamic deswelling behaviors of hydrogels at 22 °C with a sudden pH jump from 2.0 to 11.0. The hydrogels are immersed in a pH 2.0 buffer solution to the equilibrium state before the test. The deswelling rate of two hydrogels is slow due to the swelling effects of thermosensitive components at low temperature of 22 °C. Although the molar ratios of NIPAM to DMAEMA in the hydrogels are the same, the PND-50 hydrogel deswells faster than the PND-00 hydrogel. In the PND-50 hydrogel, DMAEMA segments are distributed throughout the backbone networks and the grafted chains. The pH-sensitive mobile side chains

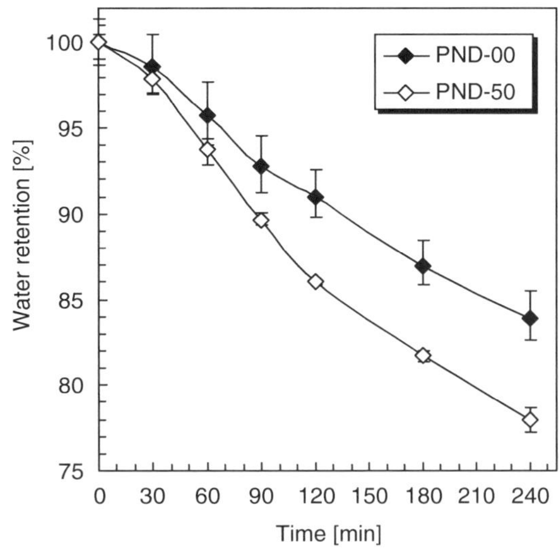

Fig. 9.24 Dynamic deswelling behaviors of hydrogels at 22 °C with a sudden pH jump from 2.0 to 11.0. The hydrogels are immersed in pH 2.0 buffer solution to the equilibrium state before the test (Reproduced with permission from Ref. [34], Copyright (2009), Elsevier)

in the comb-type PND-50 hydrogel could shrink with little restriction at alkaline solution and tow the whole networks to deswell quickly. Therefore, the shrinking rate of comb-type hydrogels in response to pH change is enhanced compared with normal-type hydrogels. In Table 9.3, the cross-linker MBA used is the same for PND-00 and PND-50, but for PND-50 half of NIPAM and DMAEMA go to the side chain. Therefore, the cross-linking density of the backbone of PND-00 should be around half that of PND-50. As a result, the mesh size of the comb-type hydrogel is much smaller than that of the normal-type one in the original state after the polymerization. Even so, the introduction of side chains with freely mobile ends in PND-50 hydrogel makes the mobility of the networks increase.

9.4.4 Dynamic Deswelling Behaviors of Hydrogels in Fixed pH Buffer Solutions with Temperature Stimuli

Figure 9.25 exhibits dynamic deswelling behaviors of hydrogels in different pH buffer solutions with a sudden temperature increase. The hydrogels before the test are immersed in the same buffer solutions to the equilibrium state at 18 °C. It is obvious that PND-50 hydrogels deswell more rapidly than PND-00 hydrogels in each pH buffer solution (from acidic pH to basic pH). Especially, the PND-50 hydrogel approaches to the smallest volume state within 5 min at pH 11.0, but the PND-00 hydrogel shrinks much slowly. Moreover, the higher pH value the buffer solution has, the larger difference between the shrinking rates the two types of hydrogels show. When the cross-linked normal-type hydrogel meets abrupt environment stimuli, the polymer networks composed of two types of thermosensitive segments near the surface of the hydrogel shrink firstly and form

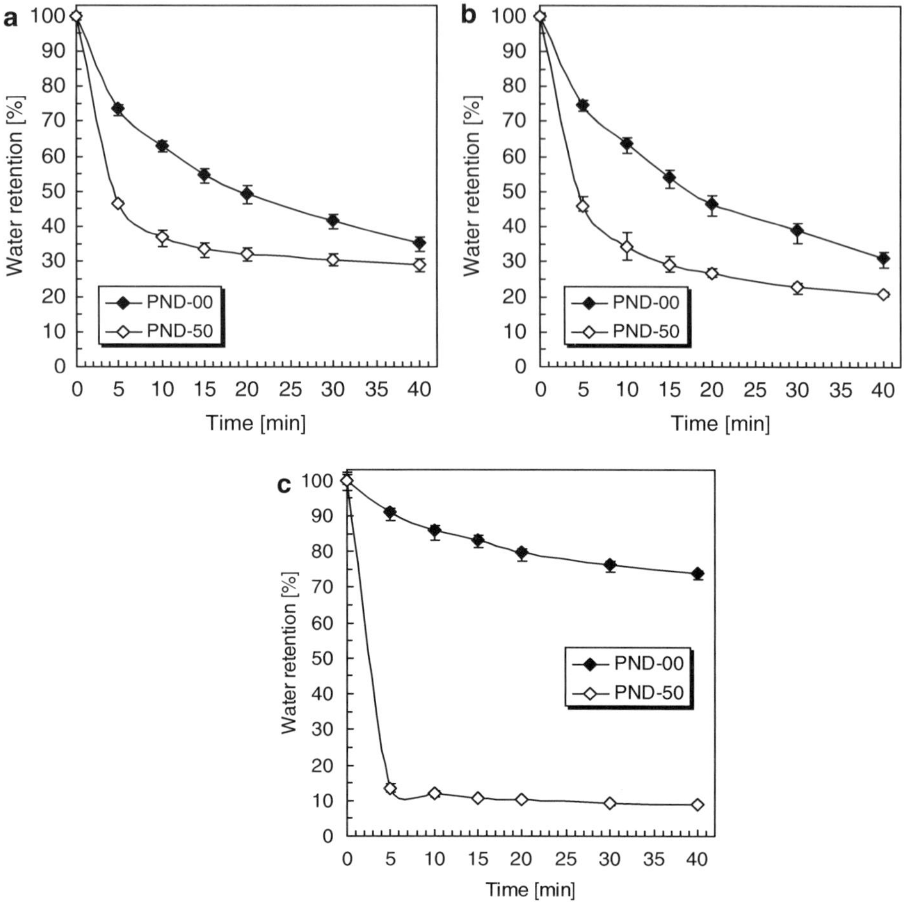

Fig. 9.25 Dynamic deswelling behaviors of hydrogels in three pH buffer solutions with a sudden temperature increase. (**a**) pH 2.0, T = 49 °C (increased suddenly from 18 °C); (**b**) pH 7.4, T = 44 °C (increased suddenly from 18 °C); (**c**) pH 11.0, T = 44 °C (increased suddenly from 18 °C). The hydrogels before the test are immersed in the same buffer solutions to the equilibrium state at 18 °C (Reproduced with permission from Ref. [34], Copyright (2009), Elsevier)

dense skin layer, which blocks further dehydrating of the hydrogel matrixes. Along with the process, the aqueous solution near the surface is vented quickly that brings anticlimactic deswelling of all hydrogels in the beginning (with surface effect). However, the swelling effect that occurs at an acidic solution resists the shrinking force caused by the temperature increase; on the contrary, the shrinking effect that occurs at a basic solution assists the shrinking force. For the PND-00 hydrogel, the aqueous solution is expulsed easily from the interior of hydrogel at pH 2.0 but is enwrapped by the dense skin layer at pH 11.0. Therefore, the response rate of PND-00 hydrogel increases with decreasing the pH value, which is mainly governed by the integrality of the dense skin layer. On the other hand, immediate dehydration of freely mobile grafted chains in the PND-50 hydrogel matrixes and the subsequent hydrophobic interactions between dehydrated grafted chains accelerate the shrinkage of the whole network [12]. Moreover, there is no integrated dense skin layer formed on the surface of comb-type grafted hydrogel; therefore, the aqueous solution could not be enwrapped tightly inside the hydrogel. As a result of the above-mentioned two reasons, the comb-type grafted hydrogel shrinks much more rapid than the normal-type hydrogel. The pH-sensitive DMAEMA units in the flexible grafted chains would counteract with the shrinking force to a certain extent because of their expanding effect in acidic solution but enhance the shrinking force in alkaline solution contrarily. Therefore, in reverse with the order of that of PND-00 hydrogels, the response rates of PND-50 hydrogels increase to a certain extent with the increase of surrounding pH.

9.4.5 Dynamic Deswelling Behaviors of Hydrogels in Buffer Solutions with Both pH and Temperature Stimuli

Figure 9.26 shows the dynamic deswelling behaviors of hydrogels in buffer solutions triggered by simultaneous temperature and pH sudden-stimuli. The shrinking trends of two types of hydrogels are similar to those in Fig. 9.25. Similarly, the comb-type grafted hydrogel shrinks more rapidly than normal-type hydrogel when both temperature and pH are increased suddenly.

Figure 9.27 shows the deswelling process of disk-shaped normal-type hydrogel (PND-00) and comb-type grafted hydrogel (PND-50) that is triggered by simultaneous temperature and pH sudden-stimuli. The pH and temperature are changed from pH 7.4 and 18 °C to pH 11.0 and 44 °C suddenly. Transparent blister formation is immediately observed on the surface of PND-00 hydrogel after an abrupt environmental change [32], and the blister breaks after a period of time. However, the PND-50 hydrogel shrinks continuously with no obvious blister. The phenomena verify the above-mentioned dense skin layer formation on the hydrogel surface.

During the shrinking process, the polymer networks near the surface of cross-linked normal-type hydrogel form the temporal dense skin layer, which is probably attributed mainly to the firstly occurred hydrophobic aggregation forces [48]. There are also anticlimactic deswelling of all hydrogels in the beginning because of the drainage near the surface (i.e., the surface effect). For the hydrogels pre-equilibrated at pH 2.0, the neutralization between the buffer solution maintained in hydrogels and the environmental buffer solution (pH 11.0 in Fig. 9.26) occurs. The actual pH surrounding is low, so the response to pH resists the shrinking force caused by the temperature increases and then makes the dense skin layer not well compacted. As a result, the well-proportioned expulsion of solution from interior of hydrogels is observed. As for the hydrogels pre-equilibrated at pH 6.0 or

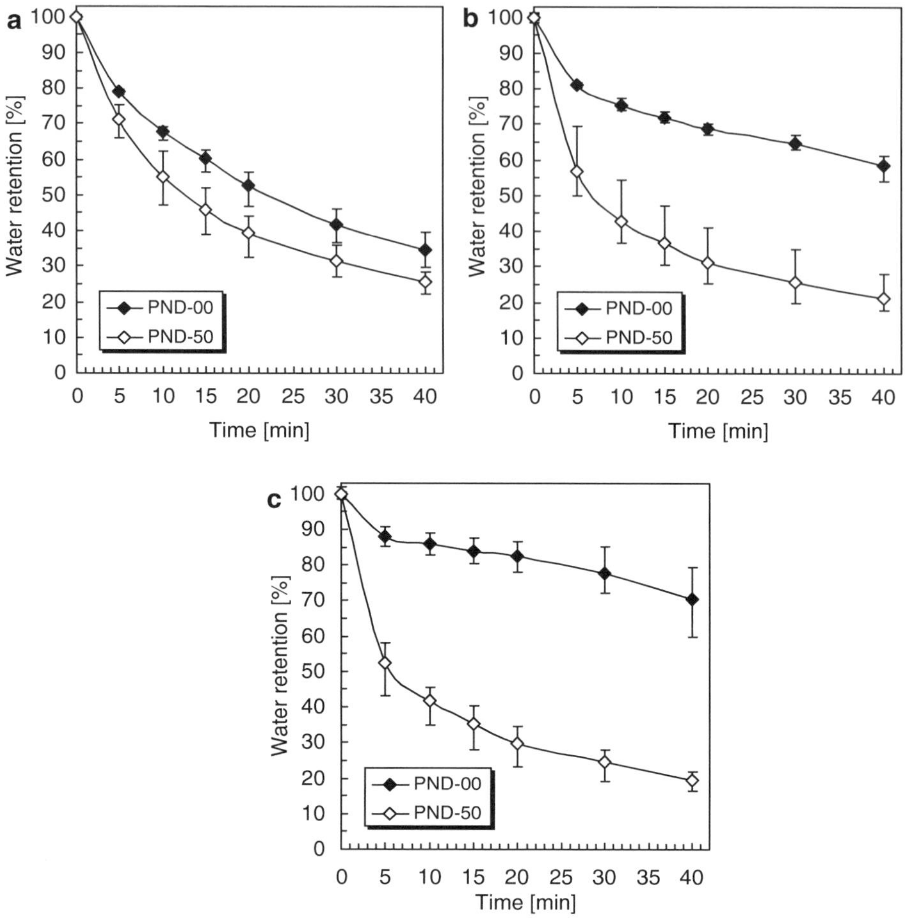

Fig. 9.26 Dynamic deswelling behaviors of hydrogels in buffer solutions triggered by simultaneous temperature and pH sudden-stimuli. The original conditions of hydrogels (in the equilibrium state) before testing are (**a**) pH 2.0, T = 18 °C; (**b**) pH 6.0, T = 18 °C; and (**c**) pH 7.4, T = 18 °C. The changed temperature and pH for the observation are 44 °C and pH 11.0, respectively (Reproduced with permission from Ref. [34], Copyright (2009), Elsevier)

pH 7.4, the pH surrounding in them is closer to the exterior pH, and the resistance to the hydrophobic aggregation would vanish and is replaced by the cooperation. Thereupon, the dense skin layer forms on the surface of the normal-type hydrogels; as a result, the solution in them is enwrapped and hardly effuses. However, the PND-50 hydrogel exhibits much faster deswelling due to the absence of dense skin layer on the surface and the acute aggregation of the dual sensitive grafted chains.

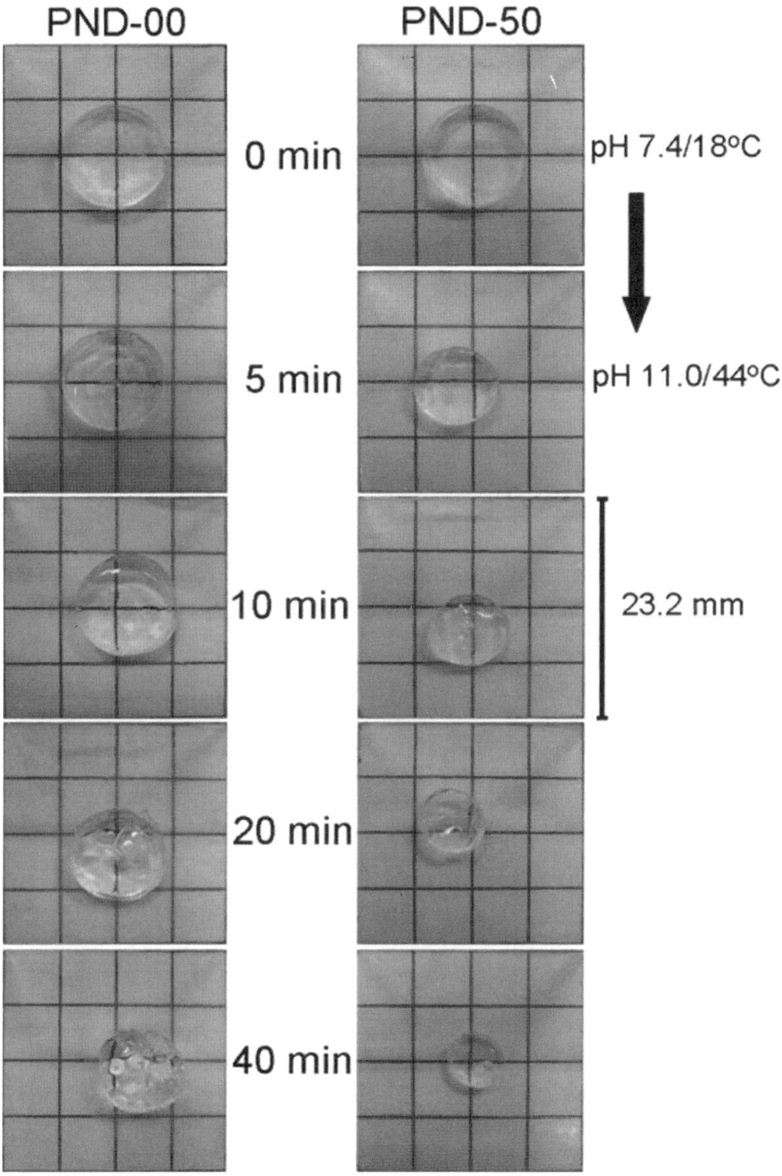

Fig. 9.27 Photographs of the deswelling process of disk-shaped normal-type hydrogel (PND-00) and comb-type grafted hydrogel (PND-50) that is triggered by simultaneous temperature and pH sudden-stimuli. The pH and temperature are changed from pH 7.4 and 18 °C to pH 11.0 and 44 °C suddenly (Reproduced with permission from Ref. [34], Copyright (2009), Elsevier)

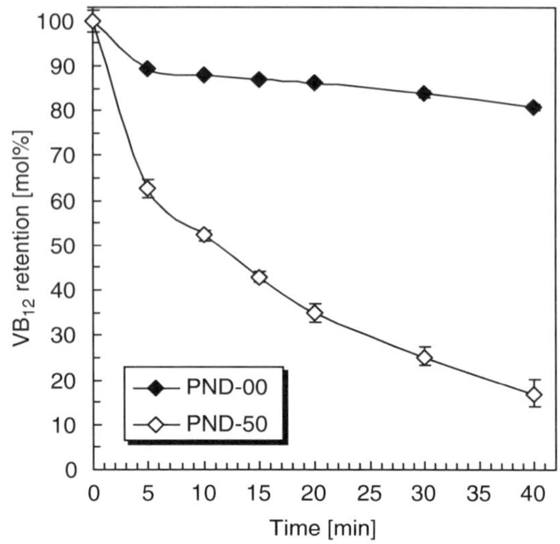

Fig. 9.28 Release of VB_{12} from the disk-shaped normal-type hydrogel (PND-00) and comb-type grafted hydrogel (PND-50) that is triggered by simultaneous temperature and pH sudden-stimuli. The pH and temperature are changed from pH 7.4 and 18 °C to pH 11.0 and 44 °C suddenly (Reproduced with permission from Ref. [34], Copyright (2009), Elsevier)

9.4.6 Drug Release from the Hydrogels

One of the current interests in stimuli-sensitive hydrogels for both biomedical and nonbiomedical uses is the possibility that the release of small molecules from the hydrogel can be controlled by environment temperature and pH stimuli. VB_{12} is selected as a model drug in this study because it is a neutral macromolecule and then there is no specific interaction between the drug and hydrogel. Dried gel disks had been immersed in 0.8 mmol L^{-1} VB_{12} buffer solutions of pH 7.4 at room temperature (18 °C) for at least 72 h before drug release experiments. The VB_{12} buffer solution is refreshed periodically to ensure the VB_{12} concentration inside the hydrogels reaches 0.8 mmol L^{-1}. Before the release experiment, hydrogels are weighted after the sample surfaces have been wiped with moistened filter paper; the total drug-loading (M_0) is defined as

$$M_0 = \frac{\left(W_s^{'} - W_d\right) \times 8 \times 10^{-4}}{\rho_L} \tag{9.3}$$

where W_s' and W_d are the weights of drug-loading hydrogel and dried gel, respectively, and ρ_L is the density of VB_{12} buffer solution. After being weighted, the drug-loading hydrogel disks are quickly transferred into pH 11.0 buffer solution without VB_{12} at a constant 44 °C using a thermostatic unit. At regular time intervals, VB_{12} concentration in the solution is measured by using a UV–vis recording spectrophotometer at a wavelength of 361 nm. VB12 release behaviors from three samples are tested and the average value is used.

The release of VB_{12} from the normal-type and comb-type grafted P(NIPAM-*co*-DMAEMA) hydrogels that is triggered by simultaneous temperature and pH sudden-stimuli is shown in Fig. 9.28. The pH and temperature are changed from

Table 9.4 Values of parameters of the kinetic model for the release rate (Eq. 9.4)

	Including surface effect[a]			Excluding surface effect[b]		
Sample	*A*	*B*	R^2	*A*	*B*	R^2
PND-00	94.185	0.0041	0.7739	90.657	0.0027	0.9845
PND-50	84.027	0.0415	0.9759	75.206	0.0374	0.9990

Note: *R* is the correlation coefficient. The closer the R^2 value approaches to 1, the better the kinetic model fits in with the experimental data

[a]Experimental data from 0 min are used for fitting the kinetic model, i.e., the surface effect during the release is included

[b]Experimental data from 5 min are used for fitting the kinetic model, i.e., the surface effect during the release is excluded

pH 7.4 and 18 °C to pH 11.0 and 44 °C suddenly. The model drug molecules are mainly squeezed out from the hydrogels together with the solution when the hydrogels meet environmental stimuli. The results show that only a small fraction of VB_{12} is released from the normal-type PND-00 hydrogel even after 40 min; on the other hand, the release rate of VB_{12} from the comb-type PND-50 hydrogel is significantly fast due to the rapid deswelling behavior of comb-type hydrogels. The release rate data are fitted to an exponential model as the following equation:

$$y = Ae^{-Bx} \tag{9.4}$$

where y is the VB_{12} retention in the hydrogel, A and B are two constants, and x is the time. The values of these parameters are listed in Table 9.4. It shows that the exponential model is accordant with the rate of release from hydrogels, especially in the condition that the surface effect is excluded. The A values for PND-00 and PND-50 hydrogels are pretty much the same thing, but the B value for the PND-50 hydrogel is over decuple larger than that for the PND-00 hydrogel. According to the characters of exponential function, the larger the constant B is, the larger the decrease degree of y along with the increase of x. That is, comparing with that of the normal-type PND-00 hydrogel, the release rate of the model drug VB_{12} from the comb-type PND-50 hydrogel is significantly faster.

9.5 Summary

By introducing comb-type grafted chains onto the hydrogel backbone networks, thermo-/pH-dual-responsive P(NIPAM-*co*-AAc) anionic hydrogels and microgels and poly(NIPAM-*co*-DMAEMA) cationic hydrogels with rapid response rate can be effectively achieved. Due to the introduction of freely mobile grafted side chains into the polymer backbone networks, the mobility of polymers in the comb-type grafted hydrogels is improved. The grafted chains inside the comb-type hydrogels could easily swell and shrink as environment pH and/or temperature changes. The grafted chains are prevented from forming dense skin layer on the surface of comb-type hydrogels; as a result, the comb-type hydrogels show acute response to temperature/pH stimuli. Fast release of model drug from the comb-type grafted hydrogels is also observed with simultaneous temperature and pH stimuli. Such rapid pH/temperature-responsive hydrogels are promising new material candidates for various applications, such as for fabricating novel sensors, actuators, and chemical/drug carriers. Based on the grafting strategy, it is possible to develop new dual stimuli-sensitive polymeric materials with rapid response to environment stimuli.

References

1. Hirokawa Y, Tanaka T (1984) Volume phase transition in a non-ionic gel. J Chem Phys 81:6379–6380
2. Dong LC, Hoffman AS (1986) Thermally reversible hydrogels: III. Immobilization of enzymes for feedback reaction. J Control Release 4:223–227
3. Bae YH, Okano T, Kim SW (1990) Temperature dependence of swelling of crosslinked poly(N, N'-alkyl substituted acrylamides) in water. J Polym Sci Part B: Polym Phys 28:923–936
4. Chen GH, Hoffman AS (1995) Graft copolymers that exhibit temperature-induced phase transitions over a wide range of pH. Nature 373:49–52
5. Qu JB, Chu LY, Yang M et al (2006) A pH-responsive gating membrane system with pumping effects for improved controlled-release. Adv Funct Mater 16:1865–1872
6. Ricks J, Tanaka T (1984) Swelling of ionic gels – quantitative performance of the Donnan theory. Macromolecules 17:2916–2921
7. Holtz JH, Asher SA (1997) Polymerized colloidal crystal hydrogel films as intelligent chemical sensing materials. Nature 389:829–832
8. Tanaka T (1978) Collapse of gels and the critical end point. Phys Rev Lett 40:820–823
9. Suzuki A, Tanaka T (1990) Phase transition in polymer gels induced by visible light. Nature 346:345–347
10. Klier J, Scranton AB, Peppas NA (1990) Self-associating networks of poly(methacrylic acid-g-ethylene glycol). Macromolecules 23:4944–4949
11. Kown LC, Bae YH, Kim SW (1991) Electrically credible polymer gel for controlled release of drugs. Nature 354:291–293
12. Yoshida R, Uchida K, Kaneko Y et al (1995) Comb-type grafted hydrogels with rapid de-swelling response to temperature changes. Nature 374:240–242
13. Kaneko Y, Sakai K, Kikuchi A et al (1996) Fast swelling/deswelling kinetics of comb-type grafted poly(*N*-isopropylacrylamide) hydrogels. Macromol Symp 109:41–53
14. Kawaguchi H, Fujimoto K (1998) Smart latexes for bioseparation. Bioseparation 7:253–258
15. Hoffman AS (2002) Hydrogels for biomedical applications. Adv Drug Deliv Rev 54:3–12
16. Hirotsu S (1993) Coexistence of phases and the nature of first-order transition in poly(*N*-isopropylacrylamide) gels. Adv Polym Sci 110:1–26
17. Bae YH, Okano T, Hsu R et al (1987) Thermosensitive polymers as on-off switches for drug release. Makromol Chem Rapid Commun 8:481–485
18. Freitas RFS, Cussler EL (1987) Temperature sensitive gels as extraction solvents. Chem Eng Sci 42:97–103
19. Okano T (1993) Responsive gels. Adv Polym Sci 110:180–195
20. Dong L, Hoffman AS (1991) A novel approach for preparation of pH-sensitive hydrogels for enteric drug delivery. J Control Release 15:141–152

21. Lee WF, Shieh CH (1999) pH-Thermoreversible hydrogels. II. Synthesis and swelling behaviors of *N*-isopropylacrylamide-*co*-acrylic acid-*co*-sodium acrylate hydrogels. J Appl Polym Sci 73:1955–1967
22. Kim SY, Cho SM, Lee YM et al (2000) Thermo- and pH sensitive behavior of graft copolymer and blend based on chitosan and *N*-isopropylacrylamide. J Appl Polym Sci 78:1381–1391
23. Serizawa T, Wakita K, Akashi M (2002) Rapid deswelling of porous poly(*N*-isopropylacrylamide) hydrogels prepared by incorporation of silica particles. Macromolecules 35:10–12
24. Zhao Y, Su HJ, Fang L et al (2005) Superabsorbent hydrogels from poly(aspartic acid) with salt- temperature- and pH-responsiveness properties. Polymer 46:5368–5376
25. Determan MD, Cox JP, Seifert S et al (2005) Synthesis and characterization of temperature and pH-responsive pentablock copolymers. Polymer 46:6933–6946
26. Krusic MK, Filipovic J (2006) Copolymer hydrogels based on *N*-isopropylacrylamide and itaconic acid. Polymer 47:148–155
27. Tanaka T, Fillmore DJ (1979) Kinetics of swelling of gels. J Chem Phys 70:1214–1218
28. Matsuo ES, Tanaka T (1988) Kinetics of discontinuous volume-phase transition of gels. J Chem Phys 89:1695–1703
29. Gutowaka A, Bae YH, Feijan J et al (1992) Heparin release from thermosensitive hydrogels. J Control Release 22:95–104
30. Yu H, Grainger DW (1993) Thermo-sensitive swelling behavior in crosslinked *N*-isopropylacrylamide networks: cationic, anionic, and ampholytic hydrogels. J Appl Polym Sci 49:1553–1563
31. Feil H, Bae YH, Feijan J et al (1993) Effect of commoner hydrophilicity and ionization on the lower critical solution temperature of *N*-isopropylacrylamide copolymers. Macromolecules 26:2496–2500
32. Zhang J, Chu LY, Li YK et al (2007) Dual thermo- and pH-sensitive poly(*N*-isopropylacrylamide-*co*-acrylic acid) hydrogels with rapid response behaviors. Polymer 48:1718–1728
33. Zhang J, Chu LY, Cheng CJ et al (2008) Graft-type poly(*N*-isopropylacrylamide-*co*-acrylic acid) microgels exhibiting rapid thermo- and pH-responsive properties. Polymer 49:2595–2603
34. Zhang J, Xie R, Zhang SB et al (2009) Rapid pH/temperature-responsive cationic hydrogels with dual stimuli-sensitive grafted side chains. Polymer 50:2516–2525
35. Annaka M, Tanaka C, Nakahira T et al (2002) Fluorescence study on the swelling behavior of comb-type grafted poly(*N*-isopropylacrylamide) hydrogels. Macromolecules 35:8173–8179
36. Yoshida R, Sakai K, Okano T et al (1994) Modulating the phase transition temperature and thermosensitivity in *N*-isopropylacrylamide copolymer gels. J Biomater Sci Polym Ed 6:585–598
37. Kaneko Y, Sakai K, Kikuchi A et al (1995) Influence of freely mobile grafted chain length on dynamic properties of comb-type grafted poly(*N*-isopropylacrylamide) hydrogels. Macromolecules 28:7717–7723
38. Jeria OM, Pizarro GDC, Marambio OG et al (2006) Synthesis of *N*-hydroxymethyl acrylamide with beta-methyl hydrogen itaconate and itaconic acid hydrogels: effects of the pH, composition, and ionic strength on the swelling behavior. J Appl Polym Sci 100:1735–1741
39. Liu Y, Velada JL, Huglin MB (1999) Thermoreversible swelling behaviour of hydrogels based on *N*-isopropylacrylamide with sodium acrylate and sodium methacrylate. Polymer 40:4299–4306
40. Wang Y, Liu ZM, Han BX et al (2004) pH Sensitive polypropylene porous membrane prepared by grafting acrylic acid in supercritical carbon dioxide. Polymer 45:855–860
41. Bokias G, Staikos G, Iliopoulos I (2000) Solution properties and phase behaviour of copolymers of acrylic acid with *N*-isopropylacrylamide: the importance of the intrachain hydrogen bonding. Polymer 41:7399–7405
42. Pich A, Tessier A, Boyko V et al (2006) Synthesis and characterization of poly(vinylcaprolactam)-based microgels exhibiting temperature and pH-sensitive properties. Macromolecules 39:7701–7707

43. Guo JT, Li L, Li XY et al (2006) Preparation and characterization of novel cationic copolymer hydrogels with pH sensitivity and thermosensitivity. J Appl Polym Sci 100:3602–3608
44. Zhang YL, Xu L, Yi M et al (2006) Radiation synthesis of poly[(dimethylaminoethyl methacrylate)-*co*-(diallyl dimethyl ammonium chloride)] hydrogels and its application as a carrier for notoginsenoside delivery. Eur Polym J 42:2959–2967
45. Şoipan D, Şen M, Klöge Z et al (2008) Adsorption of apollo reactive dyes on poly(*N*, *N*-dimethylamino ethylmethacrylate) hydrogels. Radiat Phys Chem 77:428–433
46. Kim EJ, Cho SH, Yuk SH (2001) Polymeric microspheres composed of pH/temperature-sensitive polymer complex. Biomaterials 22:2495–2499
47. Annaka M, Sugiyama M, Kasai M et al (2002) Transport properties of combtype grafted and normal-type *N*-isopropylacrylamide hydrogel. Langmuir 18:7377–7383
48. Gutowaska A, Bae YH, Feijan J et al (1992) Heparin release from thermosensitive hydrogels. J Control Release 22:95–104

Part IV
Alcohol-Responsive Hydrogel Functional Materials

Chapter 10
Smart Functional Membranes with Alcohol-Responsive Characteristics

Abstract In this chapter, the design, fabrication, and performance of smart functional membranes with alcohol-responsive characteristics are introduced. The membranes with alcohol-responsive characteristics are either flat membrane grafted with smart gates or core-shell microcapsule membrane composed of smart materials. The critical alcohol response concentrations of these membranes are regulated by simply adjusting the lower critical solution temperature (LCST) of the grafted polymeric gates in water or controlling the environmental temperature. The PNIPAM-grafted membranes with controllable critical ethanol response concentrations can be efficiently applied to various practical conditions, where the transmembrane permeability needs to be manipulated by the environmental ethanol concentration and operation temperature. The core-shell PNIPAM microcapsule membranes can function as both sensors and actuators by converting alcohol concentration variation into mechanic forces. The results in this study provide potential applications in more efficient fermentation process and may also provide opportunities in cargo delivery in alcoholic environments.

10.1 Introduction

Alcohol is one of the most common solvents for many scientific, medical, and industrial applications [1], wherein ethanol has widespread use as alcoholic beverages, solvents, feedstocks, and fuels. In organic synthesis, alcohols generally serve as versatile solvents, the concentration of which may have great effects on the reaction environment and subsequently the products. Therefore, it would be necessary to design and develop smart functional membranes with alcohol-responsive characteristics in order to meet the requirements of various applications.

Poly(*N*-isopropylacrylamide) (PNIPAM) is well known as a kind of thermo-responsive material. The nature of the thermo-responsive property of PNIPAM polymer is revealed as the balance between hydrogen bonding and hydrophobic

L.-Y. Chu et al., *Smart Hydrogel Functional Materials*,
DOI 10.1007/978-3-642-39538-3_10,

© Chemical Industry Press, Beijing and Springer Berlin Heidelberg 2013

interaction [2]. Solvents such as alcohol, which not only change the aqueous environment but also form hydrogen bonds with both water and PNIPAM molecules, can also induce the same phase transition [3]. Therefore, PNIPAM is also featured with significant alcohol-responsive characteristics [4–7]. Unlike the single critical response temperature (i.e., lower critical solution temperature; LCST), PNIPAM exhibits two critical alcohol response concentrations, i.e., the lower critical alcohol response concentration and the upper critical alcohol response concentration (C_{c1} and C_{c2}), respectively. With the alcohol concentration increasing, PNIPAM polymer will undergo the coil-to-globule transition and shrink around the lower critical alcohol response concentration, while the globule-to-coil transition and swell near the upper critical alcohol response concentration.

So far, the cononsolvency behaviors of PNIPAM polymers including linear polymers (in water or grafted on the membrane as smart gates), cross-linked macrogels, or microgels in alcohol-water mixtures with different types of alcohol and different concentrations have been focused and investigated [3, 4, 8–15]. However, the influential factors on the alcohol-responsive characteristics of PNIPAM polymers and how to use their alcohol-responsive characteristics have not been discussed in depth to the best of our knowledge.

In this chapter, the alcohol-responsive smart functional membranes, which are featured with controllable critical alcohol response concentrations, are designed and developed. The critical alcohol response concentrations of both gating membranes with grafted PNIPAM gates and core-shell PNIPAM microcapsule membranes are adjusted by simply changing the influential factors such as LCST of PNIPAM-based polymer and environmental temperature. The results in this study provide valuable guidance for designing and preparing alcohol-responsive gating membranes with adjustable critical alcohol response concentrations. Acting as a smart valve to self-regulatively adjust the transmembrane permeability, the ethanol-responsive smart gating membrane developed in this study could be used together with pervaporation membranes to remove ethanol during the fermentation process, which enables the fermentation process more efficient. The PNIPAM microcapsule membranes with encapsulated oil core can function as both sensors and actuators simultaneously. They demonstrate the conversion of alcohol concentration variation into mechanical force by ejecting the oil core out of the microcapsule membranes, which may provide opportunities in cargo delivery in alcoholic environments.

10.2 Ethanol-Responsive Hydrogels with Controllable Ethanol Response Concentration

The alcohol-responsive characteristics of either PNIPAM-grafted membranes or PNIPAM microcapsule membranes are attributable to those of PNIPAM polymer. The regulation of critical ethanol response concentrations of PNIPAM polymers by simply changing the LCST in water and operation temperatures has been practiced in our group and introduced in this section. It has been reported that the introduction

Table 10.1 The molar content of BMA or DMAA in PNIPAM-based linear polymers

Polymer	Molar ratio of BMA or DMAA to NIPAM [mol/mol][a, b]
PNB3	1: 8.6 [n_{BMA}/n_{NIPAM}]
PNB2	1: 12.1 [n_{BMA}/n_{NIPAM}]
PNB1	1: 18.9 [n_{BMA}/n_{NIPAM}]
PNIPAM	0:1 [$-/n_{NIPAM}$]
PND1	1: 7.9 [n_{DMAA}/n_{NIPAM}]
PND2	1: 6.5 [n_{DMAA}/n_{NIPAM}]
PND3	1: 3.3 [n_{DMAA}/n_{NIPAM}]

[a]The moles of NIPAM are fixed at 0.044 and 0.03 in the fabrication of PNB and PND copolymers

[b]All the molar contents are the actual molar contents which are calculated from the ^{1}H NMR data

of hydrophilic or hydrophobic comonomers into the PNIPAM-based copolymers could bring certain changes in the LCST of PNIPAM in water as well as the lower critical ethanol response concentration (C_{E1}) [4, 16]. It is attributable to the change in equilibrium between the hydrophilicity and hydrophobicity of the PNIPAM polymers since the nature of the thermo-responsive property of PNIPAM polymer is the balance between hydrogen bonding and hydrophobic interaction. As a result, the introduction of hydrophilic or hydrophobic comonomers will change the critical ethanol response concentrations of PNIPAM.

10.2.1 Preparation of PNIPAM-Based Linear Polymers

A series of PNIPAM-based linear polymers are prepared by free-radical polymerization at 63.5 °C. To investigate the effect of content of hydrophilic or hydrophobic monomers, the different feed molar ratios of hydrophobic monomer butyl methacrylate (BMA) and hydrophilic monomer *N,N′*-dimethylacrylamide (DMAA) are chosen. The actual molar content of BMA and DMAA in PNIPAM-based polymers is listed in Table 10.1. The PNIPAM-based poly(*N*-isopropylacrylamide-*co*-butyl methacrylate) and poly(*N*-isopropylacrylamide-*co*-*N,N*-dimethylacrylamide) copolymers are abbreviated as PNB and PND, respectively.

10.2.2 Adjustment of the LCST of PNIPAM-Based Copolymers in Water

Figure 10.1 shows the thermo-responsive characteristics of PNIPAM polymer, PNB copolymers, and PND copolymers in water. The transmittance of each polymer

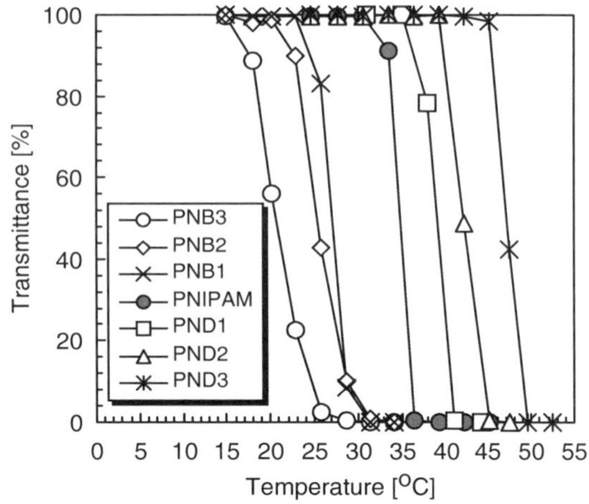

Fig. 10.1 Thermo-responsive characteristics of PNIPAM polymer and PNIPAM-based copolymers

solution (2 g/L) changes dramatically at a certain temperature, which is defined as the LCST of polymers in water. At temperature lower than the LCST of polymer in water, the polymer solution is clear and colorless, and the transmittance keeps at a high level (~100 %). However, with the temperature increases close to the LCST of polymer in water, the transmittance decreases dramatically from 100 % to 0 % and keeps unchanged as temperature increasing further, and correspondingly the polymer solution is turbid and opaque. The dramatic change in transmittance of polymer solution is attributable to the temperature-triggered coil-to-globule transition of PNIPAM in water which is due to the rupture of hydrogen bonds between the amido groups on the PNIPAM chains and the surrounding water molecules. At temperature below the LCST, PNIPAM chains extend (coil) and become hydrophilic due to the formation of hydrogen bonds, and the polymer solution is clear and colorless. However, PNIPAM chains shrink (globule) and become hydrophobic attributing to the rupture of hydrogen bonds at temperature above the LCST; thus, the polymer solution is turbid and opaque. The dramatic change in transmittance of each polymer solution is finished within a narrow temperature range near the LCST. The results clarify that a series of PNB copolymers and PND copolymers are featured with good thermo-responsive characteristics just like PNIPAM polymer.

Although the introduction of hydrophobic monomer BMA or hydrophilic monomer DMAA has no effect on the thermo-responsive characteristics, it does affect the LCST values of the PNIPAM-based copolymers. Figure 10.2 displays the effect of the molar content of hydrophobic or hydrophilic comonomers in PNIPAM-based polymers on the LCST values. The LCST values of PNB3, PNB2, PNB1, PNIPAM, PND1, PND2, and PND3 polymers in water are 20.8 °C, 25.2 °C, 27.9 °C, 34.8 °C, 39.1 °C, 42.1 °C, and 47.1 °C, respectively. The addition of hydrophobic monomer BMA leads to a lower LCST, while the hydrophilic monomer DMAA results in a higher LCST. The more the molar ratio of BMA to NIPAM is, the lower

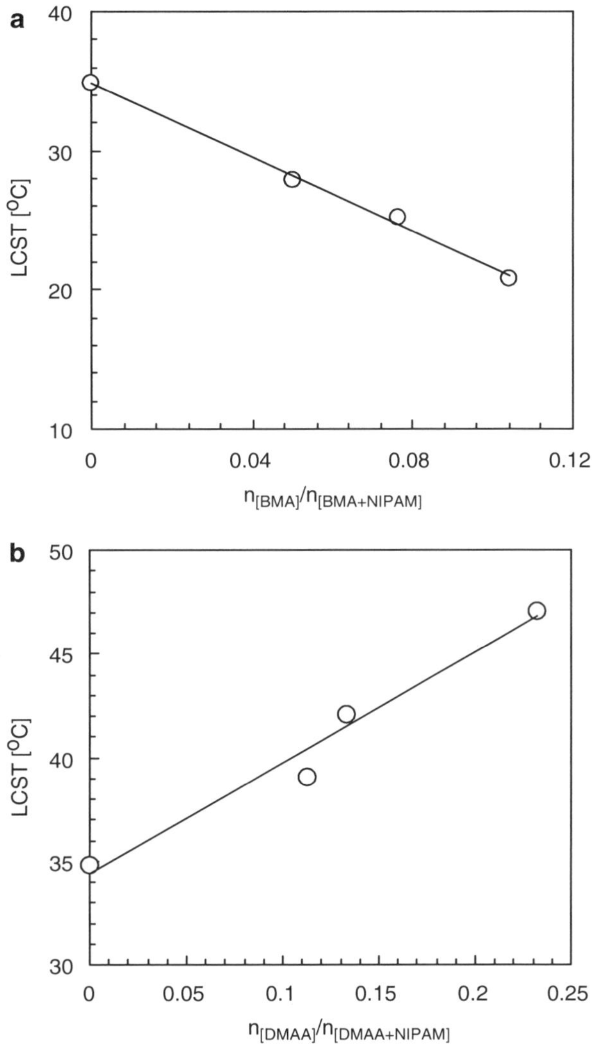

Fig. 10.2 Effect of molar ratio of BMA (**a**) or DMAA (**b**) to NIPAM in polymers on LCST values

the LCST, while the more the molar ratio of DMAA to NIPAM is, the higher the LCST. The LCST values of PNIPAM-based polymers in water decrease or increase linearly with the molar content of BMA or DMAA in polymers increasing. DMAA is more hydrophilic than NIPAM because it contains a hydrophilic amido group and lacks a hydrophobic methylene group. The more hydrophilic units in the copolymer chains, the more hydrogen bonds form, as a result the higher the LCST. On the contrary, BMA contains a high content of the hydrophobic methyl group, so it is more hydrophobic than NIPAM. The more hydrophobic units in the copolymer chains, the less hydrogen bonds form, as a result the lower the LCST in water.

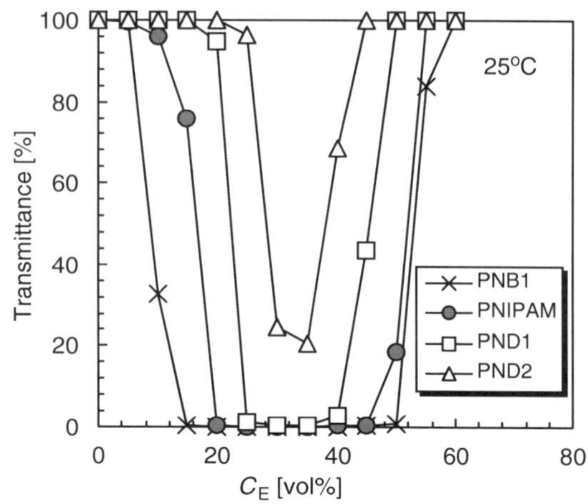

Fig. 10.3 Ethanol-responsive characteristics of PNIPAM polymers and PNIPAM-based copolymers at room temperature

10.2.3 Adjustment of the Critical Ethanol Response Concentrations of PNIPAM-Based Copolymers

The ethanol-responsive characteristics of PNIPAM polymers and PNIPAM-based copolymers at room temperature are shown in Fig. 10.3. The transmittance of ethanol solution of polymers exhibits dramatic change around two ethanol concentrations, which are defined as the lower critical ethanol response concentration (C_{E1}) and the upper critical ethanol response concentration (C_{E2}), respectively. Under ethanol concentrations below the C_{E1} or above the C_{E2}, the transmittance of polymer solutions is high (~100 %). However, the transmittance of polymer solutions decreases sharply near the C_{E1} and increases promptly close to C_{E2}, while it is much lower between the C_{E1} and C_{E2}. At ethanol concentration lower than the C_{E1}, the ethanol molecules are intended to concentrate around the PNIPAM chains [3, 5, 6], but the small amount of ethanol molecules is not enough to destroy the hydrogen bonds between the PNIPAM chains and water molecules. The polymer chains extend (coil) and are hydrophilic, and thus, the transmittance of polymer solution is high at ethanol concentration lower than C_{E1}. When the ethanol concentration increases to C_{E1}, the enrichment of ethanol molecules around PNIPAM chains might destroy the hydrogen bonds between the PNIPAM chains and water molecules. Therefore, the PNIPAM chains undergo the first phase transition from coil to globule, and the transmittance of PNIPAM polymer solution dramatically decreases. When the ethanol concentration increases further to C_{E2}, the activation energy of the ethanol solution becomes larger than the cohesive force of the PNIPAM chains [3, 5, 6]. As a result, the ethanol molecules interact with PNIPAM chains to form hydrogen bonds, and therefore the polymer chains exhibit globule-to-coil transition. Consequently, the PNIPAM chains become hydrophilic and extend again at ethanol concentration above the upper critical ethanol response concentration

C_{E2}. Obviously, the introduction of hydrophobic BMA or hydrophilic DMAA into PNIPAM polymers has no effect on the ethanol-responsive characteristics of PNIPAM-based polymers.

The C_{E1} values of the PNB1, PNIPAM, and PND1 at 25 °C are 8.7 vol. %, 16.7 vol. %, and 22.4 vol. %, respectively, while their C_{E2} values are 52.96 vol. %, 51.92 vol. %, and 45.58 vol. %, respectively (Fig. 10.3). Because BMA contains more hydrophobic groups but less hydrophilic groups than NIPAM, the introduction of hydrophobic BMA into PNIPAM polymers will bring larger influence on the hydrophilic groups than those of the hydrophobic groups. However, it is reported that the C_{E1} values of PNIPAM are mainly dependent on the hydrophilic groups, while the C_{E2} values are mainly dependent on the hydrophobic groups [7]. The molar ratio of BMA to NIPAM for PNB1 copolymers is not very large (1:18.9). Therefore, the addition of such small amount of hydrophobic BMA into the PNIPAM polymers leads to a smaller C_{E1} value while an almost unchanged C_{E2} value. Because BMA is much more hydrophobic than NIPAM, the hydration effect of PNB copolymer chains is weaker than that of PNIPAM polymer chains. Less ethanol molecules (lower ethanol concentration) are enough to break the hydrogen bonds between PNB copolymer chains and water molecules in order to initiate the coil-to-globule transition. So, the C_{E1} value of the PNB polymer is lower than that of the PNIPAM polymer. However, the introduction of hydrophilic DMAA results in larger C_{E1} but smaller C_{E2}. Because DMAA is more hydrophilic than NIPAM, more hydrogen bonds form between PND polymer and water molecules; therefore, more ethanol molecules (higher ethanol concentration) are needed to destroy the hydrogen bonds between PND polymer and water molecules in order to trigger the coil-to-globule transition. So, the C_{E1} value of the PND copolymer is higher than that of the PNIPAM polymer. On the other hand, the introduction of DMAA leads to a smaller proportion of PNIPAM in the PND copolymer. Therefore, less ethanol molecules (lower ethanol concentration) are needed to overcome the cohesion of polymers in order to trigger the globule-to-coil transition. As a result, the C_{E2} value of the PND copolymer is lower than that of the PNIPAM polymer. The response interval of ethanol concentration is defined as the difference between two critical ethanol response concentrations, i.e., the $C_{E2}-C_{E1}$ value. With the LCST of the PNIPAM-based copolymers in water increasing, the response interval ($C_{E1}-C_{E2}$) decreases. Because the response interval ($C_{E2}-C_{E1}$) of PND2 copolymer (n_{DMAA}: $n_{NIPAM} = 1{:}6.5$) is so small, the transmittance of ethanol solution of polymers cannot decrease to 0 % at ethanol concentration between the C_{E1} and C_{E2} (Fig. 10.3).

10.2.4 The Dynamic Ethanol-Responsive Characteristics of Cross-Linked Hydrogels

The poly(*N*-isopropylacrylamide-*co*-*N*,*N*-dimethylacrylamide) (PND) cross-linked hydrogels are fabricated by free-radical polymerization at 60 °C for 10 h. The feed

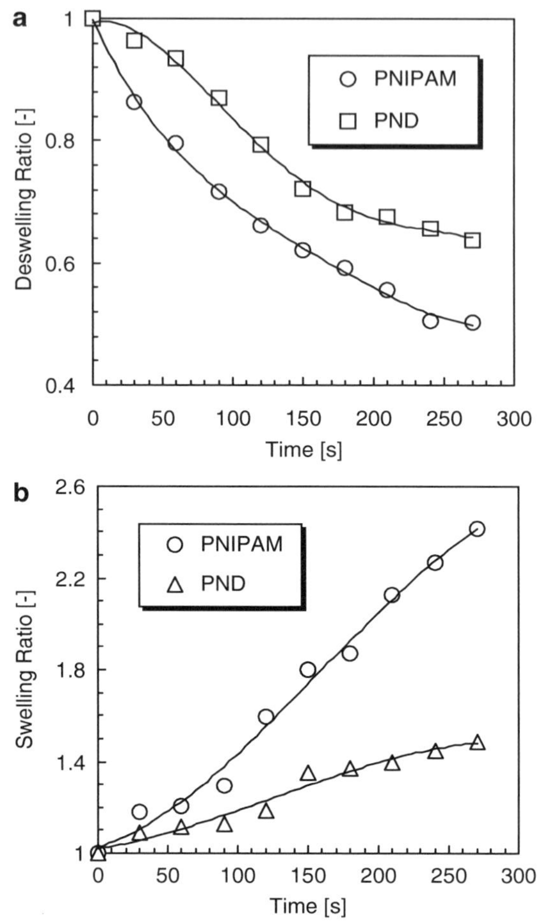

Fig. 10.4 Dynamic deswelling and swelling process of PNIPAM and PND hydrogels in ethanol solution when respectively transferred from deionized water to ethanol solution with 40 vol. % (**a**) and from ethanol solution with 40–100 vol. % (**b**) at 25 °C

molar ratio of DMAA to NIPAM is 1:8. The dynamic ethanol-responsive characteristics of PND cross-linked hydrogels are investigated by suddenly changing the ethanol concentration in the surrounding medium. The environmental temperature keeps constant at 25 °C throughout the dynamic ethanol-responsive experiment. The dynamic deswelling process of PND hydrogels is investigated by sudden transferring the PND hydrogels into ethanol solution with 40 vol. % after the hydrogels sufficiently swell in deionized water. However, the dynamic swelling process of PND hydrogels is observed by transferring the PND hydrogels into anhydrous ethanol (100 vol. %) after sufficient deswelling in ethanol solution with 40 vol. %. Figure 10.4 displays the dynamic deswelling ratio and swelling ratio change of PND hydrogels when respectively transferred into ethanol solution with 40 vol. % from deionized water and into anhydrous ethanol (100 vol. %) from ethanol solution with 40 vol. %. The deswelling ratio of PND hydrogels (R_{DE}) is defined as the volume ratio of hydrogels in ethanol solution with 40 vol. % at intermediary time during the deswelling process to that in water at the beginning. However, the swelling ratio of PND hydrogels (R_{SE}) is defined as the volume ratio of hydrogels in anhydrous ethanol at intermediary time during the swelling process to that in ethanol solution with 40 vol. %. The deswelling and swelling ratios of hydrogels are calculated by Eqs. 10.1 and 10.2, respectively:

$$R_{DE} = \frac{V_{E40,t}}{V_{W,0}} \tag{10.1}$$

$$R_{SE} = \frac{V_{E100,t}}{V_{E40,0}} \tag{10.2}$$

where $V_{E40,t}$ and $V_{E100,t}$ are the volumes of PND hydrogels in the ethanol solution with 40 vol. % and 100 vol. % at intermediary time t during the deswelling and swelling processes, respectively. $V_{W,0}$ and $V_{E40,0}$ represent the equilibrium volumes of PND hydrogels in water and ethanol solution of 40 vol. % at the beginning of deswelling and swelling processes, respectively.

As mentioned above, the lower and upper critical ethanol response concentrations (C_{E1} and C_{E2}) of PNIPAM linear polymers at 25 °C are 16.7 vol. % and 51.92 vol. %, respectively. Those critical values of PND linear polymers with molar ratio of 1:8 (DMAA: NIPAM; PND1) at 25 °C are 22.4 vol. % and 45.58 vol. %, respectively. Therefore, when the PNIPAM and PND cross-linked hydrogels are suddenly put into the ethanol solution with concentration of 40 vol. %, both hydrogels shrink because the PNIPAM-based chains undergo the coil-to-globule transition at ethanol concentrations above the lower critical ethanol response concentration C_{E1}. However, when both hydrogels are transferred from ethanol solution with concentration of 40 vol. % into anhydrous ethanol solution (100 vol. %), they swell rapidly attributing to the globule-to-coil transition of PNIPAM-based chains. Compared with PNIPAM cross-linked hydrogels, the PND hydrogels deswell and swell slower due to the introduction of non-stimuli-responsive hydrophilic segments. Moreover, the deswelling degree and swelling degree of PNIPAM cross-linked hydrogels are larger than those of PND cross-linked hydrogels.

10.3 Ethanol-Responsive Smart Gating Membranes

The traditional ethanol fermentation from biomass resources is a typical process of product inhibition [17], in which the ethanol product of an enzyme reaction inhibits enzyme activity. To achieve the most efficient and stable fermentation, there is an optimum ethanol concentration in the fermentation broth, which has been reported to be about 10 vol. %, although this value is dependent on the process conditions including tolerance of yeast [4, 18]. However, the transmembrane permeability of the existing membranes such as pervaporation membranes cannot be adjusted self-regulatively responding to the variation of ethanol concentration in the

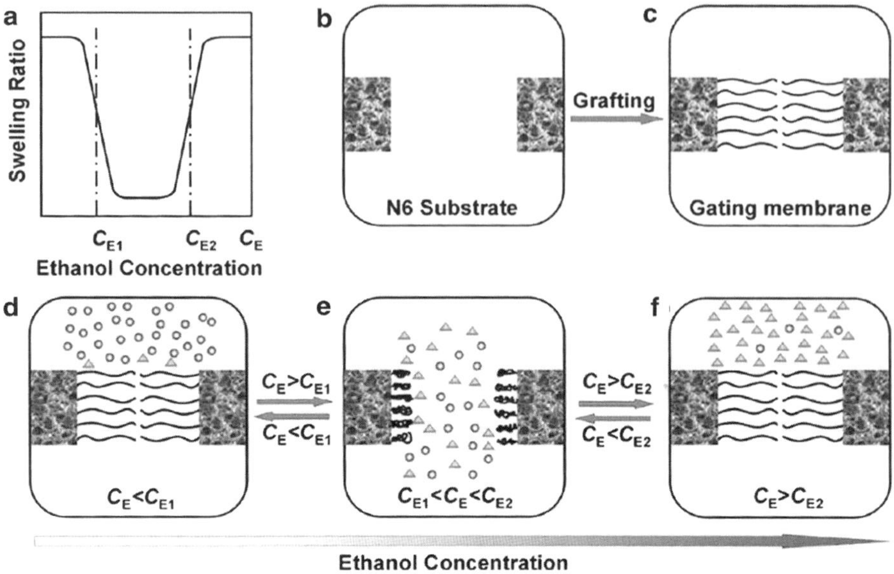

Fig. 10.5 Schematic illustration of preparation and principle of the ethanol-responsive gating membrane. (**a**) Ethanol-responsive characteristics of PNIPAM polymer, (**b**, **c**) preparation of the PNIPAM-grafted membrane, and (**d**, **e**, **f**) ethanol-responsive gating characteristics of the grafted membrane, in which the *circles* and *triangles* represent water molecules and ethanol molecules, respectively. C_E is the ethanol concentration and C_{E1} and C_{E2} ($C_{E1} < C_{E2}$) are two critical response concentrations of the PNIPAM polymer illustrated in (**a**) (Reproduced with permission from Ref. [7], Copyright (2012), American Chemical Society)

fermentation broth [18–21]. If controllable ethanol-responsive permeability across the membrane can be achieved, the ethanol concentration in the fermentation broth can be maintained in a relatively stable level, and the fermentation process is more efficient. Moreover, the adjustment of critical ethanol response concentration is very important for the membranes to be efficiently applied in various practical conditions.

On the basis of the results of ethanol-responsive polymers described in Sect. 10.2, it is possible to design ethanol-responsive smart gating membranes with controllable critical ethanol response concentrations to regulate the transmembrane permeability in various conditions. As mentioned above, PNIPAM polymer as well as PNIPAM-based copolymers exhibits the lower critical ethanol response concentration C_{E1} and the upper critical ethanol response concentration C_{E2}, as illustrated in Fig. 10.5a. The ethanol-responsive smart gating membranes have been developed by grafting PNIPAM-based copolymers onto porous Nylon-6 (N6) substrate membrane via surface-initiated atom-transfer radical polymerization (ATRP) method (as shown in Fig. 10.5b, c and Fig. 10.6) [22]. The pores of PNIPAM-grafted membrane show "closed/open" switching function when the environmental ethanol concentration changes across the aforementioned critical ethanol response concentrations (as illustrated in Fig. 10.5d–f). As a result, the permeability across ethanol-responsive membranes can be regulated, i.e., it becomes low at environmental ethanol concen-

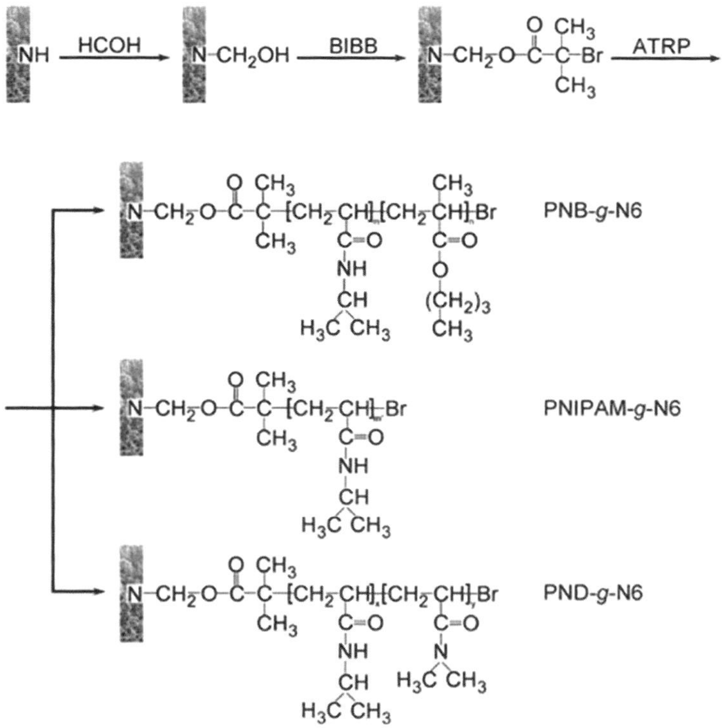

Fig. 10.6 Preparation route of the PNB-*g*-N6, PNIPAM-*g*-N6, and PND-*g*-N6 membranes (Reproduced with permission from Ref. [7], Copyright (2012), American Chemical Society)

trations lower than C_{E1} and higher than C_{E2}, while high at ethanol concentrations between C_{E1} and C_{E2}. The influential factors on the critical ethanol response concentrations of the ethanol-responsive smart gating membranes grafted with PNIPAM and PNIPAM-based copolymer gates are systematically investigated in the recent study [7]. The quantitative relationships between the LCST of PNIPAM-based polymeric gates in water and the C_{E1}, between the LCST in water and the C_{E2}, and between the LCST in water and the interval of $(C_{E1}-C_{E2})$ are ascertained for the first time. These quantitative relationships obtained in this study can be used to design and regulate the critical ethanol response concentration of grafted membranes with PNIPAM-based functional gates since the LCST values in water of PNIPAM-based materials can be easily designed and measured.

10.3.1 Preparation of Ethanol-Responsive Smart Gating Membranes

The preparation route of ethanol-responsive gating membranes by virtue of the ATRP method is schematically illustrated in Fig. 10.6. The preparation process

Table 10.2 Membrane code (Reproduced with permission from Ref. [7], Copyright (2012), American Chemical Society)

Grafted membranes	Feed molar ratio of BMA or DMAA to NIPAM [mol/mol][a]	Y [wt %]	Response temperature [°C][b]
PNB-*g*-N6-1	1: 35 [BMA/NIPAM]	71	28
PNB-*g*-N6-2	1: 40 [BMA/NIPAM]	99	31
PNIPAM-*g*-N6	0: 100 [–/NIPAM]	88	35
PND-*g*-N6-1	1: 20 [DMAA/NIPAM]	98	37.5
PND-*g*-N6-2	1: 10 [DMAA/NIPAM]	110	42.5

[a]The total molar concentration of comonomers is fixed as 0.178 mol/L

[b]The response temperatures of grafted membrane equal to the LCST values of the grafted polymer gates

included three steps, i.e., hydroxylation, acylation, and grafting. During the hydroxylation, the N6 substrate membranes are immersed into the formaldehyde solution containing 2 vol. % phosphate as catalyst at 60 °C for 12 h to generate hydroxyl groups on the pore surfaces. Later, the membranes with –OH groups reacted with 2-bromoisobutyryl bromide (BIBB) to generate –Br groups during the acylation. In the third grafting step, PNIPAM, PNB, or PND polymers are respectively grafted onto the membranes with –Br groups as functional gates by the ATRP method. The grafted membranes are coded as PNIPAM-*g*-N6, PNB-*g*-N6, and PND-*g*-N6 correspondingly. The contents of BMA and DMAA in the grafted polymeric gates are varied to adjust the response temperature of the membranes which equals to the LCST values of the grafted copolymers. The feed molar ratio of comonomers and the grafting yield of grafted membrane are listed in Table 10.2. The grafting yield of the grafted membrane is defined as the mass increase ratio after grafting polymers during ATRP and calculated by Eq. 10.3:

$$Y = \frac{m_{\mathrm{g}} - m_{\mathrm{Br}}}{m_{\mathrm{Br}}} \times 100\,\% \tag{10.3}$$

where Y stands for the grafting yield of polymers grafted on the membranes [wt %] and m_{Br} and m_{g} are the masses of the membrane with –Br groups and that with the grafted functional gates [g]. The mass of the membrane in each modification step is measured more than 4 times, and the relative errors are within 0.06 %.

10.3.2 Adjustment of the Response Temperature of Grafted Membranes

The LCST values of the polymeric gates grafted in the membranes or so-called response temperatures of the membranes in water are obtained by measuring the thermo-responsive water fluxes of grafted membranes. The response temperature of the membranes is defined as the temperature at which the water flux across the membrane changes dramatically. The experiments on water fluxes across membranes are carried out by using a microfiltration apparatus under a transmembrane pressure of 0.1 MPa. The LCST values of the grafted polymeric gates and so-called response temperatures of the membranes PNB-*g*-N6-1, PNB-*g*-N6-2, PNIPAM-*g*-N6, PND-*g*-N6-1, and PND-*g*-N6-2 in water are 28 °C, 31 °C, 35 °C, 37.5 °C, and 42.5 °C, respectively (Table 10.2). The effect of the introduction of hydrophobic monomer BMA or hydrophilic monomer DMAA into the grafted polymeric gates on the response temperatures of the grafted membranes is similar to that on the PNIPAM-based linear polymer. That is, the addition of hydrophobic monomer BMA leads to lower response temperature of membranes or LCST of grafted polymeric gates in water, and the response temperature decreases with the molar ratio of BMA to NIPAM increasing. However, the hydrophilic monomer DMAA results in higher response temperature, and the response temperature increases with the molar ratio of DMAA to NIPAM increasing.

10.3.3 Adjustment of the Critical Ethanol Response Concentrations of Grafted Membranes

The critical ethanol response concentrations of grafted membranes are investigated by measuring the fluxes of ethanol solutions with different ethanol concentrations (denoted as J_E) across the grafted membranes. The fluxes of ethanol solutions across the grafted membranes are also measured under a transmembrane pressure of 0.1 MPa. The operation temperatures are always below the LCST of the grafted polymeric gates in water to ensure that the polymeric gates are in the swollen state in water. The critical ethanol response concentration of the grafted membrane is defined as the ethanol concentration at which the flux of ethanol solution across the membrane changes dramatically.

The influential factors on the ethanol-responsive characteristics of the grafted membranes are systematically investigated. Obviously, the introduction of hydrophobic BMA or hydrophilic DMAA into the grafted polymeric gates does not eliminate the ethanol-responsive characteristics of the grafted membranes (Fig. 10.7). The fluxes of ethanol solutions through all the grafted membranes exhibit dramatic change around two critical ethanol response concentrations. Under ethanol concentrations below the C_{E1} or above the C_{E2}, the fluxes of ethanol solutions through all the grafted membranes are low because of the "closed" state of the grafted membrane pores (Fig. 10.5d, f). On the other hand, at ethanol concentrations between the C_{E1} and the C_{E2}, the fluxes become much larger due to the "open" state of the grafted membrane pores (Fig. 10.5e). The C_{E1} values of the grafted membranes PNB-*g*-N6-1, PNB-*g*-N6-2, PNIPAM-*g*-N6, PND-*g*-N6-1, and PND-*g*-N6-2 at 22 °C are 14.0 vol. %, 15.0 vol. %, 16.8 vol. %, 17.9 vol. %, and 17.5 vol. %, respectively. The corresponding C_{E2} values are 35.3 vol. %, 34.5 vol. %, 35.0 vol. %, 31.5 vol. %, and 27.0 vol. %, respectively. The effect of the introduction of BMA and DMAA into the grafted polymeric gates on the

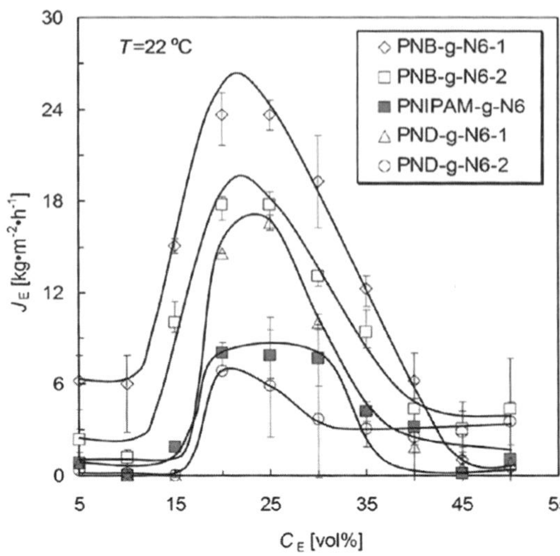

Fig. 10.7 Ethanol-responsive fluxes (J_E) of grafted membranes at 22 °C (Reproduced with permission from Ref. [7], Copyright (2012), American Chemical Society)

two critical ethanol response concentrations of grafted membranes is similar to that on those of PNIPAM-based linear polymers. That is, the addition of hydrophobic BMA into the grafted copolymer gates leads to a smaller C_{E1} value of the PNB-*g*-N6 membranes compared with that of PNIPAM-*g*-N6 membrane, while an almost unchanged C_{E2} value. The more the feed molar ratio of BMA to NIPAM is, the smaller the C_{E1}. On the contrary, the introduction of hydrophilic DMAA results in slightly larger C_{E1} of PND-*g*-N6 membrane compared with that of PNIPAM-*g*-N6 membrane but smaller C_{E2}. The more the feed molar ratio of DMAA to NIPAM is, the higher the C_{E1} value, but the lower the C_{E2} value.

As mentioned above, the response interval of ethanol concentration is defined as the difference between two critical ethanol response concentrations, i.e., the $C_{E2}-C_{E1}$ value. The results in Fig. 10.7 show that the higher the content of hydrophobic component in the grafted polymeric gates of the PNB-*g*-N6 membrane is, the wider the response interval of ethanol concentration. On the contrary, the higher the content of hydrophilic component in the grafted polymeric gates of the PND-*g*-N6 membrane is, the narrower the response interval of ethanol concentration. The aforementioned results indicate that the introduction of hydrophilic or hydrophobic component into the grafted polymeric gates can effectively regulate both C_{E1} and C_{E2} values as well as the $C_{E2}-C_{E1}$ value of the grafted membranes with PNIPAM-based functional gates.

10.3.4 Effects of Operation Temperatures on the Critical Ethanol Response Concentrations

Effects of operation temperatures on the fluxes of ethanol solutions across the grafted membranes are shown in Fig. 10.8. The operation temperatures in the

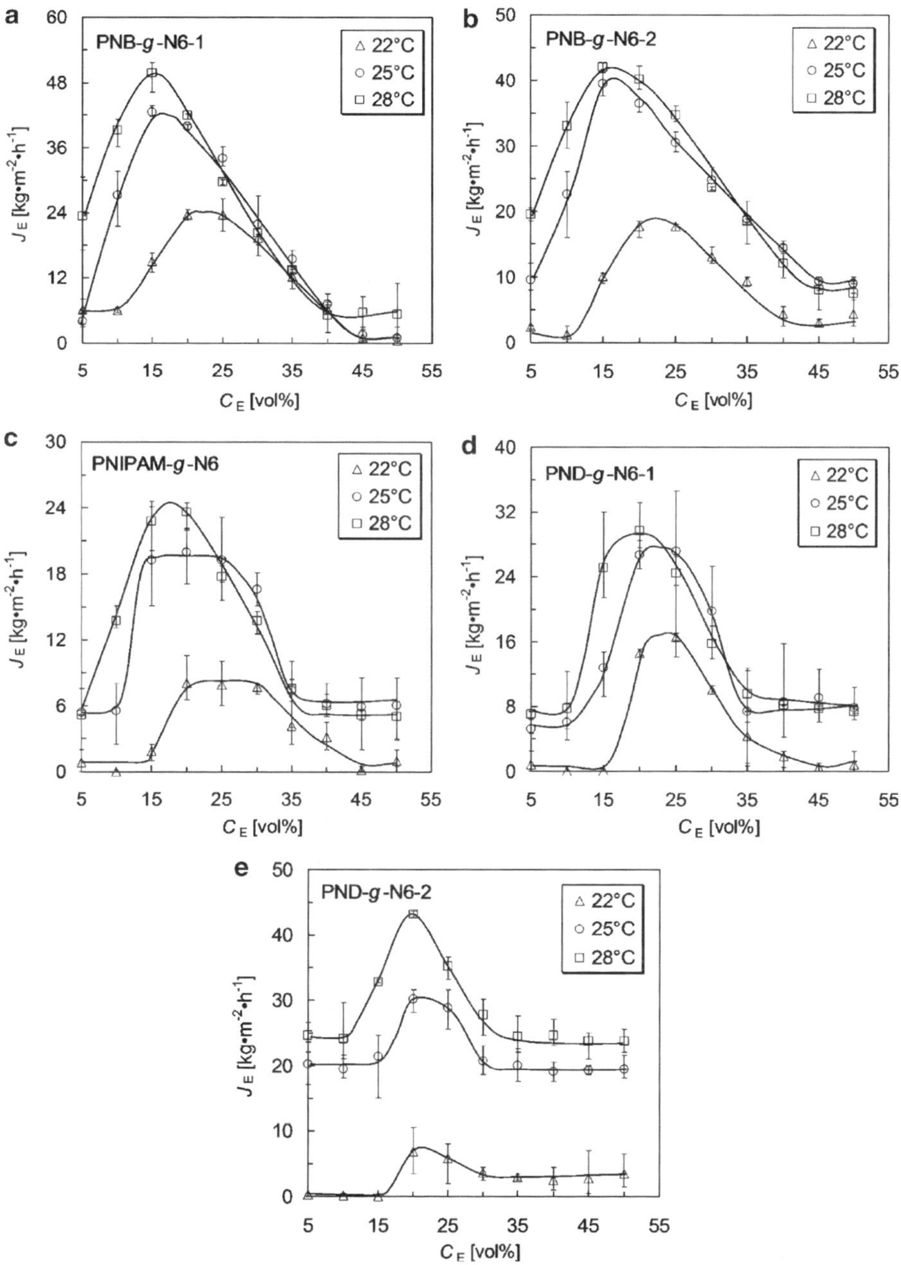

Fig. 10.8 Effects of the operation temperature on the fluxes of ethanol solutions across the grafted membranes: (**a**) PNB-*g*-N6-1 membrane, (**b**) PNB-*g*-N6-2 membrane, (**c**) PNIPAM-*g*-N6 membrane, (**d**) PND-*g*-N6-1 membrane, and (**e**) PND-*g*-N6-2 membrane (Reproduced with permission from Ref. [7], Copyright (2012), American Chemical Society)

experiments are always lower than the LCST of the grafted polymeric gates in water to ensure that the grafted gates are in the swollen state in water. With increasing the operation temperature, both the C_{E1} and C_{E2} values of all the PNB-*g*-N6-1, PNB-*g*-N6-2, PNIPAM-*g*-N6, PND-*g*-N6-1, and PND-*g*-N6-2 membranes decrease. For example, the C_{E1} values of the PND-*g*-N6-1 membrane at 22 °C, 25 °C, and 28 °C are respectively 17.9 vol. %, 15.9 vol. %, and 13.0 vol. %, and the corresponding C_{E2} values are 31.5 vol. %, 30.5 vol. %, and 28.5 vol. %, respectively (Fig. 10.8d). The increase of operation temperature destroys part of the hydrogen bonds generated between water molecules and polymers grafted inside membrane pores in ethanol solutions with concentration lower than the C_{E1}. Therefore, less ethanol molecules are needed to trigger the coil-to-globule transition of the polymeric gates grafted in the membrane pores, and thus, the C_{E1} value of the grafted membrane decreases. With increasing the environmental temperature, the cohesion degree of the grafted polymers in the membrane pore decreases to some extent. Therefore, less ethanol molecules are needed to trigger the globule-to-coil transition of the polymers; as a result, the C_{E2} value of the grafted membrane also decreases. That is, the critical ethanol response concentrations of grafted membranes can be regulated to some extent by changing operation temperatures.

10.3.5 Relationship Between the LCST in Water and the Critical Ethanol Response Concentrations

The relationships between the LCST in water and the critical ethanol response concentrations of membranes with grafted PNIPAM-based gates are illustrated in Fig. 10.9. Figure 10.9a shows that the C_{E1} value simply increases with increasing the LCST of the grafted PNIPAM-based polymeric gates in water at temperatures of 22 °C, 25 °C, and 28 °C. To achieve lower C_{E1} value, lower LCST value of the grafted PNIPAM-based polymeric gates (in water) and higher operation temperature are necessary, and vice versa. Figure 10.9b shows that the C_{E2} value keeps nearly unchanged at first and then decreases with increasing the LCST of the grafted PNIPAM-based polymeric gates in water at all temperatures of 22 °C, 25 °C, and 28 °C. To achieve a lower C_{E2} value, a higher LCST value of the grafted PNIPAM-based polymeric gates (in water) and higher operation temperature are necessary, and vice versa. Interestingly, the effect of the operation temperature on the interval of ($C_{E1}-C_{E2}$) is not remarkable (Fig. 10.9c). The response interval of ($C_{E1}-C_{E2}$) decreases with increasing the LCST of the grafted PNIPAM-based polymeric gates in water. With the relationships illustrated in Fig. 10.9, the critical ethanol response concentrations of grafted membranes with grafted PNIPAM-based gates including C_{E1}, C_{E2}, and ($C_{E1}-C_{E2}$) can be easily regulated by adjusting the LCST of the grafted PNIPAM-based gates in water and controlling the operation temperature. The optimum ethanol concentration in the fermentation broth is dependent on the process conditions and is usually considered as about 10 vol. %. For example, when

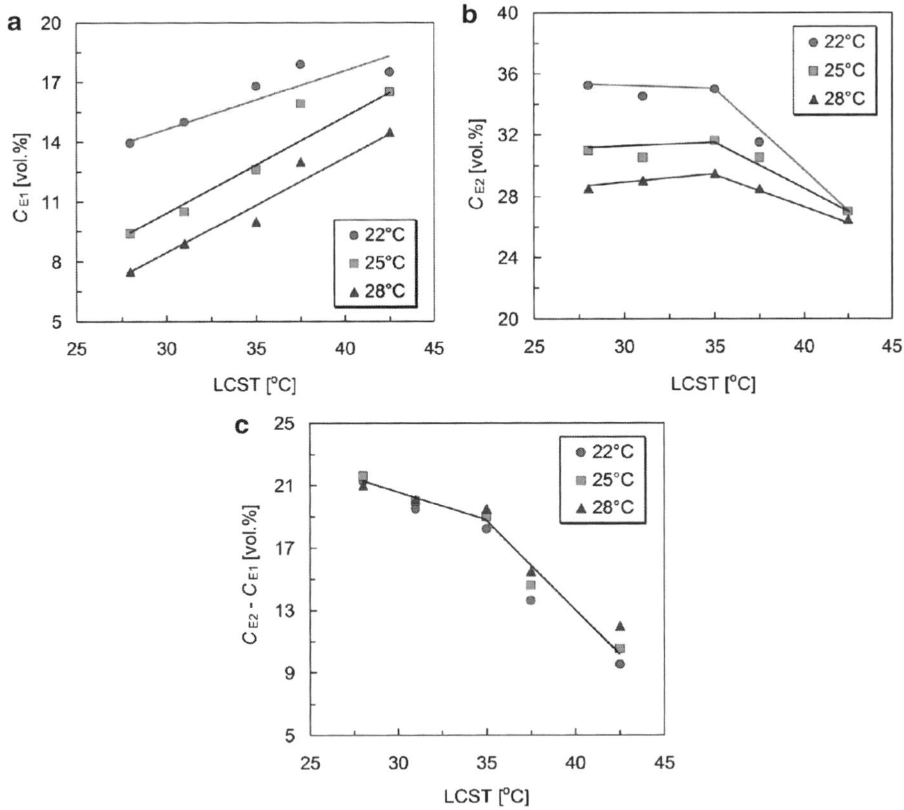

Fig. 10.9 Relationships between the LCST and the critical ethanol response concentrations of membranes with grafted PNIPAM-based gates: (**a**) C_{E1}, (**b**) C_{E2}, and (**c**) $C_{E2}-C_{E1}$ (Reproduced with permission from Ref. [7], Copyright (2012), American Chemical Society)

the operation temperature and the optimum ethanol concentration are chosen at 25 °C and 10 vol. %, respectively, the grafted membrane with the LCST of about 30 °C (i.e., the PNB-*g*-N6-1 membrane) is the most suitable membrane. However, if the operation temperature is chosen at 28 °C and the optimum ethanol concentration is still 10 vol. %, the grafted membrane with the LCST of about 34 °C (i.e., the PNIPAM-*g*-N6 membrane) is the most suitable membrane.

The ethanol-responsive smart gating membrane developed in this study could be used together with other membranes such as pervaporation membranes to remove ethanol from fermentation broths. The ethanol-responsive smart gating membrane acts as a smart valve to self-regulatively adjust the transmembrane permeability responding to ethanol concentration, and the pervaporation membrane acts as an ethanol/water separator. As a result, the ethanol concentrations in the fermentation broths can be maintained in a relatively stable level, which enables the fermentation process more efficient.

10.4 Alcohol-Responsive Microcapsule Membranes

Stimuli-responsive capsules have been considered as promising carriers in food, cosmetics, microreaction, and biomedical fields because they can change their structures and physical properties in response to external stimuli and would subsequently affect the release pattern from the reservoirs [23–25]. Recently, an alcohol-responsive PNIPAM microcapsule that can convert the variations of alcohol concentration into mechanical force has been designed as demonstrated in Fig. 10.10. The alcohol-responsive microcapsule membrane is composed of PNIPAM cross-linked hydrogels whose chemical structure is shown in Fig. 10.10. The encapsulated oil core inside the PNIPAM microcapsule membrane demonstrates the generated mechanical force by ejecting out of the microcapsule membrane upon alcohol concentration variation. When the PNIPAM microcapsule membrane remains at a swollen state, the oil core is completely encapsulated in the microcapsule membrane. However, when alcohol concentration increases to the range between the two critical alcohol response concentrations ($C_{c1} < C < C_{c2}$) (as shown in Fig. 10.10), PNIPAM microcapsule membrane becomes shrunken. Because the inner oil core is incompressible, such contraction action is hindered

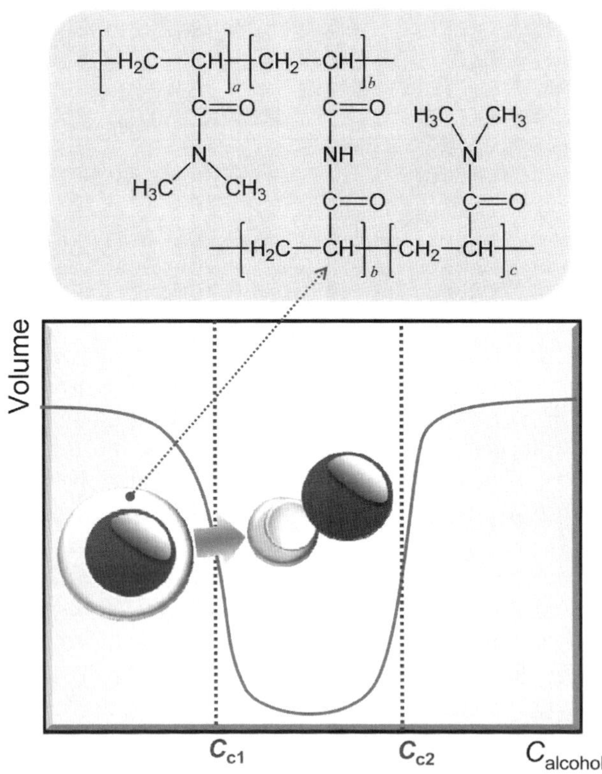

Fig. 10.10 Chemical structure of PNIPAM and schematic illustration of the concept of transition of solvent concentration variation into mechanical force (Reproduced with permission from Ref. [1], Copyright (2012), American Chemical Society)

and the pressure inside the microcapsule increases dramatically. Eventually, the accumulated pressure turns into mechanical force to rupture the microcapsule membrane and ejects the oil core into the environment. The oil core can either be an indicator itself or contain some kinds of active substances that are involved in further reaction. In addition, the quick and complete release pattern triggered by alcohol offers a potential opportunity for cargo delivery in alcoholic environments.

The core-shell PNIPAM microcapsule with encapsulated oil core is fabricated by microfluidic emulsification and subsequent UV-initiated polymerization. The inner oil core is either soybean oil or silicon oil dyed with Lumogen Red 300 (1 mg/mL). By virtue of microfluidic technique, the number of inner oil core is precisely controlled to one so that the microcapsule with a core-shell structure is successfully obtained by polymerization of the double emulsion template with one oil core.

10.4.1 Effect of Alcohol Concentration on the Thermo-responsive Characteristics of Microcapsules

The thermo-responsive size of the hollow PNIPAM microcapsules in aqueous alcoholic solutions is shown in Fig. 10.11. To eliminate the effect of oil core on the contraction of PNIPAM microcapsules, the oil core is removed with isopropyl alcohol before the experiment. The hollow PNIPAM microcapsules exhibit dramatic size change at a certain temperature which is around the VPTT of PNIPAM microcapsules. The PNIPAM microcapsules exhibit different VPTTs in aqueous solutions with different alcohol concentrations. No matter they are in a methanol solution or in an ethanol solution, the VPTT shifts of PNIPAM microcapsules show similar trends. That is, the higher the alcohol concentration is, the lower the VPTT shifts to. The VPTT values of PNIPAM microcapsules range from 12 °C to 33 °C and decrease linearly with an increase in the alcohol concentration (Fig. 10.12). Ethanol has a stronger dehydration ability than methanol at the same concentration and same temperature, which is consistent with previous studies [3]. Therefore, the VPTT value of the PNIPAM microcapsules in ethanol at the same alcoholic concentration is smaller than that in methanol.

10.4.2 Effect of Environmental Temperature on Alcohol-Responsive Characteristics of Microcapsules

The alcohol-responsive size variations of PNIPAM microcapsules at different environmental temperatures are plotted in Fig. 10.13. In the alcohol concentration range from 0 % to 40 %, the PNIPAM microcapsules change from a swollen state

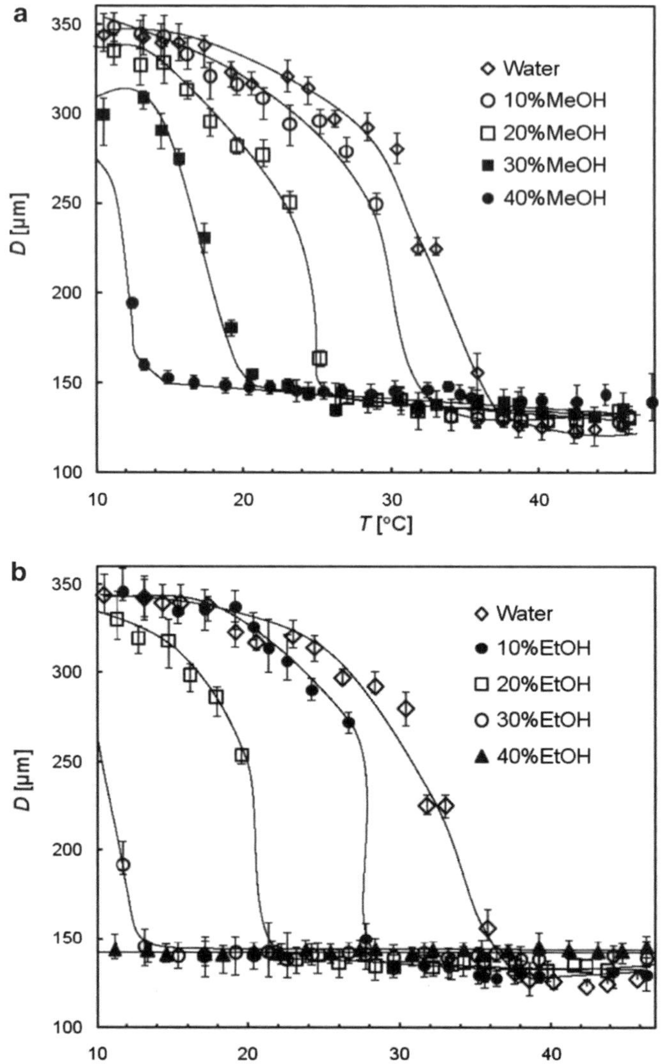

Fig. 10.11 The thermo-responsive sizes of hollow PNIPAM microcapsules in aqueous methanol solution (**a**) and aqueous ethanol solution (**b**) (Reproduced with permission from Ref. [1], Copyright (2012), American Chemical Society)

to a shrunk state and exhibit dramatic size change at a certain alcohol concentration which is the lower critical alcohol response concentration C_{c1} of PNIPAM microcapsules. In the investigated temperature range, the PNIPAM capsules reach their minimum size or shrunken equilibrium at 40 % methanol concentration or at 30 % ethanol concentration.

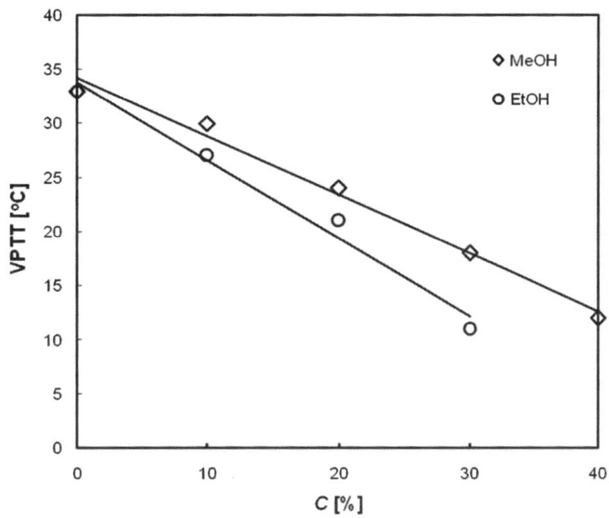

Fig. 10.12 The VPTT values of hollow PNIPAM microcapsules as a function of alcoholic concentration (Reproduced with permission from Ref. [1], Copyright (2012), American Chemical Society)

The lower critical alcohol response concentration C_{c1} values where the isothermal volume-phase transition occurs are plotted in Fig. 10.14. The C_{c1} values decrease linearly with increasing temperature, which indicates that the higher the operation temperature is, the lower the alcoholic critical concentration required for triggering the isothermal volume-phase transition. The C_{c1} values in ethanol solution are lower than those in methanol solution at the same temperature. In previous studies, it has been reported that the required concentration of alcohol solution for PNIPAM polymer to reach the minimum size shifts to a lower value as the number of carbon atoms in alcohols increases [9]. The alcohols with more carbon atoms have stronger dehydration capacity than those with fewer carbon atoms. The lower C_{c1} value can be explained in that the more carbon atoms are present, the more water molecules are required to form a clathrate structure around the alcohol molecules. As a consequence, more water molecules around the PNIPAM chains are deprived by alcohol molecules.

For the alcohol-responsive PNIPAM microcapsule, the conversion of the variations of alcohol concentration into mechanical force is usually demanded to demonstrate isothermally at room temperature. The dynamic deswelling process of PNIPAM microcapsules in alcohol solution is investigated at room temperature. In fact, the deswelling rate plays an important role in the ejection action of the oil core. If PNIPAM microcapsules cannot reach a size small enough to generate enough mechanical force to rupture the membrane before the formation of the skin layer, the PNIPAM microcapsules will never break. Generally, the higher the alcohol concentration is, the faster the PNIPAM capsules deswell and then the easier the capsule membrane breaks. The PNIPAM microcapsules shrink to their minimum sizes within 40 s or even a shorter time in alcoholic solution with a 30 % or 40 % concentration at 25 °C. Figure 10.15 shows some examples of such conversion of alcohol concentration variations into mechanical forces to eject

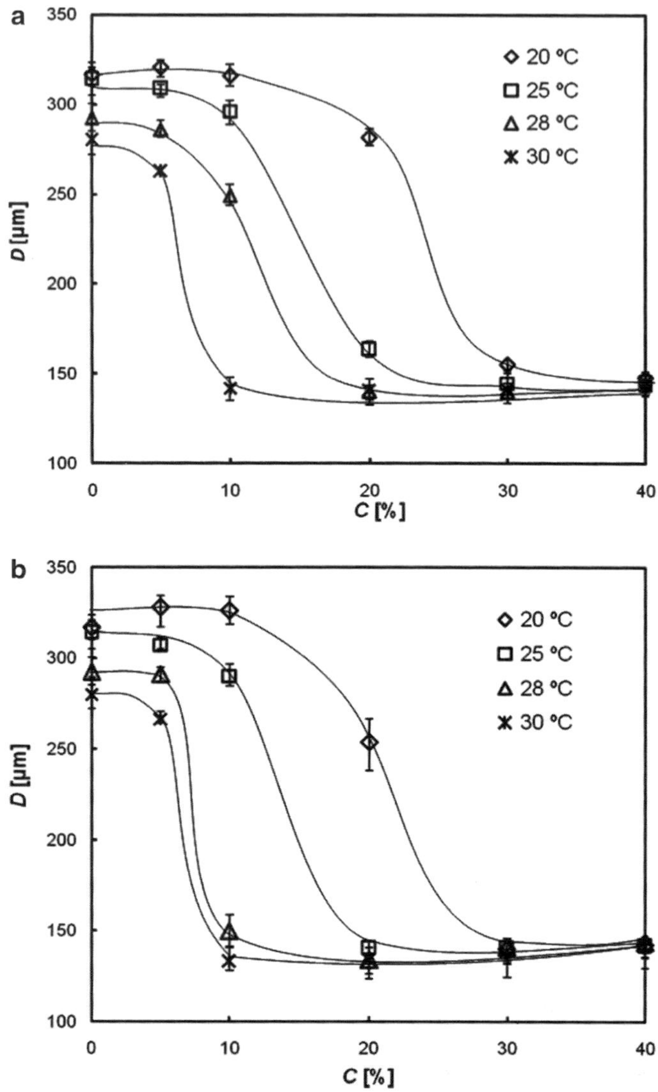

Fig. 10.13 The alcohol-responsive sizes of hollow PNIPAM capsules at different temperatures in aqueous methanol solution (**a**) and aqueous ethanol solution (**b**) (Reproduced with permission from Ref. [1], Copyright (2012), American Chemical Society)

the oil cores. Because the soybean oil is partially soluble in ethanol even at low concentration, the silicon oil, which is less soluble in ethanol, is used as the inner fluid for ethanol-responsive experiment. At 25 °C, the PNIPAM microcapsules are completely swollen in deionized water first. Then, all the environmental deionized water is removed and alcohol solution with a certain concentration at

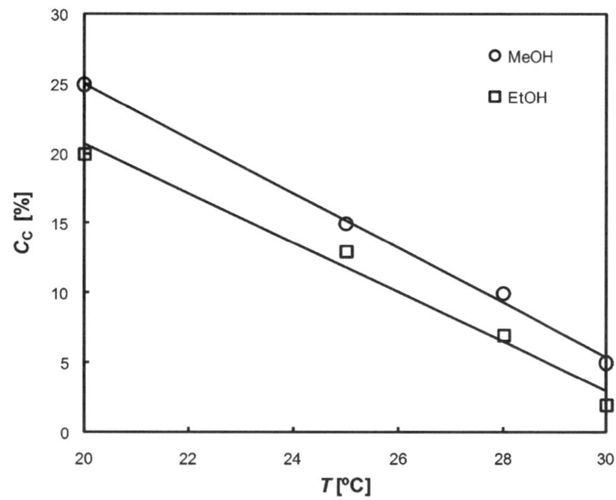

Fig. 10.14 The C_C values of hollow PNIPAM microcapsules as a function of temperature (Reproduced with permission from Ref. [1], Copyright (2012), American Chemical Society)

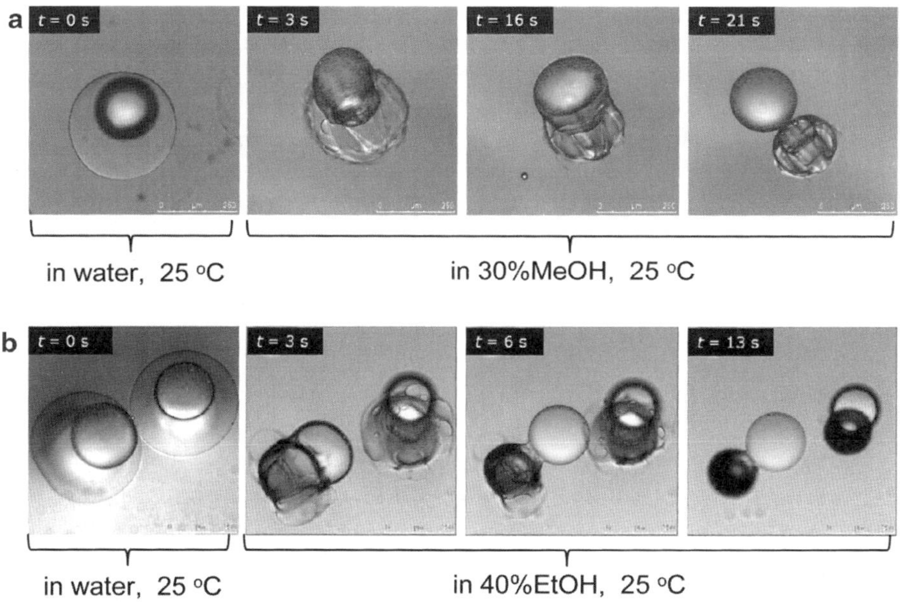

Fig. 10.15 Snapshots of the burst release process of oil cores from PNIPAM microcapsules upon adding 30 % methanol (**a**) and 40 % ethanol (**b**) at 25 °C. The PNIPAM microcapsules are in pure water at $t = 0$ s. Scale bar = 250 μm (Reproduced with permission from Ref. [1], Copyright (2012), American Chemical Society)

the same temperature is added. As the dehydration in the PNIPAM microcapsule membranes goes on, the microcapsule membranes shrink rapidly, and the oil cores are pushed aside. Eventually, the PNIPAM microcapsules shrink to the size that cannot provide enough space to trap the oil cores inside anymore. So the hydrogel microcapsule membranes rupture and eject the inner oil cores. The proposed concept of conversion of the alcohol concentration variation into mechanic force proves that core-shell PNIPAM capsules can function as both sensors and actuators of alcohol, which may also provide opportunities in cargo delivery in alcoholic environments.

10.5 Summary

In summary, alcohol-responsive smart functional membranes, either gating membranes grafted with PNIPAM gates or PNIPAM microcapsule membranes, have been designed and fabricated. Both grafted membranes and microcapsule membranes are featured with thermo-responsive characteristics and alcohol-responsive characteristics. Upon adjusting the LCST of grafted PNIPAM-based copolymer gates in water, grafted PNIPAM membranes exhibit controllable variation in critical alcohol response concentrations. The C_{E1} value simply increases with increasing the LCST of the grafted PNIPAM-based copolymer gates in water at the investigated temperatures. The C_{E2} value keeps nearly unchanged at first and then decreases with increasing the LCST of the grafted PNIPAM-based copolymer gates in water. However, with increasing the environmental temperatures, both the C_{E1} and C_{E2} values of PNIPAM-based grafted membranes and the C_c value of PNIPAM microcapsule membranes decrease. Acting as a smart valve to self-regulatively adjust the transmembrane permeability, the ethanol-responsive smart gating membrane developed in this study could be used together with pervaporation membranes to remove ethanol during the fermentation process, which enables the fermentation process more efficient. Moreover, the PNIPAM microcapsule membranes with encapsulated oil core can function as both sensors and actuators simultaneously and demonstrate the conversion of alcohol concentration variation into mechanical force by ejecting the oil core out of the microcapsule membranes, which may provide opportunities in cargo delivery in alcoholic environments.

References

1. Liu L, Song XL, Ju XJ et al (2012) Conversion of alcoholic concentration variations into mechanical force via core-shell capsules. J Phys Chem B 116:974–979
2. Pang J, Yang H, Ma J et al (2010) Solvation behaviors of *N*-isopropylacrylamide in water/methanol mixtures revealed by molecular dynamics simulations. J Phys Chem B 114:8652–8658
3. Zhu PW, Napper DH (1996) Coil-to-globule type transitions and swelling of poly(*N*-isopropylacrylamide) and poly(acrylamide) at latex interfaces in alcohol-water mixtures. J Colloid Interface Sci 177:343–352
4. Ito Y, Ito T, Takaba H et al (2005) Development of gating membrane that are sensitive to the concentration of ethanol. J Membr Sci 261:145–151
5. Xu XP, Hüther A, Maurer G (2003) The swelling equilibria of *N*-isopropylacrylamide based hydrogel in aqueous solution of ethanol. Chin J Chem Eng 11:264–268

6. Alenichev I, Sedláková Z, Ilavský M (2007) Swelling and mechanical behavior of charged poly(*N*-isopropylmethacrylamide) and poly(*N*-isopropylacrylamide) networks in water/ethanol mixtures. Cononsolvency effect. Polym Bull 58:191–199
7. Li PF, Xie R, Fan H et al (2012) Regulation of critical ethanol response concentrations of ethanol-responsive smart gating membranes. Ind Eng Chem Res 51:9554–9563
8. Winnik FM, Ringsdorf H, Venzmer J (1990) Methanol–water as a co-nonsolvent system for poly(*N*-isopropylacrylamide). Macromolecules 23:2415–2416
9. Chee CK, Hunt BJ, Rimmer S et al (2011) Time-resolved fluorescence anisotropy studies of the cononsolvency of poly(*N*-isopropyl acrylamide) in mixtures of methanol and water. Soft Matter 7:1176–1184
10. Schild HG, Muthukumar M, Tirrell DA (1991) Cononsolvency in mixed aqueous solutions of poly(*N*-isopropylacrylamide). Macromolecules 24:948–952
11. Acharya A, Goswami A, Pujari PK et al (2002) Positron annihilation studies of poly(*N*-isopropyl acrylamide) gel in mixed solvents. J Polym Sci A Polym Chem 40:1028–1036
12. Mukae K, Sakurai M, Sawamura S et al (1993) Swelling of poly(*N*-isopropylacrylamide) gels in water-alcohol (C_1-C_4) mixed solvents. J Phys Chem 97:737–741
13. Patil PN, Kathi S, Dutta D et al (2010) Understanding the swelling of poly(*N*-isopropyl acrylamide) gels through the study of free volume hole size distributions using positron annihilation spectroscopy. Polym Bull 65:577–587
14. Crowther HM, Vincent B (1998) Swelling behavior of poly*N*-isopropylacrylamide microgel particles in alcoholic solutions. Colloid Polym Sci 276:46–51
15. Scherzinger C, Lindner P, Keerl M et al (2010) Cononsolvency of poly(N, *N*-diethylacrylamide) (PDEAAM) and poly(*N*-isopropylacrylamide) (PNIPAM) based microgels in water/methanol mixtures: copolymer vs core-shell microgel. Macromolecules 43:6829–6833
16. Xie R, Li Y, Chu LY (2007) Preparation of thermo-responsive gating membranes with controllable response temperature. J Membr Sci 289:76–85
17. Lin Y, Tanaka S (2006) Ethanol fermentation from biomass resources: current state and prospects. Appl Microbiol Biotechnol 69:627–642
18. Wu Y, Xiao ZY, Huang WX et al (2005) Mass transfer in pervaporation of active fermentation broth with a composite PDMS membrane. Sep Purif Technol 42:47–53
19. Wu Y, Huang WX, Xiao ZY et al (2004) Ethanol recovery from fermentation broth by pervaporation using a composite polydimethylsiloxane membrane. Chin J Chem Eng 12:586–589
20. Lewandowska M, Kujawski W (2007) Ethanol production from lactose in a fermentation/pervaporation system. J Food Eng 79:430–437
21. Tang XY, Wang R, Xiao ZY et al (2007) Preparation and pervaporation performances of fumed-silica-filled polydimethylsiloxane-polyamide (PA) composite membranes. J Appl Polym Sci 105:3132–3137
22. Li PF, Xie R, Jiang JC et al (2009) Thermo-responsive gating membranes with controllable length and density of poly(*N*-isopropylacrylamide) chains grafted by ATRP method. J Membr Sci 337:310–317
23. Cheng R, Feng F, Meng F et al (2011) Glutathione-responsive nano-vehicles as a promising platform for targeted intracellular drug and gene delivery. J Control Release 152:2–12
24. Bagaria HG, Wong MS (2011) Polyamine-salt aggregate assembly of capsules as responsive drug delivery vehicles. J Mater Chem 21:9454–9466
25. Bae YH, Okano T, Kim SW (1991) "On-off" thermocontrol of solute transport. II. Solute release from thermosensitive hydrogels. Pharm Res 8:624–628

Part V
Glucose-Responsive Hydrogel Functional Materials

Chapter 11
Hydrogels with Rapid Response to Glucose Concentration Change at Physiological Temperature

Abstract In this chapter, a new type of glucose-responsive hydrogel with rapid response to blood glucose concentration change at physiological temperature is introduced. The hydrogel contains 3-acrylamidophenylboronic acid (AAPBA) with phenylboronic acid (PBA) groups as glucose sensors and thermo-responsive poly(*N*-isopropylacrylamide) (PNIPAM) as actuators. The response rate of the hydrogel to the change of environmental glucose concentration is significantly enhanced by introducing grafted poly(NIPAM-*co*-AAPBA) (PNA) side chains onto cross-linked PNA networks for the first time. The synthesized comb-type PNA hydrogels show satisfactory equilibrium glucose-responsive properties. Meanwhile, the hydrogels exhibit much faster response rate to glucose concentration change than normal type of cross-linked PNA hydrogels at physiological temperature. Such glucose-responsive hydrogels with rapid response rate are highly attractive for developing glucose-responsive sensors and self-regulated drug delivery systems.

11.1 Introduction

In recent years, "smart" materials that can respond to physical and chemical stimuli in the environment, such as heat [1], light [2], pH [3], magnetic field [4], electric field [5], and chemical matters [6, 7], have attracted much attention. "Smart" materials that can alter their phase or other properties when recognizing special stimuli show great potential for biomedical and pharmaceutical applications, such as drug delivery systems [1, 5], bioseparations [3], and bioactuators [4]. Among these "smart" materials, glucose-responsive materials have been widely investigated because of the increasing tendency of diabetes mellitus around the world. The investigations are aiming at developing self-regulated insulin delivery systems

L.-Y. Chu et al., *Smart Hydrogel Functional Materials*,
DOI 10.1007/978-3-642-39538-3_11,
© Chemical Industry Press, Beijing and Springer Berlin Heidelberg 2013

that can release insulin in response to blood glucose levels [6–11]. Such "smart" insulin delivery systems are expected to release insulin in time as soon as blood glucose increases [12], which can reduce the pain of patients from multiple insulin injections. To achieve such systems, Kataoka et al. [12, 13] have synthesized a novel polymeric hydrogel with glucose response at physiological pH (pH = 7.4) and collapse temperature around body temperature.

Up to now, three major kinds of glucose-responsive systems have been developed, with functional moieties involving glucose oxide [6, 7, 14, 15], concanavalin A [6, 7], and phenylboronic acid (PBA) [6, 7], respectively. However, the first two kinds of enzyme-involved and protein-involved systems usually suffer from instability, antigenicity, and expensive cost. In particular, the PBA-based systems do not contain any enzyme or protein and can reversibly form a complex with *cis*-diol such as glucose [16]. In an acid or a neutral environment, boronic acid moieties of PBA maintain plane triangle structure, which hardly form a complex with *cis*-diol. In an alkali environment, PBA turns into boronic anion tetrahedral structure by combining a hydroxyl on the boron, because the pK_a of AAPBA moiety in the copolymer of NIPAM and AAPBA is 8.2 [12, 13]. The tetrahedral structure can form a stable complex with glucose, as shown in Fig. 11.1a [17]. The PBA-based system can be made into many forms such as microgels [13, 16, 18, 19], bulk hydrogels [12, 20], microcapsules [21], and membranes [22, 23] by introducing PNIPAM as backbone. PNIPAM is a thermo-responsive polymeric material that shows reversible volume-phase transition in response to external temperature changes. The sharp phase transition of PNIPAM at the lower critical solution temperature (LCST) can be used in various applications, such as biosensors and self-regulated drug delivery systems [8, 9].

For the PBA-based hydrogel systems, their sensitivities to various sugars [24, 25], and glucose sensitivities under different conditions such as pH [12, 13, 26, 27] and salinity [18] conditions near physiological values [18, 28], have been investigated. However, the volume-phase transitions of these glucose-responsive systems are still slow, because of the inherent property of cross-linked hydrogels [29, 30]. Since rapid response rate is crucial for the glucose-responsive systems to achieve effective insulin release in time, development of PBA-based glucose-responsive systems with rapid response is essentially required.

The objective of this study is to develop a new type of glucose-responsive hydrogel with rapid response to the change of blood glucose concentration at physiological temperature. The strategy for fabricating such glucose-responsive hydrogels is to introduce grafted PNA side chains onto cross-linked PNA networks, as shown in Fig. 11.1b [17]. Cross-linked PNIPAM hydrogels containing grafted PNIPAM side chains with freely mobile ends, which are called comb-type hydrogels, exhibit fast response to environmental temperature as reported [29]. Therefore, the comb-type PNA hydrogels with grafted PNA side chains are also expected to

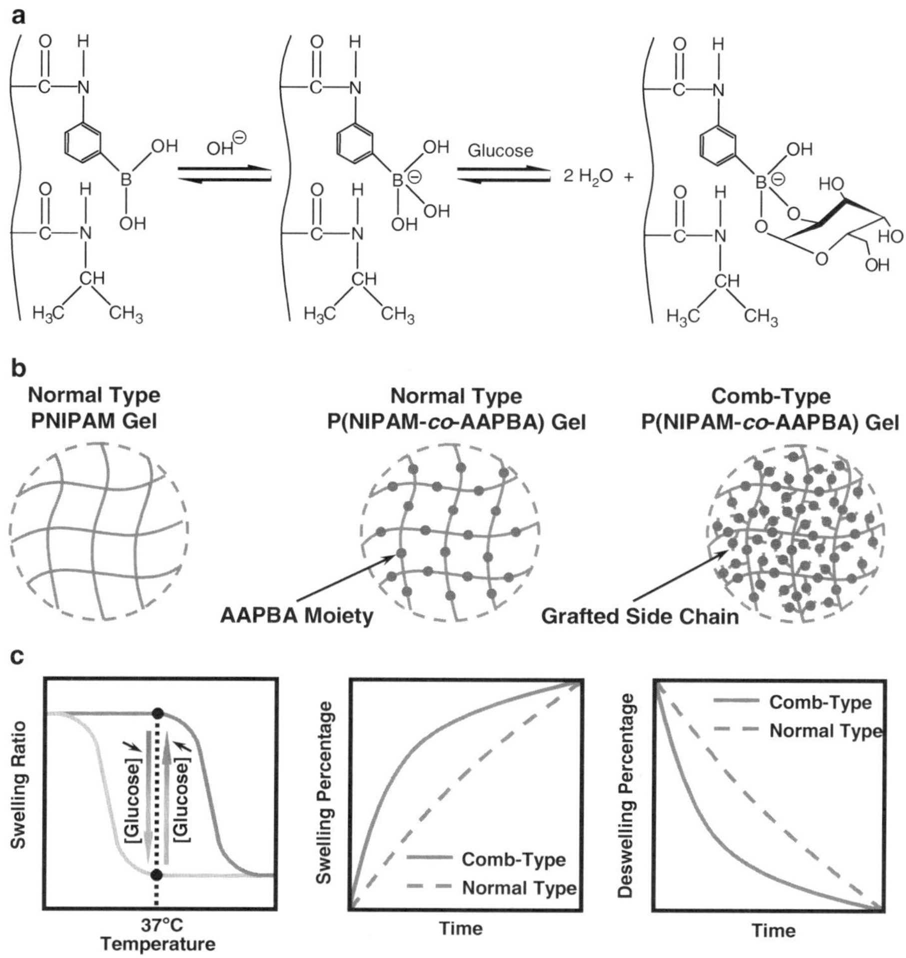

Fig. 11.1 (**a**) Complexation equilibrium between PBA moiety that grafted on PNIPAM-based polymers and glucose in alkaline environment. (**b**) Network structures of normal-type PNIPAM, normal-type PNA, and comb-type PNA hydrogels. (**c**) Swelling and shrinking behaviors of comb-type and normal-type hydrogels containing PBA moieties in response to glucose concentration change (Reproduced with permission from Ref. [17], Copyright (2008), Wiley-VCH Verlag GmbH & Co. KGaA)

present a rapid response to the environmental glucose concentrations, as illustrated in Fig. 11.1c [17]. In this chapter, the comb-type PNA hydrogels are synthesized for the first time, and their glucose-responsive swelling/deswelling transition behaviors at physiological temperature are investigated.

11.2 Preparation of Glucose-Responsive Comb-Type Hydrogels

11.2.1 Preparation of Comb-Type and Normal-Type Hydrogels Containing PBA Moieties

The reaction route for synthesizing the comb-type PNA hydrogels is shown in Fig. 11.2 [17]. The AAPBA monomer is synthesized according to previously published method [21, 24]. Intermediate macromonomer PNAH, with a terminal end group, is synthesized by free-radical polymerization method using NIPAM and AAPBA as monomers and HESH as chain transfer agent (Fig. 11.2a) [30, 31]. The comb-type PNA hydrogels are prepared by free-radical polymerization using APS as initiator and BIS as cross-linker (Fig. 11.2b) [31]. Monomers NIPAM, AAPBA, and PNAHA are dissolved in DMF, with the molar ratio of NIPAM to AAPBA at 9:1 and the weight ratio of PNAHA to the total of NIPAM and AAPBA at 3:7. A mixture solution of all the reactants is obtained by mixing the prepared DMF solution with an aqueous solution containing APS and BIS. The molar ratio of APS and that of BIS to total monomers are both 1:100. The solution is injected into a polytetrafluoroethylene tube and sealed after being bubbled with nitrogen gas for more than 15 min. Polymerization is carried out at 60 °C for 12 h. The obtained white column-type bulk hydrogel is cut into circular disks. The hydrogel disks are immersed in pure water for 6 days to remove unreacted chemicals. During this period, the water is refreshed at least 2 times per day. Normal-type PNA hydrogels are prepared in the same way without adding any PNAHA macromonomer in the recipe. The molar ratio of NIPAM to AAPBA is also 9:1 to assure that the normal-type hydrogel has the same monomer ratio as that of the comb-type hydrogel in the feed.

11.2.2 Preparation of PNAHA Macromonomer Containing PBA Moieties

The chemical structure of AAPBA is confirmed by 1H NMR spectrum, which shows results in accord with the peaks as references presented [21, 24, 29]. Because of the hydrophilicity of PBA moiety at neutral milieu (pK_a of AAPBA = 8.2), lipophilic AIBN is used as initiator [12, 13]. Chain length is able to be modulated by chain transfer agent HESH. Only one hydroxyl group is added to the terminal of a polymer chain; thus, steric hindrance is remarkable. Violent electrophilic agent acryloyl chloride is used to complete the esterifiable reaction. PNAH intermediate polymer is also confirmed by 1H NMR 400 MHz spectroscopy with DMSO [6] as solvent. The major peaks of PNAH at 9.8–10.0 ppm (acrylamide proton), 7.9–8.0 ppm (boronic hydroxyl proton), 7.2–7.9 ppm (phenyl proton), 3.8 ppm

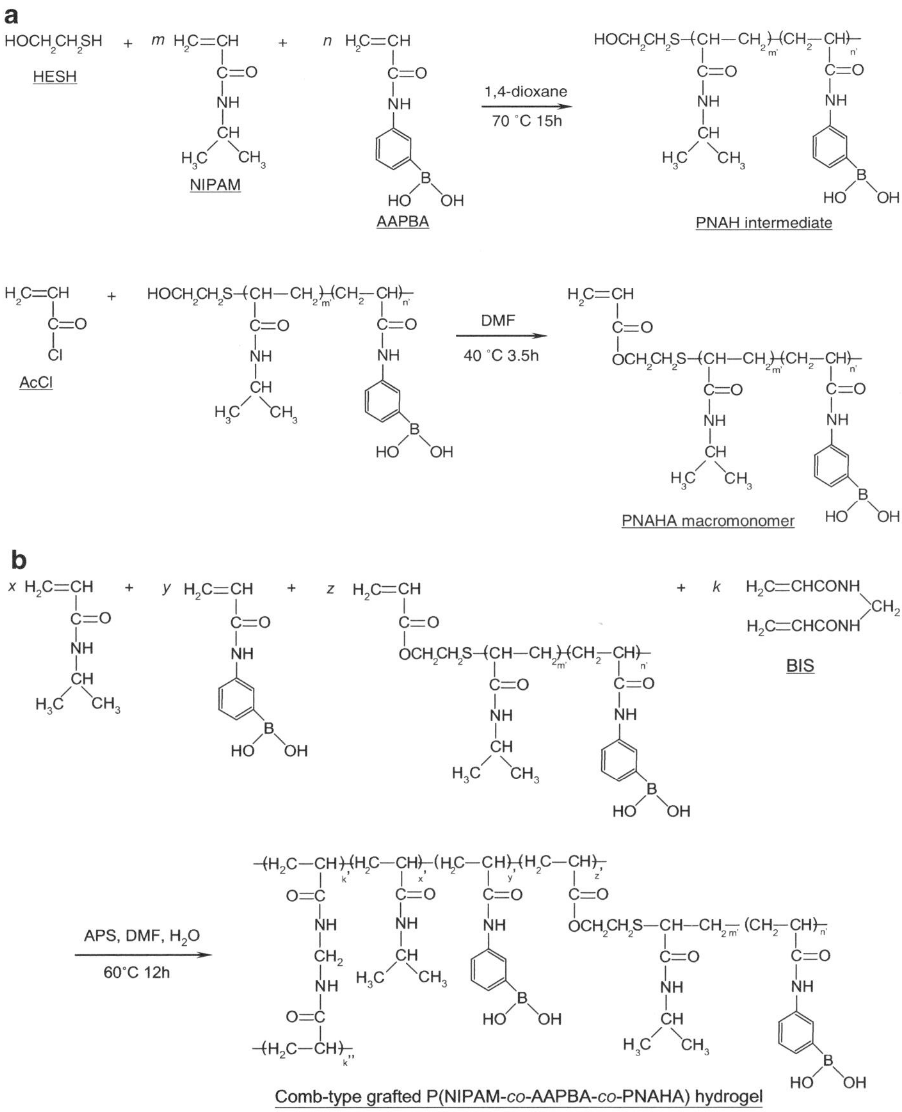

Fig. 11.2 Preparation of comb-type PNA hydrogels. (**a**) Synthesis of PNAHA macromonomers (molar ratio of NIPAM to AAPBA = 9:1). Abbreviations: HESH for 2-hydroxyethanethiol, PNAH for poly(NIPAM-*co*-AAPBA)-HESH macromolecule, AcCl for acryloyl chloride, and PNAHA for poly(NIPAM-*co*-AAPBA)-HESH-AcCl macromonomer. (**b**) Preparation of the proposed comb-type hydrogels (molar ratio of NIPAM to AAPBA = 9:1). Abbreviation *BIS N,N′*-methylenebisacrylamide, *APS* ammonium peroxide, *DMF N,N*-dimethylformamide (Reproduced with permission from Ref. [17], Copyright (2008), Wiley-VCH Verlag GmbH & Co. KGaA)

(isopropyl proton), and 2.7 ppm (terminal hydroxyl proton) are detected. In the 1H NMR spectrum of PNAHA macromonomer, other chemical shifts are nearly the same, but peaks at 5.8–6.2 ppm (vinyl proton) are detected. These results show that PNAHA macromonomer has been successfully synthesized.

The design of comb-type hydrogel is aiming at preparing hydrogel with more freely mobile chains for more rapid responsibility; this can be achieved by adding the same weight of short-chain macromonomers into the hydrogel instead of the cross-linked one. The molecular weight distribution of PNAH macromolecules is determined by gel permeation chromatography (GPC). The number average (M_n) and weight average molecular weight (M_w) are 2,016 and 4,377, respectively ($M_w/M_n = 2.17$). Thus, according to the GPC results of PNAH intermediate, we can confirm that the actual terminal group achieved is in accord with theoretical feed value. Comparing with the M_n value ($M_n = 2{,}260 \sim 3{,}800$) calculated from the feed composition (the molar ratio of HESH to total monomers is 1.5:30), the M_n value ($M_n = 2{,}016$) achieved from GPC indicates that the chain length can be well controlled by HESH.

11.2.3 Preparation of Comb-Type Poly(NIPAM-co-AAPBA) Hydrogels and Normal-Type Reference Hydrogels

Before synthesizing the comb-type PNA-grafted hydrogels, a series of normal-type PNA hydrogels are prepared. Glucose-responsive experiments are carried out to determine the optimum composition of functional monomers (AAPBA and NIPAM) for the comb-type hydrogel.

Optimal content of PBA is determined in terms of combination of two requirements. One is the glucose responsibility, and the other is the controlled release property (distinct volume change at critical temperature). The critical deswelling temperatures and thermo-responsive volume changes of normal-type PNA hydrogels that contain 10 mol % and 15 mol % PBA are investigated. As shown in Fig. 11.3a [17], the hydrogel with 10 mol % PBA displays a sharper volume-phase transition in response to temperature change than that of the hydrogel with 15 mol % PBA in pure water. However, the glucose responsibilities do not differ so much in buffer solution as shown in Fig. 11.3b [17]. This result is in accord with the data of published references [18, 19]. On one hand, the PBA moieties are not temperature-responsive, so the PNIPAM backbone of PNA hydrogel shows worse temperature responsibility with increasing PBA content. However, the result is different in alkali buffer solution. In alkali solution, the ionized PBA moieties produce electrostatic repulsion, which leads to a swelling tendency in lower temperature. Thus, the swelling behavior is enhanced in buffer solution. On the other hand, due to the ability of PBA to be reversibly combined with glucose, the glucose sensitivity of the hydrogel increases with increasing PBA content. The results show that the normal-type PNA hydrogels containing ~10 mol % of PBA moieties exhibit the optimum physical property and glucose responsiveness [18, 19].

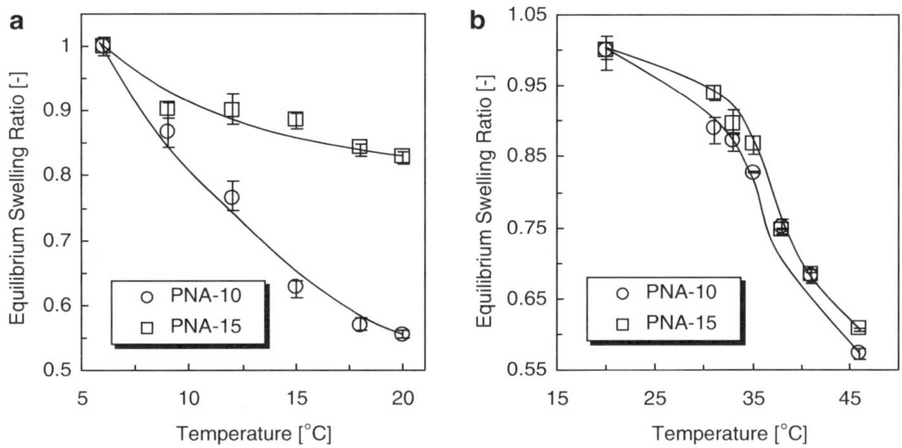

Fig. 11.3 (**a**) Temperature responsibility and volume-phase transition of normal-type PNA hydrogels with 10 and 15 mol % of PBA in pure water. (**b**) Glucose responsibility of normal-type PNA hydrogels with 10 and 15 mol % of PBA in 0.005 M glucose buffer (pH = 9.0) (Reproduced with permission from Ref. [17], Copyright (2008), Wiley-VCH Verlag GmbH & Co. KGaA)

The difference between comb-type hydrogel (copolymerized by macromolecules and other monomers) and normal-type hydrogel only lies in that the comb-type hydrogel has rapid response. It has been verified that the optimum PBA composition of normal hydrogel is 10 mol % as mentioned above. Thus, the feeding ratio of 9:1 (NIPAM to AAPBA) is considered to be optimal and it is also applicable for the comb-type hydrogel. Therefore, feed composition of 10 mol % AAPBA is fixed in the fabrication of both normal-type and comb-type hydrogels for subsequent experiments. That is, both the backbones and side chains of the comb-type hydrogels contain 10 mol % PBA moieties, the same as that in the normal-type ones. During the experiment, 30 wt % of PNAHA macromonomer is incorporated into the reaction. The normal-type hydrogels are cross-linked by using BIS to fix the polymer chains. On the other hand, linear-grafted glucose-responsive side chains with a mobile terminal are introduced on the backbones of the comb-type hydrogels.

11.3 Equilibrium Glucose-Responsive Swelling/Deswelling Behaviors of Comb-Type Hydrogels

Na_2CO_3–$NaHCO_3$ buffer solution is prepared to achieve a pH value around 9.0. 0.1 mol/L Na_2CO_3 and 0.1 mol/L $NaHCO_3$ solution are mixed at a volume ratio of 1:9. The resulting solution is then used to prepare glucose buffer solutions (GBS) with 0.4, 1.0, 2.0, and 3.0 g/L of glucose. These glucose concentrations are selected within the range of relevant physiological blood glucose concentration (0.4 g/L of glucose is the hypoglycemic limit, 0.7 ~ 1.3 g/L is the normal range, and 2.0 g/L

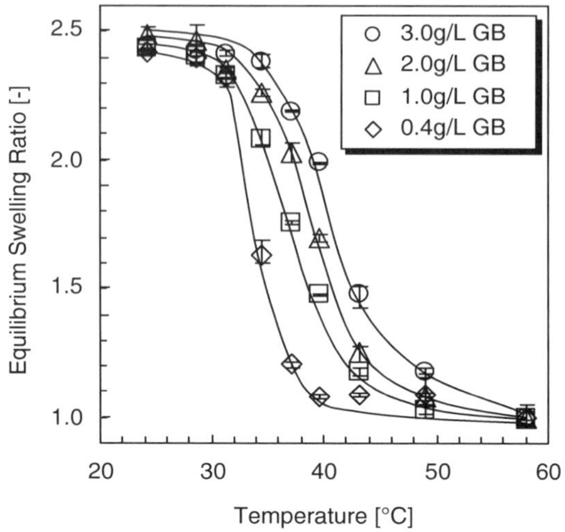

Fig. 11.4 Temperature dependence of equilibrium swelling/deswelling ratio of comb-type PNA hydrogels in buffer solutions with different glucose concentrations. Abbreviation *GB* glucose buffer, and the values before "GB" are the glucose concentrations. The pH of $NaCO_3$-$NaHCO_3$ buffer system is 8.77 at physiological temperature, which accords with pK_a of phenylboronic acid (Reproduced with permission from Ref. [17], Copyright (2008), Wiley-VCH Verlag GmbH & Co. KGaA)

is the hyperglycemic limit) [11]. The hydrogel disks are immersed in weighting bottles filled with various GBS. The bottles are placed in a water bath with constant temperature. The equilibrium swelling ratios of the hydrogels in different GBS are investigated by measuring their diameter at different temperatures that are controlled by a water bath. At each temperature, the hydrogels are maintained for at least 3 h to reach their swelling/deswelling equilibrium. The measurements are carried out at least three times, and the data are in good agreement with a standard deviation of 2 %. The value of equilibrium swelling ratio (*ESR*) is defined as the ratio of hydrogel diameter at each temperature to the minimum hydrogel diameter at the highest temperature. The temperature at which the *ESR* decreases to half of the total value change is taken as the critical deswelling temperature (*CDT*) of the hydrogel.

Figure 11.4 shows the temperature dependence of equilibrium swelling/deswelling ratio of comb-type PNA hydrogels in buffer solutions with different glucose concentrations [17]. The diameter of hydrogel disk in swollen state is about 2.5 times larger than that in shrunken state. The *CDT* of comb-type PNA hydrogel in 0.4, 1.0, 2.0, and 3.0 g/L GBS is about 34.0 °C, 36.5 °C, 38.5 °C, and 41 °C, respectively, exhibiting a directly proportional relation. This *CDT* shift indicates that the hydrogel enables a self-regulated volume change in response to glucose concentration change at a certain temperature, such as body temperature (about 37 °C).

The volume-phase transition is due to the thermo-responsive PNIPAM in the hydrogel, which can exhibit a reversible volume-phase transition as temperature changes. The LCST shift of PNIPAM is driven by an increase in entropy upon heating [32–34]. At a given temperature, the chemical potential of water in saccharide solution decreases with an increase in saccharide concentration due to the mixing free energy [35]. That is, addition of saccharides and increase of temperature have the same effect on decreasing the chemical potential of water. As a result, the decrease in the chemical potential of water molecules by adding saccharides causes saccharide-induced volume-phase transition of PNA hydrogel. Decrement of the chemical potential of water by glucose may destabilize the hydration on PNA hydrogel chains, resulting in the shrinkage of the hydrogel [35, 36]. On the other hand, the LCST of PNIPAM-based polymer can be tuned by introducing hydrophilic or hydrophobic components into the polymer. Complex of PBA moiety with glucose at a pH around nine results in change of PBA from a hydrophobic state to a hydrophilic state. Therefore, the critical deswelling temperature of the PNA hydrogel shifts to a higher temperature in GBS with higher glucose concentration.

At 37 °C (around physiological temperature), when the glucose concentration of GBS changes from 0.4 to 3.0 g/L, the *ESR* of the hydrogel increases from ~1.2 to ~2.2 (Fig. 11.4), indicating the volume change of hydrogel from a shrunken state to a swollen state. Such a hydrogel shows great potential for developing glucose-responsive sensors and self-regulated drug delivery systems.

11.4 Dynamic Glucose-Responsive Behaviors of Comb-Type Hydrogels

Glucose-responsive dynamic swelling and deswelling behaviors of comb-type and normal-type PNA hydrogels are investigated by transferring them from 0.4 to 3.0 g/L GBS (for swelling) and then transferring back to 0.4 g/L GBS (for deswelling). Before each transfer, the hydrogel disks are kept in the former solution to reach their swelling/deswelling equilibrium. All the equilibrium and dynamic processes of hydrogel disks are performed at constant physiological temperature (36.9 °C) by using a thermostatic unit. The instantaneous sizes of the hydrogel disks are obtained by measuring their diameters in GBS at fixed time interval.

The glucose-responsive dynamic behaviors of the hydrogels are investigated by measuring their swelling percentage and deswelling percentage that are respectively calculated from their swelling ratio (*SR*) and deswelling ratio (*DR*). The *SR* and *DR* are respectively defined as the ratio of instantaneous diameter at each temperature to the minimum diameter (*SR*) and that to the initial maximum diameter (*DR*). Thus, the instantaneous swelling percentage (SP_i) during the swelling process is calculated as follows:

$$SP_i = \frac{SR_i - SR_{\min}}{SR_{\max} - SR_{\min}} \times 100\,\% \tag{11.1}$$

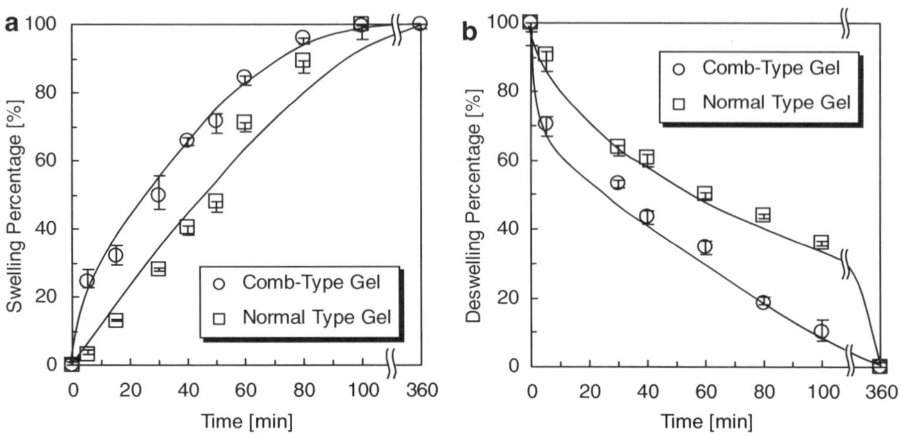

Fig. 11.5 Comparison of response rate of comb-type and normal-type hydrogels in response to glucose concentration change. Comb-type PNA hydrogels and normal-type hydrogels are both transferred from 0.4 to 3.0 g/L GBS for the swelling process after the swelling/deswelling equilibrium has been reached in the former solution. The opposite shrinking process is operated with reversed change of the glucose concentration. All the equilibrium and dynamic processes of hydrogels are performed at 36.9 °C (Reproduced with permission from Ref. [17], Copyright (2008), Wiley-VCH Verlag GmbH & Co. KGaA)

where SR_i, $SR_{\min}$, and $SR_{\max}$ are respectively the instantaneous SR, minimum SR, and maximum SR.

Similarly, the instantaneous deswelling percentage (DP_i) during the deswelling process is calculated as follows:

$$DP_i = \frac{DR_i - DR_{\min}}{DR_{\max} - DR_{\min}} \times 100\,\% \tag{11.2}$$

Figure 11.5 shows the comparison between response rates of the comb-type and normal-type PNA hydrogels in response to glucose concentration change at 36.9 °C [16]. During both swelling and deswelling processes, the comb-type PNA hydrogel exhibits faster response rate than the normal-type hydrogel, due to the presence of freely mobile side chains in the comb-type hydrogel. Meanwhile, the normal-type hydrogel can easily form a tight skin layer on the surface, which restricts the transport of water into and out of the hydrogel, resulting in slow response. By contrast, for the comb-type PNA hydrogel, its grafted side chains with freely mobile ends can prevent the hydrogel from forming tight skin-like surface to some extent [29]; as a result, the comb-type hydrogel shows faster swelling and deswelling rate than that of normal-type hydrogel.

11.5 Summary

In summary, a novel comb-type of poly(NIPAM-*co*-AAPBA) hydrogel with rapid response rate, which is highly attractive in the fields of glucose-responsive sensors and self-regulated drug delivery systems, is introduced in this chapter.

References

1. Lim YH, Kim D, Lee DS (1997) Drug releasing characteristics of thermo- and pH-sensitive interpenetrating polymer networks based on poly(*N*-isopropylacrylamide). J Appl Polym Sci 64:2647–2655
2. Beattie MS, Jackson C, Jaycox GD (1998) Azobenzene modified poly(aryl ether ketone amide)s. 2. Photo- and thermo-responsive behaviour in dilute solution. Polymer 39:2597–2605
3. Ito Y, Ochiai Y, Park YS et al (1997) pH-sensitive gating by conformational change of a polypeptide brush grafted onto a porous polymer membrane. J Am Chem Soc 119:1619–1623
4. Deng YH, Yang WL, Wang CC et al (2003) A novel approach for preparation of thermoresponsive polymer magnetic microspheres with core-shell structure. Adv Mater 15:1729–1732
5. Kim SY, Lee YM (1999) Drug release behavior of electrical responsive poly(vinyl alcohol)/poly(acrylic acid) IPN hydrogels under an electric stimulus. J Appl Polym Sci 74:1752–1761
6. Kost J, Langer R (2001) Responsive polymeric delivery systems. Adv Drug Deliv Rev 46:125–148
7. Miyata T, Uragami T, Nakamae K (2002) Biomolecule-sensitive hydrogels. Adv Drug Deliv Rev 54:79–98
8. Qiu Y, Park K (2001) Environment-sensitive hydrogels for drug delivery. Adv Drug Deliv Rev 53:321–339
9. Steil GM, Rebrin K (2005) Closed-loop insulin delivery – what lies between where we are and where we are going? Expert Opin Drug Deliv 2:353–362
10. Chu LY (2005) Controlled release systems for insulin delivery. Expert Opin Ther Pat 15:1147–1155
11. Peppas NA (2004) Is there a future in glucose-sensitive, responsive insulin delivery systems? J Drug Deliv Sci Technol 14:247–256
12. Matsumoto A, Yoshida R, Kataoka K (2004) Glucose-responsive polymer gel bearing phenylborate derivative as a glucose-sensing moiety operating at the physiological pH. Biomacromolecules 5:1038–1045
13. Matsumoto A, Ikeda S, Harada A et al (2003) Glucose-responsive polymer bearing a novel phenylborate derivative as a glucose-sensing moiety operating at physiological pH. Biomacromolecules 4:1410–1416
14. Chu LY, Li Y, Zhu JH et al (2004) Control of pore size and permeability of a glucose-responsive gating membrane for insulin delivery. J Control Release 97:43–53
15. Chu LY, Liang YJ, Chen WM et al (2004) Preparation of glucose-sensitive microcapsules with a porous membrane and functional gates. Colloids Surf B 37:9–14
16. Hoare T, Pelton R (2007) Engineering glucose swelling responses in poly(*N*-isopropylacrylamide)-based microgels. Macromolecules 40:670–678
17. Zhang SB, Chu LY, Xu D et al (2008) Poly(*N*-isopropylacrylamide)-based comb-type grafted hydrogel with rapid response to blood glucose concentration change at physiological temperature. Polym Adv Technol 19:937–943
18. Zhang YJ, Guan Y, Zhou SQ (2006) Synthesis and volume phase transitions of glucose-sensitive microgels. Biomacromolecules 7:3196–3201

19. Lapeyre V, Gosse I, Chevreux S et al (2006) Monodispersed glucose-responsive microgels operating at physiological salinity. Biomacromolecules 7:3356–3363
20. Kataoka K, Miyazaki H, Bunya M et al (1998) Totally synthetic polymer gels responding to external glucose concentration: their preparation and application to on-off regulation of insulin release. J Am Chem Soc 120:12694–12695
21. Geest BGD, Jonas AM, Demeester J et al (2006) Glucose-responsive polyelectrolyte capsules. Langmuir 22:5070–5074
22. Kikuchi A, Suvuki K, Okabayashi O et al (1996) Glucose-sensing electrode coated with polymer complex gel containing phenylboronic acid. Anal Chem 68:823–828
23. Akamatsu K, Yamaguchi T (2005) Molecular recognition gating membranes made by plasma-graft polymerization. J Photopolym Sci Technol 18:229–232
24. Shiomori K, Ivanov AE, Galaev IY et al (2004) Thermoresponsive properties of sugar sensitive copolymer of *N*-isopropylacrylamide and 3-(acrylamido)phenylboronic acid. Macromol Chem Phys 205:27–34
25. Samoei GK, Wang WH, Escobedo JO et al (2006) A chemomechanical polymer that functions in blood plasma with high glucose selectivity. Angew Chem Int Ed 45:5319–5322
26. Shiino D, Murata Y, Kubo A et al (1995) Amine containing phenylboronic acid gel for glucose-responsive insulin release under physiological pH. J Control Release 37:269–276
27. Shiino D, Murata Y, Kataoka K et al (1994) Preparation and characterization of a glucose-responsive insulin-releasing polymer device. Biomaterials 15:121–128
28. Aoki T, Nagao Y, Sanui K et al (1996) Glucose-sensitive lower critical solution temperature changes of copolymers composed of *N*-isopropylacrylamide and phenylboronic acid moieties. Polym J 28:371–374
29. Yoshida R, Uchida K, Kaneko Y et al (1995) Comb-type grafted hydrogels with rapid deswelling response to temperature changes. Nature 374:240–242
30. Annaka M, Sugiyama M, Kasai M et al (2002) Transport properties of comb-type grafted and normal-type *N*-isopropylacrylamide hydrogel. Langmuir 18:7377–7383
31. Zhang J, Chu LY, Li YK et al (2007) Dual thermo- and pH- sensitive poly(*N*-isopropylacrylamide-*co*-acrylic acid) hydrogels with rapid response behaviors. Polymer 48:1718–1728
32. Baltes T, Garret-Flaudy F, Freitag R (1999) Investigation of the LCST of polyacrylamides as a function of molecular parameters and the solvent composition. J Polym Sci Part A Polym Chem 37:2977–2989
33. Sershen S, West J (2002) Implantable polymeric systems for modulated drug delivery. Adv Drug Deliv Rev 54:1225–1235
34. Otake K, Inomata H, Konno M et al (1990) Thermal analysis of the volume phase transition with *N*-isopropylacrylamide gels. Macromolecules 23:283–289
35. Kawasaki H, Sasaki S, Maeda H (1996) Saccharide-induced volume phase transition of poly(*N*-isopropylacrylamide) gels. J Phys Chem 100:16282–16284
36. Inomata H, Goto S, Otake K et al (1992) Effect of additives on phase transition of *N*-isopropylacrylamide gels. Langmuir 8:687–690

Chapter 12
Glucose-Responsive Membranes and Microcapsules for Controlled Release

Abstract In this chapter, three glucose-responsive systems with different structures for controlled release, including flat membranes with gates in the pores, hollow microcapsules with gates in the porous shell, and hollow microcapsules with a hydrogel shell, are introduced.

12.1 Introduction

Over the past few years, sugar-responsive systems have been widely investigated in various fields such as sensors [1–5], drug delivery systems [6–8], bio-separations [9], and microreactors [10]. As one of the simple sugars, glucose is a very important target molecule for these systems, due to the key role of glucose in biological functions. The development of glucose-sensitive insulin-releasing system for diabetes therapy is a long-standing challenge for biomedical engineers [11, 12]. Although diabetes mellitus is a major cause of death in industrialized countries, periodic injections of insulin are currently the standard treatment for insulin-dependent diabetic patients. However, poor control of blood glucose level and poor patient compliance are associated with this therapy [13]. Therefore, self-regulated delivery systems that enable adapting the release rate of insulin in response to glucose concentration changes to keep the blood glucose levels within the normal range are required [14].

The glucose-responsive systems are usually based on three glucose-responsive moieties, including glucose oxidase (GOD) [7, 15], concanavalin A [6, 16, 17], and phenylboronic acid (PBA) derivatives [6, 18]. The glucose oxidase and concanavalin A can specifically interact with glucose, but these natural components are limited by the potential denature problem. The PBA derivatives, although less specific, show greater reliability and longer-term stability than the former two natural components. Based on these functional moieties, glucose-responsive systems in different forms such as membranes [19, 20], bulk hydrogels [21–23], microgels [24–27], and microcapsules [28] have been developed as potential self-regulated

L.-Y. Chu et al., *Smart Hydrogel Functional Materials*,
DOI 10.1007/978-3-642-39538-3_12,

© Chemical Industry Press, Beijing and Springer Berlin Heidelberg 2013

delivery systems. However, none of these systems could fully mimic the physiology of insulin secretion as yet [14]. Therefore, better glucose-responsive self-regulated systems are still under development, which are of both scientific and technological interests.

Recently, the authors' group has fabricated glucose-responsive flat membranes and hollow microcapsules that based on GOD enzyme for controlled release [7, 29]. GOD-immobilized polymer chains are grafted onto the pores of the porous membrane and microcapsule shell to serve as glucose-responsive gates for controlling the diffusional permeability of substances through the membrane and microcapsule shell. More recently, based on the more stable PBA system, the authors' group has developed glucose-responsive microcapsules with a PBA-based hydrogel shell for repeated glucose response under physiological temperature and glucose concentration conditions [30]. In this chapter, the design, fabrication, and performance of these three glucose-responsive systems are introduced.

12.2 Control of Pore Size and Permeability of a Glucose-Responsive Gating Membrane for Insulin Delivery

In this section, a glucose-responsive flat gating membrane with plasma-grafted poly(acrylic acid) (PAAC) gates and covalently bound GOD enzymes is introduced.

12.2.1 Preparation of Glucose-Responsive Flat Gating Membranes

Figure 12.1 schematically illustrates the preparation route of the glucose-responsive flat gating membrane and the principle for glucose-responsive control of the pore size and permeability of the gating membrane [29]. The preparation of the glucose-responsive flat gating membranes includes two steps. The first is to prepare flat gating membrane with glucose-responsive gates and the second is to immobilize GOD onto the flat gating membrane. Porous polyvinylidene fluoride (PVDF) membranes, with pore size of 0.22 μm and thickness of 62.5 μm, are used as the porous membrane substrates.

To fabricate flat gating membrane with glucose-responsive gates, PVDF membrane with linear polymer chains as gates is firstly prepared. Linear PAAC chains are grafted into the pores of the PVDF membrane substrate by plasma-graft pore-filling polymerization according to the method described previously [31–34]. Briefly, the PVDF membrane substrate is placed in a transparent glass tube, which is then filled with argon gas and evacuated to a pressure of 10 Pa. After that, the membrane substrate is treated with a radio-frequency plasma operating at 13.56 MHz,

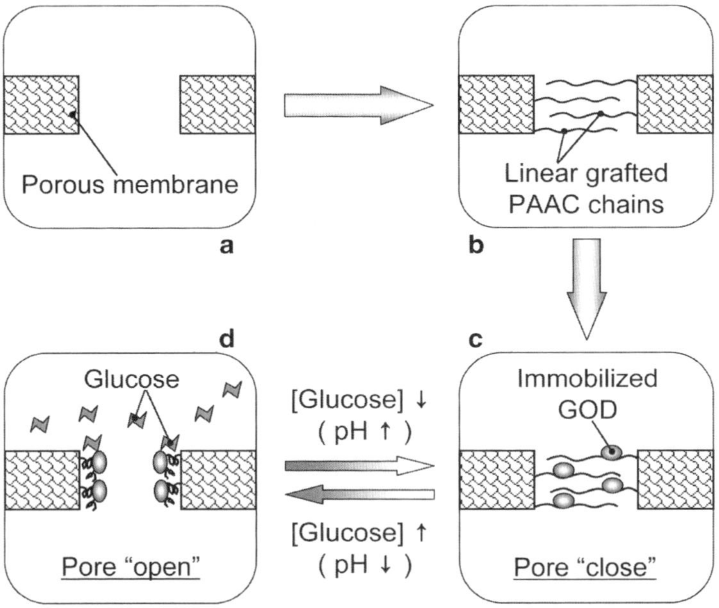

Fig. 12.1 Schematic illustration of the preparation route of the glucose-responsive gating membrane and the principle for glucose-responsive control of the pore size and permeability of the gating membrane. (**a**) Flat membrane substrate with porous structure. (**b**) Membrane substrate with grafted linear PAAC chains in the pores as the pH-responsive gates, prepared by plasma-graft pore-filling polymerization. (**c**) Glucose-responsive gating membrane fabricated by immobilization of GOD onto the grafted PAAC chains of the membrane. At neutral pH in the absence of glucose, the carboxyl groups of the grafted PAAC chains become dissociated and negatively charged. This produces repulsion force between each other to make the PAAC chains extended, so the membrane gates are "closed." (**d**) When glucose concentration increases, GOD catalyzes the oxidation of glucose into gluconic acid. This lowers the environmental pH and protonates the carboxylate groups of grafted PAAC chains, so the gates "open" due to the reduced electrostatic repulsion between grafted PAAC chains (Reproduced with permission from Ref. [29], Copyright (2004), Elsevier)

delivering 30 W for 60 s. Then, under inert atmosphere conditions, the PVDF membrane substrate is immersed into the AAC monomer solution and followed by graft polymerization under vibration in a constant-temperature bath (30 °C) for a fixed time. In the experiments, the AAC concentration in the monomer solutions is from 3 to 7 wt %, and the grafting time is from 60 to 300 min. The PAAC-grafted membranes are washed with deionized water under vibration and finally dried in a vacuum oven at 50 °C. The grafting yield (Y, %) of PAAC onto the PVDF membrane is defined as the weight increase of the membrane after the PAAC grafting.

Secondly, the GOD is immobilized onto the grafting PAAC chains of the PVDF membranes to make them glucose responsive. Immobilization of GOD is carried out by the carbodiimide method according to the method described by Ito et al. [35]. The PAAC-grafted PVDF membrane is immersed in aqueous solution with water-soluble carbodiimide (WSC), 1-(3-dimethyl-aminopropyl)-3-ethylcarbodiimide hydrochloride (10 wt %), for 1 h at 4 °C. This WSC solution is buffered at pH = 4.75 with 0.1 mol/L of 2-(*N*-morpholino)ethanesulfonic acid (MES). After being activated, the membrane is rapidly washed three times with the MES buffer solution to remove unreacted carbodiimide molecules. Then, for GOD immobilization, the activated membrane is immediately immersed in 1 wt % aqueous solution of GOD, buffered at pH = 4.75 with 0.1 mol/L MES, for 24 h at 4 °C. After that, the membrane is washed repeatedly with deionized water to remove any non-covalently bound GOD enzymes, until no further release of free GOD into the washings is detected by UV measurement.

12.2.2 Morphological Characterization of the PAAC-Grafted Membranes

Scanning electron microscope (SEM, JSM-5900LV) is used to observe the microscopic configuration of PAAC-grafted membranes with different grafting yields. The cross-sectional structures of the membranes are observed by putting the membranes into liquid nitrogen, cutting the membranes with a slice, and then gilding the membranes.

Figure 12.2 shows SEM micrographs of the cross sections of ungrafted and PAAC-grafted membranes with different grafting yields [29]. The cross sections of the ungrafted and PAAC-grafted membranes exhibit significantly different structures. As observed in Fig. 12.2a, the ungrafted membrane shows a relatively denser surface layer and a looser support layer. The cross sections of the grafted membranes, as shown in Fig. 12.2b, c, are denser than that of the ungrafted membrane, including the support layer. Comparative analysis of Fig. 12.2b with Fig. 12.2c shows that the cross section of the membrane becomes denser with the increasing grafting yield obviously, which means the porosity of the membrane becomes smaller with increasing the grafting yield. The results indicate the formation of a homogeneous graft of PAAC throughout the entire thickness of the membrane.

12.2.3 pH-Responsive Hydraulic Permeability of the PAAC-Grafted Membranes

The pH-responsive hydraulic permeability of the PAAC-grafted membranes with different grafting yields is estimated by measuring the water flux of filtration experiments. The filtration experiments of membranes are carried out with transmembrane pressure being 90 kPa. The diameter of the effective membrane area for filtration is 60 mm. The temperature of the feed solutions, buffered at pH = 4 and pH = 7,

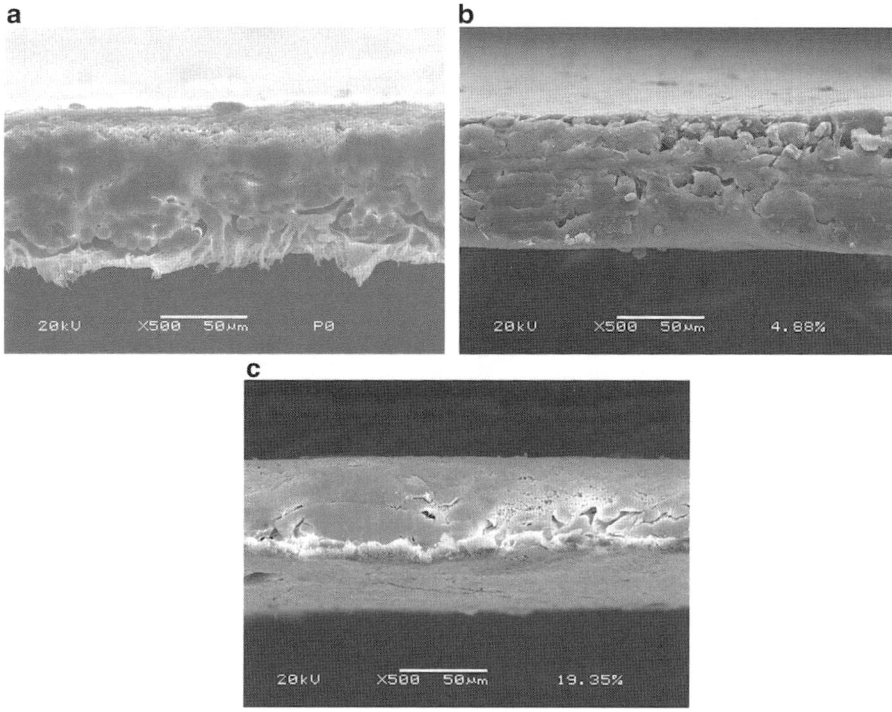

Fig. 12.2 SEM micrographs of cross sections of ungrafted and PAAC-grafted PVDF membranes. (**a**) Ungrafted PVDF membrane, (**b**) PAAC-*g*-PVDF membrane with grafting yield of 4.88 %, (**c**) PAAC-*g*-PVDF membrane with grafting yield of 19.35 % (Reproduced with permission from Ref. [29], Copyright (2004), Elsevier)

respectively, is controlled at 30 °C by using a thermostatic unit. To minimize the experimental errors, the flux measurements are carried out three to five times under each condition, and their arithmetically averaged values are taken as the results.

Figure 12.3 shows the pH-responsive hydraulic permeability of the PAAC-grafted membranes with different grafting yields [29]. The results show that the water flux of the PAAC-grafted membranes at pH = 4 is always larger than that at pH = 7. This water flux control resulted from the pH-responsive "open" and "close" functions of the grafted PAAC chains in the membrane pores. At neutral pH, the carboxyl groups of the grafted PAAC chains are dissociated and negatively charged, which produces repulsion between negative charges to make the PAAC chains extended. Therefore, the membrane gates are "closed," and the hydraulic permeability is low. Similarly, at pH lower than the pK_a of PAAC (about pH = 4.58), the carboxylate groups of the grafted PAAC chains is protonated, which reduces the electrostatic repulsion between the grafted PAAC chains in the pores. Therefore, the gates are "opened" due to the shrinking of the PAAC chains, resulting in large water flux. The grafted PAAC chains in the membrane pores act as intelligent

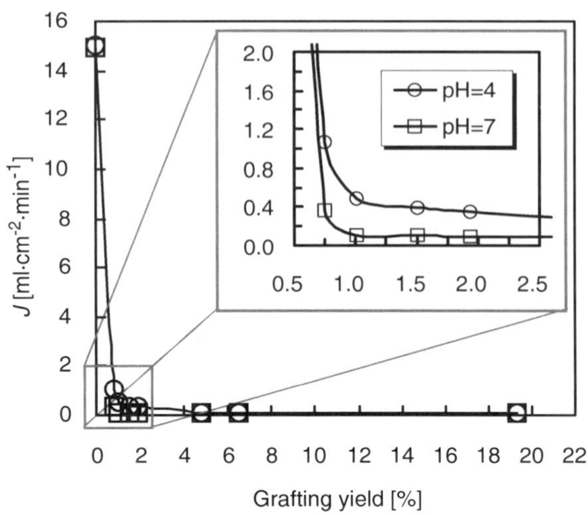

Fig. 12.3 Effect of PAAC grafting yield on the pH-controlled water flux through the PAAC-grafted membrane (Reproduced with permission from Ref. [29], Copyright (2004), Elsevier)

pH-responsive gates. On the other hand, the results show that the water flux through the membrane decreases at both pH = 4 and pH = 7 with increasing PAAC grafting yield. Because when too much PAAC is grafted into the pores (e.g., with grafting yield larger than 4.5 %), the membrane pores could be opened only a little even when the grafted PAAC chains become shrunken at pH = 4. Consequently, the water flux through the membrane becomes very small with large grafting yield of PAAC. The results indicate that the pH-responsive change of pore size is heavily dependent on the PAAC grafting yield.

12.2.4 Glucose-Responsive Controlled Release of Insulin

The diffusional permeability experiments of membranes with grafted PAAC chains and immobilized GOD are carried out using a standard side-by-side diffusion cell. Before the diffusion experiments, each test membrane is immersed in the permeant solution overnight. 0.1 mol/L Tris–HCl-buffered solution containing insulin with initial concentration of 0.1 mg/ml is used in the donor compartment, and pure 0.1 mol/L Tris–HCl-buffered solution is used in the receptor compartment. The diffusional temperature is kept constant at 30 °C by using a constant-temperature water bath, and the solutions in both the donor and the receptor compartments are magnetically stirred to enhance the mass transfer. The concentration of insulin increased in the receptor is measured using a UV-visible recording spectrophotometer at wavelength of $\lambda = 274$ nm.

The prepared glucose-responsive gating membrane is composed of porous PVDF substrate and linear-grafted PAAC chains with covalently immobilized GOD in

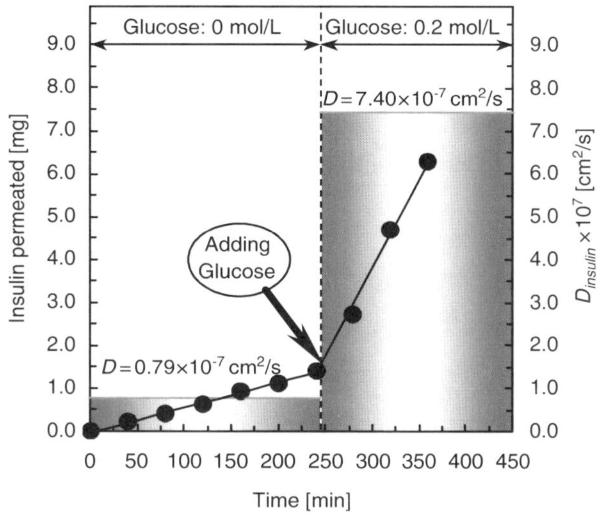

Fig. 12.4 Glucose-responsive diffusional permeation of insulin through the proposed gating membrane with PAAC grafting yield of 1.55 % (Reproduced with permission from Ref. [29], Copyright (2004), Elsevier)

the pores. The immobilized GOD, which can convert glucose into gluconic acid, acts as the glucose sensor and catalyzer. The linear-grafted PAAC chains in the membrane pores, which can extend or shrink at different pH conditions, act as the pH-responsive gates or actuators. At neutral pH conditions without glucose, the carboxyl groups of the grafted PAAC chains are dissociated and negatively charged. Therefore, the PAAC chains become extended because of the repulsion between negative charges, resulting in "closed" membrane gates. When glucose concentration increases, GOD catalyzes the oxidation of glucose into gluconic acid, resulting in lower environmental pH and protonation of the carboxylate groups in the grafted PAAC chains. This reduces electrostatic repulsion between grafted PAAC chains; therefore, the PAAC chains become shrunken and make the membrane gates "opened." Because of the glucose-responsive control of the membrane pore size, this smart gating membrane can be used for glucose-responsive controlled release of drugs such as insulin.

Figure 12.4 shows the glucose-responsive controlled release of insulin through the proposed gating membrane with PAAC grafting yield of 1.55 % [29]. In the absence of glucose, the diffusional permeation coefficient of insulin molecules across the membrane is as low as 0.79×10^{-7} cm^2/s, and the amount of insulin permeated increases linearly with time. When the environmental glucose concentration is suddenly increased from 0 to 0.2 mol/L upon glucose addition, the insulin permeation coefficient increases to 7.40×10^{-7} cm^2/s dramatically. The permeation coefficient after the glucose addition is about 9.37 times that before the addition of glucose. The results present an exciting glucose-sensitive self-regulated permeation of insulin molecules.

12.3 Glucose-Sensitive Microcapsules with a Porous Membrane and Functional Gates

In this section, glucose-sensitive microcapsules with a porous membrane and functional gates are introduced.

12.3.1 Concept of Glucose-Sensitive Microcapsules with a Porous Membrane and Functional Gates

A glucose-sensitive microcapsule with a porous membrane for encapsulation and glucose-sensitive gates for controlled release has been developed by the authors' group. Figure 12.5 schematically illustrates the concept of the proposed glucose-sensitive microcapsule [7]. The proposed microcapsule has a hollow structure

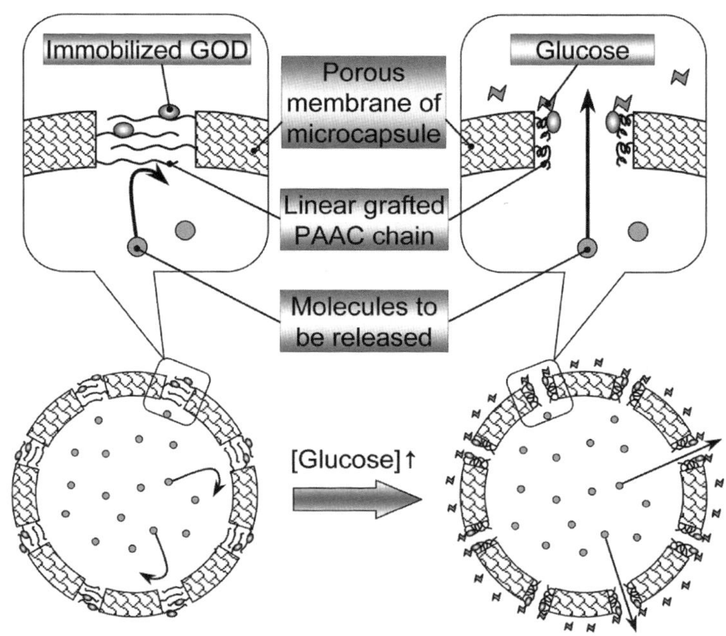

Fig. 12.5 Schematic illustration of the glucose-sensitive release principle of microcapsules with a porous membrane and functional gates. The microcapsule is composed of a core-shell porous membrane and grafted PAAC chains with covalently immobilized GOD in the pores. The substance to be released is dissolved in a solution inside the microcapsule interior. At neutral pH in the absence of glucose, the membrane pores are closed because the repulsion between negative charges makes the PAAC chains extended. When environmental glucose concentration increases, the grafted PAAC chains shrink due to the reduced electrostatic repulsion, and then the pores open (Reproduced with permission from Ref. [7], Copyright (2004), Elsevier)

comprising a porous membrane as the shell and linear-grafted PAAC chains with covalently bound GOD in the membrane pores as the functional gates. The linear-grafted PAAC chain acts as the pH-responsive actuator, and the immobilized GOD acts as glucose sensor and catalyzer. At neutral pH, the pores in the microcapsule membrane are initially closed in the absence of glucose, because of the extended PAAC chains caused by the repulsion between negative charges. Upon increasing glucose concentration, GOD catalyzes the oxidation of glucose into gluconic acid. This results in shrinking of the grafted PAAC chains because of the reduced electrostatic repulsion and then the pores are opened. The glucose-responsive “open” and “close” of the membrane gates of the microcapsule allow controlled release of the encapsulated molecules through the porous microcapsule membrane.

12.3.2 Fabrication of the Glucose-Sensitive Microcapsules

The fabrication process of the glucose-sensitive microcapsules includes two steps. The first is to prepare hollow microcapsules with a porous membrane and the second is to prepare glucose-sensitive gates in the pores of the microcapsule membrane.

Hollow microcapsules with a porous membrane are fabricated by an interfacial polymerization method as described in earlier publications [31–34]. Briefly, organic solvent mixture (10 ml) of benzene/xylene (2:1 (v/v)) containing 0.5 mol/L terephthaloyl dichloride is added to water phase (160 ml) containing 1.0 wt % sodium dodecyl sulfate as an emulsifier. After that, the mixture is mechanically agitated for 10 min with a stirring speed of 800 rpm to produce oil-in-water emulsion. The stirring speed is then reduced to 200 rpm, and both the buffer (20 ml water containing 1.18 mol/L sodium carbonate) and monomer ethylene diamine (15 ml) are added to the emulsion, followed with further agitation for 5 min. A thermostatic unit is used to keep the temperature of emulsification and interfacial polymerization constant at 10 °C. The microcapsules are separated by centrifugation and washed with deionized water, then followed with dialysis against deionized water and final freeze-drying.

To prepare glucose-sensitive gates in the membrane pores of the above prepared hollow microcapsule, linear PAAC chains are firstly grafted into the membrane pores by the plasma-graft pore-filling polymerization described previously [31–34]. Briefly, transparent glass tube containing the freeze-dried microcapsules is filled with argon gas and then evacuated to a pressure of 10 Pa. After that, the microcapsules are subjected to a radio-frequency plasma operating at 13.56 MHz, delivering 30 W for 60 s. Then, under inert atmosphere conditions, the microcapsules are immersed in AAC monomer solution, followed by graft polymerization under vibration in a constant-temperature bath (30 °C) for a fixed time. The obtained PAAC-grafted microcapsules are separated by centrifugation and washed three times with deionized water. At last, the PAAC-grafted microcapsules are dialyzed against deionized water and then freeze-dried for further immobilization of GOD.

The immobilization of GOD onto the grafted PAAC chains of the microcapsules is carried out according to the carbodiimide method [35]. Briefly, the PAAC-grafted microcapsules are immersed in an aqueous solution containing 10 wt % 1-(3-dimethyl-aminopropyl)-3-ethylcarbodiimide hydrochloride for 1 h at 4 °C. This WSC solution is buffered at pH = 4.75 with 0.1 mol/L MES. After being activated, the microcapsules are rapidly washed three times with the MES buffer solution. Then, the activated microcapsules are immediately immersed in 1 wt % aqueous solution of GOD, buffered at pH = 4.75 with 0.1 mol/L MES, for 24 h at 4 °C. After the immobilization of GOD onto the grafted PAAC chains, the microcapsules are washed repeatedly with deionized water until no further release of free GOD into the washings is detected by UV measurement. The validity of the GOD immobilization onto the PAAC-grafted microcapsule membranes is verified by using a fluorescamine method as described by Ito et al. [35].

12.3.3 Morphological Characterization of the Glucose-Sensitive Microcapsules

The microscopic morphology of the microcapsules is observed by using scanning electron microscope (SEM, JSM-5900LV), as shown in Fig. 12.6 [7]. The polyamide microcapsule shows a hollow structure with an asymmetrical and porous membrane. The membrane consists of a smooth outer surface (the water side in the interfacial polymerization) (Fig. 12.6a) and a rough inner surface (the organic side) (Fig. 12.6b). This reflects the fact that the interfacial polymerization proceeds by diffusion of the amine compounds from the aqueous phase to the organic phase. The microcapsules with porous membrane and smooth outer surface are suitable as substrates for preparing glucose-sensitive microcapsule by incorporating functional gates in the membrane pores.

12.3.4 Glucose-Sensitive Controlled-Release Behavior of the Microcapsules

The controlled-release experiments of the microcapsule are carried out using a previously published method [31–34]. Sodium chloride is selected as the model drug solute for controlled release. The freeze-dried microcapsules are dialyzed against aqueous solution containing VB_{12} with a known concentration at 30 °C under shaking for more than 3 days to load the solute inside the microcapsules. After that, a known volume of microcapsule dispersion with a known solute concentration is mixed with the same volume of deionized water. Then, the release of the solute from the microcapsules is measured by determining the increase of solute concentration in the surrounding medium with time. After a certain period of time, glucose is

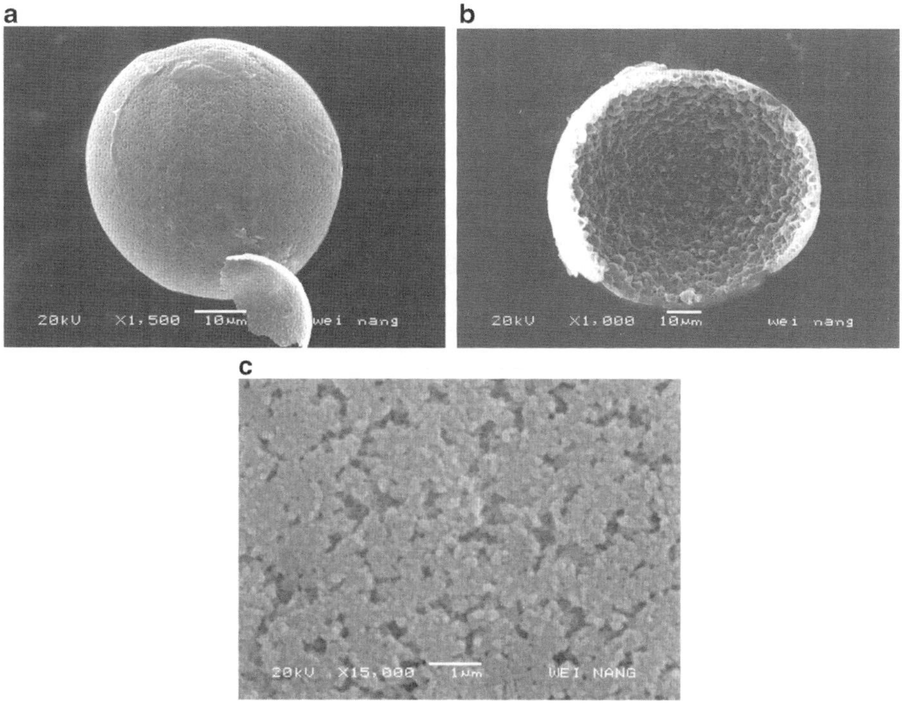

Fig. 12.6 SEM micrographs of microcapsules. (**a**) Outer surface, (**b**) cross section, and (**c**) magnified image of outer surface (Reproduced with permission from Ref. [7], Copyright (2004), Elsevier)

added to the surrounding solution at a concentration of 0.2 mol/L. During the measurements, a thermostatic unit is used to keep the temperature of the liquids constant at 30 °C. The concentration of sodium chloride is determined by measuring the electrical conductance with an electrical conductivity meter.

Figure 12.7 shows the glucose-sensitive controlled-release behaviors of sodium chloride from the ungrafted microcapsules and the glucose-sensitive microcapsules grafted with GOD-immobilized PAAC chains [7]. For the glucose-sensitive microcapsules, the release of sodium chloride molecules from the microcapsules is slow in deionized water without glucose. When the environmental glucose concentration is increased from 0 to 0.2 mol/L by adding glucose, the release rate suddenly shows a significant increase. The diffusional permeation coefficient after the addition of glucose is about 7.9 times that before glucose addition. By contrast, such a glucose-induced transition of the release rate has not been observed for the ungrafted microcapsules during the same change of the chemical environments. The grafted PAAC chains with immobilized GOD in the glucose-sensitive microcapsules play an important role as functional gates in this controlled-release process. Before adding glucose, the carboxyl groups of the grafted PAAC chains are dissociated and negatively charged, which produces repulsion between negative charges to make

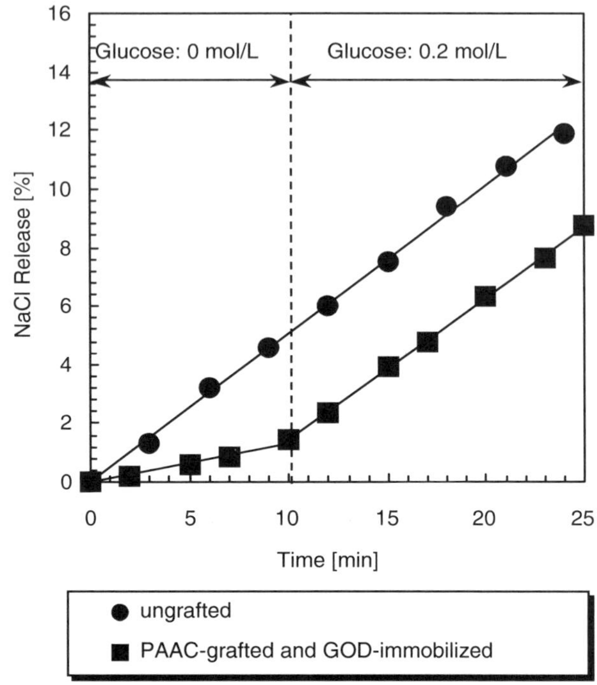

Fig. 12.7 Glucose-sensitive controlled release of sodium chloride from ungrafted microcapsules and glucose-sensitive microcapsules grafted with GOD-immobilized PAAC chains (Reproduced with permission from Ref. [7], Copyright (2004), Elsevier)

the PAAC chains extended. Therefore, the pores of the PAAC-grafted and GOD-immobilized microcapsule membrane are closed by the extended PAAC chains, resulting in a low release rate. In contrast, upon addition of glucose, the GOD immobilized on the PAAC chains catalyzes the oxidation of glucose into gluconic acid. This lowers the environmental pH and protonates the carboxylate groups of the grafted PAAC chains, which reduces the electrostatic repulsion of the grafted PAAC chains to make them shrunk. Therefore, the pores of the glucose-responsive microcapsule membrane are suddenly open due to the shrunken PAAC chains, and then a high release rate is obtained (as illustrated in Fig. 12.5). By repeatedly using the same microcapsules for VB_{12} release to verify the reversibility of the functional gates in the pores, similar glucose-sensitive controlled-release behaviors of VB_{12} are also observed [7]. These results indicate that a reversible glucose-sensitive controlled release from the fabricated hollow microcapsules is effectively achieved.

12.4 Glucose-Responsive Microcapsules with a Hydrogel Membrane

In this section, glucose-responsive microcapsules with a hydrogel membrane for repeated glucose response under physiological conditions are introduced.

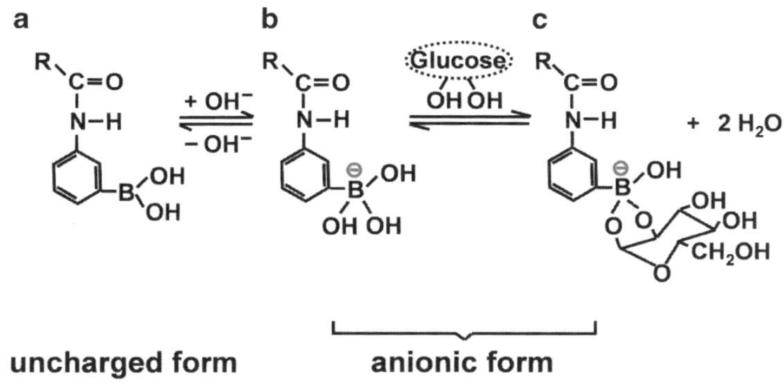

Fig. 12.8 Representation of the complex between the phenylboronic acid and glucose in aqueous solution. (**a**) Uncharged trigonal form, (**b**) charged boronate anion, and (**c**) phenylboronic acid-glucose complex (Reproduced with permission from Ref. [30], Copyright (2013), Royal Society Chemistry)

12.4.1 Concept of Glucose-Responsive Microcapsules with a Hydrogel Membrane

To construct a totally synthetic microcapsule with long-term stability and repeated glucose response, glucose-responsive 3-acrylamidophenylboronic acid (AAPBA) and thermo-responsive poly(*N*-isopropylacrylamide) (PNIPAM) are respectively employed as the glucose sensor and actuator for constructing the microcapsule shell. The sensor AAPBA can reversibly form complex with cis-diol such as glucose [26], as shown in Fig. 12.8 [30]. In an aqueous solution, PBA derivatives exist in equilibrium between an uncharged form (Fig. 12.8a) and a charged form (Fig. 12.8b), both of which can react reversibly with glucose. Especially, only the charged form can form stable complex with glucose through reversible covalent bonding (Fig. 12.8c), whereas the uncharged form is highly susceptible to hydrolysis [24]. The actuator PNIPAM is a famous thermo-responsive material that can reversibly switch between a swollen and a shrunken state via temperature changes, exhibiting a volume-phase transition temperature (VPTT) (~32 °C) close to the physiological temperature (37 °C). However, incorporation of hydrophobic PBA moiety and PNIPAM into microcapsule shell makes its VPTT lower than 32 °C. So certain amount of hydrophilic acrylic acid (AAC) is used for VPTT adjustment to make the microcapsule achieve a maximum swelling/shrinking volume change in response to glucose concentration change at 37 °C.

The concept of the proposed glucose-responsive hollow microcapsule is schematically illustrated in Fig. 12.9 [30]. In an environment with pH value close to the p*Ka* of AAPBA moiety (p*Ka* = 8.6) [36], where the PBA is supposed to present in both the uncharged and the charged forms, the glucose-responsive microcapsule is initially shrunken at 37 °C (Fig. 12.9a, c). When the glucose concentration increases, the charged form of PBA in the hydrogel shell forms stable complex with glucose

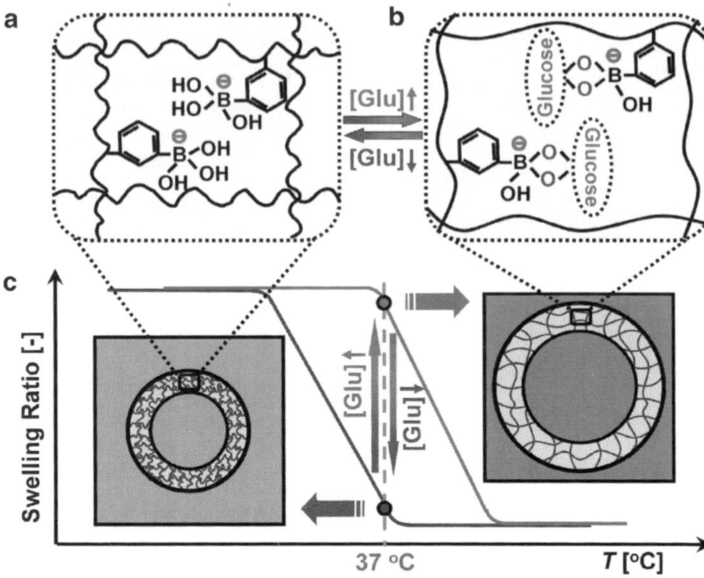

Fig. 12.9 Schematic illustration of the proposed glucose-responsive microcapsule with reversible glucose-induced swelling/shrinking behavior. Based on the complex of AAPBA with glucose, the glucose-responsive microcapsule can reversibly transit between a shrunken state (**a**) and a swollen state (**b**) by changing the glucose concentration. (**c**) Formation/decomposition of AAPBA-glucose complex induces a VPTT shift of the microcapsules, which causes reversible glucose-induced swelling/shrinking behaviors of the microcapsule (Reproduced with permission from Ref. [30], Copyright (2013), Royal Society Chemistry)

through reversible covalent bonding (Fig. 12.8b, c). The complex consumes charged PBA forms and shifts the dissociation equilibrium of PBA, which converts more hydrophobic and uncharged PBA groups into hydrophilic and charged phenylborate ions (Fig. 12.8a, b) [21]. This makes the VPTT of the microcapsule shift to a higher temperature and builds up a Donnan potential, resulting in a glucose-induced swelling of the microcapsule at 37 °C (Fig. 12.9b, c). Similarly, decrease in glucose concentration causes a glucose-induced shrinking of the microcapsule, due to the decomposition of the PBA-glucose complex. The microcapsules with reversible glucose-responsive swelling/shrinking behaviors under physiological conditions will show great potential as glucose sensors and self-regulated delivery systems for diabetes and cancer [37].

12.4.2 Template Synthesis of the Glucose-Responsive Microcapsules

The glucose-responsive microcapsules, with PNIPAM network as the actuator and AAPBA moiety as the glucose sensor, are synthesized by using monodisperse

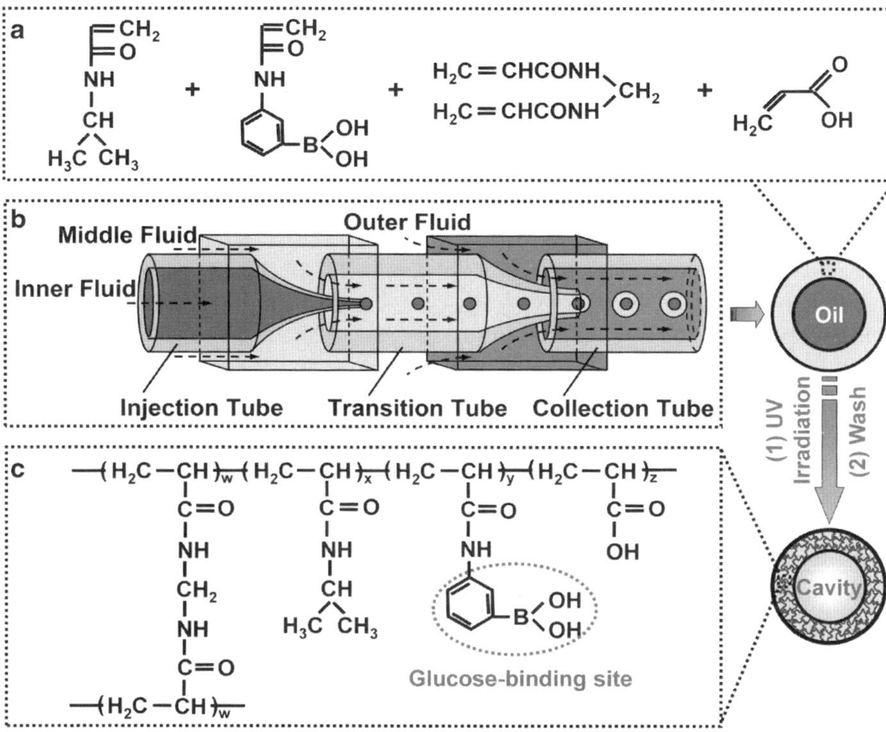

Fig. 12.10 Schematic illustration of the fabrication process of glucose-responsive microcapsule. (**a**) Middle aqueous phase containing NIPAM, AAPBA, AAC monomers, and MBA cross-linker for constructing microcapsule shell. (**b**) Capillary microfluidic device used to generate monodisperse O/W/O double emulsions. (**c**) The chemical structure of the microcapsule shell after UV-initiated polymerization (Reproduced with permission from Ref. [30], Copyright (2013), Royal Society Chemistry)

double emulsions as templates. The AAPBA is synthesized according to the procedure reported by Kitano et al. [24, 38, 39]. The glass-capillary microfluidic device (Fig. 12.10b) [30] for generating monodisperse double emulsions is fabricated according to literatures [40, 41]. The outer diameter of cylindrical capillaries and the inner dimension of square capillary tubes are both 1.0 mm. Three cylindrical capillaries, which are respectively used as the injection tube, transition tube, and collection tube, are aligned coaxially inside the square capillaries. The inner diameters of the injection tube, transition tube, and collection tube are 550, 150, and 300 μm, respectively. The end of injection tube and transition tube are tapered by a micropuller and then adjusted by a microforge.

Typically, deionized water (4 ml) containing monomer NIPAM (0.4074 g), AAPBA (0.0716 g) and cross-linker *N,N′*-methylene-*bis*-acrylamide (MBA, 0.0308 g), initiator 2,2′-azobis(2-amidinopropane dihydrochloride) (V50, 0.0217 g), glycerin (0.2 g, 5 % (w/v)), and Pluronic F127 (0.04 g, 1 % (w/v)) is used as the middle fluid to construct the glucose-responsive shell of the microcapsules.

The molar ratio of NIPAM to AAPBA is kept at 9:1. To adjust the volume-phase transition behavior of the microcapsule, different amounts of AAC are added into the middle fluid to fabricate poly(NIPAM-*co*-AAPBA-*co*-AAC) (PNAA) microcapsule with tunable VPTT (Fig. 12.10a) [30]. The amounts of AAC used in the experiment are 0, 2.5, 3, 3.5, and 4.5 mol %, with respect to the total moles of NIPAM and AAPBA.

Soybean oil containing 5 % (w/v) polyglycerol polyricinoleate (PGPR 90) and 0.2 % (w/v) $CaCO_3$ nanoparticles for emulsion stabilization is used as the outer fluids. Soybean oil containing 5 % (w/v) PGPR 90 and 1 % (w/v) 2,2-dimethoxy-2-phenylacetophenone (BDK) as photo-initiator is used as the inner fluids, respectively. O/W/O double emulsions are generated by separately pumping the inner, middle, and outer fluids into the injection tube, transition tube, and collection tube of the microfluidic device (Fig. 12.10b) and then collected in a container. The flow rates of the inner, middle, and outer fluids are respectively 500, 600, and 3,000 μL/h. The collected O/W/O double emulsions are converted into microcapsules by polymerization under UV irradiation for 20 min in an ice-water bath. Under UV light, monomers contained in the middle aqueous phase of the double emulsions are polymerized to build the glucose-responsive hydrogel shell of the microcapsule (Fig. 12.10c) [30]. A 250 W UV lamp with an illuminance spectrum of 250 ~ 450 nm is employed to produce UV light. After washing with isopropanol and deionized water for several times, these microcapsules are redispersed in deionized water for further characterization.

12.4.3 Glucose-Responsive Swelling/Shrinking Behaviors of PNAA Microcapsules with Different AAc Contents

To mimic the relevant physiological blood glucose concentrations, glucose concentration range of 0.4 ~ 3.0 g/L is selected within the range of relevant physiological blood glucose concentration [42, 43]. Na_2CO_3-$NaHCO_3$ buffer solutions (pH = 8.77, $T = 37$ °C) containing glucose with different concentrations (0.4 and 3.0 g/L), which are referred as the glucose buffer solution (GBS), are used as the medium.

To obtain a good glucose response at 37 °C, the T_{opt} of the PNAA microcapsules, defined as the temperature at which the microcapsules achieve the maximum volume change in response to glucose concentration change from 0.4 to 3.0 g/L, should be located at 37 °C. However, incorporation of hydrophobic PBA moiety into PNIPAM network of the microcapsule shell makes its VPTT lower than 32 °C, so hydrophilic AAC is incorporated into the shell for VPTT adjustment as well as the T_{opt}. To evaluate the glucose-responsive swelling/shrinking behaviors of PNAA microcapsules, a parameter defined as $R_{T,3.0}/R_{T,0.4}$ is introduced. The $R_{T,0.4}$ and $R_{T,3.0}$ are respectively the swelling ratios of PNAA microcapsules in 0.4 and 3.0 g/L GBS at temperature T. So, T_{opt} is the temperature where $R_{T,3.0}/R_{T,0.4}$ gets the maximum value. The temperature-dependent equilibrium volume changes of

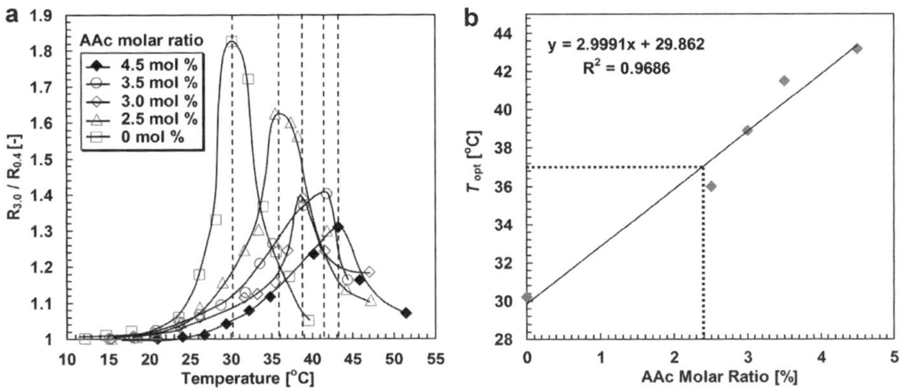

Fig. 12.11 The $R_{T,3.0}/R_{T,0.4}$ of PNAA microcapsules with different molar ratio of AAC as a function of temperature (**a**) and the T_{opt} of these microcapsules as a function of the molar ratio of AAC (**b**) (Reproduced with permission from Ref. [30], Copyright (2013), Royal Society Chemistry)

PNAA microcapsules with 2.5, 3.0, 3.5, and 4.0 mol % AAC in 0.4 g/L GBS and 3.0 g/L GBS are respectively studied within the temperature range from 12 °C to 60 °C. Optical microscope equipped with a thermostatic stage system and a CCD camera is used for temperature control and observation. Briefly, the samples are first kept in GBS at room temperature for 12 h to reach their swelling/shrinking equilibrium state. Then, the temperature-dependent equilibrium volume changes of the microcapsules in different GBS are respectively studied by stepwise increasing temperature from 12 °C to 60 °C. At each temperature, the microcapsules are equilibrated for 15 min before the measurements are made.

The $R_{T,3.0}/R_{T,0.4}$ values of PNAA microcapsules with different AAC contents (0, 2.5, 3.0, 3.5 and 4.5 mol %) in the temperature range from 10 °C to 55 °C are plotted in Fig. 12.11a [30]. At the same temperature, all PNAA microcapsules show a more swelling state in 3.0 g/L GBS than those in 0.4 g/L GBS (with $R_{T,3.0}/R_{T,0.4} > 1$), indicating a glucose-induced swelling/shrinking response. Moreover, the location of their maximum $R_{T,3.0}/R_{T,0.4}$ values indicates that the T_{opt} value increases with increasing molar ratio of AAC. Figure 12.11b shows the direct linear proportion of T_{opt} to AAC molar ratio [30], from which we can determine the AAC content required for construct microcapsule with T_{opt} at 37 °C is about 2.4 mol %.

12.4.4 Reversible Glucose-Response Behavior of the Microcapsules at Physiological Temperature

PNAA microcapsules with 2.4 mol % AAC are fabricated to achieve a good glucose-responsive swelling/shrinking volume change at physiological temperature.

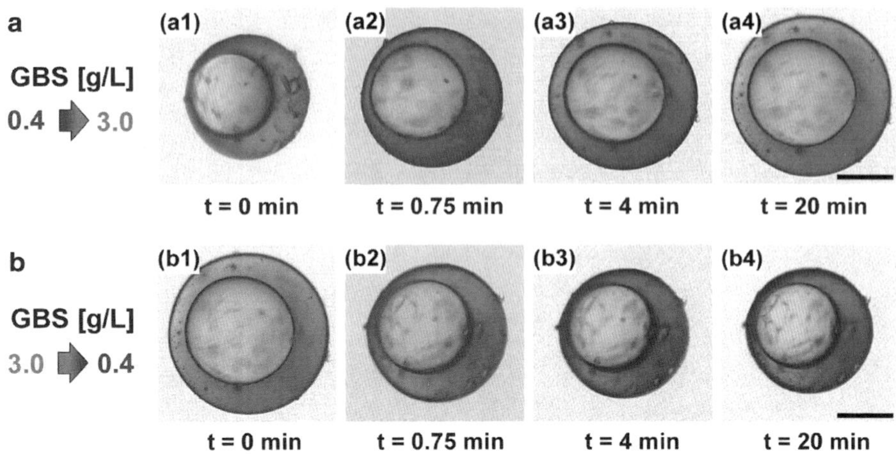

Fig. 12.12 Optical micrographs of the reversible glucose-induced swelling/shrinking behaviors of PNAA microcapsules with 2.4 mol % AAC in response to glucose concentration change between 0.4 and 3.0 g/L at 37 °C. (**a**) From 0.4 to 3.0 g/L GBS and (**b**) from 3.0 to 0.4 g/L GBS. The scale bars are 100 μm (Reproduced with permission from Ref. [30], Copyright (2013), Royal Society Chemistry)

The reversible glucose-responsive swelling/shrinking behaviors of PNAA microcapsules with 2.4 mol % AAC at 37 °C are investigated by repeatedly transferring the microcapsules between 0.4 and 3.0 g/L GBS. Their dynamic swelling volume change with increasing glucose concentration (0.4 to 3.0 g/L) and dynamic shrinking volume change with decreasing glucose concentration (3.0 to 0.4 g/L) at 37 °C are respectively studied by the same optical microscope as mentioned above.

Figure 12.12 shows the glucose-induced swelling (Fig. 12.12a) and shrinking (Fig. 12.12b) behaviors of PNAA microcapsules with 2.4 mol % AAC in response to glucose concentration changes between 0.4 and 3.0 g/L at 37 °C [30]. At 37 °C, the PNAA microcapsule is initially in a shrunken state (Fig. 12.12a1). When suddenly increasing glucose concentration from 0.4 to 3.0 g/L, the PNAA microcapsule changes from a shrunken state to a swollen state dramatically (Fig. 12.12a2–a4), because more AAPBA moieties form complex with glucose. Contrarily, when suddenly decreasing glucose concentration from 3.0 g/L back to 0.4 g/L, the PNAA microcapsule returns to the shrunken state again (Fig. 12.12b1–b4). In both of the swelling (Fig. 12.12a1, a2) and shrinking (Fig. 12.12b1, b2) processes, the microcapsule exhibits fast volume changes within $t = 0.75$ min after glucose concentration changes. These results show the good and reversible swelling/shrinking response of the PNAA microcapsules to changes in the glucose concentration at physiological temperature. These PNAA microcapsules with reversible glucose-responsive swelling/shrinking behaviors show great potential as self-regulated delivery systems for diabetes therapy and cancer treatment.

12.5 Summary

In summary, three glucose-responsive systems with different structures for controlled release, including flat membranes with gates in the pores, hollow microcapsules with gates in the porous shell, and hollow microcapsules with a hydrogel shell, are described in this chapter. The GOD-immobilized PAAC chains grafted in the pores of flat membranes can act as glucose-responsive gates for control of the solute diffusional permeability through the proposed membranes. This glucose responsivity is heavily dependent on the PAAC grafting yield, because the pH-responsive change of pore size governs the glucose-responsive diffusional permeability. The hollow microcapsules, with GOD-immobilized PAAC chains as gates in the shell pores, can achieve an increased drug release in the presence of glucose and show a reversible glucose-sensitive release characteristic. The hollow microcapsules with PNAA hydrogel shell exhibit reversible and repeated swelling/shrinking responses to glucose concentration changes within the physiological blood glucose concentration range at 37 °C. The excellent glucose sensitivity of the proposed membranes and hollow microcapsules makes these systems promising as new modes for glucose-responsive sensors and self-regulated delivery systems for diabetes therapy.

References

1. Alexeev VL, Sharma AC, Goponenko AV et al (2003) High ionic strength glucose-sensing photonic crystal. Anal Chem 75:2316–2323
2. Takahashi S, Anzai J (2005) Phenylboronic acid monolayer-modified electrodes sensitive to sugars. Langmuir 21:5102–5107
3. Egawa Y, Gotoh R, Niina S et al (2007) Ortho-azo substituted phenylboronic acids for colorimetric sugar sensors. Bioorg Med Chem Lett 17:3789–3792
4. Wang Y, Wei W, Liu X et al (2009) Carbon nanotube/chitosan/gold nanoparticles-based glucose biosensor prepared by a layer-by-layer technique. Mater Sci Eng C 29:50–54
5. Takaoka H, Yasuzawa M (2010) Fabrication of an implantable fine needle-type glucose sensor using gamma-polyglutamic acid. Anal Sci 26:551–555
6. Kost J, Langer R (2001) Responsive polymeric delivery systems. Adv Drug Deliv Rev 46:125–148
7. Chu LY, Liang YJ, Chen WM et al (2004) Preparation of glucose-sensitive microcapsules with a porous membrane and functional gates. Colloids Surf B 37:9–14
8. Tanna S, Sahota TS, Sawicka K et al (2006) The effect of degree of acrylic derivatisation on dextran and concanavalin A glucose-responsive materials for closed-loop insulin delivery. Biomaterials 27:4498–4507
9. Ito Y, Ochiai Y, Park YS et al (1997) pH-sensitive gating by conformational change of a polypeptide brush grafted onto a porous polymer membrane. J Am Chem Soc 119:1619–1623
10. Nayak S, Lyon LA (2005) Soft nanotechnology with soft nanoparticles. Angew Chem Int Ed 44:7686–7708
11. Langer R (1998) Drug delivery and targeting. Nature 392:5–10
12. Galaev IY, Mattiasson B (1999) 'Smart' polymers and what they could do in biotechnology and medicine. Trends Biotechnol 17:335–340
13. Zhang K, Wu XY (2002) Modulated insulin permeation across a glucose-sensitive polymeric composite membrane. J Control Release 80:169–178

14. Traitel T, Cohen Y, Kost J (2000) Characterization of glucose-sensitive insulin release systems in simulated in vivo conditions. Biomaterials 21:1679–1687
15. Qi W, Duan L, Li J (2011) Fabrication of glucose-sensitive protein microcapsules and their applications. Soft Matter 7:1571–1576
16. Luo Y, Liu L, Wang XB et al (2012) Sugar-installed thermoresponsive micellar aggregates self-assembled from "coil-comb-coil" triblock glycopolymers: preparation and recognition with concanavalin A. Soft Matter 8:1634–1642
17. Takahashi S, Sato K, Anzai J (2012) Layer-by-layer construction of protein architectures through avidin-biotin and lectin-sugar interactions for biosensor applications. Anal Bioanal Chem 402:1749–1758
18. Miyata T, Uragami T, Nakamae K (2002) Biomolecule-sensitive hydrogels. Adv Drug Deliv Rev 54:79–98
19. Ding ZB, Guan Y, Zhang Y et al (2009) Layer-by-layer multilayer films linked with reversible boronate ester bonds with glucose-sensitivity under physiological conditions. Soft Matter 5:2302–2309
20. Ding ZB, Guan Y, Zhang YJ et al (2009) Synthesis of glucose-sensitive self-assembled films and their application in controlled drug delivery. Polymer 50:4205–4211
21. Kataoka K, Miyazaki H, Bunya M et al (1998) Totally synthetic polymer gels responding to external glucose concentration: their preparation and application to on-off regulation of insulin release. J Am Chem Soc 120:12694–12695
22. Matsumoto A, Yoshida R, Kataoka K (2004) Glucose-responsive polymer gel bearing phenylborate derivative as a glucose-sensing moiety operating at the physiological pH. Biomacromolecules 5:1038–1045
23. Wang L, Liu M, Gao C et al (2010) A pH-, thermo-, and glucose-, triple-responsive hydrogels: synthesis and controlled drug delivery. React Funct Polym 70:159–167
24. Lapeyre V, Gosse I, Chevreux S et al (2006) Monodispersed glucose-responsive microgels operating at physiological salinity. Biomacromolecules 7:3356–3363
25. Zhang Y, Guan Y, Zhou S (2006) Synthesis and volume phase transitions of glucose-sensitive microgels. Biomacromolecules 7:3196–3201
26. Hoare T, Pelton R (2007) Engineering glucose swelling responses in poly(*N*-isopropylacrylamide)-based microgels. Macromolecules 40:670–678
27. Liu P, Luo Q, Guan Y et al (2010) Drug release kinetics from monolayer films of glucose-sensitive microgel. Polymer 51:2668–2675
28. Lapeyre V, Renaudie N, Dechezelles JF (2009) Multiresponsive hybrid microgels and hollow capsules with a layered structure. Langmuir 25:4659–4667
29. Chu LY, Li Y, Zhu JH et al (2004) Control of pore size and permeability of a glucose-responsive gating membrane for insulin delivery. J Control Release 97:43–53
30. Zhang MJ, Wang W, Xie R et al (2013) Microfluidic fabrication of monodisperse microcapsules for glucose-response at physiological temperature. Soft Matter 9:4150–4159
31. Chu LY, Niitsuma T, Yamaguchi T (2003) Thermoresponsive transport through porous membranes with grafted PNIPAM gates. AIChE J 49:896–909
32. Chu LY, Park SH, Yamaguchi T et al (2001) Preparation of thermo-responsive core-shell microcapsules with a porous membrane and poly(*N*-isopropylacrylamide) gates. J Membr Sci 192:27–39
33. Chu LY, Park SH, Yamaguchi T et al (2002) Preparation of micron-sized monodispersed thermoresponsive core-shell microcapsules. Langmuir 18:1856–1864
34. Chu LY, Yamaguchi T, Nakao S (2002) A molecular-recognition microcapsule for environmental stimuli-responsive controlled release. Adv Mater 14:386–389
35. Ito Y, Casolaro M, Kono K et al (1989) An insulin-releasing system that is responsive to glucose. J Control Release 10:195–203
36. Kataoka KMH, Okano T et al (1994) Sensitive glucose-induced change of the lower critical solution temperature of poly[*N*, *N*-(dimethylacrylamide)-*co*-3-(acrylamido)-phenylboronic acid] in physiological saline. Macromolecules 27:1061–1062

37. Manna U, Patil S (2010) Glucose-triggered drug delivery from borate mediated layer-by-layer self-assembly. ACS Appl Mat Interface 2:1521–1527
38. Kitano S, Koyama Y, Kataoka K et al (1992) A novel drug delivery system utilizing a glucose responsive polymer complex between poly (vinyl alcohol) and poly (*N*-vinyl-2-pyrrolidone) with a phenylboronic acid moiety. J Control Release 19:161–170
39. De Geest BG, Jonas AM, Demeester J et al (2006) Glucose-responsive polyelectrolyte capsules. Langmuir 22:5070–5074
40. Chu LY, Utada AS, Shah RK et al (2007) Controllable monodisperse multiple emulsions. Angew Chem Int Ed 46:8970–8974
41. Wang W, Xie R, Ju XJ et al (2011) Controllable microfluidic production of multicomponent multiple emulsions. Lab Chip 11:1587–1592
42. Peppas NA (2004) Is there a future in glucose-sensitive, responsive insulin delivery systems? J Drug Deliv Sci Technol 14:247–256
43. Zhang SB, Chu LY, Xu D et al (2008) Poly(*N*-isopropylacrylamide)-based comb-type grafted hydrogel with rapid response to blood glucose concentration change at physiological temperature. Polym Adv Technol 19:937–943

Part VI
Ion-Recognizable Hydrogel Functional Materials

Chapter 13
Preparation and Properties of Ion-Recognizable Smart Hydrogels

Abstract In this chapter, the design, fabrication, and performance of the ion-recognizable smart hydrogels with crown ether as ion-recognition receptor and thermo-responsive poly(*N*-isopropylacrylamide) (PNIPAM) as actuator are introduced. Smart responsive hydrogels capable of recognizing heavy metal ions or potassium ion are fabricated with different crown ethers for different purposes. Smart hydrogels with 18-crown-6 as ion-recognition receptor could respond to Pb^{2+} and Ba^{2+} due to the formation of 1:1 (ligand to ion) "host-guest" complex; smart hydrogels with 15-crown-5 as ion-recognition receptor could respond to K^+ due to the formation of 2:1 (ligand to ion) sandwich "host-guest" complex.

13.1 Introduction

Metal ions play important roles in the life activities. Many metal ions, such as K^+, Na^+, Mg^{2+}, Ca^{2+}, Sr^{2+}, and Cr^{3+}, are essential components of metabolism and cofactors for a variety of biological processes, including oxidative phosphorylation, gene regulation, and free-radical homeostasis [1–3]. Deficiency states of these metal ions with clinical abnormalities have been identified, whereas these essential elements can also cause toxic effects at high doses [4]. Moreover, heavy metal ions, one of the most serious environmental pollutants, are becoming a severe public health problem [5]. Some heavy metal ions, such as Pb^{2+}, Ba^{2+}, Hg^{2+}, and Cd^{2+}, have serious toxicity to the living organisms even at low concentrations [5–7]. So, the detection of ion species and ion concentration would be crucial to our life. Therefore, it is of important theoretical significance and great practical interest to study on the smart hydrogel materials with metal ion-recognition properties.

Crown ethers are a kind of heterocyclic chemical compounds and contain a ring cavity structure composed of several ether groups. The most common crown ethers are oligomers of ethylene oxide, and the repeating unit is $-CH_2CH_2O-$ group. As the first generation of synthetic host compounds, crown ethers have remarkable recognizable ability toward specific metal ions [8]. When ion diameter matches the

L.-Y. Chu et al., *Smart Hydrogel Functional Materials*,
DOI 10.1007/978-3-642-39538-3_13,

© Chemical Industry Press, Beijing and Springer Berlin Heidelberg 2013

cavity size of crown ether, the ion could be captured by the crown ether receptor, and a stable "host-guest" complex is formed. For example, 18-crown-6, which has a ring cavity structure composed of six oxygen atoms and six diethyl groups, has high affinity toward some special metal ions, such as Pb^{2+}, Ba^{2+}, and K^{+} [9]. 15-crown-5, another typical crown ether, could selectively recognize K^{+} to form a stable 2:1 (ligand to ion) sandwich "host-guest" complex [10–12].

The authors' group has developed a series of ion-recognizable smart hydrogels with crown ether as ion-recognition receptor and thermo-responsive poly(*N*-isopropylacrylamide) (PNIPAM) as actuator. The smart hydrogels with 18-crown-6 as ion-recognition receptor could respond to Pb^{2+} and Ba^{2+} due to the formation of 1:1 "host-guest" complex, while the smart responsive hydrogels with 15-crown-5 as ion-recognition receptor could respond to K^{+} due to the formation of 2:1 "host-guest" complex. In this chapter, the design, fabrication, and ion-recognition performance of these ion-recognizable smart hydrogels will be introduced.

13.2 Smart Responsive Hydrogels Capable of Recognizing Heavy Metal Ions

Water pollution with heavy metal ions is a serious problem due to their serious toxicity to human beings and other living organisms [13, 14]. For example, lead, one of the most common heavy metals, can cause damage to the central nervous system and dysfunction to the kidneys and immune systems of human beings, especially for children [15–17]. Therefore, the effective detection and elimination heavy metal ions have great scientific and practical interest.

An ion-recognizable copolymer of PNIPAM with pendent crown ether groups, poly(*N*-isopropylacrylamide-*co*-benzo-18-crown-6-acrylamide) (P(NIPAM-*co*-$B_{18}C_6$Am)), had been designed and synthesized in 1993 [18]. As shown in Fig. 13.1, these two components execute different functions. The pendent $B_{18}C_6$Am with a crown ether cavity serves as an ion-signal sensor to selectively capture metal ion, and the PNIPAM acts as actuator to induce a sudden phase transition, which exhibits a lower critical solution temperature (LCST) for phase transition. When $B_{18}C_6$Am receptors capture specific metal ions (such as K^{+}, Ba^{2+}, and Pb^{2+}) and form stable "host-guest" complexes, the LCST of P(NIPAM-*co*-$B_{18}C_6$Am) copolymer could shift to a higher temperature due to the repulsion among charged "host-guest" complexes and the hydrophilicity enhancement in the copolymer. In other words, the P(NIPAM-*co*-$B_{18}C_6$Am) copolymer could change from shrunk state to swollen state at a temperature between the two LCSTs in response to specific metal ions.

Ion-recognizable cross-linked P(NIPAM-*co*-$B_{18}C_6$Am) hydrogels are prepared via thermally initiated free-radical cross-linking copolymerization using 2,2′-azobis(2-amidinopropane dihydrochloride) (V50) as initiator and *N*,*N*′-methylenebisacrylamide (MBA) as cross-linker [19]. The synthetic scheme of the cross-linked P(NIPAM-*co*-$B_{18}C_6$Am) hydrogel is shown in Fig. 13.2. Briefly,

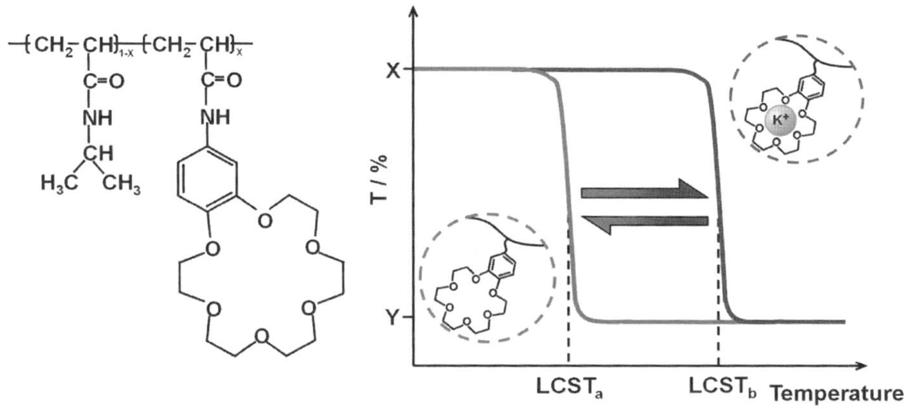

Fig. 13.1 Schematic illustration of the chemical structure and positive ion-responsive LCST shift of smart P(NIPAM-*co*-$B_{18}C_6$Am) copolymer

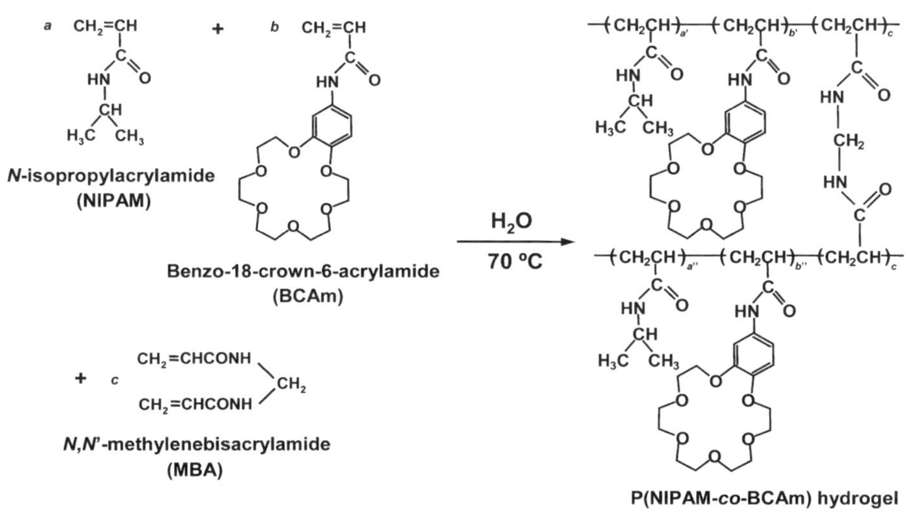

Fig. 13.2 The synthetic scheme of the cross-linked P(NIPAM-*co*-$B_{18}C_6$Am) hydrogel (Reproduced with permission from Ref. [19], Copyright (2008), American Chemical Society)

NIPAM (10 mmol) and $B_{18}C_6$Am (0.75 mmol) monomers, MBA cross-linker, and V50 initiator are dissolved in 10 mL deionized water as the reaction solution. The molar percentages of MBA and V50 in the total monomer are 1 and 0.5 mol%, respectively. After removing the dissolved oxygen by nitrogen gas, the reaction solution is immediately transferred into a small glass tube. The glass tube is sealed immediately and then immersed into a constant-temperature water bath at 70 °C. The polymerization is carried out at 70 °C for 8 h. After the gelation is completed, the prepared cylindrical hydrogel is pushed out from the glass tube and washed by

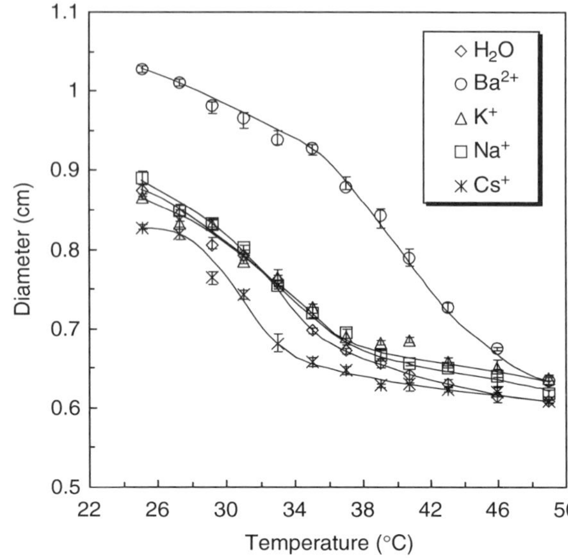

Fig. 13.3 Temperature-dependent diameter change of cross-linked P(NIPAM-*co*-$B_{18}C_6Am$) hydrogel in different ion solutions (Reproduced with permission from Ref. [19], Copyright (2008), American Chemical Society)

excess deionized water. The purified P(NIPAM-*co*-$B_{18}C_6Am$) hydrogel is cut into thin disks and then equilibrated in deionized water or various ion solutions at 25 °C.

The ion-recognition behaviors of P(NIPAM-*co*-$B_{18}C_6Am$) hydrogels are studied by evaluating the thermo-responsive volume-phase transitions in different ion solutions. The concentrations of metal ion in aqueous solutions are all 0.02 mol/L. Figure 13.3 shows the diameter changes of P(NIPAM-*co*-$B_{18}C_6Am$) hydrogel disks in different ion solutions as a function of temperature [19]. All hydrogels undergo a rapid diameter change when the ambient temperatures change across a corresponding temperature region. The diameter-change trend of P(NIPAM-*co*-$B_{18}C_6Am$) hydrogel in Ba^{2+} or Cs^{+} solution is significantly different compared with that in deionized water. In Ba^{2+} solution, the LCST of P(NIPAM-*co*-$B_{18}C_6Am$) hydrogel shifts to a higher temperature, and it turns to lower temperature in Cs^{+} solution, whereas the change of LCST in K^{+} or Na^{+} solution is not obvious.

The reason for these ion-responsive phenomena should attribute to the formation of crown ether/metal ion complexes. $B_{18}C_6Am$ receptors in the cross-linked hydrogel could selectively recognize specific ions, which bound with the cavities of crown ethers tightly and effectively. The order of the complex stability constant, log*K*, of 18-crown-6 with metal ions in water is $Ba^{2+} > K^{+} > Na^{+} > Cs^{+}$ [9].

The complex stability constant of 18-crown-6 with Ba^{2+} is the largest, so a very stable $B_{18}C_6Am/Ba^{2+}$ complex is formed. The polymer chain, attached to ionic "host-guest" complexes, behaves like an ionic polymer chain. The repulsion among charged $B_{18}C_6Am/Ba^{2+}$ groups counteracts the shrinkage of the hydrogel network with the increase of temperature, thereby resulting in the LCST changing to a higher temperature. Additionally, osmotic pressure within the hydrogel due to a Donnan potential, which arises from mobile counterions to the crown ether bound Ba^{2+},

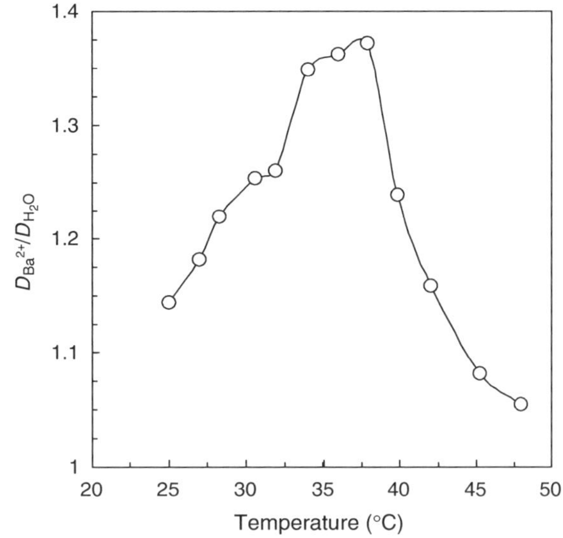

Fig. 13.4 Temperature dependence of the diameter ratio of cross-linked P(NIPAM-*co*-$B_{18}C_6Am$) hydrogel in Ba^{2+} solution to that in deionized water (Reproduced with permission from Ref. [19], Copyright (2008), American Chemical Society)

also makes the hydrogel swell more [20, 21]. So, it can be also seen in Fig. 13.3 that the P(NIPAM-*co*-$B_{18}C_6Am$) hydrogel in Ba^{2+} solution has a larger swelling degree than that in deionized water. Although Cs^+ is too large to fit into the crown ether cavity of $B_{18}C_6Am$, it could form a stable 2:1 complex with the crown ethers [22, 23]. That is, two crown ether cavities from adjacent polymer chains form a 2:1 sandwich "host-guest" complex with one Cs^+. Such complexation causes the cross-linked P(NIPAM-*co*-$B_{18}C_6Am$) hydrogel to contract, which increases the elastic restoring force of the network and decreases the LCST of the hydrogel.

Linear and linear-grafted P(NIPAM-*co*-$B_{18}C_6Am$) copolymers have been reported to present a significant increase of LCST in K^+ solution [18, 24–27]. Surprisingly, the LCST of cross-linked P(NIPAM-*co*-$B_{18}C_6Am$) hydrogel does not increase remarkably in K^+ solution in Fig. 13.3. The reason for this different phenomenon may be the result from the cross-linked network structure. Due to the cross-linked copolymer chains in the hydrogel, the cavities of crown ethers are close to each other. On the other hand, the complex stability constant of $B_{18}C_6Am/K^+$ is not high enough; therefore, the electrostatic repulsion among K^+ ions affects the formation of stable $B_{18}C_6Am/K^+$ complexes inside the cross-linked P(NIPAM-*co*-$B_{18}C_6Am$) hydrogel. Thus, the effect of K^+ on the LCST of cross-linked P(NIPAM-*co*-$B_{18}C_6Am$) hydrogel is definitely different from that of the previously reported linear P(NIPAM-*co*-$B_{18}C_6Am$) copolymer. The complex stability constant of 18-crown-6 with Na^+ is also small, so the effect of Na^+ on the LCST shift of P(NIPAM-*co*-$B_{18}C_6Am$) hydrogel is nearly negligible.

Figure 13.4 shows the effect of Ba^{2+} on the swelling behavior of cross-linked P(NIPAM-*co*-$B_{18}C_6Am$) hydrogel as a function of temperature [19]. The effect is expressed as a ratio of the diameter of hydrogel disk in Ba^{2+} solution to that in deionized water, $D_{Ba^{2+}}/D_{H2O}$. At first, the $D_{Ba^{2+}}/D_{H2O}$ value increases with the

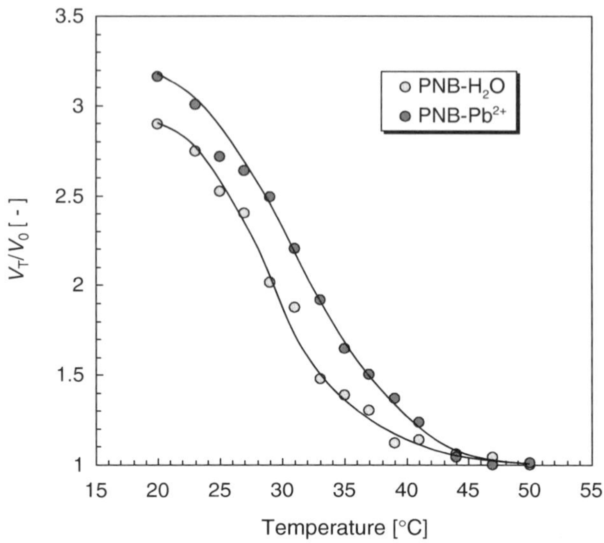

Fig. 13.5 Temperature-dependent volume changes of cross-linked P(NIPAM-*co*-$B_{18}C_6Am$) hydrogel in Pb^{2+} solution and deionized water (Reproduced with permission from Ref. [28], Copyright (2009), Elsevier)

temperature rising and then deceases. At 38 °C, the $D_{Ba}{}^{2+}/D_{H2O}$ value reaches the maximum. As mentioned above, because of the formation of $B_{18}C_6Am/Ba^{2+}$ complexes, the hydrogel swells more and has a higher LCST in Ba^{2+} solution than that in deionized water. When temperature increases above 38 °C, the $D_{Ba}{}^{2+}/D_{H2O}$ value decreases sharply as cross-linked P(NIPAM-*co*-$B_{18}C_6Am$) hydrogel in Ba^{2+} solution begins to shrink rapidly. The results indicate that the optimal operating temperature for Ba^{2+}-recognition application of cross-linked P(NIPAM-*co*-$B_{18}C_6Am$) hydrogel is around 38 °C, at which the hydrogel could spontaneously swell in response to Ba^{2+} and has the optimal ion responsibility.

As ion receptor, 18-crown-6 also exhibits high selectivity to Pb^{2+}. The authors' group applied the P(NIPAM-*co*-$B_{18}C_6Am$) hydrogel as a novel adsorbent to remove Pb^{2+} from aqueous solutions [28].

As expected, the prepared P(NIPAM-*co*-$B_{18}C_6Am$) hydrogel presents good Pb^{2+}-recognition characteristics. The trend of temperature-dependent volume change of P(NIPAM-*co*-$B_{18}C_6Am$) hydrogels in Pb^{2+} solution ($[Pb^{2+}] = 4.0$ mmol/L) is significantly different from that in deionized water as shown in Fig. 13.5 [28]. The LCST of P(NIPAM-*co*-$B_{18}C_6Am$) hydrogel shifts to a higher temperature in Pb^{2+} solution. Such positive LCST shift indicates the formation of crown ether/metal ion complexes. When $B_{18}C_6Am$ units effectively capture Pb^{2+} into their cavities through supramolecular "host-guest" complexation, the LCST of the hydrogel shifts to a higher temperature due to both the repulsion among charged $B_{18}C_6Am/Pb^{2+}$ groups and the osmotic pressure within the hydrogel, which is similar as that in Ba^{2+} solution.

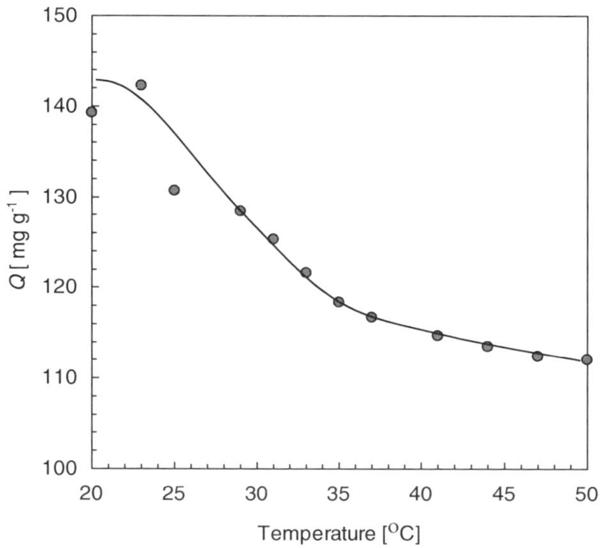

Fig. 13.6 Temperature-dependent adsorption capacity of P(NIPAM-*co*-$B_{18}C_6Am$) hydrogel for Pb^{2+} (Reproduced with permission from Ref. [28], Copyright (2009), Elsevier)

The effect of the environmental temperature on adsorption characteristics of P(NIPAM-*co*-$B_{18}C_6Am$) hydrogels toward Pb^{2+} in aqueous solution is examined [28]. Equilibrium adsorptions at various designed temperatures (in the range of 20 ~ 50 °C) are performed in batch experiments. In all experiments, the amount of dried gel sorbents is kept at approximately 37 mg for a 15 mL Pb^{2+} solution, and the initial concentrations of Pb^{2+} are all 4.0 mmol/L (828 mg/L). The weighted gel sorbents are added into a flask containing 15 mL Pb^{2+} solution, and then the flask is placed in a thermostatic water bath shaker and operated under 150 rpm at designed temperature for 3.5 h.

As shown in Fig. 13.6, Pb^{2+}-adsorption capacity of P(NIPAM-*co*-$B_{18}C_6Am$) hydrogels shows strong temperature dependence that the adsorbed Pb^{2+} amount per unit mass of the gels decreases with increasing the temperature. The results indicate that the P(NIPAM-*co*-$B_{18}C_6Am$) hydrogel has a better and higher adsorption capability toward Pb^{2+} ions at low temperature. That is to say, the prepared P(NIPAM-*co*-$B_{18}C_6Am$) hydrogels are confirmed to exhibit an interesting behavior as that "adsorption at temperature lower than the LCST and desorption at temperature higher than the LCST." The Pb^{2+} adsorption of P(NIPAM-*co*-$B_{18}C_6Am$) hydrogel mainly depends on both the complexation of Pb^{2+} with pendent $B_{18}C_6Am$ groups and the physical adsorption. The "swelling-shrinking" configuration change of P(NIPAM-*co*-$B_{18}C_6Am$) copolymer networks triggered by change in environmental temperature could influence the formation of $B_{18}C_6Am/Pb^{2+}$ complexes. At temperatures lower than the LCST, the copolymer networks stretch, which makes it easier for crown ethers to capture the guest ions

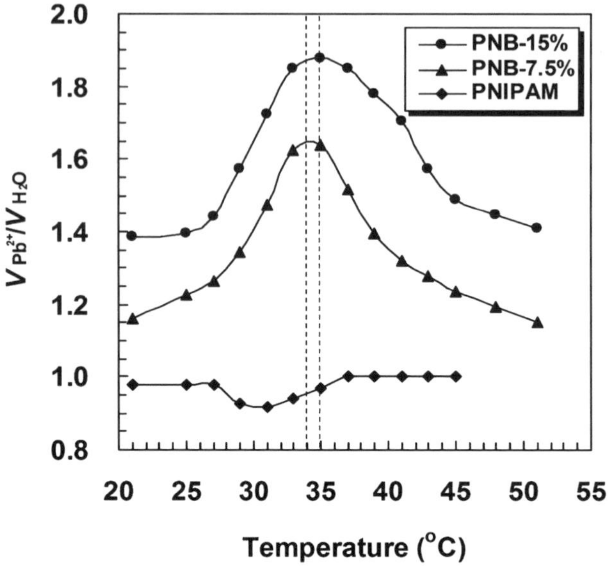

Fig. 13.7 Temperature dependence of the volume ratio of cross-linked hydrogels with different contents of $B_{18}C_6Am$ in 20 mmol/L Pb^{2+} solution to that in deionized water (Reproduced with permission from Ref. [29], Copyright (2012), American Chemical Society)

(Pb^{2+}), so that the P(NIPAM-*co*-$B_{18}C_6Am$) hydrogel exhibits a higher adsorption capacity. In contrast, at temperatures higher than the LCST, the P(NIPAM-*co*-$B_{18}C_6Am$) copolymer networks shrink, and the cavities of crown ethers are close to each other. As a result, the electrostatic repulsions among the ions affect formation of stable $B_{18}C_6Am/Pb^{2+}$ complexes inside the hydrogel, which leads to a smaller adsorbed amount of Pb^{2+}.

The results suggest that the ion-recognizable P(NIPAM-*co*-$B_{18}C_6Am$) hydrogel could serve as a novel thermo-responsive smart material for adsorption and separation of Pb^{2+}, which could be rationally achieved by simply changing the environmental temperature. The developed P(NIPAM-*co*-$B_{18}C_6Am$) hydrogel has great potential applications in environmental protections.

Systematic investigations of the effects of Pb^{2+} concentrations on the ion-responsive behaviors of P(NIPAM-*co*-$B_{18}C_6Am$) hydrogels containing different crown ether contents are also investigated recently [29]. The molar ratios of $B_{18}C_6Am$ to NIPAM are designed as 0, 7.5, and 15 mol%, and the corresponding prepared cross-linked PNIPAM and P(NIPAM-*co*-$B_{18}C_6Am$) hydrogels are labeled as PN, PNB-7.5 %, and PNB-15 %, respectively.

Figure 13.7 shows the effect of Pb^{2+} on the swelling-shrinking behaviors of cross-linked PNIPAM and P(NIPAM-*co*-$B_{18}C_6Am$) hydrogels, in which $V_{Pb}{}^{2+}$ and V_{H2O} are the hydrogel volumes in Pb^{2+} solution and deionized water, respectively. For both P(NIPAM-*co*-$B_{18}C_6Am$) hydrogels (PNB-7.5 % and PNB-15 %), the volume ratio value ($V_{Pb}{}^{2+}/V_{H2O}$) increases at first and then deceases with increase of the temperature. At a certain corresponding temperature (T °C), the $V_{Pb}{}^{2+}/V_{H2O}$ value for both P(NIPAM-*co*-$B_{18}C_6Am$) hydrogels reaches a corresponding maximum value. Below T °C, the volume of P(NIPAM-*co*-$B_{18}C_6Am$) hydrogel in deionized

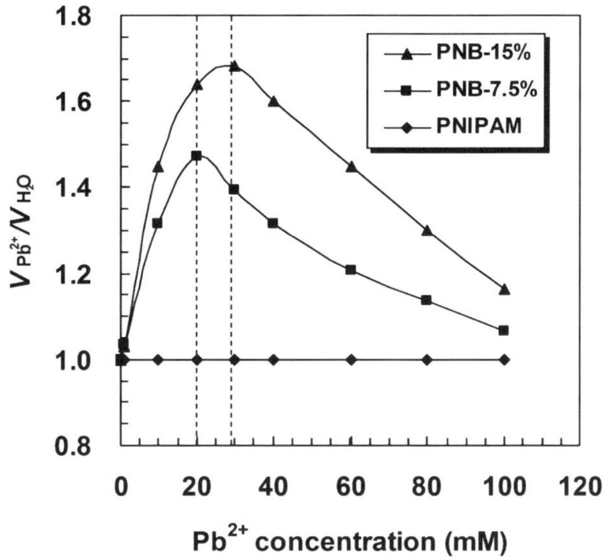

Fig. 13.8 Ion-concentration dependence of the volume ratio of cross-linked hydrogels with different $B_{18}C_6Am$ contents in Pb^{2+} solution to that in deionized water at 35 °C (Reproduced with permission from Ref. [29], Copyright (2012), American Chemical Society)

water decreases with increase of the temperature, whereas it keeps swelling in Pb^{2+} solution due to the repulsion among charged $B_{18}C_6Am/Pb^{2+}$ complexes and the osmotic pressure within the hydrogel. When temperature increases above T °C, the $V_{Pb}{}^{2+}/V_{H2O}$ value decreases as the P(NIPAM-*co*-$B_{18}C_6Am$) hydrogel in Pb^{2+} solution begins to shrink. These results indicate that the optimal operating temperature for Pb^{2+}-recognition application is around T °C, at which it has the most obvious isothermal Pb^{2+}-responsive volume change. The optimal operating temperatures for Pb^{2+} recognition of P(NIPAM-*co*-$B_{18}C_6Am$) hydrogels with different crown ether contents are different due to the different amounts of $B_{18}C_6Am/Pb^{2+}$ complexes. For PNIPAM hydrogel without $B_{18}C_6Am$ receptors, the value of $V_{Pb}{}^{2+}/V_{H2O}$ decreases slightly around 30 °C caused by the salting-out effect of nitrate on PNIPAM hydrogel [30–32].

The effect of Pb^{2+} concentration on the swelling behavior of cross-linked PNIPAM and P(NIPAM-*co*-$B_{18}C_6Am$) hydrogels with different $B_{18}C_6Am$ contents at 35 °C is shown in Fig. 13.8 [29]. With increasing the Pb^{2+} concentration, the $V_{Pb}{}^{2+}/V_{H2O}$ value of cross-linked P(NIPAM-*co*-$B_{18}C_6Am$) hydrogel also increases at first and then deceases. That is, there also exists an optimal Pb^{2+}-responsive concentration for the P(NIPAM-*co*-$B_{18}C_6Am$) hydrogel, at which the hydrogel has the largest Pb^{2+} responsivity. The optimal Pb^{2+}-responsive concentration for PNB-15 % hydrogel with larger $B_{18}C_6Am$ content is higher than that of PNB-7.5 % hydrogel.

Therefore, to achieve satisfactory Pb^{2+}-recognition characteristics of P(NIPAM-*co*-$B_{18}C_6Am$)-based hydrogel materials, both operation temperature and ion concentration should be optimized. These results provide valuable guidance for designing and applying P(NIPAM-*co*-$B_{18}C_6Am$)-based ion-recognizable hydrogel materials in various applications.

13.3 Smart Responsive Hydrogels Capable of Recognizing Potassium Ions

Among physiologically important metal ions, K^+ is the most abundant intracellular metal ion and plays an important role in biological systems. K^+ not only maintains the extracellular osmolarity with Na^+ but also regulates the concentration of other ions such as Ca^{2+} and Cl^- in the living cell [33]. The intracellular K^+ concentration is about 30 times as high as that outside the cell due to the active function of ion channels across the cell membrane [34, 35]. At some pathological sites in living organisms, serious cytoclasis or disabled K^+-Na^+ pump in cell membrane always results in abnormal increase of extracellular K^+ concentration [35]. Therefore, K^+-recognizable systems are highly attractive for various applications such as tissue engineering, targeted drug delivery systems, and sensors and/or actuators. The fabrication of K^+-recognizable materials is of both scientific and technological interests.

It has been reported that 15-crown-5, another typical crown ether, could selectively recognize K^+ to form stable 2:1 (ligand to ion) sandwich "host-guest" complex [10–12]. The authors' group designed and synthesized a novel K^+-recognizable smart copolymer, poly(*N*-isopropylacrylamide-*co*-benzo-15-crown-5-acrylamide) (P(NIPAM-*co*-$B_{15}C_5$Am)) [36]. The P(NIPAM-*co*-$B_{15}C_5$Am) copolymer exhibits a negative LCST shift for phase transition in response to K^+, which is distinctly different from the K^+-recognition behavior of P(NIPAM-*co*-$B_{18}C_6$Am) copolymer. The chemical structure and response mechanism of the P(NIPAM-*co*-$B_{15}C_5$Am) copolymer are schematically illustrated in Fig. 13.9.

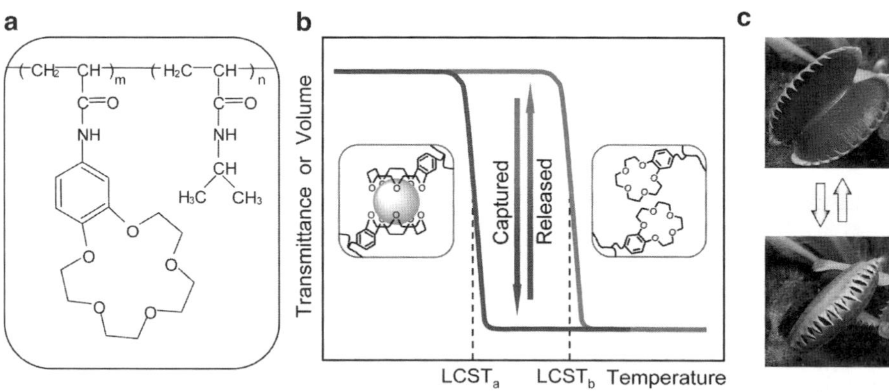

Fig. 13.9 Schematic illustration of the smart P(NIPAM-*co*-$B_{15}C_5$Am) copolymer with K^+-induced negative shift of the LCST for phase transition. (**a**) The chemical structure of P(NIPAM-*co*-$B_{15}C_5$Am) copolymer, (**b**) phase transition of the P(NIPAM-*co*-$B_{15}C_5$Am) copolymer in response to K^+, and (**c**) a similar phenomenon in nature: Venus flytrap (picture courtesy of Dr. Barry Rice, www.sarracenia.com) (Reproduced with permission from Ref. [36], Copyright (2008), Wiley-VCH Verlag GmbH & Co. KGaA)

The two components in the P(NIPAM-*co*-$B_{15}C_5$Am) copolymer execute different functions. The pendent $B_{15}C_5$Am groups serve as K^+-recognition sensors to selectively capture K^+, and the PNIPAM acts as an actuator to induce a sudden volume-phase transition in response to temperature change. Once K^+ ions are added to the environmental solution, the 15-crown-5 receptors from adjacent P(NIPAM-*co*-$B_{15}C_5$Am) copolymer chains could capture the K^+ and form stable 2:1 (ligand to ion) sandwich complexes. Such complexation would disrupt the hydrogen bonding between the oxygen atoms in the crown ether and the hydrogen atoms of water and cause the copolymer chains to contract. As a result, the hydrophobicity of the P(NIPAM-*co*-$B_{15}C_5$Am) copolymer is enhanced, and thus the LCST for phase transition of P(NIPAM-*co*-$B_{15}C_5$Am) shifts negatively to a lower value (from $LCST_a$ to $LCST_b$ in Fig. 13.9b). That is to say, when the environmental temperature is maintained between $LCST_a$ and $LCST_b$, the smart copolymer in an aqueous solution changes its physical state abruptly and reversibly from a swollen state to a shrunken state in response to the presence of K^+. The stimuli-responsive behavior of the smart copolymer induced by K^+ is similar to some phenomena in the natural world, such as the behavior of Venus flytraps shown in Fig. 13.9c. Before capturing a recognizable substance, a Venus flytrap exists in an open state, which corresponds to the copolymer in a swollen state; as soon as it captures a substance, the Venus flytrap rapidly closes, which corresponds to the copolymer in a shrunken state.

Figure 13.10a, b show the temperature-dependent phase transition behaviors of PNIPAM polymer in aqueous solution that contains different metal ions. It can be seen that the LCST of PNIPAM polymer does not change significantly in the presence of all of the selected metal ions. However, in the case of the P(NIPAM-*co*-$B_{15}C_5$Am) copolymer, as shown in Fig. 13.10c, d, the negative LCST shift induced by K^+ is extraordinarily significant in comparison with that induced by Na^+, Cs^+, or Li^+. Although Na^+ has been reported to easily form a 1:1 complex with 15-crown-5 because of their good size fit [9, 22], K^+ forms more stable 2:1 (ligand to ion) complex with two 15-crown-5 groups inter or intra the polymeric chains [10–12]. Such 2:1 complexation causes the copolymer to shrink much more effectively and the LCST to shift more significantly than that in the case of Na^+. It is also feasible for Cs^+ to form a 2:1 (ligand to ion) complex with 15-crown-5 [22, 23]. However, because the size of Cs^+ is much larger than the cavity size of 15-crown-5, the electric interaction between the Cs^+ and the oxygen atom in the crown ether is weak, which makes it difficult to form a stable complex. As a result, the LCST shift induced by Na^+ or Cs^+ is not as significant as that induced by K^+. The addition of Li^+ nearly does not affect the LCST for phase transition because the size of Li^+ is too small to effectively form stable complex with 15-crown-5.

On the other hand, when free benzo-15-crown-5 ($B_{15}C_5$) molecules are added to the aqueous solution together with K^+, the LCST of PNIPAM polymer does not significantly change, as shown in Fig. 13.10e, f. That is, the complexation of 15-crown-5 with K^+ contributes to this negative LCST shift only when the 15-crown-5 groups are chemically bonded to PNIPAM polymer chain, but does not cause a significant LCST shift if the 15-crown-5 groups are just free molecules in the solution.

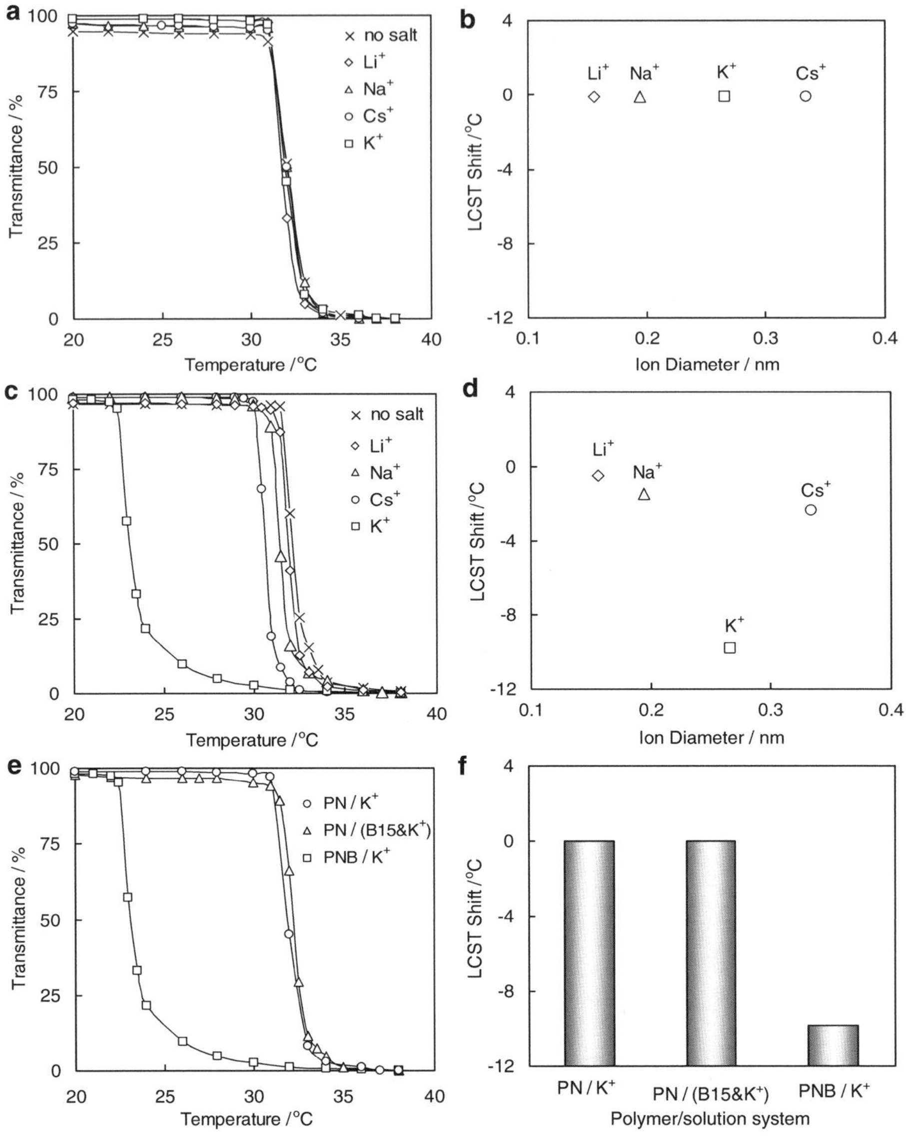

Fig. 13.10 (**a, b**) Phase transition behaviors and LCST shifts of PNIPAM polymer in different ion aqueous solutions; (**c, d**) phase transition behaviors and LCST shifts of P(NIPAM-*co*-$B_{15}C_5$Am) copolymer in different ion aqueous solutions, and the molar ratio of $B_{15}C_5$Am to NIPAM in the copolymer is 1:9; (**e, f**) comparison of phase transition behaviors and LCST shifts of PNIPAM (PN) and P(NIPAM-*co*-$B_{15}C_5$Am) (PNB) polymers in K^+ solution, in which "PN/K^+" stands for PNIPAM in K^+ solution, "PN/(B15&K^+)" for PNIPAM in K^+ solution containing free $B_{15}C_5$ molecules with the molar ratio of $B_{15}C_5$ to NIPAM being 1:9, and "PNB/K^+" for P(NIPAM-*co*-$B_{15}C_5$Am) in K^+ solution. The concentration of the polymers in all aqueous solutions is 0.5 wt%, and the metal ion concentrations are all 100 mmol/L (Reproduced with permission from Ref. [36], Copyright (2008), Wiley-VCH Verlag GmbH & Co. KGaA)

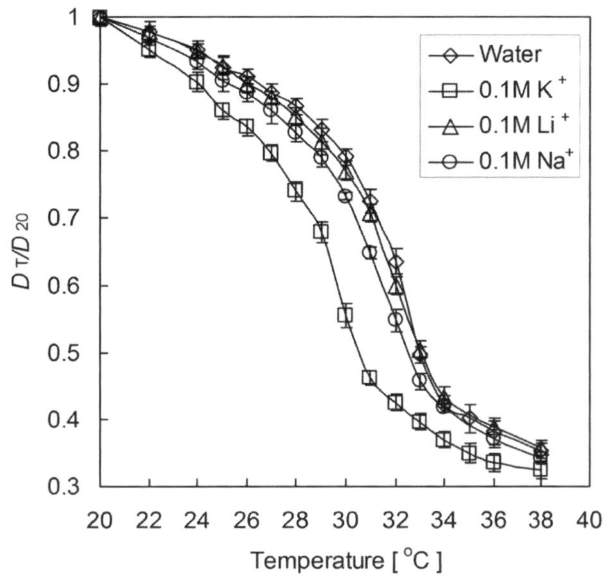

Fig. 13.11 Thermo-dependent volume change of P(NIPAM-*co*-$B_{15}C_5Am$) hydrogel in aqueous solutions containing different metal ions. D_T represents the diameter of the hydrogel disk at each test temperature *T*, and D_{20} represents that at 20 °C (Reproduced with permission from Ref. [37], Copyright (2010), Elsevier)

The authors' group also developed a K^+-recognizable smart hydrogel based on cross-linked P(NIPAM-*co*-$B_{15}C_5Am$) copolymer. The prepared hydrogel is featured with K^+-induced pulse-release mode due to its isothermal volume shrinkage induced by recognizing the increase signal of K^+ concentration in the environment [37].

The selective formation of stable 2:1 "host-guest" complex between 15-crown-5 and K^+ endows the P(NIPAM-*co*-$B_{15}C_5Am$) hydrogel with especial response to K^+. Figure 13.11 shows thermo-responsive volume change of P(NIPAM-*co*-$B_{15}C_5Am$) hydrogel in aqueous solutions containing different metal ions [37]. Just as expected, when K^+ presents in the environmental solution, the LCST of the P(NIPAM-*co*-$B_{15}C_5Am$) hydrogel shifts to a lower temperature than that in water. Whereas, when Li^+ or Na^+ appears in the solution, the LCST change is not so significant. That is, the prepared P(NIPAM-*co*-$B_{15}C_5Am$) hydrogel is especially and selectively sensitive to K^+. From the results shown in Fig. 13.11, it is expected that the prepared P(NIPAM-*co*-$B_{15}C_5Am$) hydrogel could recognize K^+ and shrink isothermally at temperature around 30 °C.

The dynamic and isothermal shrinkage behavior of P(NIPAM-*co*-$B_{15}C_5Am$) hydrogel triggered by K^+ recognition is studied by transferring the hydrogel from pure water at 30 °C to 0.2 mol/L K^+ aqueous solution at the same temperature promptly and then measuring the dynamic volume change of the hydrogel in 0.2 mol/L K^+ aqueous solution at 30 °C [37]. As shown in Fig. 13.12, the K^+-triggered dynamic volume shrinkage of P(NIPAM-*co*-$B_{15}C_5Am$) hydrogel is rapid and remarkable at 30 °C, whereas the hydrogel does not shrink in pure water. The diameter of the hydrogel disk in 0.2 mol/L K^+ aqueous solution decreases to about 50 % of that in pure water within a few minutes.

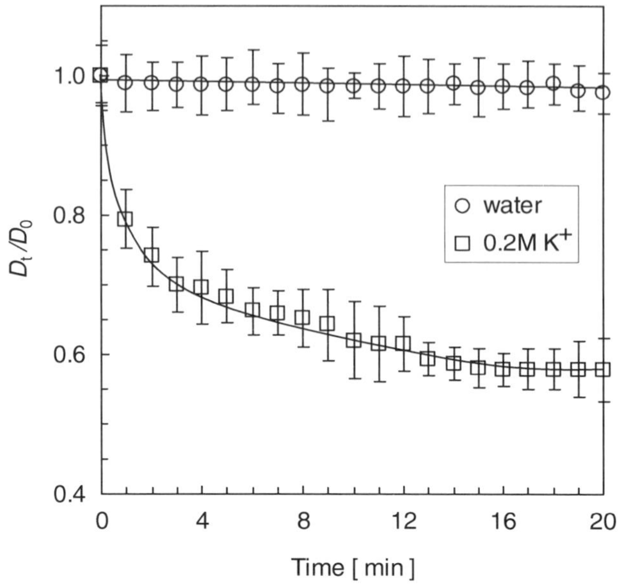

Fig. 13.12 Dynamic K^+-induced volume shrinking behavior of the P(NIPAM-*co*-$B_{15}C_5$Am) hydrogel in 0.2 mol/L K^+ aqueous solution at 30 °C. D_t represents the diameter of the hydrogel disk at time t, and D_0 represents the initial diameter of the hydrogel disk at $t = 0$ min (Reproduced with permission from Ref. [37], Copyright (2010), Elsevier)

By using such K^+-induced volume shrinking characteristics, the prepared P(NIPAM-*co*-$B_{15}C_5$Am) hydrogel could be used for sudden pulse release of drug triggered by the presence of K^+ in the surroundings. The K^+-induced isothermal release characteristics of VB_{12} as a model drug from both P(NIPAM-*co*-$B_{15}C_5$Am) hydrogel and PNIPAM hydrogel (as a reference) have been investigated [37]. In pure water, the VB_{12} releases from both hydrogels resulted from concentration-driven diffusion and have almost the same release rates, as shown in Fig. 13.13a. However, in 0.2 mol/L K^+ solution, the dynamic release behavior of VB_{12} from P(NIPAM-*co*-$B_{15}C_5$Am) hydrogel is significantly different from that from PNIPAM hydrogel (Fig. 13.13b). The release rate within the first 3 min and the total release amount of VB_{12} within 20 min from P(NIPAM-*co*-$B_{15}C_5$Am) hydrogel are both higher than those from PNIPAM hydrogel. Such drug release behaviors from P(NIPAM-*co*-$B_{15}C_5$Am) hydrogel in K^+ solution resulted from both the concentration-driven diffusion and the K^+-induced shrinkage of the hydrogel.

The dynamic behavior of K^+-induced drug release from P(NIPAM-*co*-$B_{15}C_5$Am) hydrogel by adding K^+ to the surrounding aqueous solution during the release process is also examined [37]. The results demonstrate that the addition of K^+ to the environment can trigger a pulse release of loaded drug from P(NIPAM-*co*-$B_{15}C_5$Am) hydrogel due to K^+-induced isothermal hydrogel shrinkage (Fig. 13.13c). However, PNIPAM hydrogel does not show such

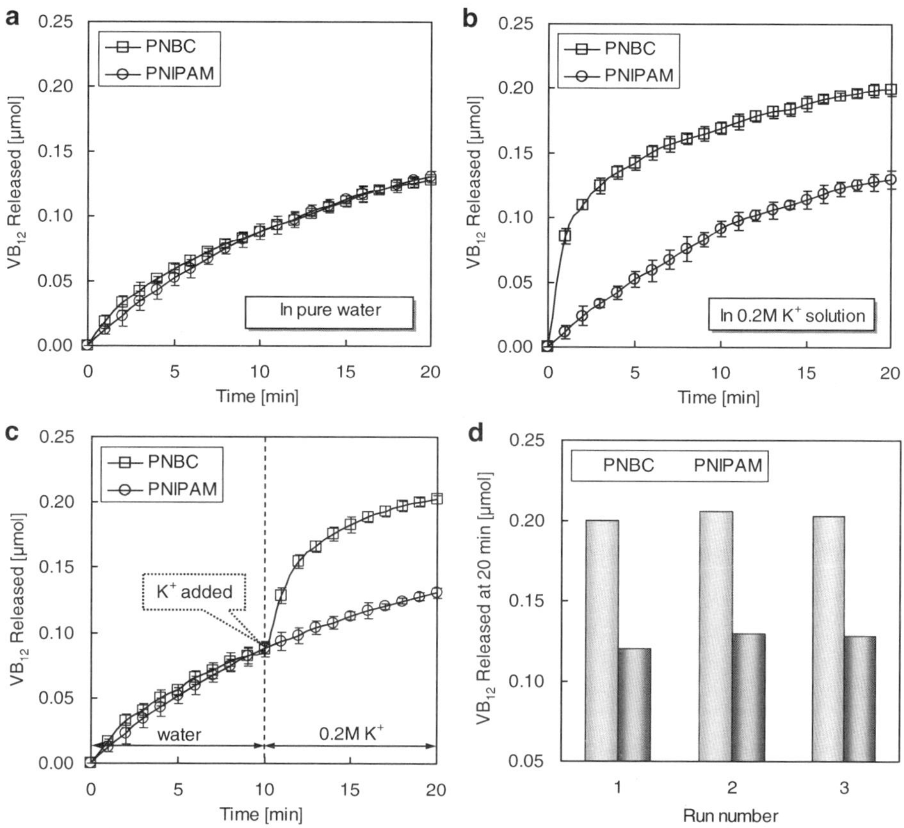

Fig. 13.13 K^+-induced controlled release of VB_{12} from P(NIPAM-*co*-$B_{15}C_5$Am) hydrogel (PNBC) and PNIPAM hydrogel. (**a**) Dynamic release behavior of VB_{12} from PNIPAM and PNBC hydrogels in pure water, (**b**) dynamic release behavior of VB_{12} from PNIPAM and PNBC hydrogels in 0.2 mol/L K^+ aqueous solution, (**c**) K^+-induced controlled release of VB_{12} from PNBC hydrogel by adding K^+ into the surrounding aqueous solution, and (**d**) reversible and repeatable characteristics of K^+-induced controlled release from PNBC and PNIPAM hydrogels. All the release experiments are carried out at 30 °C (Reproduced with permission from Ref. [37], Copyright (2010), Elsevier)

performance. This K^+-induced controlled release of model drug from P(NIPAM-*co*-$B_{15}C_5$Am) hydrogel shows reversible and repeatable characteristics, just as the PNIPAM hydrogel does due to the concentration-driven diffusion (Fig. 13.13d). Therefore, the proposed P(NIPAM-*co*-$B_{15}C_5$Am) hydrogel provides a new mode of K^+-responsive volume change for stimuli-responsive smart actuators, which is highly attractive for targeting drug delivery systems, sensors, and so on.

For responsive hydrogels, fast response to environmental stimuli is usually vital to many practical applications. For the normal cross-linked P(NIPAM-*co*-$B_{15}C_5$Am) hydrogel, the 15-crown-5 groups are randomly pendent on the polymeric networks

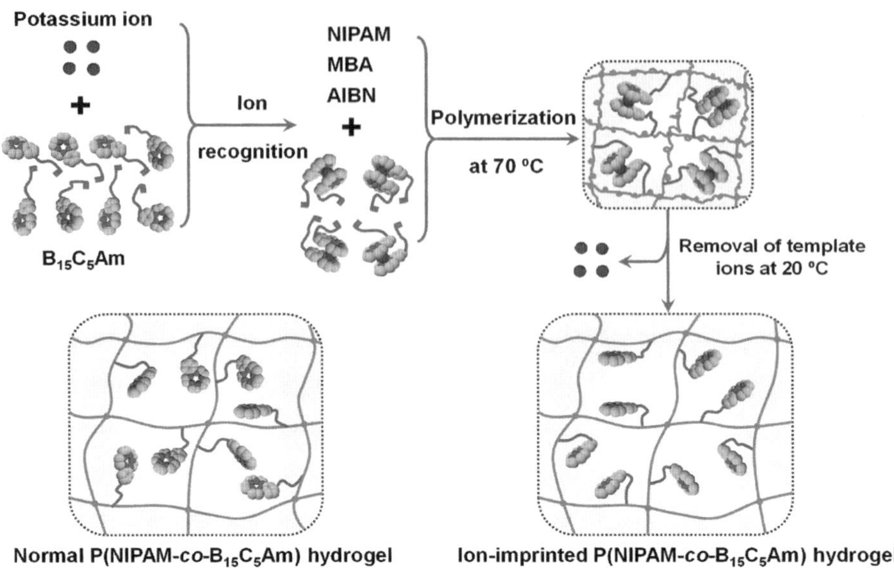

Fig. 13.14 Strategy for synthesizing the ion-imprinted P(NIPAM-*co*-$B_{15}C_5Am$) hydrogel (Reproduced with permission from Ref. [39], Copyright (2011), Wiley-VCH Verlag GmbH & Co. KGaA)

because of the random distribution of crown ethers in the polymerization. It would take more time for the randomly pendent 15-crown-5 to capture K^+ because they have to gradually adjust their positions to form stable 2:1 "host-guest" complexes. Therefore, the response rate of the normal cross-linked P(NIPAM-*co*-$B_{15}C_5Am$) hydrogel to K^+ could not be very fast in principle.

Inspired by the molecularly imprinted materials [38], the authors' group designed a novel ion-imprinted strategy for synthesizing novel P(NIPAM-*co*-$B_{15}C_5Am$) hydrogel with rapid K^+ response rate [39]. As shown in Fig. 13.14, to prepare the designed ion-imprinted P(NIPAM-*co*-$B_{15}C_5Am$) hydrogel, K^+ ions are used as ion templates in the polymerization of hydrogels. Before the polymerization, K^+ templates have been captured by $B_{15}C_5Am$ comonomers, and stable 2:1 "host-guest" complexes have been formed in monomer aqueous solution. During the polymerization, the $B_{15}C_5Am$ comonomers in the stable 2:1 "host-guest" complexes are mounted onto the copolymer networks in pairs. After polymerization, the temperature is decreased to a lower value, which is lower than the volume-phase transition temperature of P(NIPAM-*co*-$B_{15}C_5Am$), to make the hydrogel network swollen in water. So at this temperature, the K^+ templates are removed from the "host-guest" complexes by washing with excessive deionized water, and as a result, the ion-imprinted P(NIPAM-*co*-$B_{15}C_5Am$) hydrogel is obtained.

As desired, the response rate of the ion-imprinted P(NIPAM-*co*-$B_{15}C_5Am$) hydrogel to K^+ is much faster than that of normal P(NIPAM-*co*-$B_{15}C_5Am$)

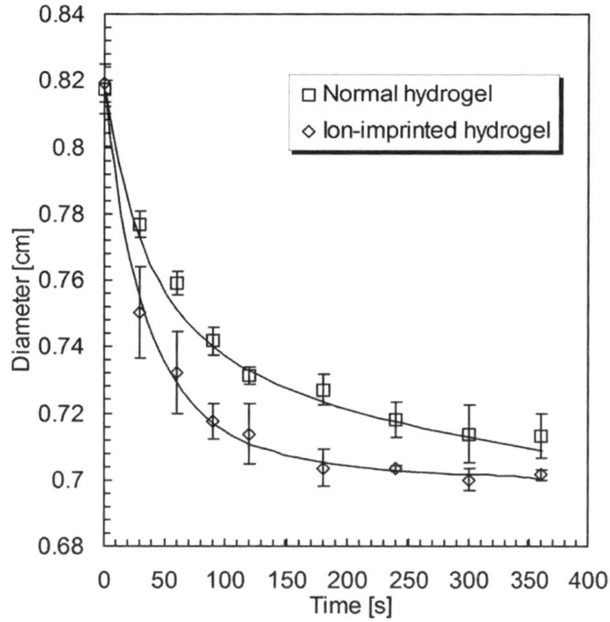

Fig. 13.15 Dynamic K^+-triggered isothermal shrinking behaviors of ion-imprinted and normal P(NIPAM-*co*-$B_{15}C_5$Am) hydrogels at 27 °C. The environmental solution is replaced immediately from deionized water to 0.2 mol/L K^+ solution just before time counting (Reproduced with permission from Ref. [39], Copyright (2011), Wiley-VCH Verlag GmbH & Co. KGaA)

hydrogel (Fig. 13.15) [39]. When environmental solution is isothermally replaced from deionized water to 0.2 mol/L K^+ solution at 27 °C, both the ion-imprinted and normal P(NIPAM-*co*-$B_{15}C_5$Am) hydrogels start to shrink due to the formation of stable 2:1 "host-guest" complexes between 15-crown-5 and K^+, whereas, within the same time intervals, the deswelling extent of the ion-imprinted P(NIPAM-*co*-$B_{15}C_5$Am) hydrogel in response to K^+ is much more significant than that of normal P(NIPAM-*co*-$B_{15}C_5$Am) hydrogel. The results in Fig. 13.15 also show that the ion-imprinted hydrogel almost reaches its equilibrium deswelling state at time $t = 200$ s; however, for the normal hydrogel, it is still far away from its equilibrium deswelling state when the time is as long as 360 s. As mentioned above, because the 15-crown-5 units in the normal hydrogel are randomly pendent on the polymer networks, it takes more time for them to adjust their positions to form stable 2:1 "host-guest" complexes with K^+. However, for the ion-imprinted P(NIPAM-*co*-$B_{15}C_5$Am) hydrogel, the 15-crown-5 units inside the hydrogel are mounted on the polymer networks in pairs. Therefore, the void between two 15-crown-5 units formed by the K^+ template provides the size-matching interaction site. So, it is much easier and faster for them to capture K^+ to form stable 2:1 "host-guest" complexes than the case of normal P(NIPAM-*co*-$B_{15}C_5$Am) hydrogel. Therefore, the response rate of the ion-imprinted P(NIPAM-*co*-$B_{15}C_5$Am) hydrogel to K^+ is significantly faster than that of normal P(NIPAM-*co*-$B_{15}C_5$Am) hydrogel.

13.4 Ion-Recognizable Monodisperse Hydrogel Microspheres

Monodisperse ion-recognizable P(NIPAM-*co*-$B_{18}C_6Am$) microspheres are successfully fabricated via precipitation copolymerization of NIPAM with $B_{18}C_6Am$ [40]. Briefly, the NIPAM (0.2 g) and $B_{18}C_6Am$ (0.05 g) monomers, MBA (0.0185 g) cross-linker, and AIBN (0.0042 g) initiator are dissolved in 20 mL of mixed solvent of H_2O/THF (99/1, v/v). The solution is bubbled with nitrogen gas for 30 min to remove dissolved oxygen and then heated to 70 °C to start the polymerization. The reaction is maintained at 70 °C for 4 h under a nitrogen atmosphere. After being cooled to room temperature, the resultant P(NIPAM-*co*-$B_{18}C_6Am$) microspheres are purified by repeating centrifugation and redispersion using deionized water to remove the residual unreacted components. PNIPAM microspheres served as the reference are also prepared and purified using the similar protocol only without any addition of $B_{18}C_6Am$.

As shown in Fig. 13.16, the prepared P(NIPAM-*co*-$B_{18}C_6Am$) microspheres exhibit nearly perfect spherical shape with good monodispersity [40]. It can be clearly seen that the size of P(NIPAM-*co*-$B_{18}C_6Am$) microspheres is larger than that of PNIPAM microspheres. The average sizes of PNIPAM and P(NIPAM-*co*-$B_{18}C_6Am$) microspheres are 390 and 570 nm, respectively. The main reason for the difference sizes is that $B_{18}C_6Am$ is a hydrophilic monomer, and introducing a hydrophilic secondary group leads to a larger particle size in the preparation of microspheres via precipitation polymerization [41]. Another reason is that the steric hindered effect of $B_{18}C_6Am$ groups also contributes to form larger microspheres. The size of P(NIPAM-*co*-$B_{18}C_6Am$) microspheres can be controlled by altering the preparation recipe. Increasing the content of NIPAM or $B_{18}C_6Am$ monomer results in P(NIPAM-*co*-$B_{18}C_6Am$) microspheres with larger size, but larger size would cause worse spherical shape due to the soft matter property of the microspheres.

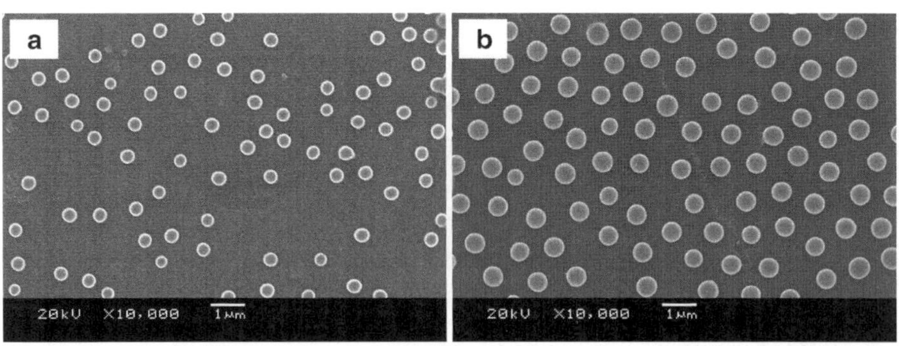

Fig. 13.16 SEM images of PNIPAM microspheres ([NIPAM] = 10 g/L) (**a**) and P(NIPAM-*co*-$B_{18}C_6Am$) microspheres ([NIPAM] = 10 g/L, [NIPAM]/[BCAm] = 4/1 (g/g)) (**b**). Scale bars are 1 μm (Reproduced with permission from Ref. [40], Copyright (2009), Elsevier)

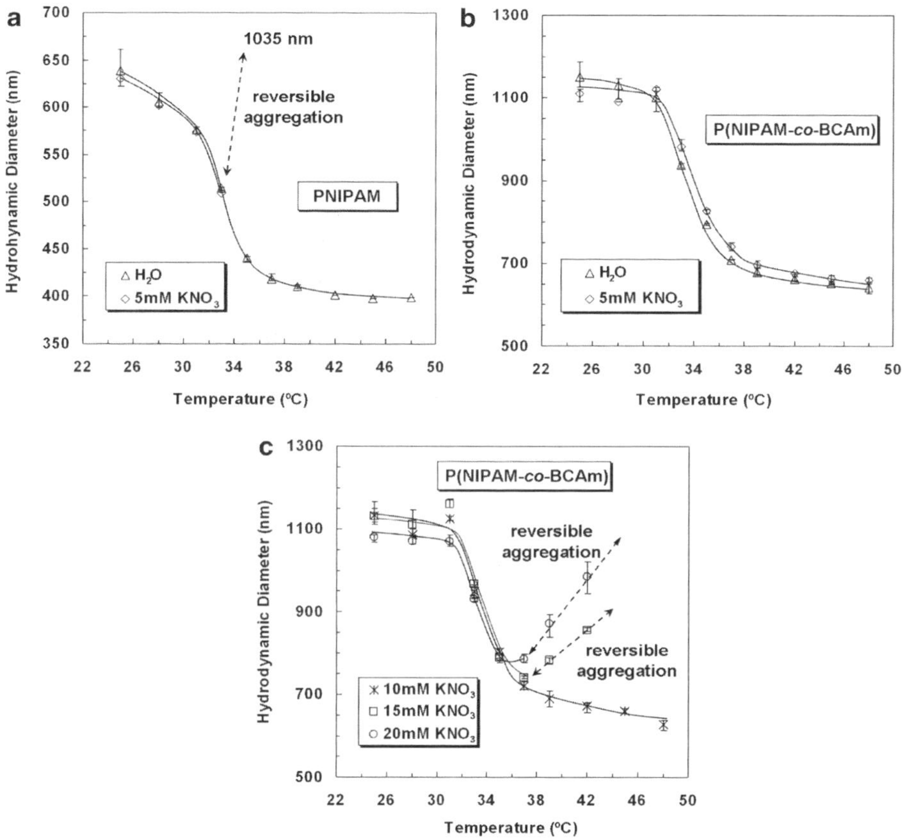

Fig. 13.17 The deswelling behaviors and dispersion stabilities of PNIPAM (**a**) and P(NIPAM-*co*-$B_{18}C_6Am$) (**b**, **c**) microspheres in the absence or presence of K^+ with different concentrations (Reproduced with permission from Ref. [40], Copyright (2009), Elsevier)

The K^+-recognition behaviors of P(NIPAM-*co*-$B_{18}C_6Am$) microspheres are investigated by dynamic light scattering (DLS) technique in K^+ solutions at different temperatures [40]. Figure 13.17 shows the deswelling behaviors and dispersion stabilities of PNIPAM and P(NIPAM-*co*-$B_{18}C_6Am$) microspheres in the absence or presence of K^+ with different concentrations. The trend of temperature-dependent diameter change of PNIPAM microspheres in 5 mmol/L K^+ solution is almost the same as that in deionized water below 33 °C (Fig. 13.17a). When the temperature increases above 33 °C, PNIPAM microspheres begin to aggregate in K^+ solution. Such aggregation phenomenon is due to the presence of electrolytes, which results in a dehydration effect to decrease the hydrophilicity of the colloidal particles and leads to microspheres' aggregation [42, 43]. In K^+ solutions, P(NIPAM-*co*-$B_{18}C_6Am$) microspheres exhibit good ion-responsive property. As expected, due to the formation of $B_{18}C_6Am/K^+$ complexes, the LCST of P(NIPAM-*co*-$B_{18}C_6Am$)

microspheres in 5 mmol/L K^+ solution shifts to a higher temperature compared with that in deionized water (Fig. 13.17b), although this increase is not so remarkable. The dispersion stabilities of P(NIPAM-*co*-$B_{18}C_6$Am) microspheres in K^+ solutions are significantly different from that of PNIPAM microspheres. When the concentration of K^+ solution is 5 mmol/L, PNIPAM microspheres begin to aggregate at temperatures above 33 °C; however, for P(NIPAM-*co*-$B_{18}C_6$Am) microspheres, the aggregation phenomenon appears only when the K^+ concentration is as high as 15 mmol/L and temperatures are above 37 °C. When the K^+ concentration is 20 mmol/L, the aggregation of P(NIPAM-*co*-$B_{18}C_6$Am) microspheres begins only when the temperature is higher than 35 °C.

The formation of crown ether/metal ion complexes and the nonuniform cross-linking density in the microsphere structure could be the reasons for the above-mentioned phenomena. It has been reported that the PNIPAM-based microspheres prepared by precipitation polymerization using MBA as cross-linker have a nonuniform structure with a solid core and a loose shell [44–47]. That is, the cross-linking density in inner part of the particle is higher than that in outer part. As a result, toward the center of the P(NIPAM-*co*-$B_{18}C_6$Am) microsphere, the cavities of crown ethers are closer to each other. Therefore, the electrostatic repulsions among K^+ affect the formation of stable $B_{18}C_6$Am/K^+ complexes in polymer chains with higher cross-linking density, which is the same as that in P(NIPAM-*co*-$B_{18}C_6$Am) hydrogels mentioned above [19]. However, in outer layer of the P(NIPAM-*co*-$B_{18}C_6$Am) microsphere, the polymer chains are hairy-like and the cavities of crown ethers are not so close. Consequently, the $B_{18}C_6$Am receptors in outer layer of microspheres can capture K^+ and form stable $B_{18}C_6$Am/K^+ complexes. The polymer chains with appended ionic "host-guest" complexes near the microsphere surface behave like ionic polymeric chains, which makes the LCST of the P(NIPAM-*co*-$B_{18}C_6$Am) microsphere shift to a higher temperature. However, due to the fact that the formation of $B_{18}C_6$Am/K^+ complexes is mainly in the outer layer of microspheres, the LCST shift is not so remarkable. Additionally, ionic "host-guest" complexes bring positive charges onto the microsphere surface; therefore, the colloid stability of dispersion is increased.

Figure 13.18 shows the isothermal volume changes of PNIPAM and P(NIPAM-*co*-$B_{18}C_6$Am) microspheres caused by different metal ions at 33 °C [40]. In the volume ratio, V_{ion}/V_{H2O}, V_{ion}, and V_{H2O} are the volumes of microspheres in ion solution and deionized water, respectively. Compared with that in deionized water, the volumes of PNIPAM microspheres in Ba^{2+} and K^+ solutions decrease slightly caused by the salting-out effect of nitrate [30–32]. On the other hand, the volumes of P(NIPAM-*co*-$B_{18}C_6$Am) microspheres in Ba^{2+} and K^+ solutions are larger than that in deionized water due to the formation of crown ether/metal ion complexes. As described above, because of the repulsion among charged crown ether/metal ion complexes and the osmotic pressure within the microspheres, the P(NIPAM-*co*-$B_{18}C_6$Am) microspheres swell more in Ba^{2+} and K^+ solutions. It can be also seen that the effects of Ba^{2+} and K^+ on the volume of P(NIPAM-*co*-$B_{18}C_6$Am) microspheres are different. The reason is that the complex stability constant of

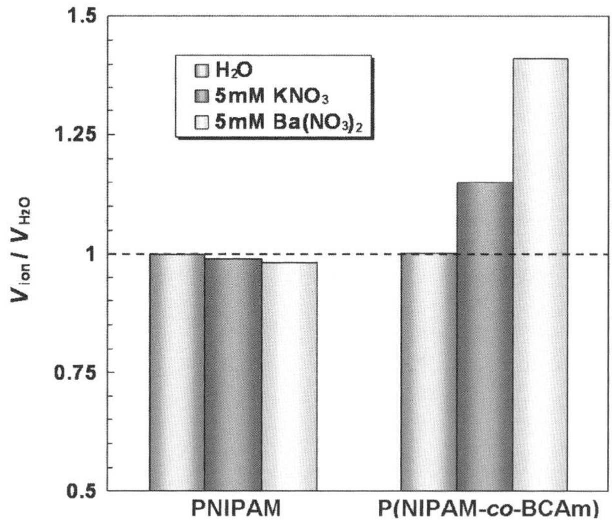

Fig. 13.18 Effect of various ions on the volume of PNIPAM and P(NIPAM-*co*-$B_{18}C_6$Am) microspheres at 33 °C (Reproduced with permission from Ref. [40], Copyright (2009), Elsevier)

18-crown-6 with Ba^{2+} is larger than that with K^+ [9]. As a result, more stable crown ether/metal ion complexes are formed inside the P(NIPAM-*co*-$B_{18}C_6$Am) microspheres in Ba^{2+} solution than that in K^+ solution, so that the P(NIPAM-*co*-$B_{18}C_6$Am) microspheres swell more in Ba^{2+} solution.

The above results demonstrate the ion-recognizable responsive capability of the P(NIPAM-*co*-$B_{18}C_6$Am) microspheres. Such smart microspheres could be utilized as new functional materials for ion-responsive systems, which are highly attractive for targeting drug delivery systems, sensors, chemical carriers, and so on.

13.5 Summary

In summary, ion-recognizable hydrogels are developed on the basis of the cooperative functions of PNIPAM as actuator and crown ether as ion-recognition receptor. Smart hydrogels with 18-crown-6 as ion-recognition receptor could respond to Ba^{2+} and Pb^{2+} due to the formation of 1:1 (ligand to ion) "host-guest" complex. In Ba^{2+} or Pb^{2+} solution, the P(NIPAM-*co*-$B_{18}C_6$Am) hydrogel swells more and has a higher LCST than that in other ion solutions, due to the repulsion among charged $B_{18}C_6$Am/ion complexes and the osmotic pressure within the hydrogel. On the other hand, the P(NIPAM-*co*-$B_{18}C_6$Am) hydrogel exhibits an interesting Pb^{2+}-adsorption behavior as that "adsorption at temperature lower than the LCST and desorption at temperature higher than the LCST" due to the "swelling-shrinking" configuration change of P(NIPAM-*co*-$B_{18}C_6$Am) chains triggered by the environmental temperature change. To achieve satisfactory ion-recognition characteristics of P(NIPAM-*co*-$B_{18}C_6$Am)-based hydrogel materials, both the operation temperature and the ion concentration should be optimized. Smart hydrogels with 15-crown-5 as ion-recognition receptor could respond to K^+ due to the formation of 2:1 (ligand to ion) "host-guest" complex. P(NIPAM-*co*-$B_{15}C_5$Am) hydrogel exhibits a K^+-triggered drug pulse release due to the isothermal volume shrinkage of the hydrogel by recognizing the increase signal of K^+ concentration in the environment. By using K^+ ions as ion templates in the polymerization of hydrogels, ion-imprinted P(NIPAM-*co*-$B_{15}C_5$Am) hydrogel with rapid K^+ response rate is prepared. The P(NIPAM-*co*-$B_{18}C_6$Am) microspheres exhibit satisfactory ion-recognition properties. In K^+ solutions, due to the formation of $B_{18}C_6$Am /K^+ complexes, the LCST of P(NIPAM-*co*-$B_{18}C_6$Am) microspheres shifts to a higher temperature, and the colloidal stability is increased. These ion-recognizable smart hydrogel materials have great potential in the field of chemical sensors and actuators, drug delivery systems, and so on.

References

1. Rolfs A, Hediger MA (1999) Metal ion transporters in mammals: structure, function and pathological implications. J Physiol (Lond) 518:1–12
2. Schnabl J, Sigel RKO (2010) Controlling ribozyme activity by metal ions. Curr Opin Chem Biol 14:269–275
3. Catterall WA (1988) Structure and function of voltage-sensitive ion channels. Science 242:50–61
4. Caussy D, Gochfeld M, Gurzau E et al (2003) Lessons from case studies of metals: investigating exposure, bioavailability, and risk. Ecotoxicol Environ Safe 56:45–51
5. Hutchinson TC, Meema KM (1988) Lead, mercury, cadmium, and arsenic in the environment. Wiley, New York
6. Vallee BL, Ulmer DD (1972) Biochemical effects of mercury, cadmium, and lead. Annu Rev Biochem 41:91–128
7. Godwin HA (2001) The biological chemistry of lead. Curr Opin Chem Biol 5:223–227
8. Pedersen CJ (1967) Cyclic polyethers and their complexes with metal salts. J Am Chem Soc 89:7017–7036
9. Izatt RM, Pawlak K, Bradshaw JS, Bruening RL et al (1991) Thermodynamic and kinetic data for macrocycle interaction with cations and anions. Chem Rev 91:1721–2085
10. Lin SY, Liu SW, Lin CM et al (2002) Recognition of potassium ion in water by 15-crown-5 functionalized gold nanoparticles. Anal Chem 74:330–335
11. Xia WS, Schmehl RH, Li CJ (1999) A highly selective fluorescent chemosensor for K^+ from a bis-15-crown-5 derivative. J Am Chem Soc 121:5599–5600
12. Yamauchi A, Hayashita T, Nishizawa S et al (1999) Benzo-15-crown-5 fluoroionophore/γ-cyclodextrin complex with remarkably high potassium ion sensitivity and selectivity in water. J Am Chem Soc 121:2319–2320
13. Nriagu JO, Pacyna JM (1988) Quantitative assessment of worldwide contamination of air, water and soils by trace metals. Nature 333:134–139
14. Mance G (1987) Pollution threat of heavy metals in aquatic environments. Elsevier Science, New York
15. Friberg L, Nordberg GF, Vouk VB (1979) Handbook on the toxicology of metals. Elsevier/North-Holland Biomedical, Amsterdam
16. Gidlow DA (2004) Lead toxicity. Occup Med (Oxford) 54:76–81

17. Boivin MJ, Giordani B (1995) A risk-evaluation of the neuropsychological effects of childhood lead toxicity. Dev Neuropsychol 11:157–180
18. Irie M, Misumi Y, Tanaka T (1993) Stimuli-responsive polymers: chemical induced reversible phase separation of an aqueous solution of poly(*N*-isopropylacrylamide) with pendent crown ether groups. Polymer 34:4531–4535
19. Ju XJ, Chu LY, Liu L et al (2008) A novel thermo-responsive hydrogel with ion-recognition property through supramolecular host-guest complexation. J Phys Chem B 112:1112–1118
20. Wang KL, Burban JH, Cussler EL (1993) Hydrogels as separation agents. Adv Polym Sci 110:67–69
21. Hirotsu S, Hirokawa Y, Tanaka T (1987) Volume-phase transitions of ionized *N*-isopropylacrylamide gels. J Chem Phys 87:1392–1395
22. Yagi K, Ruiz JA, Sanchez MC (1980) Cation binding properties of polymethacrylamide derivatives of crown ethers. Makromol Chem Rapid Commun 1:263–268
23. Kopolow S, Hogen Esch TE et al (1973) Poly(vinyl macrocyclic polyethers). Synthesis and cation binding properties. Macromolecules 6:133–142
24. Ito T, Hioki T, Yamaguchi T et al (2002) Development of a molecular recognition ion gating membrane and estimation of its pore size control. J Am Chem Soc 124:7840–7846
25. Ito T, Sato Y, Yamaguchi T et al (2004) Response mechanism of a molecular recognition ion gating membrane. Macromolecules 37:3407–3414
26. Okajima S, Sakai Y, Yamaguchi T (2005) Development of a regenerable cell culture system that senses and releases dead cells. Langmuir 21:4043–4049
27. Okajima S, Yamaguchi T, Sakai Y et al (2005) Regulation of cell adhesion using a signal-responsive membrane substrate. Biotechnol Bioeng 91:237–243
28. Ju XJ, Zhang SB, Zhou MY et al (2009) Novel heavy-metal adsorption material: Ion-recognition P(NIPAM-*co*-BCAm) hydrogels for removal of lead(II) ions. J Hazard Mater 167:114–118
29. Zhang B, Ju XJ, Xie R et al (2012) Comprehensive effects of metal ions on responsive characteristics of P(NIPAM-*co*-B18C6Am). J Phys Chem B 116:5527–5536
30. Inomata H, Goto S, Otake K et al (1992) Effect of additives on phase transition of *N*-isopropylacrylamide gels. Langmuir 8:687–690
31. Zhang YJ, Furyk S, Bergbreiter DE et al (2005) Specific ion effects on the water solubility of macromolecules: PNIPAM and the Hofmeister series. J Am Chem Soc 127:14505–14510
32. Zhang Y, Furyk S, Sagle LB et al (2007) Effects of Hofmeister anions on the LCST of PNIPAM as a function of molecular weight. J Phys Chem C 111:8916–8924
33. Berg JM, Tymoczko JL, Stryer L (2002) Biochemistry, 5th edn. W. H. Freeman, New York
34. Morth JP, Pedersen BP, Toustrup-Jensen MS et al (2007) Crystal structure of the sodium-potassium pump. Nature 450:1043–1049
35. Clausen T (2003) Na^{+}-K^{+} pump regulation and skeletal muscle contractility. Physiol Rev 83:1269–1324
36. Mi P, Chu LY, Ju XJ et al (2008) A smart polymer with ion-induced negative shift of the lower critical solution temperature for phase-transition. Macromol Rapid Comm 29:27–32
37. Mi P, Ju XJ, Xie R et al (2010) A novel stimuli-responsive hydrogel for K^{+}-induced controlled-release. Polymer 51:1648–1653
38. Yan M, Ramström O (2005) Molecularly imprinted materials: science and technology. Marcel Dekker, New York
39. Wu HG, Ju XJ, Xie R et al (2011) A novel ion-imprinted hydrogel for recognition of potassium ions with rapid response. Polym Advan Technol 22:1389–1394
40. Ju XJ, Liu L, Xie R et al (2009) Dual thermo-responsive and ion-recognizable monodisperse microspheres. Polymer 50:922–929
41. Gao J, Frisken BJ (2005) Influence of secondary components on the synthesis of self-cross-linked *N*-isopropylacrylamide microgels. Langmuir 21:545–551
42. Daly E, Saunders BR (2000) A study of the effect of electrolyte on the swelling and stability of poly(*N*-isopropylacrylamide) microgel dispersions. Langmuir 16:5546–5552

43. Rasmusson M, Vincent B (2004) Flocculation of microgel particles. React Funct Polym 58:203–211
44. Daly E, Saunders BR (2000) Temperature-dependent electrophoretic mobility and hydrodynamic radius measurements of poly(*N*-isopropylacrylamide) microgel particles: structural insights. Phys Chem Chem Phys 2:3187–3193
45. Guillermo A, Addad JPC, Bazile JP et al (2000) NMR investigations into heterogeneous structures of thermosensitive microgel particles. J Polym Sci Pol Phys 38:889–898
46. McPhee W, Tam KC, Pelton RJ (1993) Poly(*N*-isopropylacrylamide) latices prepared with sodium dodecyl sulfate. J Colloid Interface Sci 156:24–30
47. Wu X, Pelton RH, Hamielec AE et al (1994) The kinetics of poly(*N*-isopropylacrylamide) microgel latex formation. Colloid Polym Sci 272:467–477

Chapter 14
Functional Microcapsules with Ion-Recognizable Properties

Abstract In this chapter, the design, fabrication, and performance of the ion-recognizable functional microcapsules with crown ether as ion-recognition receptor and poly(*N*-isopropylacrylamide) (PNIPAM) as actuator are introduced. The microcapsules with different capsule structures and/or different crown ether receptors are developed for different purposes. For the porous microcapsule with ion-recognizable smart gates using 18-crown-6 as the ion receptor, the functional gates in the pores can close by recognizing Ba^{2+}. For the microcapsule with ion-recognizable cross-linked hydrogel membrane using 15-crown-5 as the ion receptor, the microcapsule shrinks isothermally by recognizing K^{+}. For the microcapsule with ion-recognizable cross-linked hydrogel membrane using 18-crown-6 as the ion receptor, the microcapsule swells isothermally by recognizing Ba^{2+} or Pb^{2+}. These ion-recognizable microcapsules provide new modes for smart controlled-release systems, which are highly attractive for drug delivery systems, chemical carriers, sensors, and so on.

14.1 Introduction

Microcapsules can encapsulate various active substances in their inner hollow cavity to protect them from the surrounding environment and/or to release them in a controlled manner by using appropriate microcapsule membrane materials. Since the 1980s, environmental stimuli-responsive microcapsules have been investigated widely [1]. These microcapsules control the release of their encapsulated contents in response to various environmental stimuli. They are considered to be potentially useful as controlled-release systems and especially as drug delivery systems. The target of a controlled drug delivery system is to improve drug treatment (outcome) through rate- and time-programmed and site-specific controlled release.

L.-Y. Chu et al., *Smart Hydrogel Functional Materials*,
DOI 10.1007/978-3-642-39538-3_14,

© Chemical Industry Press, Beijing and Springer Berlin Heidelberg 2013

Recently, the authors' group has developed several smart ion-recognizable functional microcapsules with crown ether as ion-recognition receptor and poly(*N*-isopropylacrylamide) (PNIPAM) as actuator. In this chapter, the design, fabrication, and performance of these ion-recognizable functional microcapsules are introduced.

14.2 Ion-Recognizable Microcapsules for Environmental Stimuli-Responsive Controlled Release

The authors' group developed a smart microcapsule with an ion-recognizable gating membrane for environmental stimuli-responsive controlled release [2]. The release of the encapsulated solutes from the proposed microcapsule is significantly sensitive to the presence of special metal ions in the environmental solution. The concept of the ion-recognizable microcapsule is schematically illustrated in Fig. 14.1. The microcapsule is composed of a porous capsule membrane and linear-grafted poly(*N*-isopropylacrylamide-*co*-benzo-18-crown-6-acrylamide)

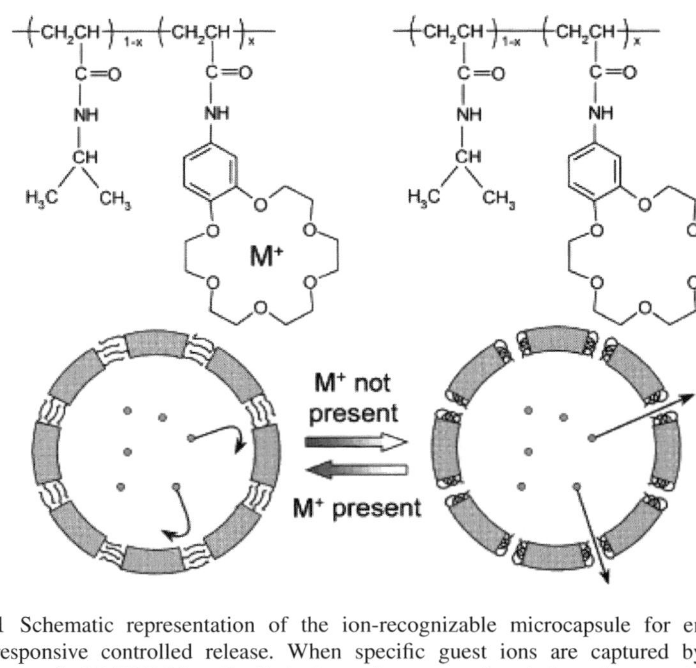

Fig. 14.1 Schematic representation of the ion-recognizable microcapsule for environmental stimuli-responsive controlled release. When specific guest ions are captured by the crown ethers, the grafted P(NIPAM-*co*-$B_{18}C_6$Am) copolymers swell and close the pores in the capsule membrane; when the guest ions are removed, the grafted P(NIPAM-*co*-$B_{18}C_6$Am) copolymers shrink and then the pores open (Reproduced with permission from Ref. [2], Copyright (2002), Wiley-VCH Verlag GmbH & Co. KGaA)

(P(NIPAM-*co*-$B_{18}C_6Am$)) copolymer chains in the pores acting as ion-recognizable gates. As described in Chap. 13, the phase transition of grafted P(NIPAM-*co*-$B_{18}C_6Am$) chains can occur isothermally as a result of the specific guest ion signal. When the specific metal ion exists in the environmental solution, the pores of the capsule membrane are closed through the swelling of P(NIPAM-*co*-$B_{18}C_6Am$) gates, thereby resulting in a low release rate of solutes from the microcapsule. On the other hand, when the specific metal ion is absent in the environmental solution, the pores open due to the shrinkage of the P(NIPAM-*co*-$B_{18}C_6Am$) gates, which leads to a high release rate. That is, the encapsulated solutes can be controlled to release only in those situations where the recognizable ions do not exist in the surrounding solution.

The porous microcapsule is prepared by the interfacial polymerization method as described in an earlier publication [3]. Plasma-graft pore-filling polymerization is employed to graft linear P(NIPAM-*co*-$B_{18}C_6Am$) copolymer chains into the pores of the microcapsule membrane according to the method described previously [3, 4]. Briefly, the freeze-dried porous microcapsules are put into a transparent glass tube, which is filled with argon gas and evacuated to a pressure of 10 Pa. And then the microcapsules are subjected to a radio-frequency plasma operating at 13.56 MHz and delivered at 30 W for 60 s. Next, the microcapsules are immersed into the monomer solution under inert atmosphere condition, and the graft polymerization is carried out in a shaking constant-temperature bath (80 °C) for a fixed period. The monomer solution is a mixture of NIPAM, $B_{18}C_6Am$, and water emulsified by sodium dodecyl sulfate (SDS). The SDS and total monomer concentrations in solution are 4 and 5 wt %, respectively, and the weight percentage ratio of NIPAM to $B_{18}C_6Am$ is 85:15. The grafted microcapsules are separated centrifugally, washed three times with deionized water, and then freeze-dried.

Figure 14.2 shows the field-emission scanning electron microscope (FE-SEM) images of the outer surface and cross section of the ungrafted and P(NIPAM-*co*-$B_{18}C_6Am$)-grafted microcapsules [2]. Comparing Fig. 14.2c with Fig. 14.2d, it can be seen that the cross-sectional structures of the ungrafted and P(NIPAM-*co*-$B_{18}C_6Am$)-grafted microcapsules are quite different. After grafting P(NIPAM-*co*-$B_{18}C_6Am$) chains onto the pore surface of the microcapsule membrane, the pore size becomes smaller. The porous structure across the cross section of the microcapsule is homogeneously covered with the grafted copolymer throughout the entire membrane thickness.

To study the ion-recognizable controlled-release behaviors of the P(NIPAM-*co*-$B_{18}C_6Am$)-grafted microcapsule, Na^+ and Ba^{2+} are chosen as the ion signals because these metal ions are important for chemical signals in biomembranes, and VB_{12} is used as the model solute [2]. The release of VB_{12} from the microcapsules is measured by determining the increase in the VB_{12} concentration of the surrounding medium with time. After 20 min, NaCl is added to the surrounding solution at a concentration of 0.2 mol/L. After another 20 min, $BaCl_2$ is also added for final NaCl and $BaCl_2$ concentrations of 0.18 and 0.02 mol/L, respectively. During the measurements, the temperature of the liquids is kept constant at 38 °C.

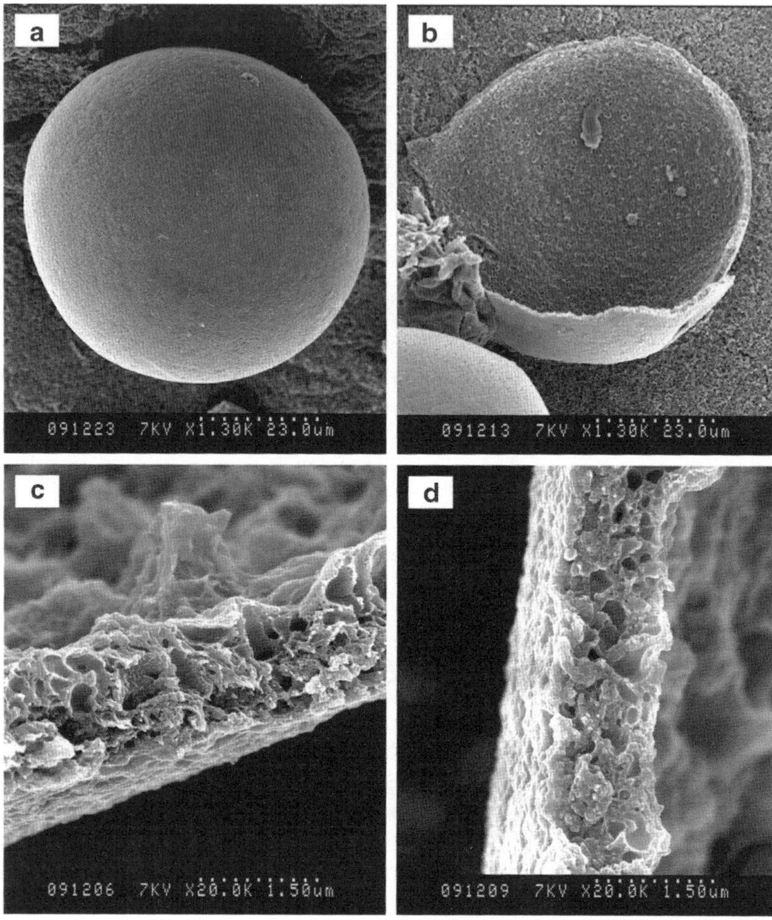

Fig. 14.2 FE-SEM images of microcapsules: (**a**) outer surface, (**b**) cross section, (**c**) cross section of an ungrafted microcapsule, and (**d**) cross section of a P(NIPAM-*co*-$B_{18}C_6$Am)-grafted microcapsule (Reproduced with permission from Ref. [2], Copyright (2002), Wiley-VCH Verlag GmbH & Co. KGaA)

Figure 14.3 shows the release results of VB_{12} from the microcapsules in different ion solutions [2]. For the P(NIPAM-*co*-$B_{18}C_6$Am)-grafted microcapsules, when deionized water or 0.2 mol/L NaCl is used as the solution, the release of VB_{12} from the microcapsules is fast; in contrast, when 0.02 mol/L $BaCl_2$ exists in the surrounding solution, the release rate suddenly drops significantly. On the other hand, the ungrafted microcapsule does not show such a sharp change of the release rate under the same experimental conditions. Because $B_{18}C_6$Am has a high stability constant for complexing with Ba^{2+}, the Ba^{2+} is selectively trapped in the cavity of crown ether, which causes the grafted P(NIPAM-*co*-$B_{18}C_6$Am) gates to swell.

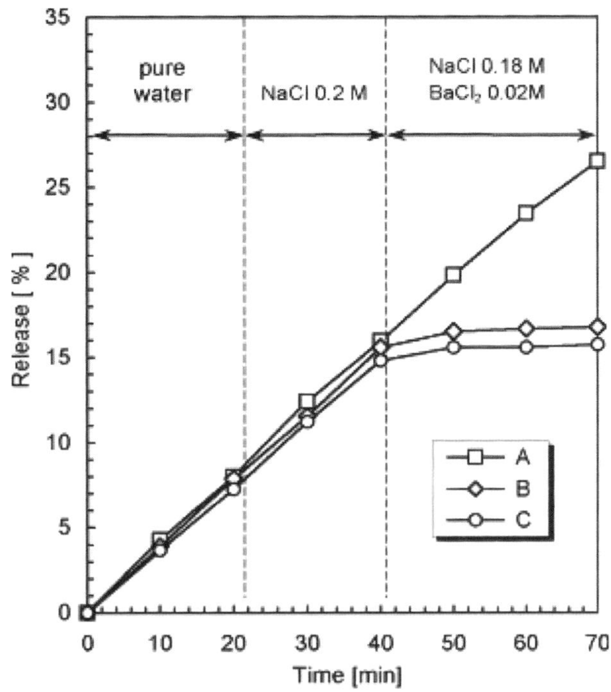

Fig. 14.3 Ion-recognition controlled-release behaviors of microcapsules. The microcapsules are *A*) ungrafted microcapsules and *B,C*) poly(NIPAM-*co*-$B_{18}C_6$Am)-grafted microcapsules with a grafting time of 113 h (*B*) and 120 h (*C*). For all experiments, the temperature is 38 °C (Reproduced with permission from Ref. [2], Copyright (2002), Wiley-VCH Verlag GmbH & Co. KGaA)

Consequently, before adding $BaCl_2$ to the solution, the pores of the P(NIPAM-*co*-$B_{18}C_6$Am)-grafted microcapsule are open because of the shrunken state of the grafted polymer gates that results in a high release rate. After adding $BaCl_2$, the pores of the P(NIPAM-*co*-$B_{18}C_6$Am)-grafted microcapsule suddenly close due to the swelling of the grafted polymer gates, and a very low release rate occurs (as illustrated in Fig. 14.1).

To verify the reversibility of ion-recognition behaviors, the VB_{12} release experiments are carried out repeatedly by using the same P(NIPAM-*co*-$B_{18}C_6$Am)-grafted microcapsules [2]. The microcapsules are washed and dialyzed against deionized water after each experiment to remove the Ba^{2+} from the crown ether receptor. Furthermore, to ensure that there is enough solutes inside the microcapsules during the entire release experiment, VB_{12} is reloaded into the microcapsules before each experiment. The reversible ion-responsive release characteristics of the P(NIPAM-*co*-$B_{18}C_6$Am)-grafted microcapsules are shown in Fig. 14.4. The ion-recognition release of the P(NIPAM-*co*-$B_{18}C_6$Am)-grafted microcapsules is found to be satisfactorily reversible and reproducible, suggesting that Ba^{2+} could be easily removed from the crown ether receptor and the grafted P(NIPAM-*co*-$B_{18}C_6$Am) gates could retain their ion-recognition properties even though they undergo repeated changes in the environmental ion signal.

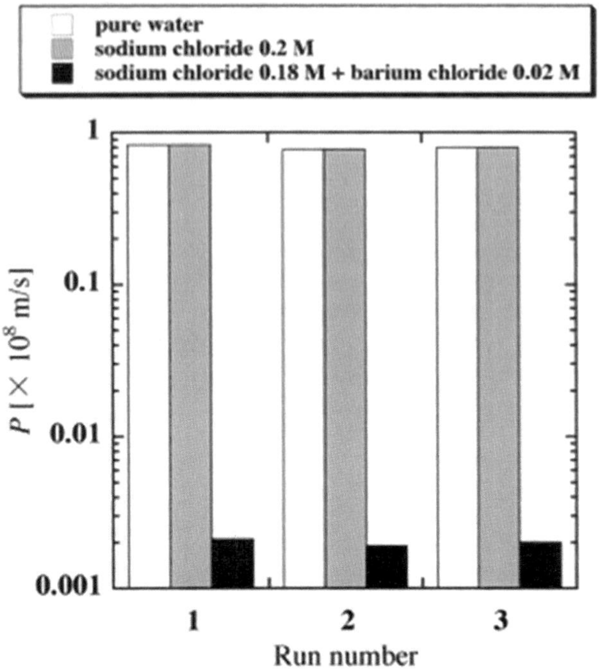

Fig. 14.4 The reversible release characteristics of the P(NIPAM-*co*-$B_{18}C_6Am$)-grafted microcapsules. For all experiments, the temperature is 38 °C (Reproduced with permission from Ref. [2], Copyright (2002), Wiley-VCH Verlag GmbH & Co. KGaA)

14.3 Monodisperse Ion-Recognizable Hydrogel Microcapsules for Burst Release of Hydrophobic Substance

A novel K^+-recognizable microcapsule that is able to directly translate the K^+ recognition into a squirting release function has been developed recently [5]. The concept of the proposed K^+-recognizable microcapsule is schematically illustrated in Fig. 14.5. The microcapsule is constructed with an ion-recognizable capsule membrane and an oil core, in which the membrane can selectively recognize K^+ and the oil core can load lipophilic substances such as tracer agents or drugs. The microcapsule membrane is composed of benzo-15-crown-5-acrylamide ($B_{15}C_5Am$) units as K^+ sensors, *N*-isopropylacrylamide (NIPAM) units as actuators, and acrylamide (AAm) units as hydrophilicity adjustors. As mentioned in Chap. 13, hydrogel based on a copolymer of NIPAM and $B_{15}C_5Am$ could isothermally change from swollen state to shrunken state due to the cooperative interaction of crown ether-based "host-guest" complexation and PNIPAM-based phase transition [6, 7]. With proper amount of AAm units, the operation temperature of K^+ recognition of the microcapsule can be adjusted from room temperature to body temperature due to the increased hydrophility of the copolymer chains [8]. Therefore, at a designed operation temperature (e.g., 37 °C), due to the K^+ recognition, the microcapsule shrinks rapidly, and the pressure on the oil core increases because the oil core

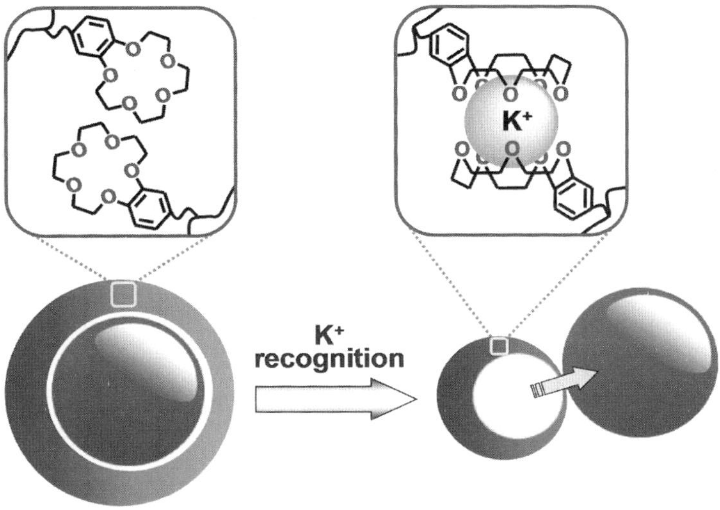

Fig. 14.5 Schematic illustration of the concept of K^+-recognizable microcapsule with a squirting release mechanism. The K^+-triggered volume shrinkage of the microcapsule membrane, which is due to the formation of a 2:1 sandwich "host-guest" complex of 15-crown-5 with K^+, results in a squirting release of the encapsulated oil core (Reproduced with permission from Ref. [5], Copyright (2011), The Royal Society of Chemistry)

is incompressible and cannot pass through the hydrogel capsule membrane via diffusion. As a result, when the internal pressure increases to a critical value, the capsule membrane ruptures suddenly due to its limited mechanical strength, and the encapsulated oil core is squirted out [9–11].

The proposed core-shell microcapsules are prepared with oil-in-water-in-oil (O/W/O) double emulsions as templates, which are generated with a microfluidic approach described previously [9–11]. In the O/W/O emulsions [5], soybean oil containing 3 % (w/v) emulsifier polyglycerol polyricinoleate (PGPR) and 0.1 % (w/v) fluorescent dye LR300 is used as the inner oil phase, and soybean oil containing 5 % (w/v) PGPR is used as the outer oil phase. For the middle aqueous phase, 4 mL deionized water containing monomers NIPAM (0.365 g), $B_{15}C_5Am$ (0.108 g), and AAm (0.023 g); cross-linker *N,N′*-methylenebisacrylamide (MBA) (0.005 g); initiator 2,2′-azobis(2-amidinopropane dihydrochloride) (V50) (0.008 g); glycerin (0.28 g); and emulsifier Pluronic F127 (0.032 g) is used typically. The ratios of $B_{15}C_5Am$ and AAm are varied for comparison. The O/W/O emulsions are collected in excess oil phase containing 1 % (w/v) initiator 2,2-dimethoxy-2-phenylacetophenone (BDK). The microcapsules are fabricated via UV-initiated polymerization in an ice bath. The O/W/O emulsions generated from microfluidic devices are highly monodisperse, from which monodisperse microcapsules are fabricated. In ion-recognition investigations, K^+, Na^+, and Ca^{2+} are chosen as the test ions because these ions are very important for signal transduction in biological systems.

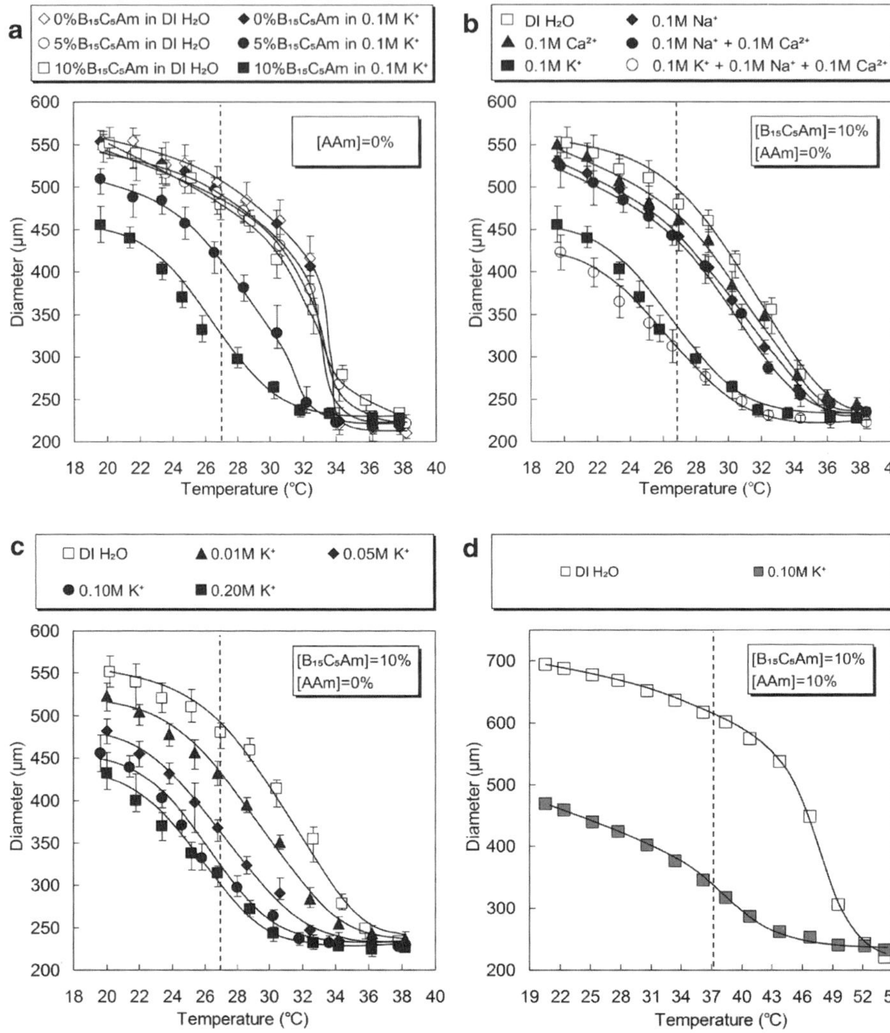

Fig. 14.6 Volume-phase transition behaviors of PNIPAM, P(NIPAM-*co*-$B_{15}C_5Am$), and P(NIPAM-*co*-AAm-*co*-$B_{15}C_5Am$) microcapsules. (**a**) Effect of $B_{15}C_5Am$ content in the microcapsule on the volume-phase transition behaviors. (**b**) Effect of ion species in the environment on the volume-phase transition behaviors. (**c**) Effect of K^+ concentration in the environment on the volume-phase transition behaviors. (**d**) Volume-phase transition behaviors of P(NIPAM-*co*-AAm-*co*-$B_{15}C_5Am$) microcapsules in deionized water (DI H_2O) and 0.1 mol/L K^+ solution (Reproduced with permission from Ref. [5], Copyright (2011), The Royal Society of Chemistry)

As shown in Fig. 14.6a, for PNIPAM microcapsules, the presence of K^+ in the environment does not stimulate any change in the phase transition behavior. As expected, all poly(*N*-isopropylacrylamide-*co*-benzo-15-crown-5-acrylamide) (P(NIPAM-*co*-$B_{15}C_5Am$)) microcapsules exhibit significant K^+-recognition behaviors (Fig. 14.6a–c). For P(NIPAM-*co*-$B_{15}C_5$Am) microcapsules, Na^+ and Ca^{2+} nearly do not cause any changes of the phase transition behaviors; however, K^+ significantly induces a negative shift of the volume-phase transition temperature (VPTT) of the P(NIPAM-*co*-$B_{15}C_5$Am) microcapsule (Fig. 14.6b). The K^+-triggered negative shift of the VPTT becomes more remarkable with increasing the $B_{15}C_5$Am amount in the microcapsule (Fig. 14.6a) or the K^+ concentration in the surrounding medium (Fig. 14.6c). If a proper operation temperature is chosen (e.g., 27 °C), the K^+-triggered isothermal volume decrease of P(NIPAM-*co*-$B_{15}C_5$Am) microcapsules is as large as 70 ~ 75 % when the molar ratio of $B_{15}C_5$Am to NIPAM is 10 mol % and the K^+ concentration is 0.1 mol/L. The VPTT of P(NIPAM-*co*-$B_{15}C_5$Am) microcapsule containing 10 mol % $B_{15}C_5$Am in deionized water is 33 °C (Fig. 14.6a), while that of P(NIPAM-*co*-AAm-*co*-$B_{15}C_5$Am) microcapsule containing 10 mol % $B_{15}C_5$Am and 10 mol % AAm increases to 46 °C due to the introduction of AAm units (Fig. 14.6d). When the environmental deionized water is replaced by 0.1 mol/L K^+ solution, the VPTT of P(NIPAM-*co*-AAm-*co*-$B_{15}C_5$Am) microcapsule shifts negatively to 35 °C (Fig. 14.6d). Therefore, the isothermal K^+-triggered shrinkage of P(NIPAM-*co*-AAm-*co*-$B_{15}C_5$Am) microcapsule can be operated at 37 °C, which is near the body temperature. When the K^+ concentration is 0.1 mol/L, P(NIPAM-*co*-AAm-*co*-$B_{15}C_5$Am) microcapsule shows a K^+-triggered isothermal volume shrinkage as large as 80 ~ 85 % at 37 °C.

Figure 14.7 shows the confocal laser scanning microscopy (CLSM) images of PNIPAM and P(NIPAM-*co*-$B_{15}C_5$Am) microcapsules when the environmental deionized water is replaced by 0.2 mol/L K^+ solution at 27 °C [5]. For PNIPAM microcapsules, the presence of K^+ at 27 °C does not induce any isothermal shrinkage (Fig. 14.7a). At the same temperature, P(NIPAM-*co*-$B_{15}C_5$Am) microcapsules containing 10 mol % $B_{15}C_5$Am exhibit a rapid and significant K^+-triggered isothermal shrinkage (Fig. 14.7b). Because the oil core is incompressible and cannot pass through the capsule membrane via diffusion, the liquid pressure inside microcapsule increases rapidly due to the membrane shrinkage. When the internal pressure increases to a critical value, the membrane ruptures suddenly and the encapsulated oil core squirts out as a result. The K^+-recognition response of the P(NIPAM-*co*-$B_{15}C_5$Am) microcapsules is very fast. The oil cores start to be burst out from the microcapsules at about 7 s, and the complete release of oil cores occurs within 25 s.

Due to the introduction of AAm units, P(NIPAM-*co*-AAm-*co*-$B_{15}C_5$Am) microcapsules are more hydrophilic than PNIPAM and P(NIPAM-*co*-$B_{15}C_5$Am) microcapsules. Therefore, P(NIPAM-*co*-AAm-*co*-$B_{15}C_5$Am) microcapsules containing 10 mol % $B_{15}C_5$Am and 10 mol % AAm are larger than PNIPAM and P(NIPAM-*co*-$B_{15}C_5$Am) ones in deionized water at room temperature, although they are all prepared using O/W/O emulsion templates with almost the same sizes, as shown in Fig. 14.8 [5]. P(NIPAM-*co*-AAm-*co*-$B_{15}C_5$Am) microcapsules are stable in deionized water at 37 °C; when the environmental deionized water is isothermally replaced by 0.2 mol/L K^+ solution, P(NIPAM-*co*-AAm-*co*-$B_{15}C_5$Am) microcapsules also exhibit very fast shrinkage and then rapidly squirt out the encapsulated oil cores.

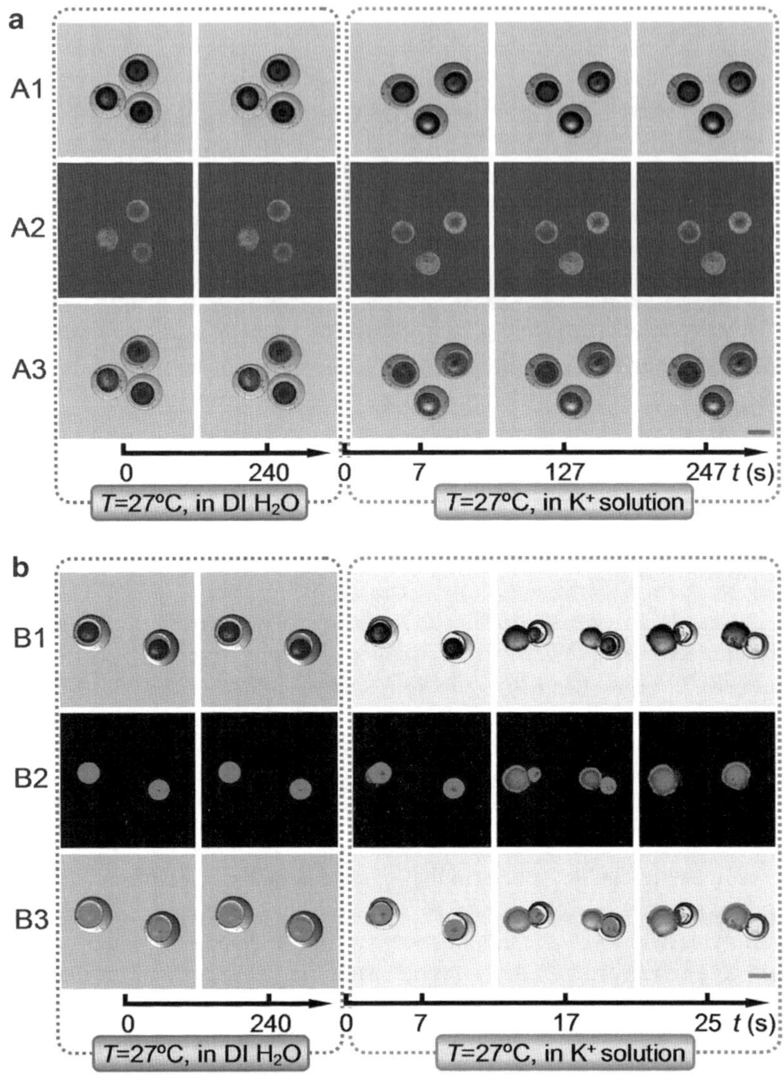

Fig. 14.7 CLSM images of PNIPAM (**a**) and P(NIPAM-*co*-$B_{15}C_5Am$) (**b**) microcapsules when the environmental deionized water (DI H_2O) is replaced by K^+ solution at 27 °C. "A1" and "B1" series show the transmission channel images, "A2" and "B2" series show the red channel images, and "A3" and "B3" series show the overlay images. The scale bars are 250 μm (Reproduced with permission from Ref. [5], Copyright (2011), The Royal Society of Chemistry)

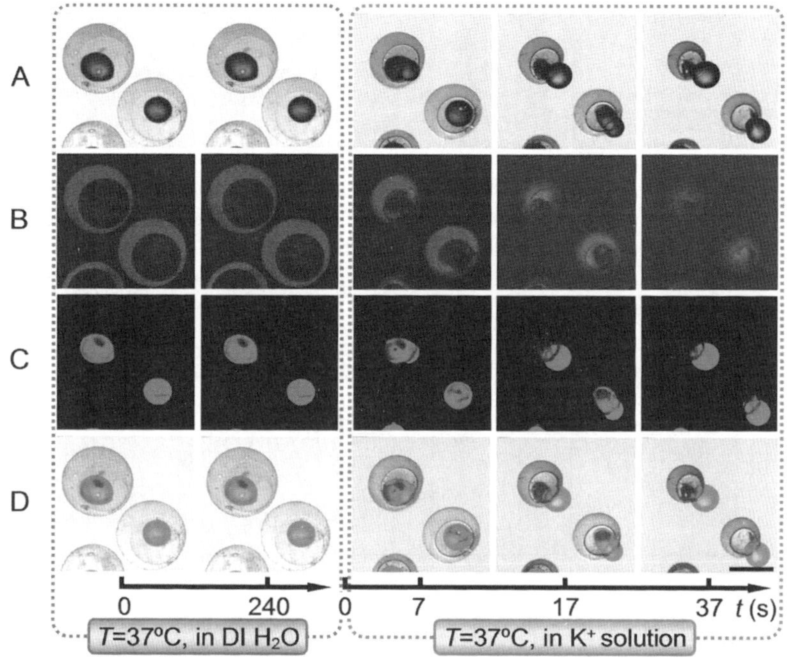

Fig. 14.8 CLSM images of K^+-triggered squirting release behavior of P(NIPAM-*co*-AAm-*co*-$B_{15}C_5$Am) microcapsules when the environmental deionized water (DI H_2O) is changed into K^+ solution at 37 °C. "A" series show the transmission channel images, "B" series show the green channel images, "C" series show the red channel images, and "D" series show the overlay images. The scale bars are 500 μm (Reproduced with permission from Ref. [5], Copyright (2011), The Royal Society of Chemistry)

14.4 Responsive Microcapsules Capable of Recognizing Heavy Metal Ions

If microcapsules could be developed to recognize heavy metal ions and then self-regulatively adjust the conformation of the microcapsule materials to increase the permeability of substances through the membranes, such smart microcapsules would serve as promising new systems for detection of heavy metals. When heavy metal ions present, such microcapsules could recognize them and automatically release the encapsulated materials that can trace the heavy metal ions to show their concentration directly or react with the heavy metal ions to eliminate their toxicity.

By using the above-mentioned P(NIPAM-*co*-$B_{18}C_6$Am) polymer materials, a smart microcapsule capable of recognizing heavy metal ions can be designed as

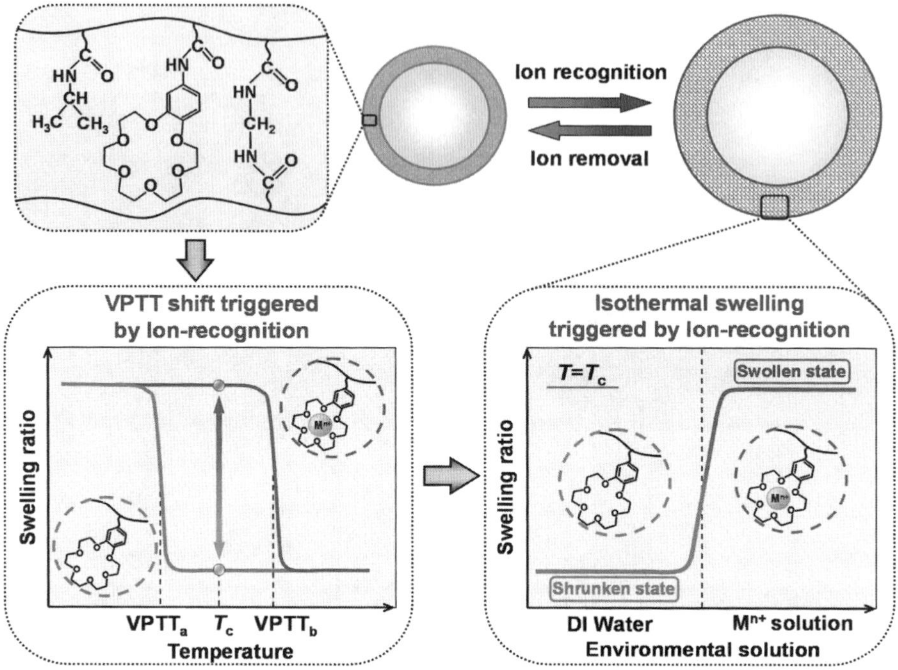

Fig. 14.9 Schematic illustration of the ion-recognizable smart microcapsule with cross-linked P(NIPAM-*co*-$B_{18}C_6Am$) hydrogel as capsule membrane (Reproduced with permission from Ref. [12], Copyright (2010), Elsevier)

illustrated in Fig. 14.9 [12]. The proposed microcapsule is featured with thin cross-linked P(NIPAM-*co*-$B_{18}C_6Am$) hydrogel as capsule membrane, which can selectively recognize special heavy metal ions such as Ba^{2+} or Pb^{2+} very well due to the "host-guest" complexation as described in Chap. 13 [13–16]. When $B_{18}C_6Am$ receptors in the cross-linked P(NIPAM-*co*-$B_{18}C_6Am$) hydrogel capture Ba^{2+} or Pb^{2+}, the VPTT of cross-linked P(NIPAM-*co*-$B_{18}C_6Am$) hydrogel shifts from a lower value ($VPTT_a$ in Fig. 14.9) to a higher one ($VPTT_b$ in Fig. 14.9) [13–16]. As a result, when the environmental temperature (T_c in Fig. 14.9) is located between $VPTT_a$ and $VPTT_b$, the proposed P(NIPAM-*co*-$B_{18}C_6Am$) microcapsule could exhibit isothermal and significant volume-phase transition from a shrunken state to a swollen state by recognizing Ba^{2+} or Pb^{2+}. As a result, the permeability of substances diffusing through the P(NIPAM-*co*-$B_{18}C_6Am$) capsule membrane increases. Such self-regulated conformational change triggered by ion recognition is reversible by removing the heavy metal ions from the crown ethers.

The proposed P(NIPAM-*co*-$B_{18}C_6Am$) microcapsules are prepared by using monodisperse O/W/O double emulsions as the polymerization templates, which are also generated by a microfluidic method described previously [8–10]. Briefly [12], in the O/W/O emulsions, soybean oil containing 3 % (w/v) PGPR is used as the

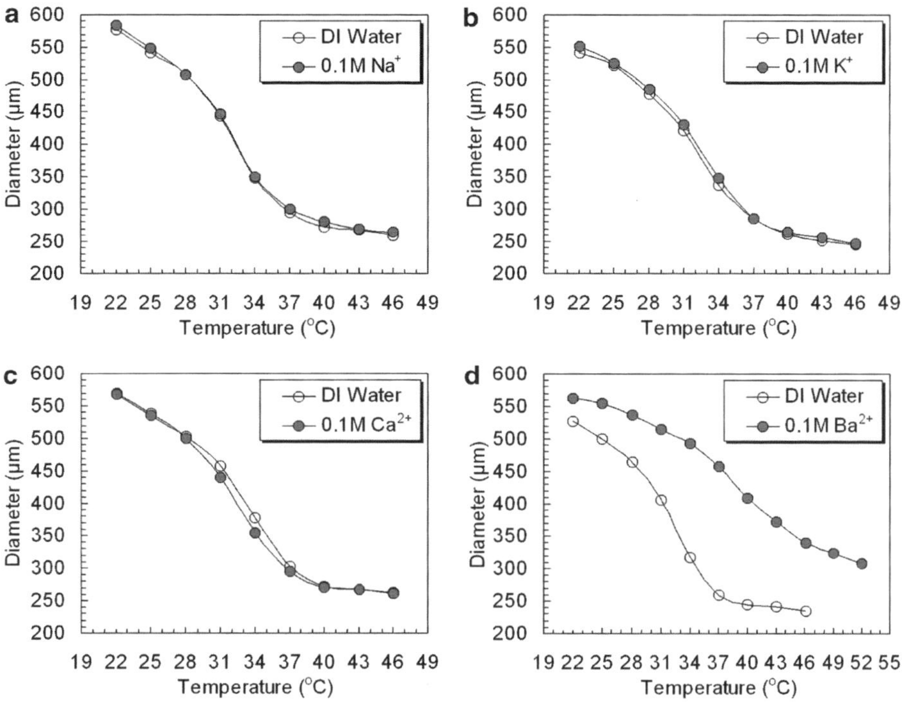

Fig. 14.10 Temperature-dependent size changes of the P(NIPAM-*co*-$B_{18}C_6Am$) microcapsules in deionized (DI) water and different ion solutions (Reproduced with permission from Ref. [12], Copyright (2010), Elsevier)

inner oil phase, and soybean oil containing 5 % (w/v) PGPR is used as the outer oil phase. For the middle aqueous phase, 5 mL deionized water containing NIPAM (0.452 g), $B_{18}C_6Am$ (0.15 g), MBA (0.006 g), V50 (0.01 g), glycerin (0.35 g), and Pluronic F127 (0.04 g) is used typically. The MBA content is also changed for comparison. The generated O/W/O emulsions are collected in excess soybean oil with another initiator 2,2-diethoxyacetophenone (DEAP). The microcapsules with cross-linked P(NIPAM-*co*-$B_{18}C_6Am$) hydrogel membrane are prepared via UV-initiated polymerization in an ice bath for 10 min.

The ion-recognition behaviors of P(NIPAM-*co*-$B_{18}C_6Am$) microcapsules are studied by evaluating their volume-phase transition behaviors in deionized water and different ion solutions [12]. Figure 14.10 shows the temperature-dependent size changes of the P(NIPAM-*co*-$B_{18}C_6Am$) microcapsules in different aqueous solutions, in which the cross-linking degree of the hydrogel capsule membrane is 1 %. All the P(NIPAM-*co*-$B_{18}C_6Am$) microcapsules exhibit excellent thermo-responsive characteristics, and their diameters decrease dramatically as the ambient temperature increases across a corresponding temperature range. It is obvious that the diameter change of the microcapsule in 0.1 mol/L Ba^{2+} solution is significantly different compared with those in deionized water and other ion solutions. The VPTT of the P(NIPAM-*co*-$B_{18}C_6$Am) microcapsule in deionized water is about 32 °C; however, in 0.1 mol/L Ba^{2+} solution, the VPTT increases about 7 °C, as shown in Fig. 14.10d, whereas the VPTT of the P(NIPAM-*co*-$B_{18}C_6$Am) microcapsule in 0.1 mol/L Na^+, K^+, or Ca^{2+} solution is almost the same as that in deionized water (Fig. 14.10a–c).

The above-mentioned phenomenon is almost the same as that of the P(NIPAM-*co*-$B_{18}C_6$Am) hydrogel described in Chap. 13, which is also caused by the formation of crown ether/metal ion "host-guest" complexes. The order of the formation constant, log*K*, of 18-crown-6 with metal ions in water is $Ba^{2+} > K^+ > Na^+ > Ca^{2+}$ [17]. This order shows a good correlation with the VPTT shifts in different ion solutions. The log*K* of 18-crown-6 with K^+, Na^+, or Ca^{2+} is small, and the polymer network of the P(NIPAM-*co*-$B_{18}C_6$Am) capsule membrane is cross-linked, so there is no significant VPTT change in Na^+, K^+, or Ca^{2+} solution (Fig. 14.10a–c). The log*K* of 18-crown-6 with Ba^{2+} is the largest; thus, $B_{18}C_6$Am can form stable "host-guest" complex with Ba^{2+}. Therefore, in Ba^{2+} solution, the VPTT of the P(NIPAM-*co*-$B_{18}C_6$Am) microcapsule shifts to a higher temperature than that in deionized water due to the repulsion among charged $B_{18}C_6Am/Ba^{2+}$ complexes and the osmotic pressure within the hydrogel capsule membrane [13]. Furthermore, the P(NIPAM-*co*-$B_{18}C_6$Am) microcapsule in Ba^{2+} solution has a larger swelling degree, and the diameter of the P(NIPAM-*co*-$B_{18}C_6$Am) microcapsule in Ba^{2+} solution is always larger than that in deionized water at any temperatures throughout the experiments (Fig. 14.10d).

Figure 14.11 shows the isothermal swelling behaviors of the P(NIPAM-*co*-$B_{18}C_6$Am) microcapsule by recognizing Ba^{2+} at 37 °C, which is set at temperature between the two VPTT values in deionized water and Ba^{2+} solution. To characterize the isothermal swelling behaviors quantitatively, three parameters indicated the swelling ratios of inner diameter (ID), outer diameter (OD), and membrane thickness of the P(NIPAM-*co*-$B_{18}C_6$Am) microcapsule caused by Ba^{2+} recognition are defined as R_{ID}, R_{OD}, and $R_{\mathrm{Thickness}}$:

$$R_{\mathrm{OD}} = \frac{\mathrm{OD}_{\mathrm{Ba}^{2+}}}{\mathrm{OD}_{\mathrm{H_2O}}}$$

$$R_{\mathrm{ID}} = \frac{\mathrm{ID}_{\mathrm{Ba}^{2+}}}{\mathrm{ID}_{\mathrm{H_2O}}}$$

$$R_{\mathrm{Thickness}} = \frac{\mathrm{OD}_{\mathrm{Ba}^{2+}} - \mathrm{ID}_{\mathrm{Ba}^{2+}}}{\mathrm{OD}_{\mathrm{H_2O}} - \mathrm{ID}_{\mathrm{H_2O}}}$$

The Ba^{2+}-triggered swelling behaviors of the P(NIPAM-*co*-$B_{18}C_6$Am) microcapsule are significant. The P(NIPAM-*co*-$B_{18}C_6$Am) microcapsule is in a shrunken state in deionized water; however, after recognizing Ba^{2+}, the microcapsule automatically switches to a swollen state. When the Ba^{2+} concentration increases to 0.2 mol/L, the values of R_{OD}, R_{ID}, and $R_{\mathrm{Thickness}}$ are slightly smaller than the corresponding values in 0.1 mol/L Ba^{2+} solution (Fig. 14.11c). Because the quantity

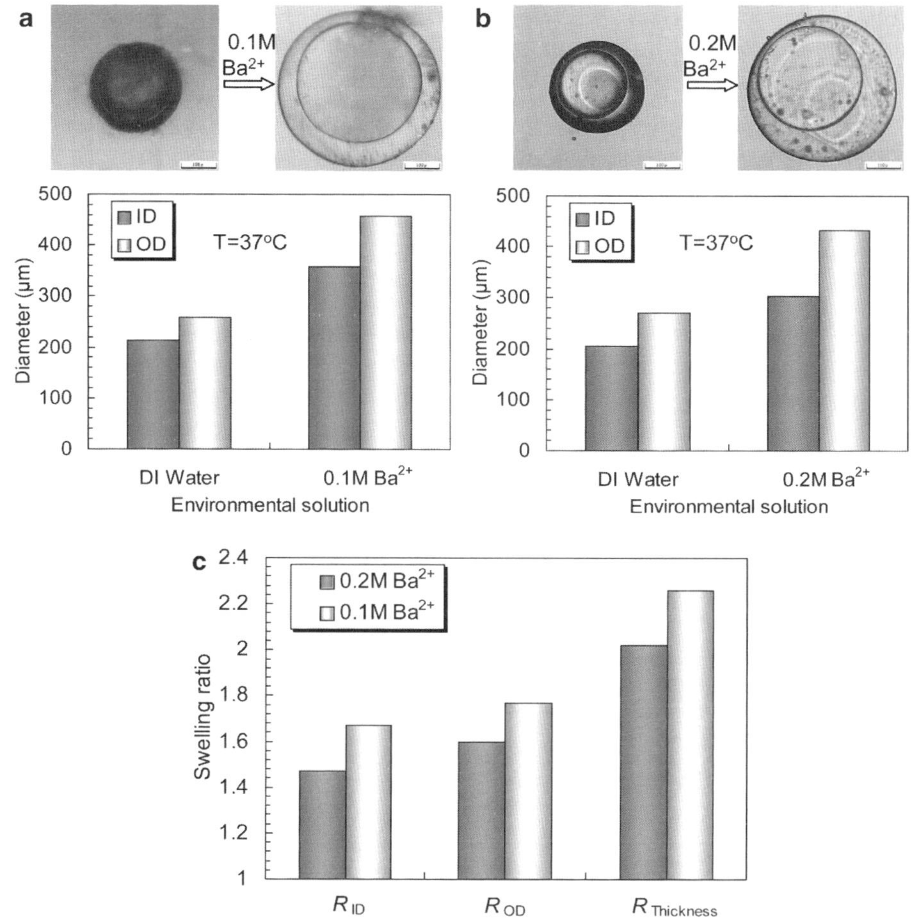

Fig. 14.11 The isothermal swelling behaviors of the P(NIPAM-*co*-$B_{18}C_6Am$) microcapsule triggered by recognizing Ba^{2+} at 37 °C when the cross-linkage is 1 %. (**a**) The size change of the microcapsule by transferring it from deionized (DI) water into 0.1 mol/L Ba^{2+} solution, (**b**) the size change of the microcapsule by transferring it from DI water into 0.2 mol/L Ba^{2+} solution, and (**c**) the comparison of R_{ID}, R_{OD}, and $R_{Thickness}$ of microcapsule caused by Ba^{2+} recognition with different concentration (Reproduced with permission from Ref. [12], Copyright (2010), Elsevier)

of $B_{18}C_6Am$ units in the P(NIPAM-*co*-$B_{18}C_6Am$) microcapsule is limited, both the 0.1 and 0.2 mol/L Ba^{2+} concentrations might provide excessive Ba^{2+} ions for the $B_{18}C_6Am$ units in the P(NIPAM-*co*-$B_{18}C_6Am$) microcapsule to form "host-guest" complexes. However, the salting-out effect from increased amount of NO^{-3} anions in solutions would counteract the swelling of the P(NIPAM-*co*-$B_{18}C_6Am$) microcapsule to some extent [18–20].

Figure 14.12 shows the swelling ratios of the inner diameter, the outer diameter, and the thickness of the prepared P(NIPAM-*co*-$B_{18}C_6Am$) microcapsule by

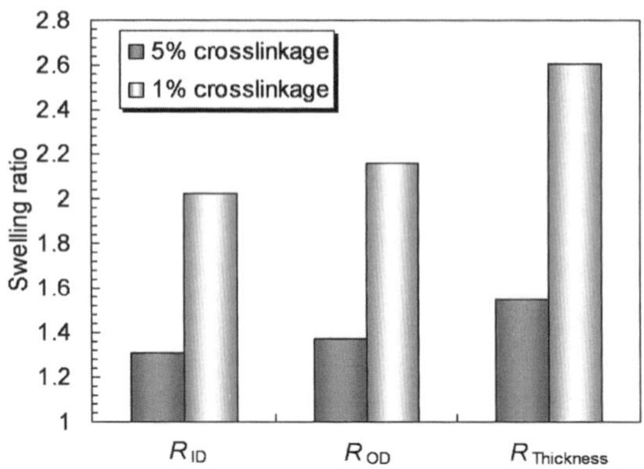

Fig. 14.12 The isothermal swelling behaviors of the prepared P(NIPAM-*co*-$B_{18}C_6Am$) microcapsule triggered by recognizing Pb^{2+} at 37 °C (Reproduced with permission from Ref. [12], Copyright (2010), Elsevier)

recognizing Pb^{2+} at 37 °C. The environmental Pb^{2+} concentration is 0.01 mol/L. When the microcapsule recognizes Pb^{2+}, all of the inner diameter, outer diameter, and the thickness increase due to the repulsion among charged $B_{18}C_6Am/Pb^{2+}$ complexes and the osmotic pressure within the hydrogel membrane, which likes that in Ba^{2+} solution as explained before [13–16].

The above results reveal that triggered by recognizing special heavy metal ions (Ba^{2+} or Pb^{2+}), the prepared P(NIPAM-*co*-$B_{18}C_6Am$) microcapsule exhibits an isothermal and significant swelling not only in outer and inner diameters but also in the thickness of capsule membrane. Such unique and dramatic isothermal swelling of the P(NIPAM-*co*-$B_{18}C_6Am$) microcapsule induced by recognizing special heavy metal ions could cause a self-regulated increase of the permeability across the microcapsule membrane, which could be highly potential for designing new materials not only as smart controlled-release systems but also as smart removers for detection and treatment of heavy metal ions.

14.5 Summary

In summary, the ion-recognizable functional microcapsules are developed with crown ether as ion-recognition receptor and PNIPAM as actuator. The porous microcapsule with linear-grafted P(NIPAM-*co*-$B_{18}C_6Am$) chains as ion-recognizable gates could control the release of encapsulated solutes only when special metal ions (such as Ba^{2+}) do not exist in surrounding solution. The core-shell microcapsule constructed with P(NIPAM-*co*-$B_{15}C_5Am$) hydrogel as the capsule membrane and an oil core exhibits a K^+-recognition controlled-release behavior. The capsule membrane could isothermally change from swollen state to shrunken state by recognizing K^+, and the rapid and significant K^+-recognition volume shrinkage of the capsule membrane results in a sudden squirting release of encapsulated oil cores. For the microcapsule with cross-linked P(NIPAM-*co*-$B_{18}C_6Am$), hydrogel as capsule membrane could swell isothermally by recognizing heavy metal ions, such as Ba^{2+} and Pb^{2+}. These ion-recognizable microcapsules provide new modes for smart controlled-release systems, which are highly attractive for drug delivery systems, chemical carriers and sensors for detection and treatment of heavy metal ions, and so on.

References

1. Breimer DD (1999) Future challenges for drug delivery. J Control Release 62:3–6
2. Chu LY, Yamaguchi T, Nakao S (2002) A molecular recognition microcapsule for environmental stimuli-responsive controlled-release. Adv Mater 14:386–389
3. Chu LY, Park SH, Yamaguchi T et al (2001) Preparation of thermo-responsive core-shell microcapsules with a porous membrane and poly(*N*-isopropylacrylamide) gates. J Membr Sci 192:27–39
4. Yamaguchi T, Nakao S, Kimura S (1991) Plasma-graft filling polymerization: preparation of a new type of pervaporation membrane for organic liquid mixtures. Macromolecules 24:5522–5527
5. Liu Z, Liu L, Ju XJ et al (2011) K^+-recognition capsules with squirting release mechanisms. Chem Commun 47:12283–12285
6. Mi P, Ju XJ, Xie R et al (2010) A novel stimuli-responsive hydrogel for K^+-induced controlled-release. Polymer 51:1648–1653
7. Wu HG, Ju XJ, Xie R et al (2011) A novel ion-imprinted hydrogel for recognition of potassium ions with rapid response. Polym Advan Technol 22:1389–1394
8. Xie R, Li Y, Chu LY (2007) Preparation of thermo-responsive gating membranes with controllable response temperature. J Membr Sci 289:76–85
9. Chu LY, Utada AS, Shah RK et al (2007) Controllable monodisperse multiple emulsions. Angew Chem Int Edit 46:8970–8974
10. Wang W, Liu L, Ju XJ et al (2009) A novel thermo-induced self-bursting microcapsule with magnetic-targeting property. Chem Phys Chem 10:2405–2409
11. Liu L, Wang W, Ju XJ et al (2010) Smart thermo-triggered squirting capsules for nanoparticle delivery. Soft Matter 6:3759–3763
12. Pi SW, Ju XJ, Wu HG et al (2010) Smart responsive microcapsules capable of recognizing heavy metal ions. J Colloid Interface Sci 349:512–518
13. Ju XJ, Chu LY, Liu L et al (2008) A novel thermo-responsive hydrogel with ion-recognition property through supramolecular host-guest complexation. J Phys Chem B 112:1112–1118
14. Ju XJ, Zhang SB, Zhou MY et al (2009) Novel heavy-metal adsorption material: Ion-recognition P(NIPAM-co-BCAm) hydrogels for removal of lead(II) ions. J Hazard Mater 167:114–118
15. Ju XJ, Liu L, Xie R et al (2009) Dual thermo-responsive and ion-recognizable monodisperse microspheres. Polymer 50:922–929
16. Zhang B, Ju XJ, Xie R et al (2012) Comprehensive effects of metal ions on responsive characteristics of P(NIPAM-*co*-B18C6Am). J Phys Chem B 116:5527–5536
17. Izatt RM, Pawlak K, Bradshaw JS et al (1991) Thermodynamic and kinetic data for macrocycle interaction with cations and anions. Chem Rev 91:1721–2085
18. Inomata H, Goto S, Otake K et al (1992) Effect of additives on phase transition of *N*-isopropylacrylamide gels. Langmuir 8:687–690
19. Zhang YJ, Furyk S, Bergbreiter DE et al (2005) Specific ion effects on the water solubility of macromolecules: PNIPAM and the Hofmeister series. J Am Chem Soc 127:14505–14510
20. Zhang Y, Furyk S, Sagle LB et al (2007) Effects of Hofmeister anions on the LCST of PNIPAM as a function of molecular weight. J Phys Chem C 111:8916–8924

Part VII
Molecular-Recognizable Hydrogel Functional Materials

Chapter 15
Preparation and Properties of Molecular-Recognizable Smart Hydrogels

Abstract In this chapter, the fabrication and performance of molecular-recognizable smart hydrogels based on poly(*N*-isopropylacrylamide) (PNIPAM) and β-cyclodextrins (CD) are introduced. The length of the monosubstituted carbon chain of modified CD as well as the categories and concentrations of guest molecules has great effect on the molecular-recognition-induced phase transition of PNIPAM polymers with pendent CD moieties. The increase in the hydrophobicity of hydrogel network leads to a considerable increase in affinity behavior toward guest molecules. Meanwhile, the association site is demonstrated to shift from original β-cyclodextrin at low temperatures to a combination of β-cyclodextrin and the PNIPAM network at high temperatures. The synthesized PNIPAM polymers and hydrogels with pendent CD moieties have the potential to be applied to the engineering of molecular-recognition sensors and switches and also the development of temperature-controlled affinity separation systems.

15.1 Introduction

β-Cyclodextrins (CD), torus-shaped molecules composed of cyclic α-1,4-oligoglucopyranosides, are host molecules possessing hydrophobic cavities and hydrophilic external surfaces. On account of a series of weak intermolecular forces, such as hydrophobic, dipole-dipole, electrostatic, van der Waals, and hydrogen-bonding interactions, cyclodextrin is able to selectively associate with guest molecules having the similar size with its cavity. Such a unique property endows cyclodextrins as well as its derivatives with valuable uses [1–5].

Combination of poly(*N*-isopropylacrylamide) (PNIPAM) and other functional groups can bring corresponding functions and/or stimuli-responsive properties to PNIPAM polymers and hydrogel in addition to the thermo-responsiveness. Due to the molecular-recognition ability of β-cyclodextrin, certain attention has been drawn to polymers and hydrogels with a combination of PNIPAM with CD which

L.-Y. Chu et al., *Smart Hydrogel Functional Materials*,
DOI 10.1007/978-3-642-39538-3_15,

© Chemical Industry Press, Beijing and Springer Berlin Heidelberg 2013

possess thermo-responsive and molecular-recognizable characteristics simultaneously [6–15]. Nozaki et al. [6, 7] have investigated the temperature dependence of CD-modified PNIPAM polymers in terms of association constant with guest molecules. The result shows that PNIPAM chain attached to the side arm of cyclodextrin has considerable influence on the association constant of cyclodextrin toward guest molecule, which is ascribed to the steric hindrance that caused by PNIPAM chain. Yamaguchi et al. [13] have developed a novel polymer system based on CD-modified PNIPAM that exhibits a coordination of molecular recognition and actuation functionalities within itself. It is found that these components affect each other. That is, the complexation between CD and the guest molecule induces a phase transition in the PNIPAM chain. Meanwhile, the phase transition in the PNIPAM chain affects the association between CD and guest molecule. Other previous studies on CD-containing PNIPAM hydrogels have mainly dealt with the controlled drug release and/or the temperature/pH sensitivity of the hydrogel [8, 9, 11, 14]. However, to the best of our knowledge, the molecular-recognition-induced phase transitions of PNIPAM polymers with pendent CD groups and the temperature dependence of affinity behavior of CD-containing PNIPAM hydrogels have not been reported.

In this chapter, the successful preparation of PNIPAM polymers and hydrogels with CD moieties will be introduced. The molecular-recognition-induced phase transition behaviors of the fabricated PNIPAM polymers with different modified CD are investigated systematically. In addition, the temperature dependence of affinity behavior of fabricated hydrogels is also investigated. The synthesized copolymers and hydrogels with combination of PNIPAM and CD are able to simultaneously respond to temperature and recognize guest molecules and could be used for the engineering of molecular-recognition sensors and switches and also for the developing temperature-controlled affinity separation systems.

15.2 Molecular-Recognition-Induced Phase Transition of Thermo-responsive Polymers with Pendent *β*-Cyclodextrin Groups

Two novel PNIPAM copolymers with pendent CD groups (PNG-ECD and PNG-HCD) are prepared via the reaction between primary amino groups of modified CD and epoxy groups of poly(*N*-isopropylacrylamide-*co*-glycidyl methacrylate) (PNG) copolymers [16]. The difference between the two fabricated copolymers is the length of the monosubstituted carbon chain of modified CD moieties. The numbers of carbon atoms in mono-6-deoxy-6-ethylene diamino-*β*-CD (ECD) and mono-6-deoxy-6-hexane diamino-*β*-CD (HCD) are 2 and 6, respectively. The 8-anilino-1-naphthalenesulfonic acid ammonium salt (ANS) and 2-naphthalenesulfonic acid (NS) are selected as the guest molecules to study the molecular-recognition-induced phase transition behaviors.

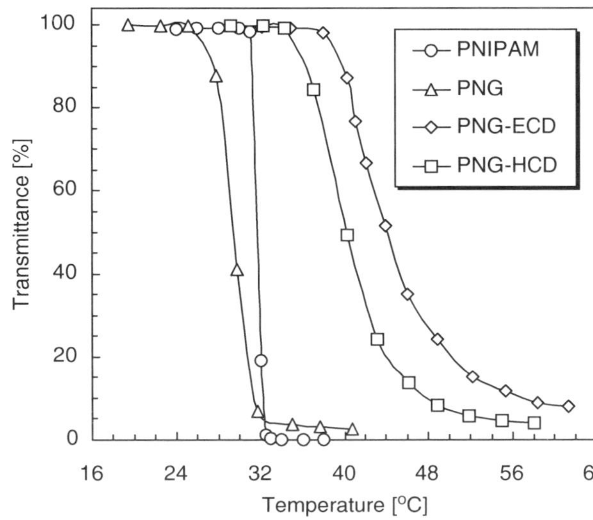

Fig. 15.1 Comparison of thermo-responsive characteristics of PNIPAM, PNG, PNG-ECD, and PNG-HCD polymers in water (Reproduced with permission from Ref. [16], Copyright (2008), Wiley-VCH Verlag GmbH & Co. KGaA)

15.2.1 The LCST of Thermo-responsive Polymers with Pendent β-Cyclodextrin Groups

To investigate the LCST values and the effect of guest molecules on the LCST values of polymers, the transmittance of PNG-ECD and PNG-HCD polymer solutions (2 wt %) is measured by a UV-visible spectrophotometer (752UV) at the wavelength of 650 nm. When the polymer chains extend (coil) and become hydrophilic, the polymer is homogeneously dispersed in solution so that the visible light can penetrate through and the transmittance of polymer solution is high. However, when the polymer chains shrink (globule) and become hydrophobic, the polymer aggregates and scatters the visible light which makes the transmittance low. Therefore, the LCST of a polymer in aqueous solution and guest molecules solution can be determined from the temperature where the transmittance of polymer solution exhibits a sharp change.

Figure 15.1 shows the thermo-responsive coil-to-globule transitions of the PNIPAM, PNG, PNG-ECD, and PNG-HCD polymers in water. According to the dramatic change in turbidity, the LCST values of PNIPAM and PNG polymer in water are found to be approximately 32.0 °C and 29.3 °C, respectively. Although the coil-to-globule transition is still significantly evident, the introduction of the hydrophobic GMA moieties results in a slight decrease in the LCST compared with that of PNIPAM polymer. The incorporation of GMA causes an increase of hydrophobic moieties in the polymer and changes the balance between the hydrophilic and hydrophobic interactions in the polymer. Consequently, the hydrogen bonds are broken and the coil-to-globule transition of polymer occurs at a lower temperature. However, the LCST values of both PNG-ECD and PNG-HCD

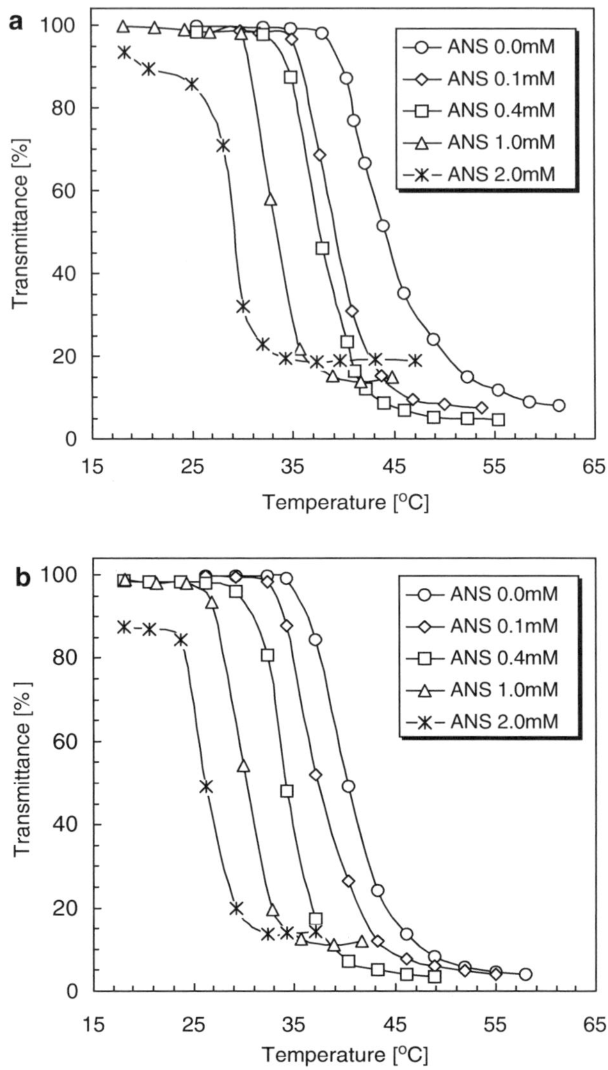

Fig. 15.2 Thermo-responsive characteristics of PNG-ECD polymers (**a**) and PNG-HCD polymers (**b**) in ANS solutions with different concentrations (Reproduced with permission from Ref. [16], Copyright (2008), Wiley-VCH Verlag GmbH & Co. KGaA)

polymers in water are much higher than that of PNG polymers, which are 43.8 °C and 39.9 °C, respectively. It is attributable to the incorporation of the hydrophilic CD moieties in polymers. Moreover, because of the longer monosubstituted carbon chain of modified CD moieties, the LCST value of PNG-HCD is slightly lower than that of PNG-ECD polymer.

Figure 15.2 shows the thermo-responsive characteristics of PNG-ECD and PNG-HCD polymers in ANS solutions with different concentrations. The LCST values of PNG-ECD in ANS solutions with different concentrations of 0.1, 0.4, 1.0, and 2.0 mM are 38.9 °C, 37.6 °C, 32.8 °C, and 29.0 °C, respectively. The LCST values

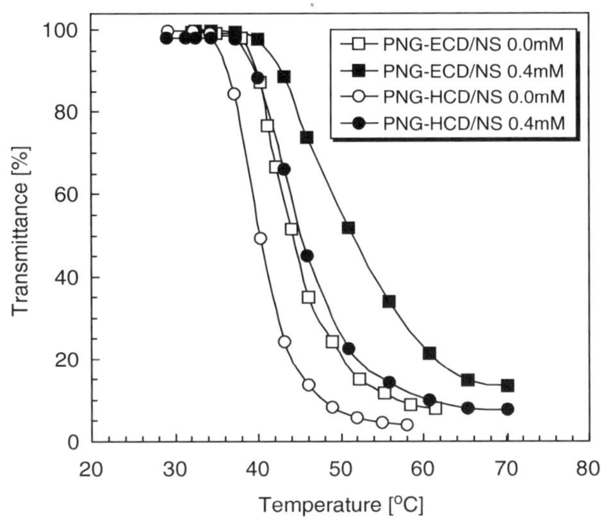

Fig. 15.3 Thermo-responsive characteristics of PNG-ECD and PNG-HCD polymers in NS solution (Reproduced with permission from Ref. [16], Copyright (2008), Wiley-VCH Verlag GmbH & Co. KGaA)

of PNG-HCD in ANS solutions with the same concentrations are 37.2 °C, 34.1 °C, 30.3 °C, and 26.0 °C, respectively. Compared with those in water, the LCST values of both PNG-ECD and PNG-HCD polymers decrease with the increase in the concentration of ANS solution. Moreover, the shift of the LCST values is larger for higher ANS concentrations. However, the introduction of ANS has almost no influence on the LCST of PNG polymer, no matter how the ANS concentration changes. The LCST shifts for the PNG-ECD and PNG-HCD polymers in ANS solutions with different concentrations are ascribed to the change of the balance between hydrophilic and hydrophobic interactions in the polymer. It is triggered by the host-guest recognition of CD moieties with guest molecules ANS. When the CD moiety recognizes the guest molecule ANS, the CD/ANS complex leaves a hydrophobic phenyl group out of the CD cavity, which enlarges the hydrophobic moiety of the polymer. The higher the concentration of ANS solution is, the more CD/ANS complexes form, and thus the more hydrophobic moieties exist in the polymer chains. As a result, the higher the concentration of ANS solution is, the lower the LCST of the polymer becomes. Moreover, with the same concentration of ANS solution, the PNG-HCD polymers are more hydrophobic than PNG-ECD polymers due to the extra four methylene groups.

Figure 15.3 shows the thermo-responsive characteristics of PNG-ECD and PNG-HCD polymers in NS solution. Unlike the result of ANS, the guest molecules NS has an opposite influence on the LCST values of PNG-ECD and PNG-HCD polymers. On adding NS molecules to the aqueous solutions, the LCST values of both PNG-ECD and PNG-HCD increase by a certain degree. When the CD moiety recognizes the guest molecules, the CD/ANS complex has a hydrophobic phenyl group out of the CD cavity, while the CD/NS complex has no side group out of the CD cavity. However, the complexation between CD and NS enlarges the hydrophilic moiety of polymers due to the dissociation of NS in water. The experimental results show that the different guest molecules have totally different effects on the thermo-responsive characteristics of PNG-ECD and PNG-HCD polymers.

15.2.2 Dynamic Thermo-responsive Characteristics of PNG-ECD and PNG-HCD in ANS Solutions

The change of absorbance is more sensitive than that of transmittance, especially for polymers in the shrunken state. Therefore, the dynamic absorbance change of PNG-ECD and PNG-HCD in ANS solutions with different concentrations responding to a "sharp temperature jump" is also observed by a UV-visible spectrophotometer at a wavelength of 650 nm. The "sharp temperature jump" referred to a range of temperature that from T_1 which below the LCST of the polymer in solution to T_2 which above the LCST of the polymer in the same solution. Because the PNG-ECD and PNG-HCD polymers in ANS solutions with different concentrations have different LCST values, the "sharp temperature jump" is different for those polymer solutions with different ANS concentrations. As shown in Fig. 15.4, the variation trends in absorbance of PNG-ECD and PNG-HCD solutions are almost the same within the "sharp temperature jump" of 10 °C. For those polymer solutions with ANS concentration lower than 2.0 mM, a mild change in absorbance is observed within the first 2 min. However, when the temperature reaches the LCST of the polymer, a sharp rise in absorbance occurs and the absorbance value is almost unchanged after the temperature is crossing the LCST. On increasing the ANS concentration in polymer solution, the rise in absorbance during the sharp temperature jump also increases.

During the "sharp temperature jump," when the environmental temperature is lower than the LCST of polymer, the polymer chains extend and become hydrophilic. The temperature is increased with time to pass through the LCST, and thus the polymer transforms from a hydrophilic to a hydrophobic structure and becomes insoluble in the solution. Once the transition is in equilibrium, the absorbance value is maximal and there is no further variation. For the polymer solution with ANS concentration of 2.0 mM, there is a moderate decrease in absorbance following the sharp rise (Fig. 15.4). The visible precipitation in the cell bottom is also observed after finishing the dynamic absorbance measurement of PNG-ECD and PNG-HCD polymers in ANS solution (2.0 mM). Elemental analysis results show that the carbon, nitrogen, and sulfur compositions of the precipitates are in the range between the compositions of ANS and PNG-CD, that is, the precipitate is verified to be the PNG-ECD/ANS or PNG-HCD/ANS complexes. This phenomenon could be useful in separations of particular molecules, such as enantiomers, proteins, and hormones, as long as the CD moieties in the synthesized polymers can recognize the target molecules.

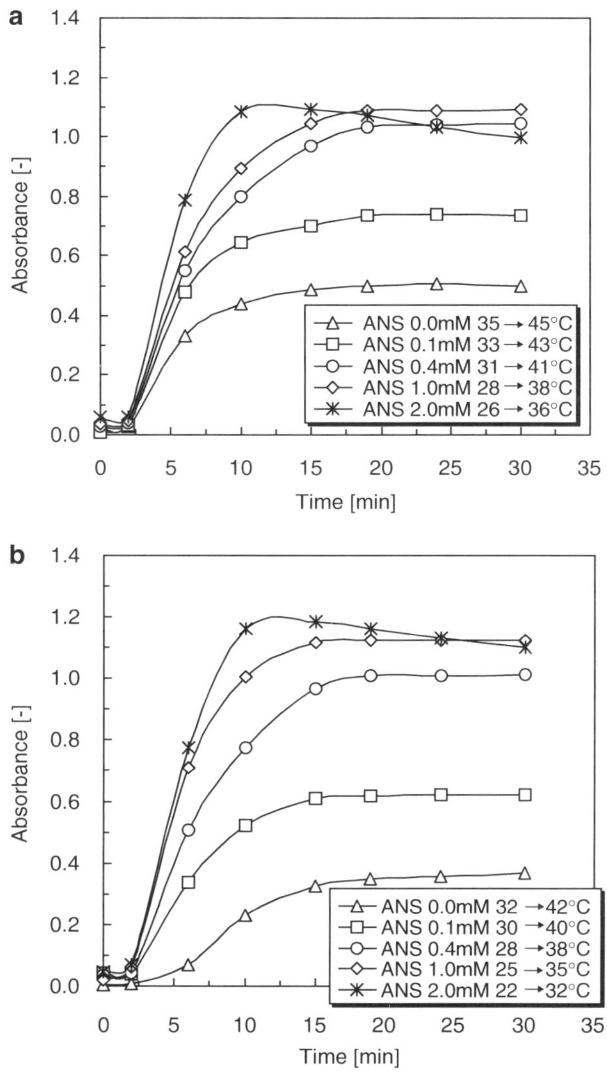

Fig. 15.4 Thermo-responsive absorbance variations of PNG-ECD (**a**) and PNG-HCD (**b**) solutions with different ANS concentrations in response to the sharp temperature jump (Reproduced with permission from Ref. [16], Copyright (2008), Wiley-VCH Verlag GmbH & Co. KGaA)

15.3 Thermo-responsive Affinity Behavior of Poly(*N*-isopropylacrylamide) Hydrogels with *β*-Cyclodextrin Moieties

The PNIPAM hydrogel with CD moieties P(NIPAM-*co*-ACD) is prepared via free-radical polymerization of *N*-isopropylacrylamide (NIPAM) with allyl-group-containing mono-(6-*N*-allylamino-6-deoxy)-*β*-cyclodextrin (ACD) at 20 °C for

24 h in a mixed solvent of DMF/H_2O [17]. Using ANS as the guest molecule, an experimental study on the temperature dependence of affinity behavior of P(NIPAM-*co*-ACD) hydrogels is carried out within a wide range of temperatures (21–60 °C).

15.3.1 Chemical and Morphological Analysis of Hydrogels

Compared with the FT-IR spectrum of PNIPAM hydrogel, a typical C-O-C stretching vibration at around 1,050 cm^{-1} as well as a conspicuously broadened peak at around 3,450 cm^{-1} (hydroxyl groups) in that of P(NIPAM-*co*-ACD) hydrogel is observed. They indicate the successful introduction of β-cyclodextrin into the PNIPAM polymer. The weight content of ACD in the P(NIPAM-*co*-ACD) hydrogel is determined to be 25.66 % through C and N content by means of elemental analysis. Moreover, the structural differences between the two hydrogels are evident through the comparison of SEM images. As shown in Fig. 15.5, both cross-sectional and surface SEM images of P(NIPAM-*co*-ACD) and PNIPAM hydrogels show porous structures. However, it is evident that PNIPAM hydrogel exhibited net-shaped structure, while P(NIPAM-*co*-ACD) hydrogel has evident flake-like structure attached to the polymeric network of the hydrogel which is due to the introduction of β-cyclodextrin.

15.3.2 Volume-Phase Transition Temperature (VPTT) Shift of Hydrogels Induced by Molecular Recognition

The volume-phase transition temperature (VPTT) of hydrogel could be determined by the temperature, at which a dramatic volume change begins to occur during equilibrium swelling ratio (ESR) measurement. Or it could be determined by the intersection of the baseline and leading edge of the endothermic peak during differential scanning calorimetry (DSC) measurement. The temperature dependence of ESR of P(NIPAM-*co*-ACD) hydrogel as well as PNIPAM hydrogel in deionized water and dilute aqueous ANS solution (1.0 mM) is investigated in the temperature range from 20 °C to 50 °C. There is nearly no variation in VPTT values of PNIPAM hydrogel in either deionized water or dilute aqueous ANS solution. It implies no apparent influence of dilute aqueous ANS solution on the hydrophilic/hydrophobic balance of neutral PNIPAM polymeric network. In regard to P(NIPAM-*co*-ACD) hydrogel, both deionized water and dilute aqueous ANS solution could give rise to a conspicuous shift in VPTT. It displays an uplifted VPTT which results from the incorporation of ACD with a hydrophilic external surface. In the case of dilute aqueous ANS solution, the VPTT of P(NIPAM-*co*-ACD) hydrogel descends from nearly 35 °C observed in deionized

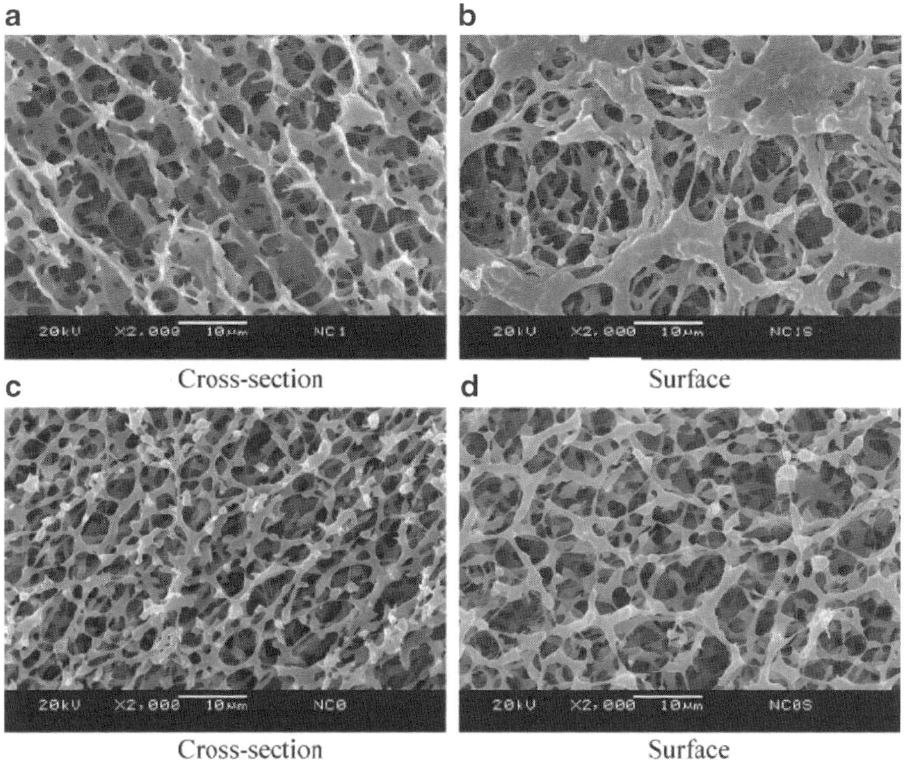

Fig. 15.5 SEM images of cross sections and surfaces of P(NIPAM-*co*-ACD) (**a**, **b**) and PNIPAM hydrogels (**c**, **d**) (Reproduced with permission from Ref. [17], Copyright (2007), American Chemical Society)

water to around 30 °C. It indicates an augment of hydrophobicity of hydrogel polymeric network. It is the association of β-cyclodextrin and guest molecule ANS that accounts for the VPTT descent. The thermo-responsive behavior of the P(NIPAM-*co*-ACD) hydrogel described above indicates that β-cyclodextrin, as well as its complexation with guest molecule ANS, plays an important role in the hydrophilic/hydrophobic balance of hydrogel in the solution, which in turn affects the VPTT behavior.

DSC thermograms of hydrogels under investigation are schematically illustrated in Fig. 15.6 to exhibit the endothermic behavior of hydrogels on heating, which are in agreement with the ESR results. P(NIPAM-*co*-ACD) hydrogel shows increased and decreased VPTTs in deionized water (around 34.4 °C) and dilute aqueous ANS solution (around 30.2 °C), respectively. However, the VPTT of PNIPAM hydrogel nearly remains the same, i.e., 32.6 °C in deionized water and 32.2 °C in dilute aqueous ANS solution. The VPTT shift of P(NIPAM-*co*-ACD) hydrogel is attributable to the introduction of ACD and host-guest association.

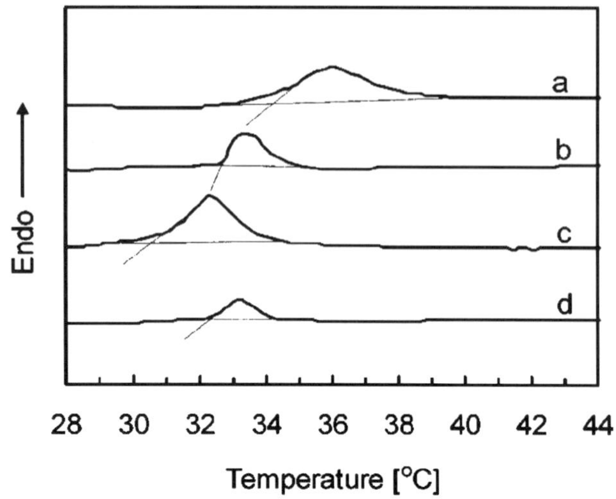

Fig. 15.6 DSC thermograms of P(NIPAM-*co*-ACD) hydrogel in deionized water (**a**), PNIPAM hydrogel in deionized water (**b**), P(NIPAM-*co*-ACD) hydrogel in dilute aqueous ANS solution (1.0 mM) (**c**), and PNIPAM hydrogel in dilute aqueous ANS solution (1.0 mM) (**d**) (Reproduced with permission from Ref. [17], Copyright (2007), American Chemical Society)

15.3.3 Temperature Dependence of Affinity Behavior and Temperature-Induced Shift of Association Sites

The association constant (K_s) of P(NIPAM-*co*-ACD) and PNIPAM hydrogels with ANS at each given temperature is determined using the Benesi-Hildebrand equation [18] through UV spectrophotometry. With regard to P(NIPAM-*co*-ACD) hydrogel (Table 15.1), the association constant K_s is relatively low at lower temperatures (below VPTT), and rise dramatically when temperature increases above 35 °C. As for PNIPAM hydrogel, the association constant K_s is difficult to be accurately determined at low temperatures due to rather weak interaction between hydrogel and ANS molecule. However, similar to P(NIPAM-*co*-ACD) hydrogel, a marked increase in K_s for PNIPAM hydrogel is observed when temperature rises above VPTT. As temperature goes up from below VPTT to above, the polymeric network of PNIPAM undergoes a coil-to-globule transition [19–23], shifting from a hydrophilic state to a hydrophobic state. Therefore, in view of the hydrophobicity of the guest molecule ANS, we make a hypothesis that the hydrophilic/hydrophobic balance of polymeric network of the hydrogel plays an important role in the affinity behavior of P(NIPAM-*co*-ACD) and PNIPAM hydrogels. Thereby, a satisfactory elucidation might be reached. Accordingly, it is necessary to ascertain the temperature dependence of the association constants of PNIPAM polymer and β-cyclodextrin toward ANS, respectively.

Using hydrophobic molecule ANS as fluorescence probe, which is highly sensitive to environmental changes, the variations in fluorescence intensity of ANS in the presence of PNIPAM and natural β-cyclodextrin at 20 °C and 45 °C during the fluorimetric titration experiments are schematically illustrated in Figs. 15.7 and 15.8, respectively.

Table 15.1 Association constants for P(NIPAM-*co*-ACD) and PNIPAM hydrogels with ANS at different temperatures[a] (Reproduced with permission from Ref. [17], Copyright (2007), American Chemical Society)

	P(NIPAM-*co*-ACD) hydrogel		PNIPAM hydrogel	
T (°C)	Ks (M^{-1})	R^2	Ks (M^{-1})	R^2
21	10,350	0.9915	[b]	
28	8,123	0.9943	[b]	
30	10,100	0.9951	440	0.9889
35	39,680	0.9852	507	0.9807
40	39,630	0.9743	822	0.9714
50	53,800[c]	0.9498	2,650[c]	0.9403
60	55,000[c]	0.9367	3,214[c]	0.9278

[a]Measured by UV spectrophotometry

[b]No apparent UV variations observed to determine K_s accurately

[c]K_s cannot be accurately determined due to relatively poor correlation coefficient

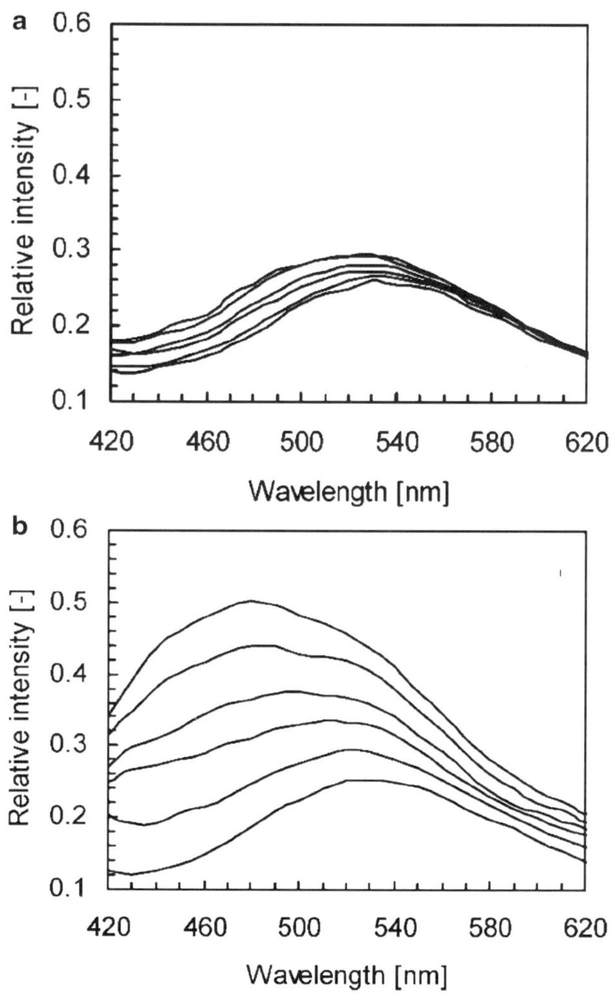

Fig. 15.7 Fluorescence spectral changes of ANS (15 μM) upon the regular addition of PNIPAM polymer at (**a**) 20 °C and (**b**) 45 °C. [PNIPAM polymer]: 0, 300, 450, 600, 750, and 900 (μM, from the *bottom* to *top*). Excitation wavelength: 350 nm (Reproduced with permission from Ref. [17], Copyright (2007), American Chemical Society)

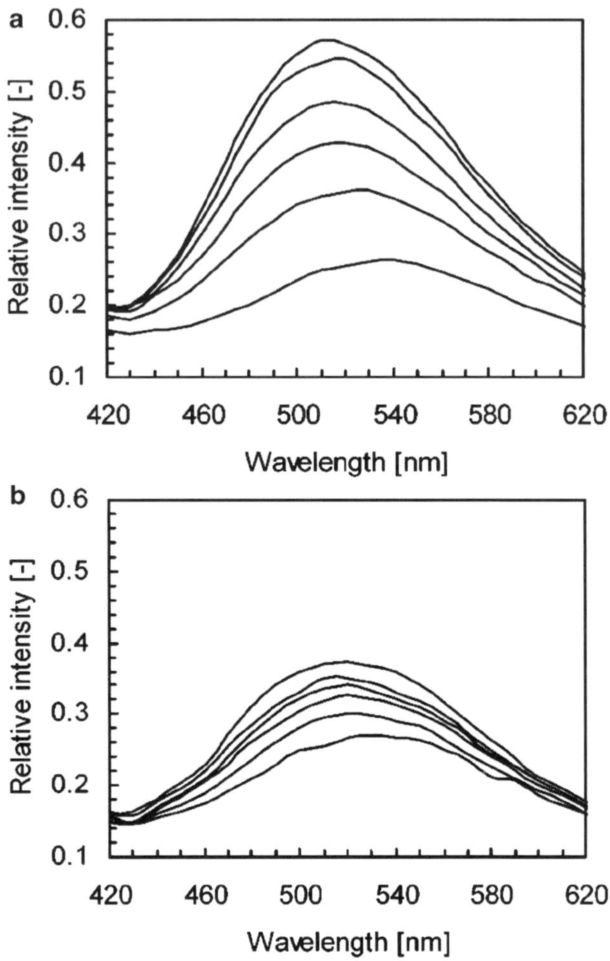

Fig. 15.8 Fluorescence spectral changes of ANS (15 μM) upon the regular addition of *β*-cyclodextrin at (**a**) 20 °C and (**b**) 45 °C. [*β*-cyclodextrin]: 0, 300, 450, 600, 750, and 900 (μM, from the *bottom* to *top*). Excitation wavelength: 350 nm (Reproduced with permission from Ref. [17], Copyright (2007), American Chemical Society)

As shown in Fig. 15.7, the increase in fluorescence intensity of ANS upon the regular addition of PNIPAM aqueous solution is nearly undetectable at 20 °C but is considerably enhanced with a blue shift by 40 nm at 45 °C. It indicates a sharp hydrophilic/hydrophobic transition with a marked temperature increase across VPTT. In contrast, Fig. 15.8 exhibits that the fluorescence intensity of ANS in the presence of *β*-cyclodextrin is much higher at 20 °C compared with that observed at 45 °C. It suggests that much more ANS molecules reside in the hydrophobic cavities of *β*-cyclodextrins at 20 °C than at 45 °C. Though utterly opposite in tendency, they are identical in themselves: high fluorescence intensity implies a more hydrophobic environment and vice versa. As can be seen from Table 15.2, the association constants for PNIPAM/ANS cannot be accurately calculated at temperatures below VPTT due to no apparent spectral variations in fluorimetric titration experiments. However, when temperature increases above VPTT, the association constant keeps

Table 15.2 Association constants for PNIPAM linear polymer and β-cyclodextrin with ANS at different temperatures[a] (Reproduced with permission from Ref. [17], Copyright (2007), American Chemical Society)

	PNIPAM polymer		β-cyclodextrin	
T (°C)	Ks (M^{-1})	R^2	Ks (M^{-1})	R^2
20	[b]		102	0.9914
28	[b]		112	0.9956
35	35	0.9983	108	0.9996
45	84	0.9976	86	0.9930
55	126	0.9998	61	0.9959

[a]Measured by fluorimetry
[b]No apparent UV spectrum changes observed to determine Ks accurately

increasing noticeably. At temperatures below the VPTT, the hydrophilic PNIPAM has little influence on the fluorescence intensity of probe ANS, while hydrophobic PNIPAM at high temperatures greatly impacts the surroundings of ANS, leading to a considerable increase in fluorescence intensity. The association constants of β-cyclodextrin/ANS maintain a decrease with a temperature increase due to the fact that the β-cyclodextrin/ANS complex is more thermodynamically stable at low temperatures than at high temperatures.

Considering the data of association constants lists in Table 15.2, the hypothesis made previously holds that the hydrophilic/hydrophobic balance of polymeric network of the hydrogel plays an important role in the affinity behavior. At temperatures below the VPTT, the network of P(NIPAM-*co*-ACD) and PNIPAM hydrogels is highly hydrophilic. Therefore, the inclusion toward guest ANS is mostly attributed to β-cyclodextrin with respect to P(NIPAM-*co*-ACD) hydrogel. As temperature goes up across VPTT, the increase in the hydrophobicity of PNIPAM network gives rise to a much more enhanced affinity toward ANS. Meanwhile, with regard to β-cyclodextrin, there is a decrease in association ability with ANS. Thus, at temperatures above VPTT, the enhanced affinity toward ANS is attributable to the combination of host-guest complexation and hydrophobic interaction.

Along with K_s listed in Table 15.1 is the correlation coefficient of linear regression of experimental data R^2. As for P(NIPAM-*co*-ACD) hydrogel, it is noteworthy that an excellent curve fitting with correlation coefficient greater than 0.99 is observed at temperatures below VPTT, compared with a relatively poor fitting when temperature rises above VPTT. The derivation process of the Benesi-Hildebrand equation is based on the 1:1 stoichiometry of host-guest complexation. In our work, at low temperatures, β-cyclodextrin makes the greatest contribution to the inclusion of ANS and makes a 1:1 complexation. A 1:1 stoichiometry is evidenced by excellent linearity shown in Table 15.2, which meets the precondition of the Benesi-Hildebrand equation quite well. Accordingly, a desirable curve fitting is acquired. At high temperatures, hydrophobicity is predominant, and thus, a combination of host-guest association as well as hydrophobic interaction accounts for the increased affinity toward ANS. Therefore, there is a shift of association sites from original β-cyclodextrin to a combination of β-cyclodextrin and PNIPAM polymeric network, which accordingly leads to a less desirable curve fitting. As the temperature increases above 50 °C, a further decrease in association ability of β-cyclodextrin with ANS and a further considerable increase in hydrophobicity of PNIPAM network eventually result in a poor curve fitting. Thus, accurate association constants are not available.

15.4 Summary

In summary, the molecular-recognizable hydrogel functional materials based on PNIPAM and CD have been successfully synthesized. Both the PNIPAM polymers and hydrogels with pendent CD moieties are featured with thermo-responsive characteristics and molecular-recognizable characteristics. In regard to the PNIPAM copolymers with different lengths of monosubstituted carbon chain of modified CD (PNG-ECD and PNG-HCD), both of them can perfectly recognize the guest ANS molecules and NS molecules. However, the two guest molecules have opposite influence on the LCSTs of PNG-ECD and PNG-HCD polymers. With an increase in the concentration of ANS solution, the LCST values of PNG-ECD and PNG-HCD polymers both decrease, and the shift of the LCST values is larger for higher ANS concentrations. However, the LCST values of both PNG-ECD and PNG-HCD increase by a certain degree on adding NS molecules to the aqueous solutions. Moreover, the quantitative research on the thermo-responsive affinity behavior of P(NIPAM-*co*-ACD) hydrogel toward ANS reveals that a considerable increase in the affinity of P(NIPAM-*co*-ACD) hydrogel is attributable to a drastic increase in the hydrophobicity of hydrogel network with temperature rising across the VPTT. Meanwhile, the association site is demonstrated to shift from original β-cyclodextrin at low temperatures to a combination of β-cyclodextrin and the PNIPAM network at high temperatures. It is noteworthy to draw attention to an upshift in the VPTT of P(NIPAM-*co*-ACD) hydrogel in deionized water and a downshift in dilute aqueous ANS solution. The former is ascribed to the result of the incorporation of β-cyclodextrin onto the hydrogel network, and the latter is ascribed to the complexation of β-cyclodextrin and ANS. The synthesized PNIPAM polymers and hydrogels with pendent CD moieties have the potential to be applied to the engineering of molecular-recognition sensors and switches and also the development of temperature-controlled affinity separation system.

References

1. Murakami H, Kawabuchi A, Kotoo K et al (1997) A light-driven molecular shuttle based on a rotaxane. J Am Chem Soc 119:7605–7606
2. Breslow R, Zhang BL (1992) Very fast ester hydrolysis by a cyclodextrin dimer with a catalytic linking group. J Am Chem Soc 114:5882–5883
3. Stella VJ, Rajewski RA (1997) Cyclodextrins: their future in drug formulation and delivery. Pharm Res 14:556–567

4. Hirasawa T, Maeda Y, Kitano H (1998) Inclusional complexation by cyclodextrin-polymer conjugates in organic solvents. Macromolecules 31:4480–4485
5. Krieg HM, Lotter J, Keizer K et al (2000) Enrichment of chlorthalidone enantiomers by an aqueous bulk liquid membrane containing β-cyclodextrin. J Membr Sci 167:33–45
6. Nozaki T, Maeda Y, Ito K et al (1995) Cyclodextrins modified with polymer chains which are responsive to external stimuli. Macromolecules 28:522–524
7. Nozaki T, Maeda Y, Kitano H (1997) Cyclodextrin gels which have a temperature responsiveness. J Polym Sci A Polym Chem 35:1535–1541
8. Liu YY, Fan XD, Hu H et al (2004) Release of chlorambucil from poly(*N*-isopropylacrylamide) hydrogels with *β*-cyclodextrin moieties. Macromol Biosci 4:729–736
9. Liu YY, Fan XD (2002) Synthesis and characterization of pH- and temperature-sensitive hydrogel of *N*-isopropylacrylamide/cyclodextrin based copolymer. Polymer 43:4997–5003
10. Wei HL, Yu HQ, Zhang AY et al (2005) Synthesis and characterization of thermosensitive and supramolecular structured hydrogels. Macromolecules 38:8833–8839
11. Zhang JT, Huang SW, Gao FZ et al (2005) Novel temperature-sensitive, β-cyclodextrin-incorporated poly(*N*-isopropylacrylamide) hydrogels for slow release of drug. Colloid Polym Sci 283:461–464
12. Yanagioka M, Kurita H, Yamaguchi T et al (2003) Development of a molecular recognition separation membrane using cyclodextrin complexation controlled by thermosensitive polymer chains. Ind Eng Chem Res 42:380–385
13. Ohashi H, Hiraoka Y, Yamaguchi T (2006) An autonomous phase transition-complexation/decomplexation polymer system with a molecular recognition property. Macromolecules 39:2614–2620
14. Liu YY, Fan XD, Zhao YB (2005) Synthesis and characterization of a poly(*N*-isopropylacrylamide) with β-cyclodextrin as pendant groups. J Polym Sci A Polym Chem 43:3516–3524
15. Liu YY, Fan XD, Gao L (2003) Synthesis and characterization of *β*-cyclodextrin based functional monomers and its copolymers with *N*-isopropylacrylamide. Macromol Biosci 3:715–719
16. Yang M, Chu LY, Xie R et al (2008) Molecular-recognition induced phase-transition of two thermo-responsive polymers with pendent beta-cyclodextrin groups. Macromol Chem Phys 209:204–211
17. Wang HD, Chu LY, Yu XQ et al (2007) Thermo-sensitive affinity behavior of poly(*N*-isopropylacrylamide) hydrogels with *β*-cyclodextrin moieties. Ind Eng Chem Res 46:1511–1518
18. Benesi HA, Hildebrand JH (1949) A spectrophotometric investigation of the interaction of iodine with aromatic hydrocarbons. J Am Chem Soc 71:2703–2707
19. Maeda Y, Higuchi T, Ikeda I (2001) FTIR spectroscopic and calorimetric studies of the phase transitions of *N*-isopropylacrylamide copolymers in water. Langmuir 17:7535–7539
20. Maeda Y, Higuchi T, Ikeda I (2000) Change in hydration state during the coil-globule transition of aqueous solutions of poly(*N*-isopropylacrylamide) as evidenced by FTIR spectroscopy. Langmuir 16:7503–7509
21. Shan J, Chen J, Nuopponen M et al (2004) Two phase transitions of poly(*N*-isopropylacrylamide) brushes bound to gold nanoparticles. Langmuir 20:4671–4676
22. Ramon O, Kesselman E, Berkovici R et al (2001) Attenuated total reflectance/Fourier transform infrared studies on the phase separation process of aqueous solutions of poly(*N*-isopropylacrylamide). J Polym Sci B Polym Phys 39:1665–1677
23. Tiktopulo EI, Uversky VN, Lushchik VB et al (1995) Domain coil-globule transition in homopolymers. Macromolecules 28:7519–7524

Chapter 16
Functional Membranes with Molecular-Recognizable Properties

Abstract In this chapter, the design, fabrication, and performance of smart functional membranes with molecular-recognizable properties are introduced. The membranes are prepared by suspending β-cyclodextrin (CD) host molecules onto the freely mobile ends of thermo-responsive poly(*N*-isopropylacrylamide) (PNIPAM) chains grafted in the membrane pores. By virtue of the interaction between the lower critical solution temperature (LCST) of PNIPAM chains and association constant of CD toward guest molecules, the membranes based on PNIPAM and CD are demonstrated to chiral separate enantiomers, adsorb/desorb guest molecules, and controlled-release model drug molecules. During the chiral separation, the PNG-ECD-grafted membranes separate *D,L*-tryptophan enantiomers with a high selectivity and regenerate with a high decomplexation ratio by simply changing the operation temperature. By simply adjusting the operation temperature, the membranes adsorb guest molecules at lower temperature and desorb at higher temperature. Moreover, the membranes with the same functional gate are featured with triple stimuli-responsive gating functions and release the model drug molecules in a controlled manner. The as-prepared membranes based on PNIPAM and CD are highly attractive for chiral separation, affinity separation, and controlled-release systems responding to the environmental temperature and guest molecules.

16.1 Introduction

Recently, a novel molecular-recognizable membrane is prepared by suspending molecular-recognizable host molecules onto the freely mobile ends of environmental-stimuli-responsive polymers grafted in the membrane pores [1–6]. The molecular-recognizable host molecules could recognize special molecules or ions and thus act as "sensors," while linear-grafted stimuli-responsive polymers could change its configuration after "sensors" recognize molecules or

L.-Y. Chu et al., *Smart Hydrogel Functional Materials*,
DOI 10.1007/978-3-642-39538-3_16,

© Chemical Industry Press, Beijing and Springer Berlin Heidelberg 2013

ions and thus act as "actuators." The cooperation of "sensors" and "actuators" realizes the switching function of molecular-recognizable gating membranes. The molecular-recognizable host molecules suspended on polymer chains are usually macromolecules such as crown ether and cyclodextrin.

It has been reported that the association constants of β-cyclodextrin (CD) with guest molecules are influenced dramatically by temperature variation across the lower critical solution temperature (LCST) [7, 8]. By taking advantage of such a property, previous investigations on temperature-dependent molecular-recognizable membranes based on PNIPAM and CD have been focused on the performances of autonomously adsorbing and desorbing guest molecules at temperatures across the LCST [6, 9]. Moreover, in our previous study (Chap. 15), the thermo-responsive and molecular-recognizable characteristics of PNIPAM polymers and hydrogels with pendent CD moieties, as well as their thermo-responsive affinity toward guest molecules, are investigated systematically. The results clarify that the categories and concentrations of guest molecules have great effect on the molecular-recognition-induced phase transition of PNIPAM polymers with pendent CD moieties. Moreover, the hydrophilic/hydrophobic balance of polymeric network of the hydrogel based on PNIPAM and CD plays an important role in the affinity behavior. By virtue of PNIPAM polymers with CD moieties as smart gates in the membrane pores, the proposed membranes have great potential to be used in thermo-responsive chiral separation, thermo-responsive affinity separation, and thermo-responsive and molecular-recognition-induced controlled release.

The proposed membranes based on PNIPAM and CD are prepared by grafting poly(*N*-isopropylacrylamide-*co*-glycidyl methacrylate/mono-6-deoxy-6-ethylene diamino-β-cyclodextrin) (PNG-ECD) with the combination of plasma-induced pore-filling graft polymerization and chemical reaction (Fig. 16.1). First, poly(*N*-isopropylacrylamide-*co*-glycidyl methacrylate) (PNG) linear chains are grafted onto porous substrate membrane via plasma-induced pore-filling graft polymerization. The substrate membranes are either nylon-6 (N6) membranes or polyethylene terephthalate (PET) track-etched membranes. Subsequently, the mono-6-deoxy-6-ethylene diamino-β-cyclodextrin (ECD) is appended onto the PNG polymer chains by the reaction between the epoxy groups of PNG chains on the membrane and the primary amino groups on the CD.

16.2 Molecular-Recognizable and Thermo-responsive Membranes for Chiral Resolution

The PNG-ECD-grafted N6 membranes are demonstrated to achieve the thermo-responsive chiral separation with a simple and efficient process for membrane regeneration [10]. The proposed PNG-ECD-grafted membranes simultaneously exhibit chiral selectivity, based on the molecular recognition

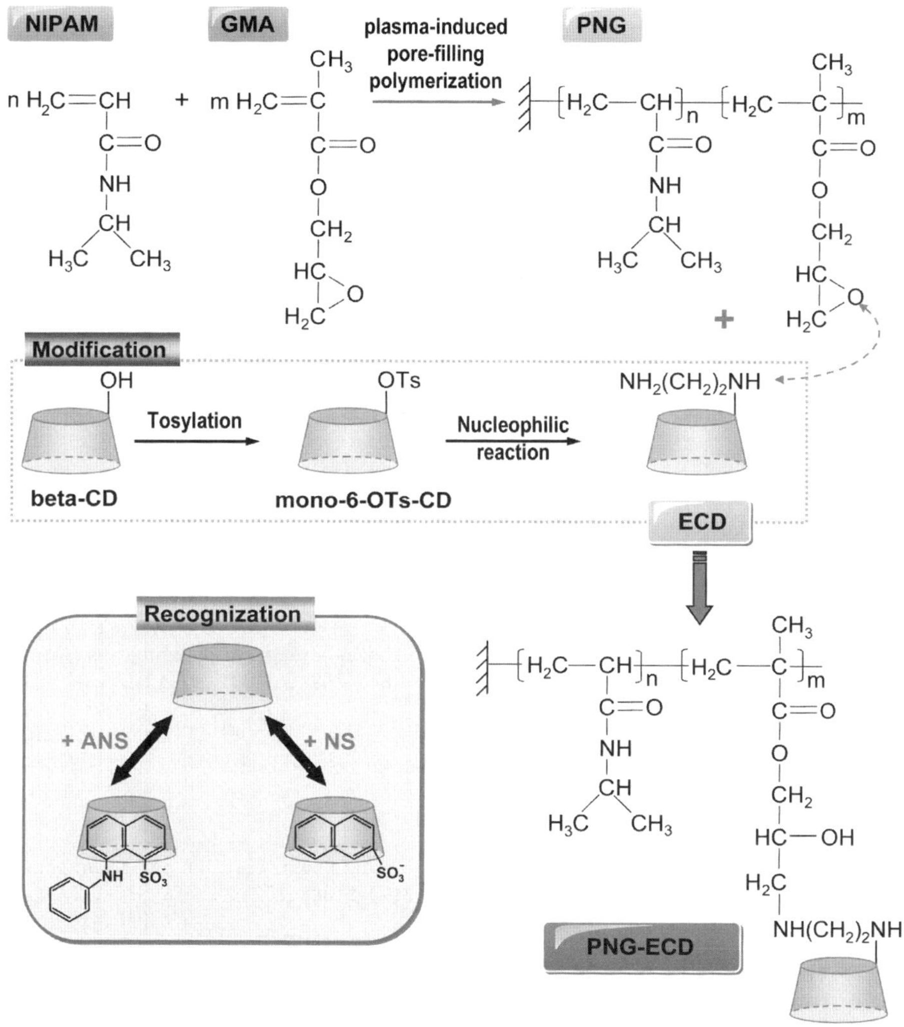

Fig. 16.1 The synthesis route of molecular-recognizable and thermo-responsive membranes based on PNIPAM and CD

of CD, and thermo-responsiveness, based on the phase transition of poly(*N*-isopropylacrylamide) (PNIPAM). It has been reported that the phase transition of PNIPAM responding to temperature change could affect the association constant between CD and guest molecules in a PNIPAM/CD polymer system [6–8, 11, 12]. Therefore, by simply changing the operation temperature, chiral resolution with high selectivity and efficient membrane regeneration are achieved (Fig. 16.2). At a temperature below the LCST, the association constant of CD toward guest molecules is large. As a result, one of the two enantiomers is captured by CD

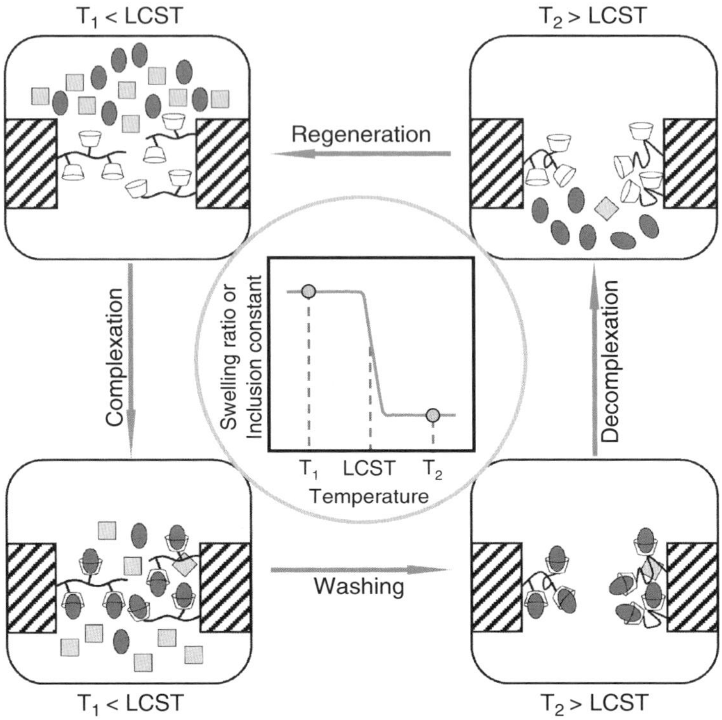

Fig. 16.2 Schematic illustration of the proposed molecular-recognizable and thermo-responsive membranes for chiral resolution and membrane regeneration (Reproduced with permission from Ref. [10], Copyright (2008), Wiley-VCH Verlag GmbH & Co. KGaA)

molecules because of chiral recognition during the permeation of racemates, while the other is permeated. When the complexation between CD and the captured enantiomers in the membrane reaches equilibrium, a wash process is carried out to remove the free molecules. However, the captured molecules desorb from the CD cavities due to the decreased association constant at the operation temperature increased above the LCST. Compared with currently existing affinity-membrane processes, the proposed membrane, especially the decomplexation process, is completely environment friendly and can be easily operated.

By the virtue of scanning electron microscopy (SEM), X-ray photoelectron spectroscopy (XPS), and contact-angle measurements, the PNG-ECD chains are confirmed to be grafted on both the outer membrane surfaces and the inner surfaces of the membrane pores. Chiral resolution and decomplexation are carried out to investigate the selectivity and decomplexation ratio of the grafted membranes. Effects of the operation temperature and grafting yield of membranes on the aforementioned performance are also studied.

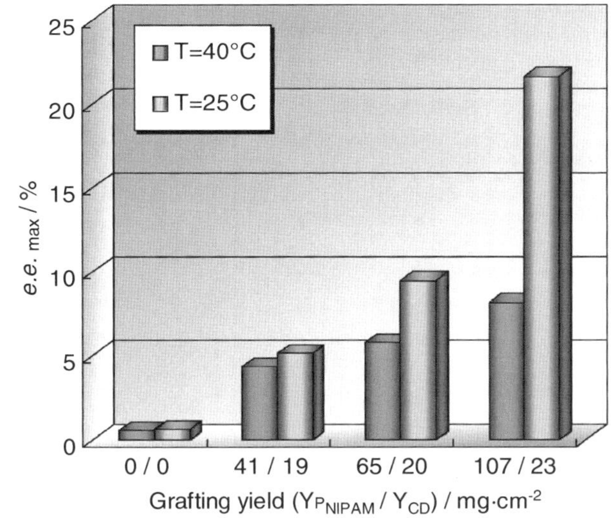

Fig. 16.3 Effect of grafting yields on the maximal *e.e.* value of *D,L*-tryptophan chiral resolution through the membranes (Reproduced with permission from Ref. [10], Copyright (2008), Wiley-VCH Verlag GmbH & Co. KGaA)

16.2.1 *Effects of Grafting Yield and Operation Temperature on Chiral Resolution of* D,L-*Tryptophan Through Grafted Membranes*

For all the PNG-ECD-grafted N6 membranes with different grafting yields, the enantiomeric excess (*e.e.* given in %, Eq. 16.1) value during the resolution process has a peak at both 25 °C and 40 °C:

$$e.e. = \frac{A_{\mathrm{D}} - A_{\mathrm{L}}}{A_{\mathrm{D}} + A_{\mathrm{L}}} \times 100 \tag{16.1}$$

where A_{D} and A_{L} stand for the peak areas of *D*-enantiomer and *L*-enantiomer in the permeate, respectively.

The higher the grafting yield of CD on the membrane is, the larger the maximal *e.e.* value is, no matter at the environmental temperature of 25 °C or 40 °C (Fig. 16.3). If the grafting yield of CD on the membrane increases, the chiral selectivity is improved. However, once recognition sites on membranes are saturated by guest molecules, the selectivity would gradually decrease. That is, the peak is the critical point of saturation for recognition sites. By considering both permeability and selectivity, an optimum grafting yield in this study is obtained as $Y_{\mathrm{PNIPAM}} = 107$ mg·cm^{-2}/$Y_{\mathrm{CD}} = 23$ mg·cm^{-2} for subsequent investigations.

Moreover, the maximal *e.e.* values of the PNG-ECD-grafted N6 membranes with different grafting yields at 25 °C are always higher than those at 40 °C. When the environmental temperature is higher than the LCST of PNIPAM, the

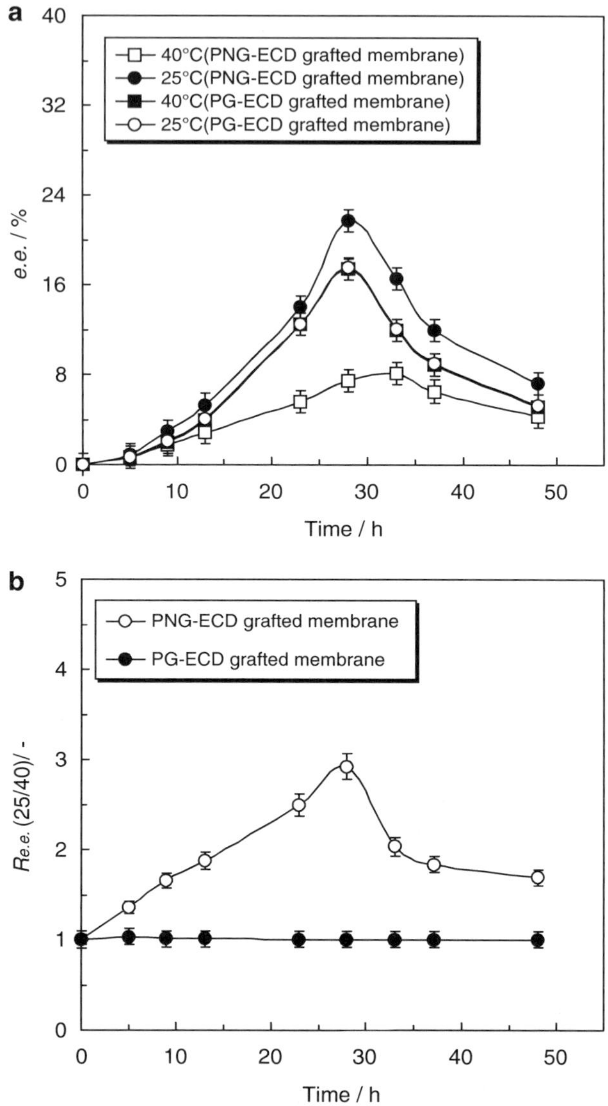

Fig. 16.4 Effect of temperature on the *e.e.* values of *D,L*-tryptophan chiral resolution through membranes. (**a**) *e.e.* values at 25 °C and 40 °C, (**b**) the thermo-responsive gating coefficient of grafted membranes (Reproduced with permission from Ref. [10], Copyright (2008), Wiley-VCH Verlag GmbH & Co. KGaA)

PNIPAM chains are in a shrunken state. Thus, the steric hindrance caused by the shrinking of the chains makes the association constant between CD and tryptophan molecules decrease; as a result, the chiral selectivity decreases. To clearly show the effect of operation temperature on chiral selectivity, the grafted membranes with and without PNIPAM is compared in Fig. 16.4. Both of PNG-ECD-grafted N6 membrane ($Y_{PNIPAM} = 107$ mg·cm^{-2}/$Y_{CD} = 23$ mg·cm^{-2}) and poly(glycidyl methacrylate/mono-6-deoxy-6-ethylene diamino-β-cyclodextrin)

(PG-ECD) grafted N6 membrane ($Y_{PNIPAM} = 0$ mg·cm^{-2}/$Y_{CD} = 23$ mg·cm^{-2}) have a chiral selectivity for *D,L*-tryptophan with a peak in the *e.e.* value (Fig. 16.4a). The thermo-responsive gating coefficient $R_{e.e.(25/40)}$ is defined as the ratio of *e.e.* values at environmental temperature of 25 °C to that at 40 °C. For the PG-ECD-grafted membrane, there is very little effect of environmental temperature on chiral selectivity since the chiral selectivity at 40 °C is almost the same as that at 25 °C ($R_{e.e.(25/40)}$ is 1.0). However, for the PNG-ECD-grafted membrane, the difference between the chiral selectivity at 25 °C and 40 °C is remarkable. The thermo-responsive coefficient $R_{e.e.(25/40)}$ is approximately 3 at 28 h (Fig. 16.4b). In conclusion, it is found that both the grafting yield and environmental temperature of PNG-ECD-grafted N6 membrane affect the chiral selectivity significantly.

16.2.2 Chiral Resolution Performance of D,L*-Tryptophan Through Grafted Membranes*

Figure 16.5 shows the chiral resolution performance of *D,L*-tryptophan through a PNG-ECD-grafted N6 membrane ($Y_{PNIPAM} = 107$ mg·cm^{-2}/$Y_{CD} = 23$ mg·cm^{-2}) at the temperature below and above the LCST. At both 25 °C and 40 °C, the concentration of *D*-tryptophan (*D*-Trp) and *L*-tryptophan (*L*-Trp) enantiomers in the receptor cell increases with time. The concentration of *D*-Trp is more than that of *L*-Trp because the CD molecule has a higher association constant with *L*-Trp in neutral environments [13]. Most of *L*-Trp enantiomers absorb in the membrane, while *D*-Trp enantiomers pass through the membrane. For the chiral resolution through a PNG-ECD-grafted membrane, the separation efficiency of two enantiomers is reduced with increasing temperature from 25 °C to 40 °C since the shrinking of PNIPAM chains at 40 °C hinders the association between CD and *L*-Trp enantiomers.

16.2.3 Decomplexation of Tryptophan Enantiomers from Grafted Membranes

The decomplexation performance of tryptophan enantiomers from PNG-ECD-grafted ($Y_{PNIPAM} = 107$ mg·cm^{-2}/$Y_{CD} = 23$ mg·cm^{-2}) and PG-ECD-grafted ($Y_{PNIPAM} = 0$ mg·cm^{-2}/$Y_{CD} = 23$ mg·cm^{-2}) N6 membranes is compared in Fig. 16.6. The decomplexation of tryptophan enantiomers from membranes after chiral resolution is carried out at 40 °C, which is above the LCST of PNIPAM. The absolute amounts of *L*-Trp and *D*-Trp released from the PNG-ECD-grafted membrane are much higher than those from the PG-ECD-grafted membrane. The decomplexation ratio η (see Eq. 16.2) of the PNG-ECD-grafted membrane is nearly 3 times as that of the PG-ECD-grafted membrane. The above-mentioned

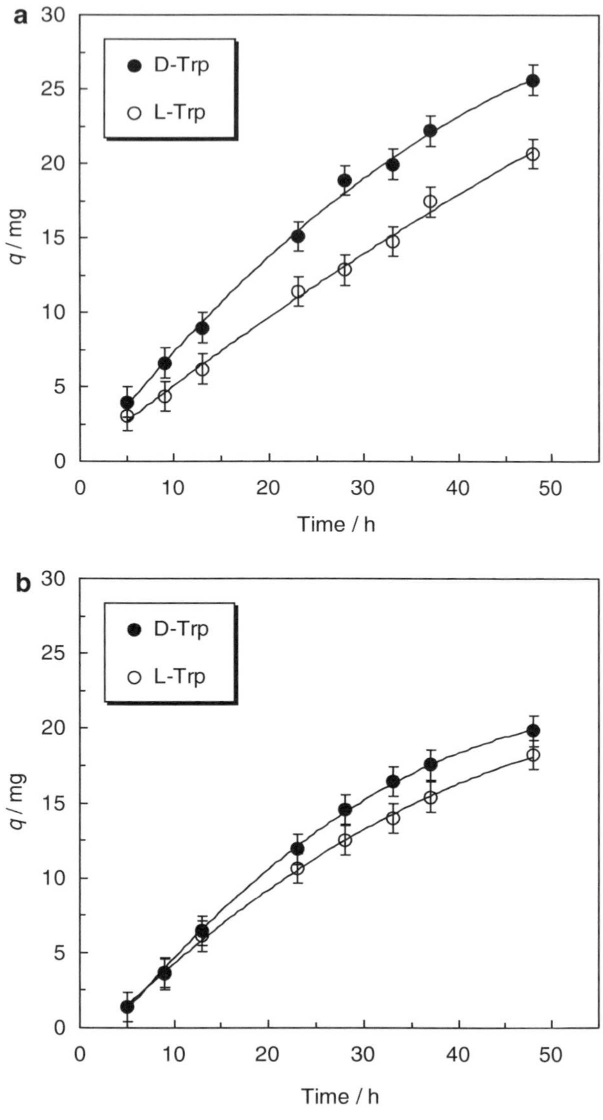

Fig. 16.5 Chiral resolution of *D,L*-tryptophan through PNG-ECD-grafted N6 membrane at 25 °C (**a**) and 40 °C (**b**) (Reproduced with permission from Ref. [10], Copyright (2008), Wiley-VCH Verlag GmbH & Co. KGaA)

results demonstrate the promising regeneration capability of the proposed thermo-responsive chiral membranes:

$$\eta = \frac{q}{q_0} \times 100 \tag{16.2}$$

where q and q_0 stand for the amount [mg] of *D,L*-tryptophan desorbed from a membrane and that adsorbed in the membrane before decomplexation, respectively.

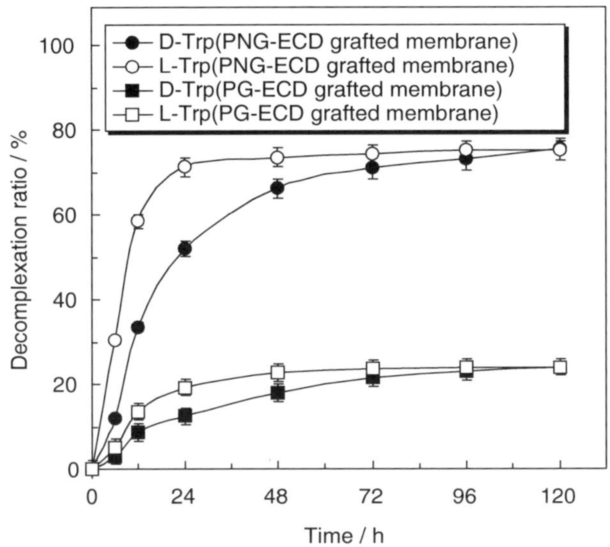

Fig. 16.6 Decomplexation ratio of tryptophan enantiomers from membranes after chiral resolution. The operation temperature for decomplexation is 40 °C (Reproduced with permission from Ref. [10], Copyright (2008), Wiley-VCH Verlag GmbH & Co. KGaA)

16.3 Molecular-Recognizable and Thermo-responsive Membranes for Affinity Separation

The combination of plasma-induced pore-filling graft polymerization and chemical reaction is also demonstrated to successfully graft PNG-ECD polymers on the PET track-etched membranes [14]. The PET membranes are featured with straight transmembrane pores and narrow size distribution. The fabricated molecular-recognizable and thermo-responsive membranes based on PNIPAM and CD are applied to affinity separation toward model guest molecule (i.e., 8-anilino-1-naphthalenesulfonic acid ammonium salt, ANS). By virtue of the effect of configuration change of PNG chains grafted on the membrane pores on the binding ability toward guest molecules, the thermo-responsive affinity separation process toward ANS is achieved (Fig. 16.7). ANS molecules are adsorbed on the PNG-ECD-grafted PET membrane at lower temperature while desorbed at higher temperature. SEM images display that the pores of the PNG-ECD-grafted PET membrane with grafting yield $Y_{PNIPAM} = 8.1\ \%$ and $Y_{CD} = 14.5\ \mu g{\cdot}cm^{-2}$ are filled uniformly and completely with polymer chains. Fourier transform infrared spectra (FT-IR) verify that ECD is immobilized on the grafted membranes via the chemical reaction between primary amino groups of ECD and epoxy groups of PNG chains.

As references, the adsorption ability of ungrafted, PNG grafted ($Y_{PNIPAM} = 6.21\ \%$), and PG-ECD grafted ($Y_{GMA} = 5.33\ \%$ and $Y_{CD} = 11.4\ \mu g{\cdot}cm^{-2}$) PET

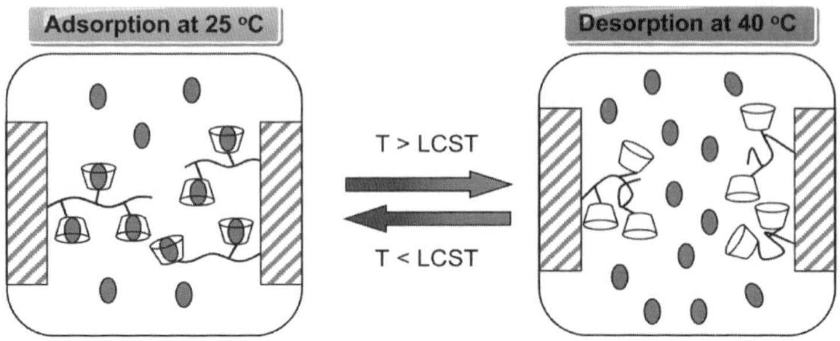

Fig. 16.7 Schematic illustration of thermo-responsive adsorption and desorption of ANS on the PNG-ECD-grafted PET membrane

membranes toward ANS is investigated first. At 25 °C, the unit adsorption amount of ANS (Q_m) on the three membranes increases at beginning of adsorption experiments and then stays the same even when the temperature increases above 40 °C. Moreover, the unit equilibrium adsorption amount Q'_m on the CD-grafted PG-ECD membrane is much higher (~3.5 μg·cm^{-2}) than the ungrafted and PNG grafted membranes (~0.5 μg·cm^{-2}), which is attributable to stronger recognition capability of CD toward ANS. The dynamic adsorption experiment on the PNG-ECD-grafted PET membrane ($Y_{PNIPAM} = 11.8$ % and $Y_{CD} = 4$ μg·cm^{-2}) toward ANS in three low-/high-temperature cycles shows that ANS molecules are adsorbed onto the membrane at 25 °C and desorbed at 40 °C (Fig. 16.8). Compared with the results of other three membranes mentioned above, it is evident that the molecular-recognizable and thermo-responsive characteristics of PNG-ECD-grafted PET membrane are attributable to the effect of configuration change of PNG chains on the binding ability of CD toward ANS at temperatures across the LCST. At temperatures below the LCST, the swollen PNG grafted chains provide enough space for stable complexation of CD and ANS, and much ANS molecules are adsorbed on the PNG-ECD-grafted PET membrane. However, at temperatures above the LCST, the shrunken polymer chains make the complexation unstable, and the adsorbed ANS molecules are desorbed from the membrane. The cooperation of the swelling-shrinking configuration change of PNG grafted chains and the stronger recognition of CD toward ANS endow the PNG-ECD-grafted PET membrane with remarkable molecular-recognizable and thermo-responsive characteristics. The equilibrium Q'_m values on the PNG-ECD-grafted membrane show good repeatability in the investigated three low-/high-temperature cycles (Fig. 16.8b).

In addition, as CD content grafted on the PNG chains increases, the difference of ANS adsorption amount on the PNG-ECD-grafted PET membrane between 25 °C and 40 °C becomes larger. The difference of equilibrium Q'_m values on the PNG-ECD-grafted PET membrane with $Y_{CD} = 4$ μg·cm^{-2} between 25 °C and 40 °C is around 0.7 μg·cm^{-2}. It is much smaller than that of PNG-ECD-grafted PET membrane with $Y_{CD} = 10.5$ μg·cm^{-2} (about 1.2 μg·cm^{-2}).

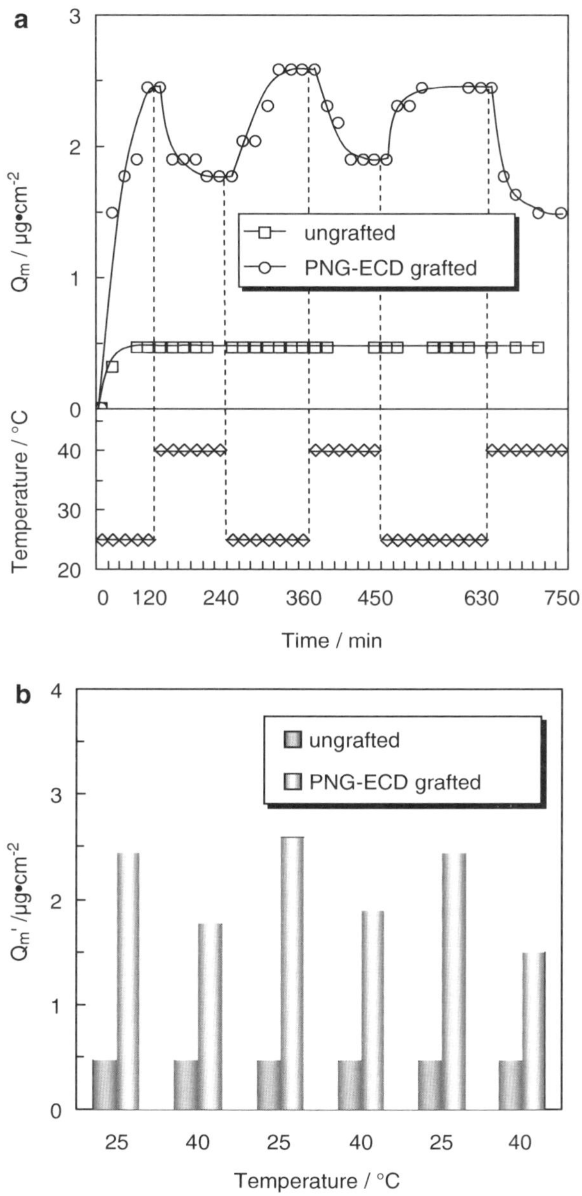

Fig. 16.8 Temperature-responsive adsorption of ANS on (**a**) ungrafted membrane and (**b**) PNG-ECD-grafted PET membranes ($Y_{PNIPAM} = 11.8$ %, $Y_{CD} = 4$ μg·cm^{-2}) in the three low-/high-temperature cycles (Reproduced with permission from Ref. [14], Copyright (2009), Elsevier)

16.4 Gating Characteristics of Thermo-responsive and Molecular-Recognizable Membranes

In our previous study, it is found that the different guest molecules have opposite effect on the LCST values of PNG-ECD polymers [15]. If the PNG-ECD polymers are grafted on the membrane, the grafted membrane may exhibit different gating

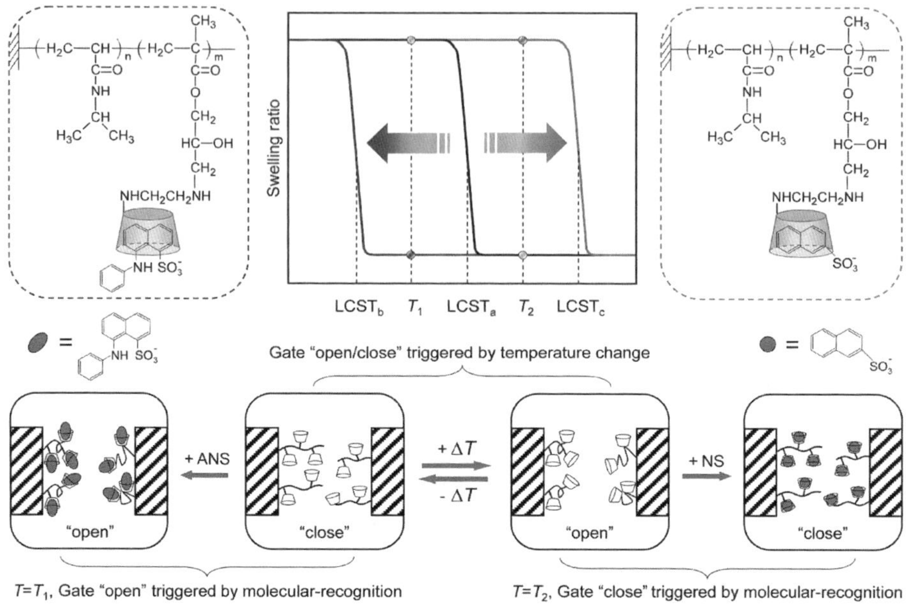

Fig. 16.9 Schematic illustration of the concept of the proposed thermo-responsive and molecular-recognizable gating membrane with the same functional gate exhibiting triple stimuli-responsive gating functions (Reproduced with permission from Ref. [16], Copyright (2010), Elsevier)

functions. In this section, the PNG-ECD-grafted N6 membranes are successfully fabricated by the aforementioned method [16]. The effect of the grafting yields of both PNIPAM and CD as well as the molar ratio of PNIPAM in the grafted chains on the thermo-responsive and molecular-recognizable gating characteristics of the as-prepared membranes is investigated systematically. Due to the thermo-responsive phase transition of PNIPAM and molecular recognition of CD, the as-designed membranes with the same functional gate can achieve triple stimuli-responsive gating functions (Fig. 16.9): (1) gate "open/close" function triggered by temperature change, (2) gate "open" function isothermally triggered by recognition of ANS guest molecules with a hydrophobic side group at T_1, and (3) gate "close" function isothermally triggered by recognition of NS guest molecules with a hydrophilic side group or without side group at T_2.

16.4.1 Thermo-responsive Diffusional Permeability and Gating Characteristics

Both the thermo-responsive and molecular-recognizable gating characteristics of the PNG-ECD-grafted N6 membranes with different grafting yields are systematically investigated. The diffusional permeability of vitamin B12 (VB12) molecules through membranes in different conditions with changing environmental temperatures and guest molecules is examined. Compared with the ungrafted membrane, the diffusional coefficient through the PNG-ECD-grafted membranes changes greatly when environmental temperature increases from 37 °C to 50 °C. The diffusion thermo-responsive gating coefficient $R_{D,H2O(50/37)}$ is defined as the ratio of diffusional coefficient of VB12 through membranes in pure water at 50 °C to that at 37 °C. The thermo-responsive gating coefficient of grafted membranes increases at first due to the thermo-responsive swelling-shrinking configuration change of the grafted PNG-ECD gates. However, the length and density of the grafted PNG-CD chains are larger and larger, while the grafting yield is increasing, resulting in the pores being finally choked by the over-grafted polymers. The diffusion thermo-responsive gating coefficient $R_{D,H2O(50/37)}$ reaches to the maximal value 3.0 when the grafting yield is $Y_{PNIPAM} = 33$ mg·cm^{-2}/$Y_{CD} = 10$ mg·cm^{-2}.

16.4.2 Molecular-Recognizable Diffusional Permeability and Gating Characteristics

The ANS-recognition gating coefficient (at 37 °C) $R_{D,ANS/H2O(37)}$ of grafted membranes is defined as the ratio of diffusional coefficient of VB12 passing through membranes at 37 °C in 1.0 mM ANS solution to that in pure water. The ANS-recognition gating coefficient $R_{D,ANS/H2O(37)}$ of the ungrafted membrane is slightly smaller than 1.0 because the adsorbed ANS molecules slightly hinder the diffusion of VB12 molecules. However, the ANS-recognition gating coefficient $R_{D,ANS/H2O(37)}$ of PNG-ECD membranes with $Y_{PNIPAM} = 33$ mg·cm^{-2}/$Y_{CD} = 10$ mg·cm^{-2} increases to as high as 1.6 (Fig. 16.10a). The significant increase of $R_{D,ANS/H2O(37)}$ value demonstrates that the membrane pores switch from "close" state to "open" state to some extent. It results from the ANS-recognition shrinking conformational change of the polymer chains grafted in membrane pores. When ANS guest molecules are present, CD moieties recognize and form complexes. Therefore, the LCST of grafted copolymers in ANS aqueous solution shifts to a lower value ($LCST_b$) due to the hydrophobic phenyl group of ANS out of the CD cavity [12, 15]. Because of molecular recognition of ANS, the phase transition of the grafted chains can occur isothermally at a temperature between the $LCST_a$ and $LCST_b$ (e.g., $T_1 = 37$ °C). The membrane gates switch from "close" state in water to "open" state in ANS aqueous solution isothermally at 37 °C.

The NS-recognition gating coefficient (at 50 °C) in Fig. 16.10b, $R_{D,NS/H2O(50)}$, is defined as the ratio of diffusional coefficient of VB12 passing through membranes at 50 °C in 1.0 mM NS solution to that in pure water. Compared with the $R_{D,NS/H2O(50)}$ of ungrafted membrane which is slightly smaller than 1.0, that of the PNG-ECD-grafted membrane with $Y_{PNIPAM} = 33$ mg·cm^{-2}/$Y_{CD} = 10$ mg·cm^{-2} reduces to 0.85. The significant decrease of the NS-recognition gating coefficient $R_{D,NS/H2O(50)}$ indicates that the membrane pores switch from "open" state to "close" state to some extent. It is because the NS-recognition swelling conformational

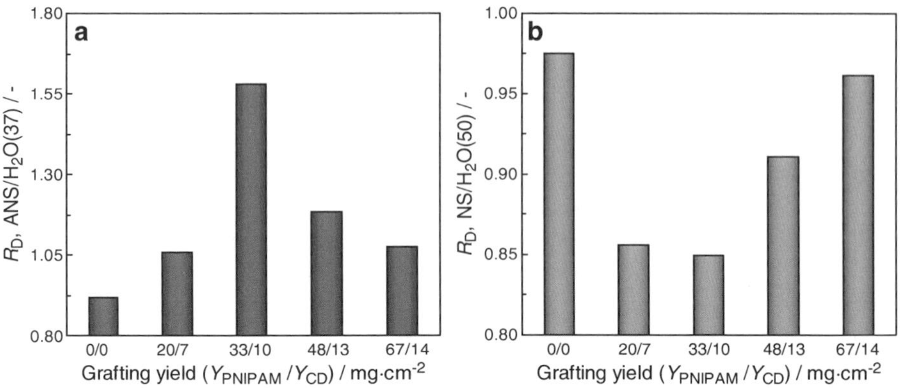

Fig. 16.10 Gating characteristics of membranes in terms of diffusional permeability of solute VB12 molecules passing through PNG-ECD-grafted N6 membranes with different grafting yields: (**a**) ANS-recognition gating coefficient. (**b**) NS-recognition gating coefficient (Reproduced with permission from Ref. [16], Copyright (2010), Elsevier)

change of the grafted polymer chains in membrane pores. When NS molecules exist in the surrounding solution, CD/NS complexes form, and thus the LCST of grafted copolymers shifts to a higher temperature (LCST$_c$). This is attributable to the complexation between CD and NS, which slightly enlarges the hydrophilic moiety of copolymers due to the dissociation of NS molecules in the water [15]. Therefore, the molecular-recognition ability of NS leads to the isothermal phase transition of grafted chains at a temperature between the LCST$_a$ and LCST$_c$ (e.g., $T_2 = 50$ °C). The membrane gates switch from "open" state in water to "close" state in NS aqueous solution isothermally at 50 °C.

The amount of both CD/ANS and CD/NS complexes increases with the increasing of grafting yield of PNIPAM chains on PNG-ECD-grafted membranes, due to more and more ECD attached on PNG chains when the grafting yields increase. As a result, the membrane pores open or close to a larger extent, and the ANS-recognition gating coefficient increases while NS-recognition gating coefficient decreases. There is an optimum grafting yield (e.g., $Y_{PNIPAM} = 33$ mg·cm^{-2}/$Y_{CD} = 10$ mg·cm^{-2}) for obtaining an optimum ANS-recognition gating coefficient and NS-recognition gating coefficient.

16.4.3 Adjustment of Thermo-responsive and Molecular-Recognizable Gating Characteristics

As the actuator for the functional gate, the molar ratio of PNIPAM in the grafted gates may adjust the gating characteristics of the as-prepared membranes. A series of membranes with similar amount of CD but different molar ratios of NIPAM to GMA (α) in the grafted chains are prepared. The effect of α value on thermo-responsive and molecular-recognizable gating characteristics is investigated experimentally. The diffusion coefficients of VB12 passing through grafted membranes at 50 °C are much higher than that at 37 °C. The diffusion thermo-responsive gating coefficient $R_{D,H2O(50/37)}$ increases with increasing the α value. When the α value increases, i.e., the PNIPAM component in grafted chains increases, as a result, the thermo-responsive swelling-shrinking volume phase transition of the grafted PNG-ECD gates is more significant. When the α value is 8, the thermo-responsive gating coefficient is improved to be about 5.0 compared with 3.0 at $\alpha = 1$. In addition, the α value also affects the characteristics of ANS-recognition gating significantly. When the α value is 4, the ANS-recognition gating coefficient is improved to be about 2.0 compared with 1.58 at $\alpha = 1$. The molecular-recognizable gating function of membranes with similar amount of CD but different molar ratios is achieved by the cooperation between PNIPAM and CD. When the PNIPAM component in grafted chains increases, the swelling-shrinking volume phase transition of the grafted PNG-ECD gates in the membranes becomes more significant, and the above-mentioned cooperation is enhanced. However, the steric hindrance caused by the significant phase transition of grafted chains could affect the association constant of CD toward guest molecules ANS. As a result, when the α value increases further to 8, the ANS-recognition gating coefficient decreases to a certain degree. Therefore, there are optimum molar ratios α for both the thermo-responsive and the molecular-recognizable gating characteristics of the membranes.

16.5 Summary

In summary, the thermo-responsive and molecular-recognizable PNG-ECD-grafted membranes have been fabricated successfully by plasma-induced pore-filling graft polymerization and chemical reaction. By virtue of the adjustment of association constant between CD and guest molecules via the configuration change of grafted PNIPAM chains, the as-designed membranes are demonstrated to chiral separate enantiomers with a high selectivity and regenerate with a high decomplexation ratio by simply changing the operation temperature. The maximal thermo-responsive coefficient $R_{e.e.(25/40)}$ of about 3 is achieved by the grafted membrane with the optimum grafting yield of $Y_{PNIPAM} = 107$ mg·cm^{-2}/$Y_{CD} = 23$ mg·cm^{-2} in the investigated range. Moreover, the PNG-ECD-grafted membranes are also used to adsorb and desorb ANS molecules by simply adjusting the operation temperature due to the swelling-shrinking configuration change of PNIPAM-grafted chains around the LCST and the molecular recognition of CD toward ANS. The higher the CD content in the grafted chains on the membrane, the larger the difference between the adsorption amount of ANS onto PNG-ECD-grafted PET membrane at 25 °C and that at 40 °C. In addition, taking advantage of the opposite effect of different guest molecules on the LCST values of PNG-ECD chains, the prepared membranes with the same functional gate are featured with triple stimuli-responsive gating functions. The thermo-responsive and molecular-recognizable gating characteristics of the

PNG-ECD-grafted membranes can be adjusted by changing the grafting yields of both PNIPAM and CD on the membranes as well as the molar ratio of PNIPAM to CD in the grafted chains. The as-prepared membranes based on PNIPAM and CD are highly attractive for chiral separation, affinity separation, and controlled-release systems responding to the environmental temperature.

References

1. Ito T, Hioki T, Yamaguchi T et al (2002) Development of a molecular recognition ion gating membrane and estimation of its pore size control. J Am Chem Soc 124:7840–7846
2. Ito T, Sato Y, Yamaguchi T et al (2004) Response mechanism of a molecular recognition ion gating membrane. Macromolecules 37:3407–3414
3. Ito T, Yamaguchi T (2004) Osmotic pressure control in response to a specific ion signal at physiological temperature using a molecular recognition ion gating membrane. J Am Chem Soc 126:6202–6203
4. Ito T, Yamaguchi T (2006) Controlled release of model drugs through a molecular recognition ion gating membrane in response to a specific ion signal. Langmuir 22:3945–3949
5. Yamaguchi T, Ito T, Sato T et al (1999) Development of a fast response molecular recognition ion gating membrane. J Am Chem Soc 121:4078–4079
6. Yanagioka M, Kurita H, Yamaguchi T et al (2003) Development of a molecular recognition separation membrane using cyclodextrin complexation controlled by thermosensitive polymer chains. Ind Eng Chem Res 42:380–385
7. Nozaki T, Maeda Y, Ito K et al (1995) Cyclodextrins modified with polymer chains which are responsive to external stimuli. Macromolecules 28:522–524
8. Wang HD, Chu LY, Yu XQ et al (2007) Thermo-sensitive affinity behavior of poly(*N*-isopropylacrylamide) hydrogels with β-cyclodextrin moieties. Ind Eng Chem Res 46:1511–1518
9. Ohashi H, Shimada T, Yamaguchi T (2009) Development of molecular recognition membrane showing autonomous adsorption-desorption phenomenon. J Photopolym Sci Technol 22:473–476
10. Yang M, Chu LY, Wang HD et al (2008) A novel thermo-responsive membrane for chiral resolution. Adv Funct Mater 18:652–663
11. Nozaki T, Maeda Y, Kitano H (1997) Cyclodextrin gels which have a temperature responsiveness. J Polym Sci A Polym Chem 35:1535–1541
12. Ohashi H, Hiraoka Y, Yamaguchi T (2006) An autonomous phase transition-complexation/decomplexation polymer system with a molecular recognition property. Macromolecules 39:2614–2620
13. Tabushi I, Kuroda Y, Mizutani T (1986) Artificial receptors for amino acids in water. local environmental effect on polar recognition by 6A-Amino-6B-carboxy- and 6B-amini-6A-carboxy-β-cyclodextrins. J Am Chem Soc 108:4514–4518
14. Xie R, Zhang SB, Wang HD et al (2009) Temperature-dependent molecular-recognizable membranes based on poly(N-isopropylacrylamide) and β-cyclodextrin. J Membr Sci 326:618–626
15. Yang M, Chu LY, Xie R et al (2008) Molecular-recognition induced phase-transition of two thermo-responsive polymers with pendent β-cyclodextrin groups. Macromol Chem Phys 209:204–211
16. Yang M, Xie R, Wang JY et al (2010) Gating characteristics of thermo-responsive and molecular-recognizable membranes based on poly(*N*-isopropylacrylamide) and β-cyclodextrin. J Membr Sci 355:142–150

Index

L.-Y. Chu et al., *Smart Hydrogel Functional Materials*,
DOI 10.1007/978-3-642-39538-3,

© Chemical Industry Press, Beijing and Springer Berlin Heidelberg 2013

D

E

Printed by Printforce, the Netherlands